P9-CDG-690

ELASTIC WAVES AND ULTRASONIC NONDESTRUCTIVE EVALUATION

ELASTIC WAVES AND ULTRASONIC NONDESTRUCTIVE EVALUATION

Proceedings of the IUTAM Symposium on Elastic Wave Propagation and Ultrasonic Evaluation University of Colorado, Boulder, Colorado, U.S.A. July 30 - August 3, 1989

Edited by

S.K. DATTA

Department of Mechanical Engineering
University of Colorado
Boulder, Colorado, U.S.A.

J.D. ACHENBACH

Department of Civil Engineering
Northwestern University
Evanston, Illinois, U.S.A.

Y.S. RAJAPAKSE

Mechanics Division
Office of Naval Research
Arlington, Virginia, U.S.A.

1990

NORTH-HOLLAND
AMSTERDAM•NEW YORK•OXFORD•TOKYO

ELSEVIER SCIENCE PUBLISHERS B.V.
Sara Burgerhartstraat 25
P.O. Box 211
1000 AE Amsterdam, The Netherlands

Distributors for the United States and Canada:

ELSEVIER SCIENCE PUBLISHING COMPANY, INC.
655, Avenue of the Americas
New York, N.Y. 10010, U.S.A.

ISBN: 0 444 87485 2

Printed in The Netherlands

PREFACE

The IUTAM Symposium on Elastic Wave Propagation and Ultrasonic Nondestructive Evaluation was held during July 30 - August 3, 1989, at the Clarion Hotel in Boulder, Colorado, USA. The purpose of the symposium was to highlight the role of the mechanics of wave propagation in the development of Ultrasonic Nondestructive Evaluation (NDE) techniques. Many fundamental problems of elastic wave propagation as they relate to ultrasonic material and defect characterization were discussed. There were 57 oral presentations, of which two were general review lectures of 45 minutes duration each, six were for 30 minutes each, and the rest were for 20 minutes. In addition to the oral presentations there was a session of 33 poster presentations. With 138 registrants, the symposium was somewhat larger than anticipated, indicating the importance of the topic in mechanics and engineering.

During the last decade, emergence of ultrasonic techniques for nondestructive evaluation of materials has provided a strong impetus for the study of propagation and diffraction of elastic waves. Elastic wave propagation studies have led to quantitative techniques for evaluation and service life prediction of structures. Advances towards the development of ultrasonic quantitative nondestructive evaluation (QNDE) have been made possible by the interaction between the mechanicians, material scientists, electrical engineers and applied physicists. The purpose of this IUTAM Symposium was to provide an opportunity for researchers in the general area of solid mechanics, who are active in the interdisciplinary QNDE field, and scientists from other related areas to come together to exchange ideas, to assess the progress that has been made in recent years and to delineate areas of future research. It was clear from the presentations that a considerable amount of work in the area has been done in the last ten years, and important advances have been made in the numerical, analytical and experimental techniques for studying elastic wave propagation in the context of detection and characterization of flaws, characterization of materials and inhomogeneous stress field and evaluation of progressive failures or phase changes. Leading researchers from various parts of the world presented state-of-the-art lectures, focusing attention on an important technological area in which wave propagation studies have made and will continue to make important contributions toward advancing the technology.

The Symposium is believed to have achieved its goal of a clear presentation of the current state of analytical, numerical, and experimental techniques as well as future research needs. It has provided evidence of important advances that have been made in the detailed analysis and experimental verification of ultrasonic scattering by single and multiple defects in isotropic medium. It was shown that some progress has been made in extracting relevant fracture parameters from ultrasonic scattered field data. There is no doubt that our understanding of the effects of stress, strain, material inhomogeneities, texture, anisotropy, and layering on wave propagation has been enhanced by recent theoretical works. Some important areas that are beginning to get attention are: localization in periodic or random media, interfaces and their role in failure mechanisms, ultrasonic characterization of damage and life expectancy, supercomputer use in modeling, image processing and enhancement techniques, and laser ultrasonics and acoustic microscopy. Judging by the progress during the last decade it is anticipated that ultrasonic techniques will find more and more use in quantitative nondestructive evaluation. In that regard it is hoped that this volume will be a valuable reference to researchers and practitioners in ultrasonic NDE.

The organizers of the Symposium gratefully acknowledge financial support for the organization of the conference and for travel expenses for conferees from the International Union of Applied Mechanics, the University of Colorado (Cooperative Institute for Research in Environmental Sciences, Graduate School, and College of Engineering and Applied Science), the Office of Naval Research, the National Science Foundation, Martin Marietta Laboratories, Elsevier Science Publishers, Pergamon Press, and Springer-Verlag. Our special thanks go to Dr. A.R. Seebass, Dean of the College of Engineering and Applied Science, Dr. R.E. Sievers, Director of the Cooperative Institute for Research in Environmental Sciences, Dr. Risa Palm, Associate Vice Chancellor for Research and Dean of the Graduate School, and Dr. B. Djordjevic of the Martin

Marietta Laboratories. Without their generous support it would not have been possible to organize this international conference.

We are indebted to the members of the Scientific Committee who provided valuable assistance in the planning of the scientific program in the selection of the contributions. We thank the members of the Organizing Committee members for their hard work that made the symposium not only technically rewarding but also socially enjoyable. We are particularly thankful to Ms. Julie McKie, who worked uncounted hours and looked after every little detail from the very beginning of the Symposium planning to the completion of this Proceedings volume. Her good spirits, meticulous management and organizational skills made the Symposium such an enjoyable one.

Our thanks go also to the Clarion Hotel conference management staff, particularly to Linda Montgomery and Jackie Prendergast, for being so helpful and cooperative before and during the conference. Many students (Mr. Y. Al-Nassar, Mr. S. Liu, Mr. T. Ju, Mr. J. Fox, Dr. P. Xu, Mr. T. Kohl, Dr. K. Khair, Mr. M. Bouden), especially Mr. Robert Bratton, gave their untiring help during the symposium. Our special thanks go to them all. We also express our appreciation to the staff of Elsevier Science Publishers B.V. for their cooperation.

S.K. Datta, Boulder, Colorado
J.D. Achenbach, Evanston, Illinois
Y.S. Rajapakse, Arlington, Virginia

November, 1989

TABLE OF CONTENTS

COMMITTEES

Scientific Committee:

J.D. Achenbach, Co-chairman, USA
S.K. Datta, Co-chairman, USA
P.M. van den Berg, The Netherlands
L.J. Bond, England
H. Fukuoka, Japan
P. Höller, FRG
J. Hult, Sweden
U. Nigul, USSR
G. Quentin, France
C.M. Sayers, The Netherlands
R.B. Thompson, USA

Organizing Committee:

S.K. Datta, Chairman
H.M. Ledbetter
R. Pak
Y. Rajapakse
H. Spetzler
S. Sture
J. McKie

SESSION CHAIRMEN:

G.A. Alers
D. Barnett
A.V. Clark
R.J. Clifton
A.K. Gautesen
W.A. Green
R.D. Gregory
L.M. Keer
G. Matzkanin
J.J. McCoy
A.N. Norris
S. Rokhlin
A. Tucker
G.R. Wickham

LIST OF PARTICIPANTS

AUSTRIA
P. Borejko
H. Irschik

CANADA
D. Jarman
W. Karunasena
R. Paskaramurthy
A.H. Shah

FEDERAL REPUBLIC OF GERMANY
M. Braun
P. Fellinger
P. Hagedorn
P. Höller
T. Kreutter
K.J. Langenberg
J. Lenz
J. Wauer

FRANCE
B. Hosten
G. Quentin

INDIA
S.K. Bose
A.K. Rao
K. Viswanathan

IRELAND
P.M. O'Leary

ITALY
G. Baronio
L. Binda
P.P. Delsanto

JAPAN
Y. Arai
M. Enoki
H. Fukuoka
M. Hirao
Y. Iwashimizu
M. Kitahara
O. Kobori
E. Matsumoto
S.G. Nomachi
M. Ohtsu
T. Oshima
T. Sawano
Y. Shindo
H. Toda

MEXICO
F.J. Sabina

THE NETHERLANDS
P.M. van den Berg
A.T. de Hoop
M. Lorenz
C.M. Sayers
S.M. de Vries
L.F. van der Wal

SWEDEN
A. Boström
P. Olsson

SWITZERLAND
J. Dual

TAIWAN
M.K. Kuo

UNITED KINGDOM
I.D. Abrahams
P.H. Albach
L.J. Bond
G.A.D. Briggs
E.R. Green
W.A. Green
R.D. Gregory
P.A. Lewis
P.A. Martin
J.A. Ogilvy
N. Scott
M. Somekh
A. Temple
G.R. Wickham

USSR
V. Babeshko

UNITED STATES

J.D. Achenbach
L. Adler
G.A. Alers
R.H. Atkinson
D. Barnett
N. Bleistein
M. Bouden
A.M.B. Braga
R. Bratton
C.J. Brown
O. Buck
D.E. Budreck
T.C.T. Ting
E.J. Chern
H.C. Choi
A.V. Clark
R.J. Clifton
S.K. Datta
A.J. Devaney
L.B. Felsen
D.W. Fitting
C.M. Fortunko
J. Fox
C.K. Frederickson
J.M. Gary
A.K. Gautesen
S.M. Handley
J.G. Harris
M.R. Holland
G.C. Johnson
T.H. Ju
L.M. Keer
K. Khair
V.K. Kinra
T. Kohl
R. Kranz
R.D. Kriz
H.M. Ledbetter
J.W. Lee
S.W. Liu
J.H. Loewenherz
P.L. Marston
T. Mase
G. Matzkanin
J. McCoy
D. Mendelsohn
J.G. Miller
H. Murakami
R.J. Nagem
Y. Al·Nassar
A. Nayfeh
A.N. Norris
S.J. Norton
J. Paffenholz
R. Pak
K.C. Park
G.E. Phillips
C. Pierre
A.M. Porter
F.J. Rizzo
R.A. Roberts
T.R. Rogge
S.I. Rokhlin
W. Sachse
K. Salama
P. Sheng
H. Spetzler
S. Sture
R.B. Thompson
A. Tucker
A. Tverdokhlebov
R.L. Weaver
P.C. Xu
Y. Yang
Z. You

A

ULTRASONIC SCATTERING, INVERSE PROBLEMS AND LOCALIZATION

Elastic Waves and Ultrasonic Nondestructive Evaluation
S.K. Datta, J.D. Achenbach and Y.S. Rajapakse (Editors)

FROM ULTRASONICS TO FAILURE PREVENTION

J. D. Achenbach

Center for Quality Engineering and Failure Prevention
Northwestern University
Evanston, IL. 60208-3020

Ultrasonic tests provide travel times of signals and time traces of amplitudes. By the use of appropriate data processing techniques, this information can be used to obtain, for example, elastic constants, the location and dimensional characteristics of single inhomogeneities and certain features of distributions of inhomogeneities. Such results are related to the residual strength of a component, and hence they play an essential role in failure prevention. In this paper we will consider examples of strength considerations based on ultrasonic results.

1. INTRODUCTION

The strength of a structural component is conventionally determined by placing it in a testing machine and increasing the load until the component fails. A preferable way to obtain strength information is in a non-destructive manner, for example by an ultrasonic technique. This is certainly true if the component is expensive or if it is part of a structural system that cannot easily be disassembled. Unfortunately it is generally not possible to measure strength directly by ultrasonic methods. A component whose strength is controlled by the presence of flaws is, however, an exception. For such a component it is possible to obtain ultrasonic data, which when properly processed and interpreted, will yield relevant information on the component's strength.

Fracture mechanics and, in a more general sense, failure mechanics have made great strides in the prediction of the integrity of structural components. For a component made of a material of known properties subjected to a given set of loads, it is possible to calculate the critical size of a crack at a specified location. The component is judged to be safe if the crack is smaller than a critical size and when it may be expected that it will not grow to that size during the service life or prior to the next inspection. Naturally, to tolerate cracks of subcritical size, reliable methods must be available to detect and characterize cracks. Quantitative non-destructive evaluation provides the techniques to detect a crack (or more generally a flaw), and to determine its location, size, shape and orientation. It is also possible that once a crack has been detected its characterization can circumvent sizing and can directly yield the magnitude of stress intensity factors under service loads. This paper will discuss some of the ultrasonic methods, particularly with regard to strength assessment of bodies that may contain cracks. An earlier discussion has been given in Ref.[1].

By an ultrasonic technique the time of travel of a signal as well as its amplitude can be measured, the latter as a function of time. For a homogeneous material the time of travel yields a wave speed and hence an elastic constant. For a material with a known wave speed, a signal arrival time which is inconsistent with the geometry of an unflawed body and the placement of source and receiver (pulse-echo or pitch-catch), is an indication of the presence of an inhomogeneity. For some simple cases, for example edge cracks in plates, time of flight measurements can be used to size a crack.

A compact flaw acts as a reflector of ultrasonic wave motion. A bigger flaw reflects more sound, and hence the amplitude of the voltage produced by a piezoelectric transducer that is exposed to the reflected sound is related to the dimensions of the crack. Unfortunately, quantitative characterization of a flaw based on amplitude measurements may be very inaccurate unless the measurement is carefully calibrated. The reason is that part of the incident energy that insonifies the flaw is scattered rather than specularly reflected. A calculation of the total fields which includes scattering effects is rather complicated. Another problem with amplitude considerations is that damping in the material and scattering by microstructure, microcracks, and voids and inclusions tend to reduce the measured signals, particularly as the frequency increases. Hence the productive use of amplitude considerations requires careful adjustment of the results for these secondary effects.

In general terms, two approaches to ultrasonic flaw detection and characterization have been taken. The imaging approach seeks to process the scattered field in such a manner that a visual outline of the object is produced on a display. The inverse-scattering approach attempts to infer geometrical characteristics of a flaw from either the angular dependence of its far-field scattering amplitude at fixed frequency, or from the frequency dependence of its far-field scattering amplitude at fixed angles.

Imaging is conceptually simple, but difficult to implement with ultrasonic signals, see e.g.[2]. The basic idea is to collect signals scattered by the flaw into an experimentally accessible aperture and then to recombine these signals such that the ones scattered from a particular point on the flaw add coherently at a unique point on the image. At other points in the image, these same signals should not add in phase. Not surprisingly, the main difficulty with imaging is to obtain an acceptable level of resolution. A number of innovative instruments and signal processing methods have been developed. An important example is the acoustic microscope, [3]-[4]. This instrument was initially developed in the gigahertz frequency regime, where the wavelength is on the order of a micrometer and resolution approaches that of optical microscopes. The acoustic microscope can, however, also "hear" under the surface of a solid, and it can image both microstructure and flaws. Imaging is an important method of detection and characterization, but it falls somewhat outside the scope of this paper. For an interesting discussion of its potential, but also of the current problems with imaging, we refer to the review paper by Thompson and Thompson [5].

In experimental work on quantitative flaw definition by the scattered-field method, either the pulse-echo method with one transducer or the pitch-catch method with two transducers is used. The transducer(s) may be either in direct contact with the specimen, or transducer(s) and specimen may be immersed in a water bath. Most experimental setups include instrumentation to gate out and spectrum analyze the signal diffracted by a flaw. The raw scattering data generally need to be corrected for transducer transfer functions and other characteristics of the system, which have been obtained on the basis of appropriate calibrations. After processing, amplitudes and phase functions, are available as functions of the frequency and the scattering angle. These experimental data can then be directly compared with theoretical results. For the inverse scattering method, the experimental data can be interpreted with the aid of analytical methods, to characterize the scatterer.

In recent years several methods have been developed to investigate scattering of elastic waves by interior cracks as well as by surface-breaking cracks, in both the high- and the low-frequency domains. The appeal of the high-frequency approach is that the probing wavelength is of the same order of magnitude as the length-dimensions of the crack. This gives rise to interference phenomena which can be detected. The advantage of the low frequency approach is that useful approximations can be based on static results. A large body of numerical results has been developed for the direct problem of elastodynamic scattering by an inhomogeneity. In particular, several numerical programs based on the use of the T-matrix method and the boundary element method have been developed.

The solution to the direct scattering problem, that is, the computation of the field generated when an ultrasonic wave is scattered by a known flaw, is a necessary preliminary to the solution of the inverse problem, which is the problem of inferring the geometrical characteristics of an unknown flaw from either the angular dependence of the amplitude of the scattered far-field at fixed frequency, or from the frequency dependence of the far-field amplitude at fixed angle. In recent years solutions to the inverse problem have been obtained by the use of nonlinear optimization methods.

Other applications of ultrasonic wave methods, e.g., to acoustic emission techniques and distributed property measurements are not discussed here, due to length limitations. More complete discussions of ultrasonic QNDE methods can be found in recent review papers by Thompson and Thompson [5], Fu [6], and Thompson [7]. The role of elastodynamic scattering problems in quantitative non-destructive evaluation has also been reviewed by, e.g., Gubernatis [8] and Bond et al. [9]. Interesting applications have been discussed by Coffey and Chapman [10].

A single crack is only one kind of material discontinuity. Cracks are particularly objectionable since they are very obvious causes of catastrophic failure, but voids, cavities, inclusions, interfaces, distribution of cracks, or in general terms damaged regions of a material, may have equally deleterious effects on the strength of components. By the use of appropriate non-destructive evaluation methods it should be possible to discriminate between a broad spectrum of flaws and to determine the relevant characteristics of each kind. For purposes of specificity within the allotted length of this paper, the attention will, however, be restricted to components containing cracks.

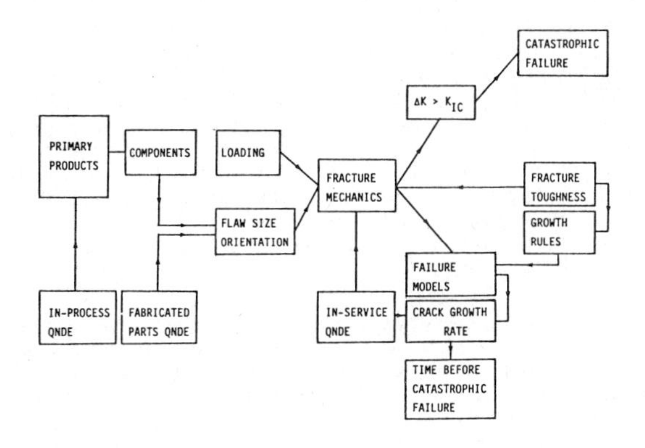

FIGURE 1
Role of QNDE in the life cycle of a component.

The role of quantitative non-destructive evaluation during the various stages of the life cycle of a structural component is illustrated in Fig. 1. The schematic depiction is meant to apply to metal components which are subjected to cyclic loading, and hence may sustain metal fatigue. As indicated in Fig. 1, Quantitative NDE methods should enter in the material processing stage, to play a role in maintaining material quality of primary products. At this stage QNDE methods ensure that primary products such as plates, sheets and strips do not contain cracks or other flaws whose dimensions exceed a certain specified level. In the next stage, QNDE methods should be applied to characterize flaws that have been induced in the process of fabricating components. The maximum dimensions of such cracks, in conjunction with the magnitude of the cyclic load, can be used to calculate the maximum values of corresponding stress intensity factors. Naturally these maximum values, ΔK, should be less than the fracture toughness. If this is indeed the case, then a crack may still propagate, but at a controlled rate which in principle is predictable. Hence within the framework of the "damage tolerant" philosophy, a part containing a macroscopic flaw is acceptable if it can be shown that at the predicted stress levels, the flaw will not grow to critical size during the design lifetime. Reliable quantitative methods of non-destructive in-service inspection of parts, in conjunction with the concept of "probability of detection," [5], are clearly essential for a successful implementation of the damage tolerant philosophy. Flaws which at the time of in-service inspection are greater in size than consistent with the design lifetime, must be detected and characterized. On the other hand the QNDE procedure should not reject components that contain only smaller size flaws. A part should be returned to service if no flaws are found, or if it can be shown that the size of a detected flaw is small enough that it will propagate to failure only over a period substantially larger than the next inspection interval. Very considerable life cycle cost savings can be achieved with this "retirement for cause" procedure. Retirement for cause procedures have been discussed in considerable detail in the literature, see e.g. Refs.[11]-[12].

2. SCATTERING BY A PENNY-SHAPED CRACK

We start by considering a penny shaped crack in an infinite, homogeneous, isotropic and linearly elastic solid, as shown in Fig. 2. An incident time-harmonic, plane longitudinal wave of the form

$$u_i^{in} = u_o d_i^{L+} \exp(ik_L p_j^{L+} x_j) \tag{2.1}$$

interacts with the crack and generates scattered waves. The amplitude of the incident wave is u_o and its propagation direction is taken in the x_1x_3 plane. Also, θ is the angle of incidence and k_L is the wave number of the incident wave:

$$k_L = \omega/c_L \ , \quad c_L^2 = (\lambda + 2\mu)/\rho \quad , \tag{2.2a,b}$$

$$d_i^{L+} = p_i^{L+} = (\sin\theta, 0, \cos\theta) \quad . \tag{2.3a,b}$$

The fields are time-harmonic with angular frequency ω, but the time factor exp(-$i\omega t$) is omitted throughout this paper.

It is assumed that the faces of the crack will not touch. The boundary conditions on the crack faces then are

$$\sigma_{3q}^{in}(\underset{\sim}{x}) + \sigma_{3q}^{sc}(\underset{\sim}{x}) = 0, \quad x \in A^{\pm} \ , \tag{2.4}$$

where σ_{3q}^{in} represent the stress components due to the incident wave in the absence of the crack, σ_{3q}^{sc} denote the stress components of the scattered wave induced by the interaction of the incident wave with the crack, and A^+ and A^- are the insonified and shadow sides of the crack.

The scattered displacement field for a crack of general orientation can be expressed by the following representation integral

$$u_k^{sc}(\underset{\sim}{x}) = \int_{A^+} \sigma_{ijk}^{G}(\underset{\sim}{x}-\underset{\sim}{\xi})\Delta u_i(\underset{\sim}{\xi}) n_j dS(\underset{\sim}{\xi}) \ , \quad x \notin A^+ \tag{2.5}$$

in which $\underset{\sim}{x}$ denotes the position vector of the observation point, $\underset{\sim}{\xi}$ denotes the position vector of the source point, n_j is the unit normal vector of A^+, and Δu_j are the crack opening displacements (displacement jumps across the crack faces) defined by

$$\Delta u_j(\underset{\sim}{\xi}) = u_j\Big|_{\underset{\sim}{\xi}\in A^+} - u_j\Big|_{\underset{\sim}{\xi}\in A^-} \ , \tag{2.6}$$

and σ_{ijk}^{G} is the stress Green's function of the uncracked medium. The function σ_{ijk}^{G} denotes the stress components at position $\underset{\sim}{x}$ due to a time-harmonic unit point force applied at position $\underset{\sim}{\xi}$ in the direction ξ_k.

By substituting (2.5) into Hooke's law

$$\sigma_{pq} = C_{pqk\ell} u_{k,\ell} \quad , \tag{2.7}$$

and by using the boundary conditions (2.4), the following system of boundary integral equations can be obtained

$$\sigma^{in}_{3q}(\underset{\sim}{x}) = -C_{3qk\ell} \frac{\partial}{\partial x_\ell} \int_{A^+} \sigma^G_{i3k}(\underset{\sim}{x}-\underset{\sim}{\xi}) \Delta u_i(\underset{\sim}{\xi}) n_3 dS(\underset{\sim}{\xi}) \quad , \underset{\sim}{x} \in A^+ \quad . \tag{2.8}$$

Here $C_{pqk\ell}$ is the elastic tensor which for an isotropic material is given by

$$C_{pqk\ell} = \lambda\delta_{pq}\delta_{k\ell} + \mu(\delta_{pk}\delta_{q\ell} + \delta_{p\ell}\delta_{qk}) \quad , \tag{2.9}$$

where λ and μ are Lamé's elastic constants and δ_{pq} is the Kronecker delta. Closed form solutions to Eq.(2.8) can not be obtained, and in general a suitable numerical method must be employed. Among many studies of this problem we mention the works by Martin and Wickham [13], Lin and Keer [14], Budreck and Achenbach [15], and Nishimura and Kobayashi [16]. Earlier studies of elastic wave scattering by a penny-shaped crack have been presented by Robertson [17], Mal [18], Garbin and Knopoff [19] and Krenk and Schmidt [20], who used integral transform and dual integral equations techniques.

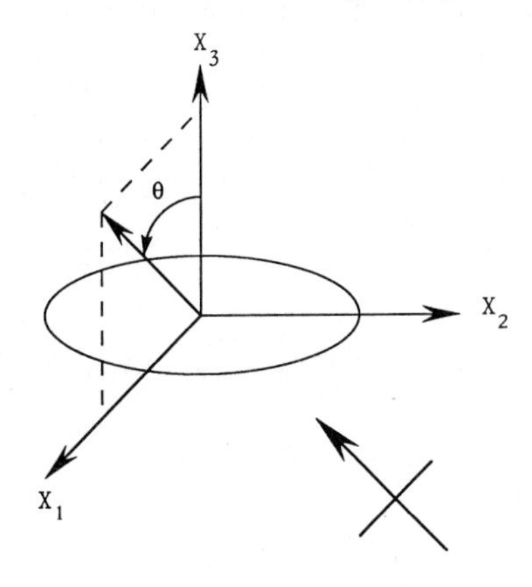

FIGURE 2
Longitudinal wave incident on a penny-shaped crack.

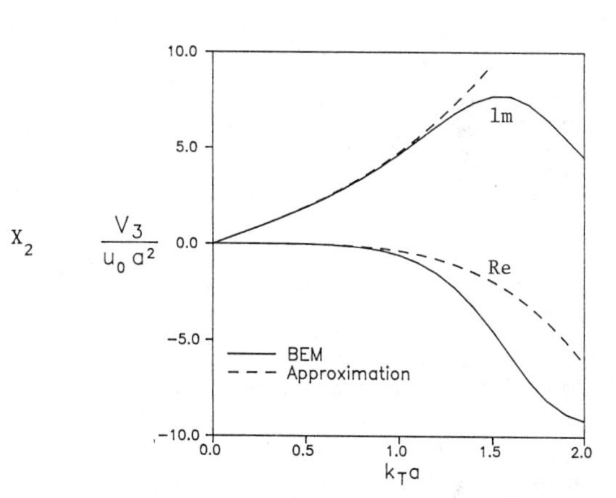

FIGURE 3
Crack-opening volume as a function of $k_T a$.

In an alternative approach, approximate solutions to Eq.(2.8) can be obtained in the low-frequency range. This can be done, for example, by the perturbation technique proposed by Roy [21], who investigated the elastic wave scattering by an elliptical crack. In this procedure, the known and unknown quantities of Eq.(2.8) are expanded into power series of ik_T as

$$\sigma^{in}_{3q}(\underset{\sim}{x}) = \mu \sum_{m=0}^{\infty} \frac{(ik_T)^m}{m!} \sigma^{(m)}_{3q} (x_1 \sin\theta)^m \quad , \tag{2.10}$$

$$\sigma^G_{ijk}(\underset{\sim}{x}-\underset{\sim}{\xi})\ ) = \mu \sum_{m=0}^{\infty} \frac{(ik_T)^m}{m!} \sigma^{(m)}_{ijk}(\underset{\sim}{x};\underset{\sim}{\xi}) \quad , \tag{2.11}$$

$$\Delta u_i(\underset{\sim}{x}) = \sum_{m=0}^{\infty} \frac{(ik_T)^m}{m!} \Delta u^{(m)}_i(\underset{\sim}{x}) \quad , \tag{2.12}$$

where $\sigma^{(m)}_{3q}$, $\sigma^{(m)}_{ijk}$ and $\Delta u^{(m)}_i$ are expansion coefficients, and k_T is the wavenumber of transverse waves

$$k_T = \omega/c_T \;, \quad c_T^2 = \mu/\rho \tag{2.13a,b}$$

Substituting (2.10) - (2.12) into (2.8) and collecting terms of the same order in k_T yields a system of integral equations which can be solved analytically.

When the crack-opening displacement has been calculated, the crack-opening volume follows by integration over the area of the crack:

$$V_i = \int_A \Delta u_i(\underset{\sim}{x})dS \tag{2.14}$$

Keeping three terms in the expansions, the result for normal incidence is

$$V_3 = V_{30} + V_{32}(k_Ta)^2 + V_{33}(k_Ta)^3 + \cdots, \tag{2.15}$$

where a is the radius of the crack, and

$$V_{30} = \frac{4}{3}\frac{\kappa^4}{\kappa^2-1}\,a^3ik_Lu_o \tag{2.16}$$

$$V_{32} = \frac{14}{90}\frac{\kappa^2[3(\kappa^2-1)^2+2\kappa^2]}{(\kappa^2-1)^2}\,a^3ik_Lu_o \tag{2.17}$$

$$V_{33} = -\frac{4}{135}\frac{\kappa[32-40\kappa^2+15\kappa^4+8\kappa^5]}{\pi(\kappa^2-1)^2}\,a^3k_Lu_o \tag{2.18}$$

In these expressions

$$\kappa = c_L/c_T = k_T/k_L = \left[\frac{2(1-\nu)}{1-2\nu}\right]^{1/2}, \tag{2.19}$$

where ν is Poisson's ratio. It may be verified that V_{30} as given by Eq.(2.16) is just the static crack-opening volume induced by surface tractions on the crack faces which are opposite in sign to the ones corresponding to the incident wave given by Eq.(2.1), i.e., corresponding to

$$\sigma_{33} = -ik_L(\lambda+2\mu)u_o \tag{2.20}$$

It is of interest to check the accuracy of the approximation given by Eq.(2.15), as the dimensionless wave number k_Ta increases. Figure 3 shows a comparison of V_3/a^2u_o with numerically exact results which were obtained by a boundary element solution of Eq.(2.8). It is noted that a very adequate approximate solution is obtained for values of $k_Ta = 0$ to $k_Ta = 1$.

Equation (2.5) forms the basis for an inverse method to determine the location of the crack and its crack opening volume. The method was discussed in some detail in Ref.[22]. In the approach of Ref.[22], Eq.(2.5) is written in the form

$$u_k^{sc}(\underset{\sim}{x}) = \int_{A^+} C_{ijpq}\frac{\partial u_{pk}^G(\underset{\sim}{x}-\underset{\sim}{\xi})}{\partial\xi_q}\Delta u_i(\underset{\sim}{\xi})n_j(\underset{\sim}{\xi})dA(\underset{\sim}{\xi}) \tag{2.21}$$

For a crack in the x_1x_2-plane and for normal incidence, i.e., the incident wave given by Eq.(2.1) propagates in the x_3-direction ($\theta=0$), Eq.(2.21) may be rewritten as

$$u_k^{sc}(\underset{\sim}{x}) = \int_A C_{33\ell m}D_{k\ell m}(\underset{\sim}{x}-\underset{\sim}{\xi})\Delta u_3(\xi)dA(\underset{\sim}{\xi}) \tag{2.22}$$

where

$$D_{k\ell m}(\underset{\sim}{x}-\underset{\sim}{\xi}) = \frac{\partial}{\partial x_m}u_{k\ell}^G(\underset{\sim}{x}-\underset{\sim}{\xi}) \tag{2.23}$$

Since $u_{k\ell}^G(\underset{\sim}{x}-\underset{\sim}{\xi})$ is the displacement at position x in the direction x_k due to a unit force applied in the direction x_ℓ at $\underset{\sim}{x}=\underset{\sim}{\xi}$, Eq.(2.23) implies that $-D_{k\ell m}(\underset{\sim}{x}-\underset{\sim}{\xi})$ is the displacement produced at position $\underset{\sim}{x}$ in the x_k-direction, by a double force applied at $\underset{\sim}{x}=\underset{\sim}{\xi}$ with forces in the x_ℓ-direction and moment arm in the x_m-direction.

In the far-field ($k_\alpha|\underset{\sim}{x}-\underset{\sim}{\xi}| >> 1$, α = L,T) and for relatively low frequencies ($k_\alpha a << 1$, where a is a characteristic length dimension of the crack), Eq.(2.22) may be simplified to

$$u_k^{Sc}(\underset{\sim}{x}) = C_{33\ell m} D_{k\ell m}(\underset{\sim}{x}) V_3 \tag{2.24}$$

where V_3 is the crack-opening volume defined by Eq.(2.14). Physically, the scattered field of Eq.(2.24) can be thought of as being produced by three double forces (force and arm in the same direction) in the x_1, x_2 and x_3 directions, located at the centroid of the crack, and of strength λV_3, λV_3 and $(\lambda+2\mu)V_3$, respectively.

Now suppose that the three displacement components of the scattered field have been measured at a point of observation $\underset{\sim}{x}$. If the origin of the coordinate system is placed on the crack, then the coordinates of the point of observation become unknown, since the location of the crack is unknown. The displacements are complex valued, and hence Eq.(2.24) defines a set of six nonlinear equations for five unknowns, namely, for the 3 components of $\underset{\sim}{x}$ and the real and imaginary parts of V_3. It follows that Eq.(2.24) is an overdetermined set of nonlinear equations. The system can, however, be solved numerically, as has been discussed in detail in Ref.[22].

As discussed earlier in this paper, the ultimate aim of quantitative non-destructive evaluation is to obtain information on the residual strength of components. By combining the information on the crack opening volume, obtained in the manner described in this section, with some concepts of fracture mechanics, it is possible to estimate the maximum stress intensity factor. The approach is based on a formula stated by Budiansky and O'Connell [23] which relates the static crack opening volume, V_3^{st}, to the static Mode-I stress intensity factor, for the case of a planar crack of arbitrary shape in a uniform field of tensile stresses, σ_o, directed normal to the crack faces:

$$V_3^{st} = \frac{1-\nu}{3\mu}\sigma_o \int_\Gamma \ell_c \left(\frac{K_I^{st}}{\sigma_o}\right)^2 d\Gamma \tag{2.25}$$

where Γ is the edge of the crack, K_I^{st} is the Mode-I stress intensity factor and ℓ_c is a length parameter. An approximate expression for K_I^{st} in terms of V_3^{st} has been given by Budiansky and Rice [24] as:

$$\left(\frac{K_I^{st}}{\sigma_o}\right)_{max} = \left\{\frac{24\mu}{(1-\nu)\pi^3}\frac{V_3^{st}}{\sigma_o}\right\}^{1/6} \tag{2.26}$$

Equation (2.26) is actually an exact result for a penny-shaped crack, as can easily be verified. It should be noted that (2.26) expresses the stress-intensity factor per unit applied tensile stress in terms of the crack opening volume per unit stress. The observation of interest now is that the latter ratio can actually be obtained from ultrasonic information in the limit of vanishing frequency. This is true because in the zero-frequency limit an ultrasonic test reduces to a quasi-static problem. For example for the scattering problem discussed earlier in this section we have

$$\frac{V^{st}}{\sigma_o} = \lim_{k_L\to 0} \frac{V_3}{ik_L(\lambda+2\mu)u_o} \tag{2.27}$$

Hence, if V_3 can be measured as a function of the frequency, Eqs.(2.26) and (2.27) will yield the Mode I stress intensity factor for any service-load induced tensile field, without detailed crack sizing. This procedure does require a low frequency limit of the crack-opening volume. Such a low frequency limit may be difficult to obtain directly from experimental data. The general dependence on frequency as the limit of zero frequency is approached is, however, known from analytical results as given by Eq.(2.15), and this should help to determine the required limit.

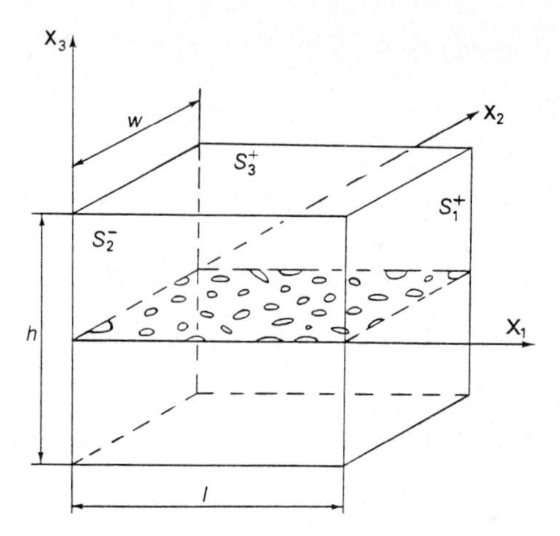

FIGURE 4
Cell containing a distribution of cracks in the x_1x_2-plane

3. REFLECTION BY A PLANARY DISTRIBUTION OF CRACKS

Next we consider a plane in a homogeneous, isotropic, linearly elastic solid which is permeated by a distribution of in-plane cracks. The geometry is shown in Fig. 4. An incident plane longitudinal wave (L-wave) whose propagation vector $\underset{\sim}{p}^{L+}$ is in the x_1x_3-plane, is of the form given by Eq.(2.1). The interaction of the plane incident wave with the plane containing the cracks gives rise to a complicated pattern of reflected and transmitted waves. Sufficiently far from the cracked plane, and at sufficiently low frequencies, it may however be assumed that the reflected and transmitted waves are homogeneous plane waves. The dominance of these reflected and transmitted homogeneous plane waves has been shown rigorously by Achenbach and Kitahara [25] for interaction of an incident wave with a distribution of spherical cavities whose centers are periodically distributed over a plane, and by Angel and Achenbach [26] for the corresponding two-dimensional case of a periodic distribution of line cracks. For these periodic cases the incident wave gives rise to reflected and transmitted wave modes with the same periodicity as the geometrical configuration, each with its own cut-off frequency. At frequencies under its cut-off frequency, a mode will be evanescent, while it will be propagating at frequencies above that cut-off frequency. Under the lowest cut-off frequency only the zeroth order mode will be propagating, and that mode is a homogeneous plane wave, i.e., a plane wave which is uniform over its wavefronts. Even though it has not been proven that a similar system of wave modes exist for an arbitrary distribution of scatterers over a plane, it seems very reasonable to assume that the far field still consists of homogeneous plane waves when the frequency is not too high.

By virtue of these arguments, the wave motion far from the cracked layer may then be written in the form of reflected and transmitted longitudinal and transverse waves as:

$$x_3 < 0:\ u_i(\underset{\sim}{x}) = u_o d_i^{L+} \exp(ik_L p_j^{L+} x_j) + u_o R^L d_i^{L-} \exp(ik_L p_j^{L-} x_j) + u_o R^T d_i^{T-} \exp(ik_T p_j^{T-} x_j), \tag{3.1}$$

$$x_3 > 0:\ u_i(\underset{\sim}{x}) = u_o T^L d_i^{L+} \exp(ik_L p_j^{L+} x_j) + u_o T^T d_i^{T+} \exp(ik_T p_j^{T+} x_j) \tag{3.2}$$

where R^L, R^T and T^L, T^T are the reflection and transmission coefficients of the reflected and transmitted longitudinal and transverse waves, and

$$d_i^{L\pm} = p_i^{L\pm} = (\sin\theta, 0, \pm\cos\theta) \quad , \tag{3.3}$$

$$p_i^{T\pm} = (\sin\theta_T, 0, \pm\cos\theta_T) \quad , \tag{3.4}$$

$$d_i^{T\pm} = (\mp\cos\theta_T, 0, \sin\theta_T) \quad , \tag{3.5}$$

$$\theta_T = \arcsin(\eta\sin\theta) \ , \ \eta = \kappa^{-1} = c_T/c_L = [(1-2\nu)/2(1-\nu)]^{\frac{1}{2}} \tag{3.6a,b}$$

Once it has been assumed that the far-fields consist of plane reflected and transmitted waves, the corresponding reflection and transmission coefficients can be obtained rigorously by an application of the Betti-Rayleigh reciprocal theorem. This approach was first used by Achenbach and Kitahara [25] and Sotiropoulos and Achenbach [27]. For a cell defined by $|x_1| \le \ell/2$, $|x_2| \le w/2$ and $|x_3| \le h/2$, shown in Fig. 4, the reciprocal theorem relates two distinct elastodynamic states, A and B, of the same body

$$\int_S (t_i^A u_i^B - u_i^A t_i^B) dS = 0 \tag{3.7}$$

where

$$S = S_1^{\pm} + S_2^{\pm} + S_3^{\pm} + \sum_{m=1}^{N} A_m^{\pm} \quad . \tag{3.8}$$

Here, t_i are the traction components acting on S, $S_m^{\pm}$ are the exterior bounding surfaces of the cell, and $A_m^{\pm}$ are the insonified side and the shadow side of the m-th crack. Also, body forces are not considered in Eq.(3.8), h is assumed to be sufficiently large, and N is the number of cracks contained in the cell. For $\{u_i^B, t_i^B\}$ one of the following auxiliary elastodynamic states is taken

$$u_{i(R)}^{BL} = - u_o d_i^{L-} \exp(-ik_L p_j^{L-} \cdot x_j) \quad , \tag{3.9}$$

$$u_{i(R)}^{BT} = - u_o d_i^{T-} \exp(-ik_T p_j^{T-} \cdot x_j) \quad , \tag{3.10}$$

where d_i^{L-}, p_i^{L-}, d_i^{T-} and p_i^{T-} are defined by Eqs.(3.3)-(3.5). The auxiliary elastodynamic states (3.9)-(3.10) are opposite in phase and propagate in directions opposite to the reflected longitudinal and transverse waves.

Substituting the displacements and tractions corresponding to Eq.(2.1) for state A, and those corresponding to (3.9) for state B, and considering the traction-free boundary conditions on the faces of the cracks, the following expression is obtained for the reflection coefficient R^L

$$R^L = - F_L \sum_{m=1}^{N} \int_{A_m^+} t_{i(R)}^{BL} \Delta u_i^m dS \quad , \tag{3.11}$$

where the quantities Δu_i^m, i=1,2,3, denote the crack opening displacements of the m-th crack, and

$$F_L = [2ik_L u_o^2 (\lambda+2\mu) \ell w \cos\theta]^{-1} \quad . \tag{3.12}$$

Similarly, by substituting the displacements and tractions corresponding to Eq.(2.1) for state A, and those corresponding to Eq.(3.10) for state B, the reflection coefficient R^T can be obtained as

$$R^T = - F_T \sum_{m=1}^{N} \int_{A_m^+} t_{i(R)}^{BT} \Delta u_i^m dS \quad , \tag{3.13}$$

where

$$F_T = [2ik_T u_o^2 \mu \ell w \cos\theta_T]^{-1} \quad . \tag{3.14}$$

In these equations, $t_{i(R)}^{BL}$, and $t_{i(R)}^{BT}$ are the traction components induced by the auxiliary displacement fields (3.9)-(3.10), acting on the insonified side of the cracks.

In principle, the reflection and transmission coefficients can be computed, once the crack opening displacements have been calculated for all N cracks. However, this procedure is impractical since we are dealing with a large number of cracks. In what follows, we assume that all cracks are identical (with radius a) and parallel to each other. We consider low frequency and low crack density, and we further assume that the cracks are randomly distributed so that all positions of the cracks are equally probable. Under these assumptions, the amplitudes of the crack opening displacements for all cracks are approximately the same, except for a difference in phase due to the incident wave, i.e,

$$\Delta u_i^m(\underset{\sim}{x}) \simeq \Delta u_i(\underset{\sim}{x}^o)\exp(ik_L p^{L+} x_j) \quad , \tag{3.15}$$

where $\Delta u_i(\underset{\sim}{x}^o)$ are the crack opening displacements of a reference crack centered at the origin of the coordinate system. Substitution of Eq.(3.15) and the expression for $t_{i(R)}^{BL}$, corresponding to Eq.(3.9), into Eq.(3.11) results in

$$R^L = F_L\left\{-V_1\sin2\theta + \left(\frac{1-2\eta^2}{\eta^2} + 2\cos^2\theta\right)V_3\right\} ik_L\mu u_o N \ , \tag{3.16}$$

where η is defined by Eq.(3.6b), and V_i are the crack opening volumes of the reference crack, which are defined by

$$V_i = \int_A \Delta u_i(\underset{\sim}{y})dS \quad . \tag{3.17}$$

By substituting Eq.(3.12) into Eq.(3.16), the final result for R^L is obtained as

$$R^L = \frac{\eta^2\bar{N}}{2u_o\cos^2\theta}\left\{-V_1\sin2\theta + \left(\frac{1-2\eta^2}{\eta^2} + 2\cos^2\theta\right)V_3\right\} \quad . \tag{3.18}$$

where $\bar{N} = N/\ell\omega$ denotes the number of cracks per unit area. In a similar manner, an expression for R^T is obtained as

$$R^T = -\frac{\bar{N}}{2u_o\cos\theta_T}\left\{V_1\cos2\theta_T + V_3\sin^2\theta_T)\right\} \quad , \tag{3.19}$$

It should be noted that the present approximation does not take account of the interaction effects between individual cracks. Thus, the validity of the present analysis is limited to crack densities that are not too large.

The time-averaged power flow of an elastodynamic state is defined by

$$< P > = -\frac{1}{2}\,\omega\ \mathrm{Im}\int_S t_i u_i^* dS \quad , \tag{3.20}$$

where t_i defines the traction components corresponding to the displacement field u_i, and u_i^* denotes the complex conjugate of u_i. Also S is defined by Eq.(3.7). Since no energy is created or destroyed inside the domain enclosed by S, we must evidently have

$$< P > = 0. \tag{3.21}$$

Substitution of the appropriate displacements and tractions in Eq.(31) yields by virtue of Eq.(3.21):

$$|T^L|^2 = 1 - |R^L|^2 - \eta\,\frac{\cos\theta_T}{\cos\theta}\,(|R^T|^2 + |T^T|^2) \ . \tag{3.22}$$

The reflection and transmission coefficients for a layer permeated by distributed 2-D cracks (line cracks) have exactly the same expressions as for penny shaped cracks, except that $\bar{N}$ should be interpreted as the number of cracks per unit length along the x_1-axis. The reflection coefficient for a special two-dimensional case, namely, the one of periodically spaced cracks of equal width, can be obtained by a numerically rigorous method, see Ref.[26]. This coefficient provides a convenient result to test the approximations that must be used to make Eq.(3.18) useful. Suppose the cracks are of width 2a, and their centers are spaced a distance D apart. We then have $\bar{N} = 1/D$, and for normal incidence ($\theta = 0$) the reflection coefficient given by Eq.(3.18) becomes

$$R_L = V_3/2u_o D \tag{3.23}$$

A considerable effort is required to determine a crack-opening volume if the effects of neighboring cracks are taken into account. Hence in first approximation these effects are often neglected and the crack opening volume is approximated by the one for a single crack in an unbounded solid. For the two-dimensional case it is not difficult to check the accuracy of this approximation in the limit of vanishing frequence. A static crack-opening volume, which would correspond to the zero-frequency limit of a dynamic problem can be calculated from the following energy relation, which is well-known in fracture mechanics

$$\frac{1}{2}\sigma_{33}V_3 = \frac{1-\nu}{\mu}\int_o^a K_I^2 da \tag{3.24}$$

For a single line crack of width 2a in an unbound solid we have $K_I = \sigma_{33}(\pi\alpha)^{1/2}$, and hence if σ_{33} is defined by Eq.(2.20), Eq. (3.24) yields

$$V_3^{sin} = \pi(1-\nu)\frac{\lambda+2\mu}{\mu} ik_L u_o a^2 \tag{3.25}$$

For the periodic case of cracks of width 2a whose centers are spaced a distance D apart along the x_1-axis, Ref.[28] gives the Mode-I stress intensity factor as $K_I = \sigma_{33}[D\tan(\pi a/D)]^{1/2}$. Substitution of this expression in Eq.(3.24) yields

$$V_3^{per} = -\frac{2}{\pi}(1-\nu)\frac{\lambda+2\mu}{\mu} ik_L u_o D^2 \ \ell n[\cos\frac{\pi a}{D}] \tag{3.26}$$

Substitution of either (3.25) or (3.26) would give the zero frequency slope of the R_L versus $k_T D$ curve. The ratio of the slopes follows as

$$\text{ratio} = -\frac{2}{\pi^2}\left(\frac{D}{a}\right)^2 \ell n\left[\cos\frac{\pi a}{D}\right] \tag{3.27}$$

In the limit as a/D approaches zero this limit approaches unity. The next term in an expansion for small values of $\pi a/D$ is $(1/6)(\pi a/D)^2$, which should be sufficiently smaller than unity in order that interaction between the cracks may be neglected. For example for an error of 5% we can allow $a/D \simeq 0.17$, which would correspond to about 2 crack widths of intact material between cracks.

On the basis of the preceding discussion it may be assumed that for a distribution of penny-shaped cracks, the crack-opening volume for a single crack may be substituted in Eq.(3.18) provided that the centers of the crack are spaced say some six crack radii apart. At low frequencies Eq.(2.15) may be used, but at higher frequencies the numerical results shown in Fig. 3 should be used for the crack-opening volume.

If for normal incidence the reflection coefficient has been measured as a function of the frequency, then the inverse problem requires the solution of two quantities, namely, $\bar{N}$ and V_3. At low frequencies only the slope of the reflection coefficient will be known, which would not provide enough information. If higher frequency information will be available, particularly the frequency at which R_L achieves a maximum value, then, in principle, both $\bar{N}$ and a can be determined. It should, however, be kept in mind that a maximum value of R_L is not necessarily related to the crack radius a, but may alternatively be related to the crack spacing D.

As discussed in Ref.[27], low frequency information may be sufficient to obtain a reasonable estimate of the relevant stress intensity factor for a service load condition, provided that the number of cracks per unit area is known a-priori within reasonable bounds, say 25%. Then by expressing V_3 in terms of R_L, and substituting the result in Eqs.(2.27) and (2.26) quite good K_I estimates are obtained, due to the error reducing effect of the 1/6 root in Eq.(2.26). Details have been discussed in Ref.[27].

4. ATTENUATION DUE TO DISTRIBUTED PENNY-SHAPED CRACKS

Finally we will consider the attenuation of ultrasonic waves which propagate through a solid containing distributed penny-shaped cracks. This is a complicated problem. In general both the phase velocity and the amplitude are affected when an harmonic wave propagates through such a solid. Here we will briefly consider the long wavelength limit for the special case of parallel penny-shaped cracks of equal size which are sufficiently widely spaced not to interact with each other.

The usual argument for the coefficient of attenuation is based on the use of the scattering cross section of a single crack. In physical terms the scattering cross section represents the ratio of the time-averaged rate of energy scattered in all directions to the time averaged intensity of the incident wave. Usually the scattering cross section is obtained from its relations to the far-field amplitude of the scattered field on the incident ray behind the obstacle. Since the scattered field can be considered as being generated by crack tractions, one would expect that the scattering cross section can also be obtained by a rate of energy calculation for loading of the crack faces. This has been shown rigorously by Hudson [29]. It is easily verified that for normal incidence a very simple expression is obtained which involves the crack-opening volume and the normal stress of the incident wave on the plane of the crack:

$$Q_L = \frac{\frac{1}{2}\mathrm{Re}(i\omega\sigma_{33}V_3^*)}{I}, \qquad (4.1)$$

where the asterisk indicates the complex conjugate, and I is the intensity of the incident wave

$$I = \frac{1}{2}\rho c_L u_o^2 \omega^2 \qquad (4.2)$$

Upon substitution of σ_{33} from Eq.(2.20) and V_3 from Eq.(2.15) we obtain

$$Q_L = \frac{4}{135\pi} \frac{8+15\eta - 40\eta^3 + 32\eta^5}{(1-\eta^2)^2 \eta^5} \frac{\omega^4 a^6}{c_L^4} \qquad (4.3)$$

where η is defined by Eq.(3.6b). It may be verified that this is the same result as calculated by Piau [30, Eq.(21)] by a different method.

Now if a unit volume of the solid contains n cracks, then the time-averaged rate of energy scattering per unit volume may be written as

$$D = -n\, Q_L I \qquad (4.4)$$

The attenuation factor α_L may be introduced by considering a plane wave of the general form

$$u_3 = u_o \exp(-\alpha_L x_3)\exp(i\bar{k}_L x) \qquad (4.5)$$

For this wave we have $dI/dx_3 = -2\alpha_L I$, which must just equal $-n\, Q_L I$. Hence in first approximation we have $\alpha_L = (1/2)nQ$.

It is generally possible to measure the coefficient of attenuation. Unfortunately the useful quantity V_{30} does not enter in Q_L, and hence not in α_L. Consequently it is not possible to establish a connection with a stress intensity factor, at least not from an attenuation measurement.

A more advanced analysis of the relation between phase velocity and attenuation and certain strength parameters has been given by Evans et al [31].

ACKNOWLEDGMENT

This work was carried out in the course of research for Contract N00014-85-K-0401 with the Office of Naval Research, Mechanics Program. The author is pleased to acknowledge a helpful discussion with Prof. S. K. Datta of the material of Section 4.

REFERENCES

1. J.D. Achenbach, in Advances in Fracture Research, 1989, vol. 5 (Proceedings ICF7, eds. K. Salama, K. Ravi-Chandar, D.M.R. Taplin and P. Rama Rao), Pergamon Press, New York.
2. K. Wang (ed): Acoustical Imaging, 1980, vol. 9, Plenum Press, New York, NY.
3. A. Briggs: An Introduction to Scanning Acoustic Microscopy, 1985, Oxford University Press, Oxford, UK.
4. C.F. Quate, A. Atalar and H.K. Wickramasinghe: Proc. IEEE, 1979, vol. 67, p. 1092.
5. R.B. Thompson and D.O. Thompson: Proc. of the IEEE, 1985, vol. 73, p. 1716.
6. L.S. Fu: Applied Mechanics Reviews, 1982, vol. 35, p. 1047.
7. R.B. Thompson: J. Appl. Mech., vol. 50, (50th Anniversary Issue), 1983, p. 1191.
8. J.E. Gubernatis, in Review of Progress in Quantitative Nondestructive Evaluation, vol. 5 (eds. D.O. Thompson and D.E. Chimenti), p. 21, Plenum, New York, 1986.
9. L.J. Bond, M. Punjani and N. Saffari: IEEE Proceedings, 1984, vol. 131, p. 265.
10. J.M. Coffey and R.K. Chapman: Nucl. Energy, 1983, vol. 22, p. 319.
11. J.E. Allison, J.M. Hyzak and W.H. Reimann, in Preventions of Structural Failure (ed. W. Lewis), p. 240, Am. Soc. of Metals, Metals Park, OH., 1978.
12. C.G. Annis, Jr., M.C. Van Wanderham, J.A. Harris, Jr. and D.L. Sims: J. Eng. for Power, 1981, vol. 103, p. 198.
13. P.A. Martin and G.R. Wickam, Proc. Roy. Soc. London, 1983, vol. A390, p. 91.
14. W. Lin and L.M. Keer: J. Acoust. Soc. Am., 1987, vol. 82, p. 1442.
15. D.E. Budreck and J.D. Achenbach: J. Appl. Mech., 1988, vol. 55, p. 405.
16. N. Nishimura and S. Kobayashi, Comp. Mech, 1989, vol. 4, p. 319.
17. I.A. Robertson: Proc. Camb. Phil. Soc., 1967, vol. 63, p. 229.
18. A.K. Mal: Int. J. Engng. Sc., 1970, vol. 8, p. 389.
19. H.D. Garbin and L. Knopoff: Quart. Appl. Math, 1973, vol. 30, p. 453.
20. S. Krenk and H. Schmidt, Phil. Trans. R.Soc. London, 1982, vol. A308, p. 167.
21. A. Roy: Int. J. Engng.Sc., 1987, vol. 25, p. 155.
22. J.D. Achenbach, D.A. Sotiropoulos and H. Zhu: J. Appl. Mech, 1987, vol. 54, p. 754.
23. B. Budiansky and R.J. O'Connell: Int. J. Solids and Structures, 1976, vol. 12, p. 81.
24. B. Budiansky and J.R. Rice: J. Appl. Mech, 1978, vol. 45, p. 453.
25. J.D. Achenbach and M. Kitahara: J. Acoust. Soc. Am, 1986, vol. 80, p. 1209.
26. Y. Angel and J.D. Achenbach: J. Appl. Mech., 1985, vol. 52, p. 33.
27. D.A. Sotiropoulos and J.D. Achenbach: J. Nondestr. Eval., 1988, vol. 7, p. 123.
28. H. Tada, P. Paris and G. Irwin: The Stress Analysis of Cracks Handbook, 1973, Del Research Co., St. Louis, MO.
29. J.A. Hudson: Geophys. J.R. Astr. Soc., 1981, vol. 64, p. 133.
30. M. Piau: Int. J. Engng. Sc., 1980, vol. 18, p. 549.
31. A.G. Evans, B.R. Tittmann, L. Ahlberg, B.T. Khuri-Yakub and G.S. Kino: J. Appl. Phys. 1978, vol. 49, p. 2669.

Elastic Waves and Ultrasonic Nondestructive Evaluation
S.K. Datta, J.D. Achenbach and Y.S. Rajapakse (Editors)

NUMERICAL TECHNIQUES AND THEIR USE TO STUDY WAVE PROPAGATION AND SCATTERING - A REVIEW

Leonard J. BOND

NDE Centre, Department of Mechanical Engineering
University College London
Torrington Place
LONDON, WC1E 7JE, United Kingdom

Various numerical modelling techniques which are applied to ultrasonic wave propagation and scattering are reviewed. The factors which are important in the selection of a modelling technique, to solve a specific problem, are discussed. A comparison of the capabilities and limitations of the various models is attempted and current trends and developments outlined, with interest focused on techniques which are applied to problems in the mid-frequency scattering regime.

1. INTRODUCTION.

It was recognised in the mid-1970's that if ultrasonic NDT were to become quantitative there was the need to develop an adequate science base for the subject. The provision of an adequate theoretical analysis for ultrasonic wave propagation and scattering is essential for the optimization of current NDT techniques and also for the development of new techniques, especially where novel problems are encountered.

Practical ultrasonic measurements usually produce an A-scan or 'RF' trace which may be combined with scanning techniques to give either a B-scan or a C-scan [1]. In addition there are quantitative NDT techniques which are seeking to give flaw characterisation from the solution of experimental inverse problems. These are, by their very nature, often ill posed and with very restricted data sets and the solution of such problems is difficult. It however remains the solution of such practical inverse problems, which give material and/or flaw characteristics, that is at the heart of quantitative ultrasonic NDT. Before adequate solutions for the inverse problem can be developed it is necessary to understand the relevant range of corresponding forward scattering problems.

Elastic wave propagation and scattering in solids has been under investigation in a wide range of fields of study, including seismology, geophysics, electronics and applied mechanics for over 50 years. There is now a well established text book literature on 'elasto-dynamics' [e.g. 2], as well as a range of monographs and detailed research literature which was reviewed by Pao [3]. The range of problems for which solutions are available, although extensive, remains limited [2,3,4] and it is quickly seen from the literature that many important practical ultrasonic problems are analytically intractable. To overcome these limitations with theory a variety of approximate solutions are employed [3,4,5,6]; for example, Born approximation for weak scattering and the Kirchhoff approximation for strong scattering in the high frequency regime. It has however been found that the various approximate models also fail to provide an adequate description for some important problems involving interactions especially in the mid-frequency scattering regime. To provide the necessary theory a range of numerical models have been developed. The role of modelling and its important place in NDT and NDE was recently discussed by Coffey [7]. A review of all types of mathematical modelling employed in NDT is provided by Georgiou and Blakemore [8] and that paper is currently being revised and expanded [9].

For modelling, as with other aspects of ultrasonic NDT, many of the techniques employed were originally developed to solve problems encountered in other fields of study, such as seismology [10]. Various reviews of modelling for NDT, including ultrasonics, have now been given [5,8,9,11]. The need for improved NDT, especially in the nuclear and aero-space industries, combined with the growing capability of computers has resulted in a major growth in the application of numerical modelling to NDT problems [6,7].

This paper seeks to provide an overview of various numerical techniques which are used to model wave propagation and scattering for ultrasonic NDT. Interest is focused on problem analysis and a comparison of the methods, especially for problems which involve mid-frequency scattering, where feature dimensions are of the same order as the wavelengths of the ultrasound involved.

2. DEVELOPMENTS IN MODELLING CAPABILITY.

Over the past decade the vast increase in computer capability has revolutionised the capabilities and potential for numerical modelling. To illustrate these developments finite difference methods, applied to elastic wave problems are considered here [11], but corresponding growth is also found with other methods.

This family of techniques was first employed in 1967/68 by Alterman [12] in Israel and Bertholf [13] in the USA and both groups considered a range of pulsed wave problems with a 2-D grid of nodes. By 1974 the models run using a CDC 7600 computer involved about 200 by 100 nodes (20,000 grid points), with two components of displacement calculated at each node. Such a grid corresponds to a 2-D physical region with dimensions of about 10 by 5 wavelengths and the model was then iterated to give displacements over 200 time steps. Such a program results in the calculation of $8x10^6$ components of displacement and can be used to model either near-field or far-field scattering problems.

By 1988/89 finite difference models run on a CRAY computer have been reported using grids of 2,000 by 2,000 nodes ($4x10^6$ grid points), again with two components of displacement calculated at each node. Such a grid corresponds to a region of 100 wavelengths square. In this case the number of time iterations corresponds to about the maximum grid dimensions and the result is a program which calculates up to $16x10^9$ components of displacement.

In addition to 2-D problems full 3-D finite difference models are also under development and are being run on CRAY 2 machines. These have reported grid dimensions, in each of the three dimensions, in excess of 100 nodes (over $1x10^6$ grid points) with all three components of displacement calculated at all nodes. These 3-D models require similar size calculations to those for the large 2-D systems [14].

It is therefore observed that numerical modelling for ultrasonics is now in many cases no longer hardware limited. The modelling community do not have to try and shoe-horn their program into a machine and then leave it to run as a batch job for a day or more. They are not now faced with data interpretation based on simple line printer or graph plotter output. The current range of work stations and graphics packages, combined with the availability of computational power are all making revolutionary changes in modelling, including data display and manipulation.

3. PROBLEM ANALYSIS FOR MODELLING IN ULTRASONIC NDT.

In quantitative ultrasonic NDT practical measurements on signals seek to determine various flaw parameters which may be those for an isolated feature or more general material properties.

For the case of an isolated flaw in an otherwise homogeneous matrix the interest is in the determination of;

i. Detection capability and limitations. Probability of detection (POD) for various classes of flaws.
ii. Flaw location, within the part geometry.
iii. Flaw type and its shape.
iv. Orientation.
v. Size.
vi. Distribution (if more than one)

The ultrasonic measurement is required to give the scatterers characterisation, and this should be provided in such a form that the data obtained can then be used in damage tolerance or fracture mechanics calculations.

Any theoretical model or description of ultrasonic NDT is required to give data in a form where experimental comparisons are valid, such as with;

i. Simulated 'A-scans' or 'RF' data. (to give arrival time data)
ii. Reflection coefficients. (spectrum or frequency domain)
iii. Complete wave field visualisation.
iv. Far-field radiation pattern. (directivity)
v. Surface point displacements.

The analysis of a complete ultrasonic NDT problem for modelling becomes a matter of identification of key features, knowing what input data is available and the required form of output. The features selected for modelling can then be compared with the capabilities of the various modelling techniques. Such analysis may involve the use of a simple linear system or the more general "Total Wave Model"[4]. Some models may well be required to consider all the elements in a system, whereas others exclude consideration of the transducers and the electrical circuits. The central aspect of a model for ultrasonic NDT is the description of the physical acoustics. In this paper only techniques which are used to calculate displacements are considered.

A key factor in numerical modelling is the completeness of the model; i.e. does it include all the necessary physics? It is then required that the validity of the model is demonstrated. The range of validity and level of output accuracy to be achieved, with specific data sets are also required.

4. REVIEW OF NUMERICAL METHODS.

Modelling now has a major role to play in NDT and a wide range of techniques have been employed to solve the forward or direct scattering problems [1,2,3,7,9,11,16]. Examples of most modelling techniques that have been employed to study ultrasonic NDT are to be found reported in the series "Review of Progress in Quantitative NDE" [15] and various other publications [16], including these proceedings.

To seek to review any extensive literature is never an easy task and this is especially true for a subject like numerical modelling for ultrasonic NDT where so many of the methods are also techniques that are being developed in other fields of study.

In this review seven groups of modelling techniques are considered and if the single property of ka range (where k is the wave number and a the flaw dimension) is used they can be grouped together in terms of scattering regime [4,5,6]. All the methods calculate components of displacement in the wave-field. In the high ka regime there are some analytic solutions and both elastic Kirchhoff and Geometrical Theory of Diffraction (GTD) methods apply to give far-field solutions. In the low ka regime other analytic techniques are available together with the Born approximation which again gives far-field solutions. The three remaining techniques considered are finite difference, T-matrix and the Boundary Integral Equation (BIE) or Boundary Element Methods (BEM) and these can all be applied to either far-field or near-field solutions in the mid-frequency scattering regime.

The properties of the various numerical methods considered are summarised as Table 1. In addition the various techniques that are related to finite difference methods are summarised in Table 2. Each group of techniques is now considered.

Standard problems.

In many cases the available analytical solutions are found in the various textbooks [2,3]. In addition there are some problems which are of particular interest to NDT, such as the semi-infinite crack which was considered by Achenbach et al [17]. There are also integral equation solutions available for problems such as wave interaction with the penny-shaped crack [18], the infinite strip, the cylinder and sphere [19], and some wedges.

Low ka or small flaws. (ka $\ll$ wavelength)

This is the scattering regime where flaws have overall dimensions much smaller than wavelength and it tends to be of interest in the aero-space industry, where it is required to inspect critical rotating components such as turbine discs. In this regime two techniques are in common use and these are the Born and the quasi-static approximations.

The Born approximation has been extensively used to describe weak scattering inhomogeneity. Much of the early work was reported by Gubernatis et al [20].

Technique.	Analytical methods	Finite differences	Elastic Kirchhoff	Born Approx.	Geometrical theory of diffraction	T-Matrix.	BIE or BEM.
Property.							
ka restriction.	$ka \ll 1$ $ka \gg 1$	$0.1 < ka < 20$	$ka \gg 1$	$ka < 1$	$ka \gg 1$	$0 < ka < 15$	$0 < ka < 6$
Field region	Far	Near or Far	Far	Far	Far	Either	Near or Far
Dimension	3-D	2-D (3-D)	3-D	3-D	2.5-D	3-D	Most 3-D
Shape of scatterer	circle, cylinder sphere.	limited (square.) *	small surface slope.	good range **	crack-like	ellipsoidal cavity.	Arbitrary.
Included material	restricted	most ***	**	weak	X	***	Strong or weak
Mode conversion	**	***	***	**	*	***	***
Incident wave.	plane or spherical	arbitrary 2-D	plane or spherical	**	**	**	Arbitrary
Short pulse.	**	***	*	+	*	X	FFT used.
Multiple scattering. (2 Body)	+	***	+	+	*	**	**.

Rating system: *** Very good, ** Good, * Copes, + poor, X Very Poor. (after [6])

Table 1. Comparison of various modelling techniques applied to ultrasonic wave propagation and scattering.

Technique.	Finite difference	FE/FD	Hybrid FD/GTD.	Lumped Mass.	FD + perturbation.
Property.					
ka restriction.	$0.1 < ka < 20$	$0.1 < ka < 20$	$0.1 < ka < 20$	$0.1 < ka < 20$	$0.01 < ka < 20$ **
Field region	Near or Far.	Near or Far.	Near and Far. ***	Near or Far.	Near or Far.
Dimension	2-D & 3-D	2-D	2-D	2-D	2-D
Shape of scatterer	limited (square.) *	***	Limited	**	**
Included material	most	most	most	most	most
Mode conversion	***	***	***	***	***
Incident wave.	arbitrary 2-D	arbitrary 2-D	arbitrary 2-D	arbitrary 2-D	arbitrary 2-D
Short pulse.	***	***	***	***	***
Multiple scattering. (2 Body)	***	***	***	***	***

Rating system: *** Very good, ** Good, * Copes, + poor, X Very Poor. (after [6])

Table 2. Comparison of various finite difference and related modelling techniques.

The quasi-static approximation employs a Green's integral for scattering by flaws that are small compared with the wavelengths involved. The method is described in detail by Gubernatis and Domany [21].

For both these methods many examples of applications are found in the literature [e.g. 15].

High ka or large flaws (ka >> wavelength).

This is the regime where flaws are large compared with the wavelength employed and it tends to be considered in NDT of large thick section structures such as those found in the power generation industry. It is also a regime where the theory applies to many problems where acoustic imaging techniques are employed. The most useful theories in this regime are those developed from classical optical scattering theory.

One of the most commonly used approximations is the Kirchhoff approximation which is applied to wave interactions with large, strong scattering features. This formulation is based on the Green's integral [22]. This approach has been applied to a number of problems and good agreement with experimental data has been obtained [5,15,23].

What is probably the most widely used and among the most powerful elastic wave scattering model is the elasto-dynamic Geometrical Theory of Diffraction (GTD) which is derived directly from optical ray theory. The classic paper in this field is by Keller [24] and there are numerous applications of the technique [5,15,17,22].

Mid-frequency scattering or intermediate size flaws (ka = wavelength).

It is this regime where the flaws are of the same order of magnitude, in size, as the wavelength of the ultrasound used. It is in this size range where many problems occur and it is also in this regime where the theory is the most difficult. Four main families of techniques are commonly employed in this regime and as seen from Tables 1 and 2 each have their own particular strengths and limitations.

i. T-Matrix.

This is a method based on the solution of the systems integral equation which combines the incident and scattered fields that have been formulated as a set of basis functions. The technique has been fully described in several places [25]. The major disadvantages of this scheme are in the effort which may be required in the formulation of the initial basis functions and even then the scheme may well be computationally intensive. It is however a powerful modelling tool which has been demonstrated for a range of applications including scattering by simple volumetric scatterers, cracks of arbitrary shape [26] and for obstacles near a boundary [27].

ii. Boundary Integral Equation or Boundary Element Methods.

This is a family of techniques which are similar to the T-Matrix in that this technique involves the solution of an integral equation, which in this case is obtained by dividing the surface of the scatterer into elements each of which then has an associated unknown displacement [28]. This is a technique that is increasingly used and which has already been used for scattering by arbitrarily shaped voids and also a range of both volumetric [30] and surface features [31,32]. The major limitations would appear to be due to the complexity which can be encountered in the formulation of the initial equations and then solution of the equation set involved.

iii. Finite Difference and Finite Element Methods.

In numerical modelling applied to all types of elastic waves much effort has been concentrated on explicit finite difference schemes [11,33,5] and more recently finite element, mixed finite element/finite difference [34] schemes and various forms of hybrid schemes. These types of models all considered mid-frequency scattering where resonant scattering and complex mode-conversion phenomena are encountered. Such models give a complete description of the time development of the wave-field of interest. They can be applied to consider pulsed and tone-burst problems for inputs of many types of waves including Rayleigh, Lamb, shear and compression, which can in turn provide sources of various shapes. They can also model mixed solid-fluid problems and include a wide range of material properties.

Results have been produced for a range of 2-D canonical geometries with pulsed plane wave excitation [35] and the results obtained agree well with experiments. In addition immersion inspection systems have been modelled where there is propagation in both an elastic solid and a viscous-fluid [33,35]. Finite difference model results have also been produced for pulsed wave scattering by a semi-infinite 2-D slit and these compared with those obtained with GTD. The two sets of data give good agreement for many features in the data and also some interesting differences, between the single frequency GTD model results and finite difference model [36].

Special nodal formulations have been produced to model particular effects such as partial crack closure and qualitative agreement with other theories has been obtained [36]. Most recently large scale finite difference models for 3-D wave propagation in inhomogeneous anisotropic media have been produced by Temple [14].

For conventional finite difference models the major limitations are those due to the limited range of nodal formulations, the restrictions imposed with simple grid coordinate systems and the shear size of the computational effort required to give accurate solutions for models that consider wave propagation over large distances.

To overcome the various limitations imposed by conventional finite difference techniques a series of related models have been developed and the properties of these are outlined in Table 2.

An early development was by Harumi and his co-workers who produced various 'lumped mass' schemes which appear to overcome many of the limitations on scatterer shape imposed with a conventional Cartesian coordinate scheme. This group's models have been used to produce a wide range of results which they have compared with experiments [37]. Not all details for this work are easily obtained.

A number of groups have produced schemes which employ a finite element formulation for the spatial dimensions and a finite difference formulation to give time propagation [34]. Lord and his co-workers have produced a number of such models and these are reported in several papers [15], as is similar work by various other groups. Models using this method are also reported in other papers (these proceedings). The major advantage of this approach is the flexibility that it gives in terms of the geometries that can be modelled.

One limitation on conventional finite difference modelling is its inability to consider small surface perturbations. One scheme which combines a finite difference model with a perturbation has been proposed by Hong et al [38] and applied to various problems in seismology.

A final area of limitation for finite difference models is found when seeking to model both the near-field and the far-field for a scattering problem. The shear size of grid involved can limit the domain that is modelled. This has been overcome by various groups who have combined GTD with finite difference methods. The finite difference scheme considers the scattering, including mode-conversion and the GTD technique is then used to give propagation to the far-field [16,39].

5. CURRENT TRENDS AND CONCLUSIONS.

The main objective in practical ultrasonic NDT will remain the detection and characterisation of flaws using various forms of sparse data. Modelling can be expected to be increasingly used to understand more complex forward scattering problems and in the formulation of inversion schemes for flaw characterisation.

A wide range of numerical modelling techniques are in current use and many examples of application are to be found in the literature. No one technique is the most suitable for all problems and it is important to match technique to the problem. All numerical methods have limitations, as well as strengths, and in any evaluation or comparison of techniques this should be known and acknowledged.

There is a clear need to ensure that the numerical model can provide data in a form for comparison with experimental results e.g. reflection coefficients. Such data for specific geometries can be made independent of the measurement system used and they are less subject to variability than the prediction of data such as the A-scan or RF signals which in most cases clearly include transducer and instrumentation effects.

With the growing use of models to predict results for systems where experimental data is not available, it is necessary to ensure the validation of model results, within some specified range of accuracy.

In terms of the mathematics involved there is a shift from consideration of 2-D scattering to full three dimensional problems. The range of geometries involved is becoming increasingly complex and more realistic flaws and material properties are being considered, such as stainless steel welds and composites. For single flaws in metals partial crack closure, roughness, and matrix material, that is inhomogeneous and anisotropic, are all on the agenda. There are also challenges which result from the use of new manufacturing processes, an example of which is modelling inspection of diffusion bonding.

The growth in the application of the various mathematical modelling techniques employed in ultrasonic NDT can be expected to continue as the provision of an adequate theoretical analysis for ultrasonic wave propagation and scattering is essential for the optimisation of current NDT techniques and also for the development of new techniques, especially where novel problems are encountered.

A range of signal processing techniques are also under development to try and extract more features and all such data is being considered for processing by artificial intelligence (AI) schemes. Model data is important in testing the capabilities of such signal processing techniques. There is also the major change in the move from modelling the physics of the forward scattering problem to the direct solution of the inverse problem. A range of inversion schemes are being developed which perform a "Model

Based Inversion" that assumes that some characteristics of the problem are known.

ACKNOWLEDGEMENTS.

Thanks are due to all my colleagues and students who have worked with me over the past 15 years on a wide range of the modelling techniques discussed in this paper. Thanks are especially due to Dr George Georgiou (now Topexpress Ltd) and Dr Richard Blake (now SERC Daresbury) for the contributions that they have made to the groups modelling activity over many years. The helpful discussions with Dr Andrew Temple, (Harwell Laboratory) and Professor Frank Rizzo (Iowa State University) which resulted in improvements to this paper are also gratefully acknowledged.

REFERENCES.

[1] Bray D E and Stanley R K (1989) NonDestructive Evaluation. McGraw Hill (New York).
[2] Graff K F (1975) Wave motion in elastic solids. Oxford University Press (Oxford).
[3] Pao Y-H (1983) Elastic waves in solids. J Appl Mechanics, Trans. ASME 50 pp 1152-1164.
[4] Bond L J and Saffari N. (1984) Mode-conversion Ultrasonic Testing. In "Research Techniques in NDT, Vol 7, Ed. R S Sharpe, Academic Press (London)." pp 145-189.
[5] Harker A H (1988) Elastic waves in solids, with applications to NDT of pipelines. Adam Hilger, in association with British Gas. (Bristol)
[6] Temple J A G (1987) European Developments in Theoretical Modelling of NDE for pipework. Int. J Pres. Ves. and Piping. 28 pp 227-267.
[7] Coffey J M (1988) Mathematical modelling of NDT techniques. In "Non-Destructive Testing", J M Farley and R W Nichols (eds) Pergamon Press (Oxford) Vol 1, pp 79-94.
[8] G A Georgiou and M Blakemore (1987) Mathematical Modelling and NDT; A state of the art. Proceedings 6th OMAE, (preprint)
[9] Georgiou, G.A. Blakemore M. and Bond L J (1989) A review of theoretical modelling for NDT. NDT International. (In Preparation)
[10] Richards P G (1979) Theoretical seismic wave propagation. Reviews of Geophysics and Space Physics 17 (2) pp 312-328.
[11] Bond L J (1982) Methods for the Computer Modelling of Ultrasonic Waves in Solids. In "Research Techniques in NDT, Vol 6, Ed. R S Sharpe, Academic Press (London)." pp 107-150.
[12] Alterman Z S (1968) Finite difference solutions to geophysical problems. J of Physics of the Earth. Vol 16 pp 113-128
[13] Bertholf L D (1976) Numerical solution for two-dimensional elastic wave propagation in finite bars. J Applied Mechanics 34 (3) 725-734.
[14] Temple J A G (1988) Modelling the propagation and scattering of elastic waves in inhomogeneous anisotropic media. J Phys D: Appl Phys. 21 pp 859-873.
[15] Thompson D O and Chimenti D (eds) (1980-on) Proceedings, Review of Progress in Quantitative Non-Destructive Evaluation, Vols 1-8. Plenum. (New York) Annual meeting proceedings
[16] Blakemore M and Georgiou G A. Editors. (1988) Mathematical Modelling in NDT. Clarendon Press (Oxford)
[17] Achenbach J D, Gautesen A K, and McMaken H (1982) Ray Methods for waves in elastic solids. Pitman (London)
[18] Martin P A and Wickham G R (1983) Diffraction of elastic waves by a penny-shaped crack: analytical and numerical results. Proc Roy Soc Lond. A390 pp 91-129.

[19] Pao Y-H and Mow C-C (1973) Diffraction of elastic waves and dynamic stress concentrations. Adam Hilger (New York)
[20] Gubernatis J E, Domany E, Krumhansl J A and Hubernam M (1977) The Born Approximation in the theory of the scattering of elastic waves by flaws. J Appl Phys. 48 pp 2812-2819.
[21] Gubernatis J E and Domany E (1979) Rayleigh wave scattering of elastic waves from cracks.
[22] Ogilvy J A (1988) Ultrasonic propagation in anisotropic welds and cast materials. In "Mathematical Modelling in NDT". (eds) M Blakemore and G A Georgiou. Clarendon Press (Oxford) pp 191-208.
[23] Haines N F and Langston D B (1980) The reflection of ultrasonic pulses from surfaces. J.A.S.A. 67 pp 1443-1454.
[24] Keller J B (1957) Diffraction by an aperture. J Appl Phys 28 pp 426-444.
[25] Varadan V K and Varadan V V (1980) Acoustic, electromagnetic and elastic wave scattering - focus on the T-Matrix approach. Pergamon Press (New York)
[26] Visscher W M (1983) Theory of scattering of elastic waves from cracks of arbitrary shape. Wave Motion 5 pp15-32.
[27] Varadan V V and Varadan V K (1981) Scattering of waves by obstacles in the presence of nearby boundaries- a self-consistent T-matrix. In "Elastic wave scattering and propagation." (eds) Varadan V K and Varadan V V. Ann Arbour Science (Michigan, USA)
[28] Manolis G D and Beskos D E (1988) Boundary element methods in elasto-dynamics. Unwin Hyman (London)
[29] Schafbuch P J, Thompson R B, Rizzo F J and Rudolphi T J (1989) Elastic wave scattering by arbitrarily shaped voids. In " Review of Progress in Quantitative Non-Destructive Evaluation, Vol 8. (eds) D O Thompson and D Chimenti. Plenum. (New York) pp 15-22.
[30] Rizzo F J, Shippy D J and Rezayat M (1985) A boundary integral equation method for radiation and scattering of elastic waves in three dimensions. Int. J. Numerical Methods in Eng. Vol 21 pp 115-129.
[31] Albach P A and Bond L J (1989) Numerical simulation of elastic wave scattering from three dimensional axisymmetric surface features. In " Review of Progress in Quantitative Non-Destructive Evaluation, Vol 9. (eds) D O Thompson and D Chimenti. Plenum(New York) (In Press)
[32] Sanchez-Sesma F J (1987) Site effects on strong ground motion. Soil Dynamics and Earthquake Engineering 6 (2) 124-132.
[33] Bond L J, Punjani M and Saffari N (1988) Ultrasonic wave propagation and scattering using explicit finite difference methods. In "Mathematical Modelling in NDT". (eds) M Blakemore and G A Georgiou. Clarendon Press (Oxford) pp 81-124.
[34] Blake R J and Bond L J (1989) Rayleigh wave scattering from surface features: wedges and down steps. Ultrasonics. (In Press)
[35] Saffari N and Bond L J (1987) Body to Rayleigh wave mode-conversion at steps and slots. J Nondestructive Evaluation 6 (1) pp 1-21.
[36] Punjani M and Bond L J (1986) Scattering of plane waves by a partially closed crack. In " Review of Progress in Quantitative Non-Destructive Evaluation, Vol 5A (eds) D O Thompson and D Chimenti. Plenum. (New York) pp 61-71.
[37] Harumi K, Ogura Y and Uchida M (eds) (1989) Ultrasonic Defect Sizing. Japanese Society for NDI.
[38] Hong M, Bond L J and Murrell S A F (1986) Computer modelling of seismic wave propagation and interaction with discontinuities in rock masses. In " Rock Mechanics: Key to Energy Production." (eds) H L Hartman. Society of Mining Engineers (Littleton, Colorado, USA) pp 488-495.
[39] Punjani M (1987) Crack characterisation using ultrasonic elastic waves. Ph.D. Thesis, University College London.

Elastic Waves and Ultrasonic Nondestructive Evaluation
S.K. Datta, J.D. Achenbach and Y.S. Rajapakse (Editors)

ELASTIC WAVE INVERSE SCATTERING

Anthony J. Devaney*

Department of Electrical and Computer Engineering
Northeastern University
Boston, Mass. 02115

The mathematical foundations of the inverse scattering problem at fixed frequency are presented for viscoelastic systems consisting of an isotropic, semi-transparent scatterer embedded in a homogeneous isotropic background. The inverse scattering problem is shown to reduce to the problem of estimating certain off-diagonal elements of an elastic wave scattering tensor ("T matrix") from measurements of the compressional and shear wave scattering amplitudes of the scatterer. In the weak scattering limit (Born approximation) the components of the T matrix tensor are found to be equal to certain linear combinations of the spatial Fourier transforms of the Lamé parameters and density of the scatterer and the inverse scattering problem reduces to estimating the density and elastic parameters from specification of these linear combinations over sets of "Ewald spheres" in Fourier space. This latter problem is discussed in some length for the special case of an elastic scatterer embedded in a fluid background and it is shown that for this case the two elastic parameters and density can be separately reconstructed if scattering data is available at multiple frequencies and if the bulk dispersion in the background fluid and scatterer can be neglected.

1. INTRODUCTION

In the following paper† the inverse scattering problem is formulated for linear viscoelastic systems consisting of one or more elastic inhomogeneities embedded in a homogeneous, isotropic elastic matrix. For the sake of simplicity it is also assumed that the inhomogeneities (scatterer) are isotropic and are thus characterized by three space and frequency dependent parameters (the two Lamé parameters and density). Because of the possible frequency dependence of the elastic parameters and density of the matrix and scatterer the inverse scattering problem is formulated in the frequency domain and thus consists of the determination of the elastic parameters and density of the scatterer at some given frequency from sets of scattered field measurements performed at that frequency.

The inverse scattering problem as defined above is often referred to as inverse scattering at "fixed energy" since it first arose in quantum mechanical inverse scattering applications involving conservative potentials probed by fixed energy particles [1]. However, it is clear that

* Also with A.J. Devaney Associates, 26 Edmunds Rd., Wellesley, MA 02181

†A preliminary account of the research reported here was presented at the *Workshop on Recent Techniques on Wave Propagation and Scattering*, Ohio State University, Oct. 1982.

by employing many probing frequencies (or, equivalently, wide bandwidth probing wavefields) that the inverse problem as formulated here is quite general and, in particular, is not inherently less general than time domain formulations of the inverse scattering problem and, in fact, has obvious advantages over time domain formulations for viscoelastic applications.

The measurement geometries employed in inverse scattering can be roughly divided into two categories: (i) tomographic type measurement geometries that employ near field measurements that are often performed in forward scattering directions and (ii) X-ray crystallographic type geometries that employ far field measurements over a full 4π steradians surrounding the scatterer. Although the applications of inverse scattering theory divide into the above two categories the inverse scattering *theory* required for these two categories is essentially identical and differs only in how one interprets and, possibly, preprocesses the scattered field data before inputting this data into an inverse scattering reconstruction algorithm. For this reason, and because the theory underlying category (ii) is more "pure" and less encumbered with non-essential details of the scattering geometry, only the second category will be considered in this paper.

The fundamental scattering model employed in the paper is presented in Section 2. The wave equation satisfied by the elastic displacement vector is expressed in terms of an "elastic scattering potential" that is a differential tensor operator that plays the same role in elastic wave scattering that the scalar valued scattering potential plays in quantum mechanical scattering [2]. A Lippmann Schwinger integral equation [2] relating the elastic displacement vector and the elastic scattering potential is introduced and shown to lead directly to an "elastic transition operator" that is completely analogous to the transition operator of quantum scattering theory [2]. The "T matrix" is defined in the usual way [2,3] as the matrix representation of the (elastic) transition operator in momentum space (Fourier space). Finally, the elastic wave scattering amplitudes are introduced and shown to be proportional to certain (on-shell) components of the T matrix.

The elastic wave inverse scattering problem is formulated and discussed in Section 3. The inverse problem is defined, in complete analogy to the inverse scattering problem in quantum mechanical scattering, to consist of determining the elastic scattering potential from knowledge of the elastic wave scattering amplitudes at a given frequency and for all incident and scattering directions. This problem is shown to reduce ultimately to determining the T matrix from its on-shell components as specified by the elastic wave scattering amplitudes. For the class of weakly inhomogeneous scatterers (weak scatterers) the scattering amplitudes are shown to specify certain linear combinations of the spatial Fourier transforms of the Lamé parameters and density of the scatterer. Fourier inversion of these linear combinations then leads to reconstructions of linear functionals of the Lamé parameters and density directly.

The special case of weak scatterers embedded in a fluid background is treated as an example of the general theory in Section 3.2. An algorithm is presented that generates a linear functional of the Lamé parameters and density of the scatterer from the compressional wave scattering amplitude specified for all incident and scattering directions. For the case where the scatterer and fluid background are weakly viscoelastic it is found that the Lamé parameters and density can be *separately* determined from scattered field measurements performed at three distinct frequencies. The final section summarizes the major results reported in the paper.

2. ELASTIC WAVE SCATTERING MODEL

We consider an elastic scattering system consisting of a semitransparent inhomogeneity (scatterer) embedded in a homogeneous background having Lamé constants λ_0, μ_0 and density

ρ_0. The scatterer will be characterized in terms of its elastic moduli tensor C_{ijkl} and density ρ. The Lamé parameters of the background medium as well as the elastic moduli tensor of the scatterer will be allowed to depend on frequency ω (the system can be viscoelastic) and both the elastic moduli tensor and density of the scatterer will, in general, depend on position $\mathbf{r}$. We will restrict our attention to isotropic media so that the elastic moduli tensor can be expressed in terms of Lamé parameters $\lambda(\mathbf{r})$ and $\mu(\mathbf{r})$† via the equation [4]

$$C_{ijkl} = \lambda(\mathbf{r})\delta_{ij}\delta_{kl} + \mu(\mathbf{r})(\delta_{ik}\delta_{jl} + \delta_{il}\delta_{jk}), \tag{2.1}$$

where δ_{ij} is the Kronecker delta function (= 1 if $i = j$ and zero otherwise).

We will characterize an elastic wavefield propagating in the composite system comprised of background with embedded inhomogeneity by an elastic displacement vector $\mathbf{u}(\mathbf{r})$ which represents the material displacement within the medium at point $\mathbf{r}$ and at frequency ω. This quantity satisfies the elastic wave equation [4,5]

$$(\lambda_0 + 2\mu_0)\nabla(\nabla \cdot \mathbf{u}) + \mu_0[\nabla^2\mathbf{u} - \nabla(\nabla \cdot \mathbf{u})] + \omega^2\rho_0\mathbf{u} = -\vec{V} \cdot \mathbf{u} \tag{2.2}$$

where $\vec{V}$ is the dyadic (tensor) operator‡

$$\vec{V} = V_{ik} = \frac{\partial}{\partial x_j}\delta C_{ijkl}(\mathbf{r})\frac{\partial}{\partial x_l} + \omega^2\delta\rho(\mathbf{r})\delta_{ik}. \tag{2.3}$$

Here, $\delta\rho = \rho - \rho_0$ is the deviation between the density of the medium at point $\mathbf{r}$ from that of the background and δC_{ijkl} is the deviation of the elastic moduli tensor C_{ijkl} from the elastic moduli tensor of the homogeneous isotropic background; i.e.,

$$\begin{aligned}\delta C_{ijkl}(\mathbf{r}) &= C_{ijkl}(\mathbf{r}) - [\lambda_0\delta_{ij}\delta_{kl} + \mu_0(\delta_{ik}\delta_{jl} + \delta_{il}\delta_{jk})] \\ &= \delta\lambda(\mathbf{r})\delta_{ij}\delta_{kl} + \delta\mu(\mathbf{r})(\delta_{ik}\delta_{jl} + \delta_{il}\delta_{jk}),\end{aligned} \tag{2.4}$$

where $\delta\lambda = \lambda - \lambda_0$ and $\delta\mu = \mu - \mu_0$

Besides the differential equation (2.2) we also require that the scattered field component of the displacement vector $\mathbf{u}$ satisfy the Sommerfeld radiation condition [6] which can be expressed in the form

$$\mathbf{u}^{(s)}(r\mathbf{s}) \sim \mathbf{A}^{(1)}\frac{e^{ik_1r}}{r} + \mathbf{A}^{(2)}\frac{e^{ik_2r}}{r} \tag{2.5}$$

as k_1r and k_2r tend to infinity along the fixed direction $\mathbf{s}$. Here, $\mathbf{u}^{(s)}$ denotes the scattered wave component of the elastic wavefield and

$$k_1 = \omega\sqrt{\frac{\rho_0}{\lambda_0 + 2\mu_0}} \tag{2.6a}$$

$$k_2 = \omega\sqrt{\frac{\rho_0}{\mu_0}} \tag{2.6b}$$

†We will not explicitly display the frequency ω in the arguments of the various field quantities since the theory will be developed in the frequency domain and ω can, thus, be regarded as a fixed parameter.

‡We will denote dyadics by an overline arrow and will refer to the dyadic operator $\vec{V}$ as the "elastic scattering potential" since it plays the same role in elastic wave scattering as does the usual scalar scattering potential in quantum mechanical scattering [2].

are the compressional and shear wavenumbers, respectively, of the background medium.

The asymptotic form of the scattered wavefield given in Eq.(2.5) is seen to consist of a pair of outgoing spherical waves, one of which is a compressional wave having amplitude $\mathbf{A}^{(1)}$ and the second is a shear wave having amplitude $\mathbf{A}^{(2)}$. We will in the following be primarily interested in the special case where the wavefield incident to the inhomogeneity is either a compressional or shear plane wave propagating with unit wavevector $\mathbf{s}_0$. In these cases the amplitudes $\mathbf{A}^{(j)}$, (j=1,2) of the two spherical waves attain special significance and are known as *scattering amplitudes* and will be denoted by $\mathbf{A}^{(j)}(k_j\mathbf{s}, k_v\mathbf{s}_0)$ where $j = 1, 2$ indexes the scattered wave and $v = 1, 2$ indexes the incident wave ($v = 1$ for a compressional wave and $v = 2$ for a shear wave).

The Sommerfeld radiation condition as stated in Eq.(2.5) together with the wave equation (2.2) lead directly to a Lippmann Schwinger integral equation [2] for the elastic displacement vector $\mathbf{u}$. This equation is readily derived using standard Green function methods and is found to be given by [5]

$$\mathbf{u}(\mathbf{r}) = \mathbf{u}_0(\mathbf{r}) + \int d^3r'\, \vec{\vec{G}}(\mathbf{r} - \mathbf{r}') \cdot \vec{\vec{V}} \cdot \mathbf{u}(\mathbf{r}') \tag{2.7}$$

where $\mathbf{u}_0$ is the incident wave and $\vec{\vec{G}}$ is the Green dyadic for the background media that satisfies the Sommerfeld radiation condition [5]:

$$\vec{\vec{G}}(\mathbf{R}) = \frac{-1}{4\pi\rho_0\omega^2}[k_2^2 \frac{e^{ik_2R}}{R}\vec{\vec{I}} - \nabla\nabla(\frac{e^{ik_1R}}{R} - \frac{e^{ik_2R}}{R})] \tag{2.8}$$

where $\vec{\vec{I}}$ is the unit dyadic and $R = |\mathbf{R}|$.

2.2 Elastic Transition Operator and T Matrix

The Lippmann Schwinger equation (2.7) can be expressed in the concise operator form†

$$\mathbf{u} = \mathbf{u}_0 + \hat{\vec{\vec{G}}} \cdot \hat{\vec{\vec{V}}} \cdot \mathbf{u} \tag{2.9}$$

where $\mathbf{u}$ and $\mathbf{u}_0$ are now regarded as vectors defined in an abstract vector space (e.g., the space of square integrable functions in R^3) and $\hat{\vec{\vec{G}}}$ and $\hat{\vec{\vec{V}}}$ are operators defined on this vector space. Following the standard method employed in quantum scattering theory [2] we are lead to define an *transition operator* $\hat{\vec{\vec{T}}}$ according to the equation

$$\hat{\vec{\vec{T}}} \cdot \mathbf{u}_0 = \hat{\vec{\vec{V}}} \cdot \mathbf{u} \tag{2.10}$$

so that Eq.(2.9) can be expressed in the form

$$\mathbf{u} = \mathbf{u}_0 + \hat{\vec{\vec{G}}} \cdot \hat{\vec{\vec{T}}} \cdot \mathbf{u}_0. \tag{2.11}$$

The transition operator $\hat{\vec{\vec{T}}}$ is an integral operator in coordinate space so that in this particular representation Eq.(2.10) becomes

$$\hat{\vec{\vec{T}}} \cdot \mathbf{u}_0(\mathbf{r}') = \int d^3r''\, \vec{\vec{T}}(\mathbf{r}', \mathbf{r}'') \cdot \mathbf{u}_0(\mathbf{r}'') \tag{2.12}$$

†We will denote abstract operators by placing a caret ^ over the symbol representing the operator.

where we have denoted the matrix elements of $\hat{\vec{T}}$ in the coordinate representation by $\vec{T}(\mathbf{r}', \mathbf{r}'')$. We now assume that the incident wave is a compressional or shear plane wave:

$$\mathbf{u}_0(\mathbf{r}) = \mathbf{a}_v e^{ik_v \mathbf{s}_0 \cdot \mathbf{r}}, \tag{2.13}$$

with $v = 1, 2$ and where $\mathbf{a}_v = \mathbf{s}_0$ for $v = 1$ (a compressional wave) and $\mathbf{a}_v = \mathbf{l}_0$ with $\mathbf{l}_0 \cdot \mathbf{s}_0 = 0$ for $v = 2$ (a shear wave). We then conclude from Eq.(2.12) that

$$\hat{\vec{T}} \cdot \mathbf{u}_0(\mathbf{r}') = \vec{T}(\mathbf{r}', k_v \mathbf{s}_0) \cdot \mathbf{a}_v \tag{2.14}$$

where

$$\vec{T}(\mathbf{r}', k_v \mathbf{s}_0) = \int d^3 r'' \, \vec{T}(\mathbf{r}', \mathbf{r}'') e^{ik_v \mathbf{s}_0 \cdot \mathbf{r}''} \tag{2.15}$$

is the mixed matrix element of the Transition operator $\hat{\vec{T}}$ between the coordinate and momentum representations.

The scattering amplitudes can be expressed directly in terms of matrix elements of the transition operator $\hat{\vec{T}}$ in the momentum representation. To obtain the desired expressions we first express Eq.(2.11) in the coordinate representation and make use of Eq.(2.14). We then find that the scattered field component of the elastic displacement field is given by

$$\mathbf{u}^{(s)}(\mathbf{r}; v) = \int d^3 r' \, \vec{G}(\mathbf{r} - \mathbf{r}') \cdot \vec{T}(\mathbf{r}', k_v \mathbf{s}_0) \cdot \mathbf{a}_v \tag{2.16}$$

where we have explicitly displayed the dependence of the scattered field on the incident wave ($v = 1$ for incident compressional wave and $v = 2$ for incident shear wave).

The scattering amplitudes are computed from Eq.(2.16) by making use of the asymptotic form of the Green tensor (2.8) [5]:

$$\vec{G}(\mathbf{r} - \mathbf{r}') \sim \frac{-1}{4\pi \rho_0 \omega^2} [k_1^2 \mathbf{s}\mathbf{s} e^{-ik_1 \mathbf{s} \cdot \mathbf{r}'} \frac{e^{ik_1 r}}{r} + k_2^2 (\vec{I} - \mathbf{s}\mathbf{s}) e^{-ik_2 \mathbf{s} \cdot \mathbf{r}'} \frac{e^{ik_2 r}}{r}] \tag{2.17}$$

where $\mathbf{s} = \frac{\mathbf{r}}{r}$ is the unit vector in the $\mathbf{r}$ direction. On substituting Eq.(2.17) into Eq.(2.16) and comparing the result with Eq.(2.5) we conclude that

$$\mathbf{A}^{(1)}(k_1 \mathbf{s}, k_v \mathbf{s}_0) = -\frac{k_1^2}{4\pi \rho_0 \omega^2} \mathbf{s}\mathbf{s} \cdot \vec{T}(k_1 \mathbf{s}, k_v \mathbf{s}_0) \cdot \mathbf{a}_v \tag{2.18a}$$

$$\mathbf{A}^{(2)}(k_2 \mathbf{s}, k_v \mathbf{s}_0) = -\frac{k_2^2}{4\pi \rho_0 \omega^2} \mathbf{l}\mathbf{l} \cdot \vec{T}(k_2 \mathbf{s}, k_v \mathbf{s}_0) \cdot \mathbf{a}_v \tag{2.18b}$$

where $\mathbf{l}$ is a unit vector perpendicular to $\mathbf{s}$ and

$$\begin{aligned} \vec{T}(k_j \mathbf{s}, k_v \mathbf{s}_0) &= \int d^3 r' \, e^{-ik_j \mathbf{s} \cdot \mathbf{r}'} \vec{T}(\mathbf{r}', k_v \mathbf{s}_0) \\ &= \int d^3 r'' \int d^3 r' \, e^{-ik_j \mathbf{s} \cdot \mathbf{r}'} \vec{T}(\mathbf{r}', \mathbf{r}'') e^{ik_v \mathbf{s}_0 \cdot \mathbf{r}''} \end{aligned} \tag{2.19}$$

is the matrix representation of the transition operator in the momentum representation and is known as the *T matrix* [3].

3. ELASTIC WAVE INVERSE SCATTERING

We turn now to the inverse scattering problem which can be loosely defined to be the problem of determining the elastic parameters and density of the inhomogeneity (scatterer) from a set of scattered field measurements. In the following we will formulate the inverse scattering problem in terms of the elastic scattering potential $\vec{V}$ rather than directly in terms of the elastic parameters of the scatterer. However, since the elastic parameters and density of the scatterer uniquely specify the elastic scattering potential $\vec{V}$ and vice versa it is clear that these two formulations are equivalent.

We will assume in this paper that we are concerned with the ideal case where the data is not corrupted with noise and the maximum allowable information obtainable from scattered field measurements is available. Even within this "ideal" framework the inverse scattering problem can be defined in a number of ways all of which, however, are more-or-less equivalent. For our purposes we will define the inverse scattering problem as follows:

INVERSE SCATTERING PROBLEM. *Determine the elastic scattering potential* $\vec{V}$ *at some given frequency* ω *from knowledge of the scattering amplitudes* $\mathbf{A}^{(j)}(k_j\mathbf{s}, k_v\mathbf{s}_0)$, *(j,v=1,2) specified for all incident and scattering directions* $\mathbf{s}_0$ *and* $\mathbf{s}$ *and at one or more frequencies.*

Referring to Eqs.(2.18) we see that the above definition is equivalent to the statement that the elastic scattering potential be determined from certain components of the T matrix of the scatterer. These components can be determined experimentally by performing a sequence of scattering experiments employing incident compressional and shear plane waves and measuring the scattering amplitudes using detectors that are sensitive to compressional (pressure) and shear waves. Alternatively, the amplitudes are directly obtainable from measurements of the elastic displacement vector in the far field.

The T matrix is related to the scattering potential through an integral equation that can be derived from the operator form of the Lippmann Schwinger equation (2.9). In particular, by making use of the definition of the transition operator given in Eq.(2.10) one finds that this latter quantity satisfies the operator equation

$$\hat{\vec{T}} = \hat{\vec{V}} + \hat{\vec{V}} \cdot \hat{\vec{G}} \cdot \hat{\vec{T}}. \tag{3.1}$$

The equation satisfied by the T matrix is then simply obtained by expressing Eq.(3.1) in the momentum representation (in Fourier space). We find that

$$\vec{T}(\mathbf{p}_1, \mathbf{p}_2) = \vec{V}(\mathbf{p}_1, \mathbf{p}_2) + \int d^3p' \vec{V}(\mathbf{p}_1, \mathbf{p}') \cdot \vec{G}(p') \cdot \vec{T}(\mathbf{p}', \mathbf{p}_2) \tag{3.2}$$

where $\vec{V}(\mathbf{p}_1, \mathbf{p}_2)$ is the momentum space representation of the scattering potential $\hat{\vec{V}}$ and where we have made use of the fact that the Green dyadic is diagonal in the momentum representation [5]. We note for future reference that

$$\begin{aligned}\vec{V}(\mathbf{p}_1, \mathbf{p}_2) &= \int d^3r \int d^3r'\, e^{-i\mathbf{p}_1\cdot\mathbf{r}} \vec{V}(\mathbf{r}) \delta(\mathbf{r} - \mathbf{r}') e^{i\mathbf{p}_2\cdot\mathbf{r}'} \\ &= \int d^3r e^{-i\mathbf{p}_1\cdot\mathbf{r}} \left\{ \frac{\partial}{\partial x_j} \delta C_{ijkl}(\mathbf{r}) \frac{\partial}{\partial x_l} + \omega^2 \delta\rho(\mathbf{r}) \delta_{ik} \right\} e^{i\mathbf{p}_2\cdot\mathbf{r}} \\ &= -\mathbf{p}_1\mathbf{p}_2 \delta\tilde{\lambda}(\mathbf{p}_1 - \mathbf{p}_2) - (\mathbf{p}_1 \cdot \mathbf{p}_2 + \mathbf{p}_2\mathbf{p}_1)\delta\tilde{\mu}(\mathbf{p}_1 - \mathbf{p}_2) \\ &\quad + \omega^2 \delta\tilde{\rho}(\mathbf{p}_1 - \mathbf{p}_2)\vec{I}\end{aligned} \tag{3.3}$$

where $\delta\tilde{\lambda}$, $\delta\tilde{\mu}$ and $\delta\tilde{\rho}$ are the spatial Fourier transforms of $\delta\lambda$, $\delta\mu$, and $\delta\rho$, respectively.

The difficulty of the inverse scattering problem is apparent from Eq.(3.2). In particular, the scattering data only specifies certain components of the T matrix and, in particular, only those matrix elements for which $|\mathbf{p}_1|$ and $|\mathbf{p}_2|$ are equal to either k_1 or k_2. The integral equation, on the other-hand, relates the entire T matrix, and not only those "on-shell" components, to the unknown scatterer as defined by the matrix $\vec{V}(\mathbf{p}_1, \mathbf{p}_2)$. Thus, in order to "solve" the above integral equation for the scattering potential in momentum space we evidently need more information than is available or, alternatively, we require some means of inferring the off-shell elements of the T matrix from the (known) on-shell components.

This same problem arises in scalar wave inverse scattering and various methods have been proposed to effectively determine the off-shell components of the T matrix from the observable on-shell components [7-12] in these applications. . Alternatively, this problem can be avoided by employing simplifying approximations such as the Born [2] or Rytov approximations [12]. Recently, a new approach to this problem has been employed in quantum mechanical inverse scattering where the single T matrix equation (3.2) is replaced by a coupled pair of equations that cleanly separate out the on-shell and off-shell components of the T matrix [13].

Due to space limitations we will restrict our attention here to the use of the Born approximation which consists of approximating the T matrix by the first term in Eq.(3.2); i.e.,

$$\vec{T}(\mathbf{p}_1, \mathbf{p}_2) \approx \vec{V}(\mathbf{p}_1, \mathbf{p}_2). \tag{3.4}$$

On setting $\mathbf{p}_1 = k_j\mathbf{s}$ and $\mathbf{p}_2 = k_v\mathbf{s}_0$ with $j, v = 1, 2$ we conclude, on using Eq.(3.3), that

$$\begin{aligned}\vec{T}_B(k_j\mathbf{s}, k_v\mathbf{s}_0) = &- k_jk_v\mathbf{s}\mathbf{s}_0\delta\tilde{\lambda}(k_j\mathbf{s} - k_v\mathbf{s}_0) - k_jk_v(\mathbf{s}_0 \cdot \mathbf{s}\vec{I} + \mathbf{s}_0\mathbf{s})\delta\tilde{\mu}(k_j\mathbf{s} - k_v\mathbf{s}_0)\\ &+ \omega^2\delta\tilde{\rho}(k_j\mathbf{s} - k_v\mathbf{s}_0)\vec{I}\end{aligned} \tag{3.5}$$

where $\vec{T}_B$ is the Born approximation to the T matrix.

Within the Born approximation we conclude that the inverse scattering problem reduces to estimating the Lamé parameters and density of the scatterer from certain linear combinations of the spatial Fourier transforms of these quantities as specified by the scattering amplitudes and Eq.(3.5). These linear combinations are obtained from performing sets of scattering experiments using incident compressional and shear plane waves and measuring the scattering amplitudes of the resulting compressional and shear scattered wavefields. By making use of Eqs.(2.18) and Eq.(3.5) we obtain the following results for the various possible of scattering experiments:

Compressional Incident Wave/ Compressional Scattered Wave

$$\begin{aligned}&\mathbf{s} \cdot \vec{T}_B(k_1\mathbf{s}, k_1\mathbf{s}_0) \cdot \mathbf{s}_0 =\\ &- k_1^2\delta\tilde{\lambda}[k_1(\mathbf{s} - \mathbf{s}_0)] - 2k_1^2(\mathbf{s}_0 \cdot \mathbf{s})^2\delta\tilde{\mu}[k_1(\mathbf{s} - \mathbf{s}_0)] + \omega^2(\mathbf{s}_0 \cdot \mathbf{s})\delta\tilde{\rho}[k_1(\mathbf{s} - \mathbf{s}_0)]\end{aligned}$$

Shear Incident Wave/ Compressional Scattered Wave

$$\begin{aligned}&\mathbf{s} \cdot \vec{T}_B(k_1\mathbf{s}, k_2\mathbf{s}_0) \cdot \mathbf{l}_0 =\\ &- 2k_1k_2(\mathbf{s}_0 \cdot \mathbf{s})(\mathbf{s} \cdot \mathbf{l}_0)\delta\tilde{\mu}(k_1\mathbf{s} - k_2\mathbf{s}_0) + \omega^2(\mathbf{s} \cdot \mathbf{l}_0)\delta\tilde{\rho}(k_1\mathbf{s} - k_2\mathbf{s}_0)\end{aligned}$$

Compressional Incident Wave/ Shear Scattered Wave

$$\mathbf{l}\cdot\vec{T}_B(k_2\mathbf{s},k_1\mathbf{s}_0)\cdot\mathbf{s}_0 = \\ -2k_1k_2(\mathbf{s}_0\cdot\mathbf{s})(\mathbf{s}\cdot\mathbf{l}_0)\delta\tilde{\mu}(k_2\mathbf{s}-k_1\mathbf{s}_0)+\omega^2(\mathbf{s}\cdot\mathbf{l}_0)\delta\tilde{\rho}(k_2\mathbf{s}-k_1\mathbf{s}_0)$$

Shear Incident Wave/ Shear Scattered Wave

$$\mathbf{l}\cdot\vec{T}_B(k_2\mathbf{s},k_2\mathbf{s}_0)\cdot\mathbf{l}_0 = \\ -k_2^2[(\mathbf{s}_0\cdot\mathbf{s})(\mathbf{l}\cdot\mathbf{l}_0)+(\mathbf{l}\cdot\mathbf{s}_0)(\mathbf{s}\cdot\mathbf{l}_0)]\delta\tilde{\mu}[k_2(\mathbf{s}-\mathbf{s}_0)]+\omega^2(\mathbf{l}\cdot\mathbf{l}_0)\delta\tilde{\rho}[k_2(\mathbf{s}-\mathbf{s}_0)]$$

3.2 Fluid Background

As an example of the theory developed above we consider the special case where the background medium is a fluid ($\mu_0 = 0$). In this case the only scattering amplitude that is non-zero is that corresponding to an incident compressional wave and a scattered compressional wave. By making use of the vector identity $\mathbf{s}\cdot\mathbf{s}_0 = 1-\frac{1}{2}(\mathbf{s}-\mathbf{s}_0)^2$ we can write this scattering amplitude in the form

$$\mathbf{s}\cdot\vec{T}_B(k_1\mathbf{s},k_1\mathbf{s}_0)\cdot\mathbf{s}_0 = \tilde{F}[k_1(\mathbf{s}-\mathbf{s}_0)] \tag{3.6}$$

where

$$\tilde{F}(\mathbf{K}) = \omega^2\tilde{F}_1(\mathbf{K})+\tilde{F}_2(\mathbf{K})+\frac{1}{\omega^2}\tilde{F}_3(\mathbf{K}) \tag{3.7}$$

with

$$\tilde{F}_1(\mathbf{K}) = \frac{1}{C_1^2}[\delta\tilde{\lambda}(\mathbf{K})+2\delta\tilde{\mu}(\mathbf{K})]-\delta\tilde{\rho}(\mathbf{K}) \tag{3.8a}$$

$$\tilde{F}_2(\mathbf{K}) = 2K^2\delta\tilde{\mu}(\mathbf{K})+\frac{C_1^2}{2}K^2\delta\tilde{\rho}(\mathbf{K}) \tag{3.8b}$$

$$\tilde{F}_3(\mathbf{K}) = \frac{C_1^2}{2}K^4\delta\tilde{\mu}(\mathbf{K}) \tag{3.8c}$$

and where ω is the frequency and C_1 the compressional wave velocity in the background.

The scattering data then specifies the spatial Fourier transform of the scalar function

$$F(\mathbf{r}) = \frac{1}{(2\pi)^3}\int d^3K\tilde{F}(\mathbf{K})e^{i\mathbf{K}\cdot\mathbf{r}} \tag{3.9}$$

over the set of spherical surfaces $\mathbf{K} = k_1(\mathbf{s}-\mathbf{s}_0)$ in Fourier space. These set of surfaces are the well known *Ewald spheres* from X-ray crystallography [14] and that play a central role in linearized optical, acoustical and quantum mechanical inverse scattering [15-17]. If a *complete* set of scattering experiments are performed then the unit vectors $\mathbf{s}$ and $\mathbf{s}_0$ vary over the entire unit sphere and $\tilde{F}(\mathbf{K})$ will be completely determined from Eq.(3.6) throughout the interior of the so-called *Ewald Limiting Sphere* [14] $|\mathbf{K}| \leq 2k_1$. It then follows that a low pass filtered version of $F(\mathbf{r})$ can be determined, bandlimited to the interior of this sphere. This reconstruction can, for example, be generated directly from Fourier inversion of the scattered field data or, alternatively, can be obtained from an integral transform [18] that has been employed with success in diffraction tomography [16,17,19].

The function $F(\mathbf{r})$ that is generated by inversion of the scattering data within the Born approximation is a linear combination of $\delta\lambda$, $\delta\mu$, and $\delta\rho$. In particular, it follows immediately from Eqs.(3.7) and (3.8) that

$$F_1(\mathbf{r}) = \frac{1}{C_1^2}[\delta\lambda(\mathbf{r}) + 2\delta\mu(\mathbf{r})] - \delta\rho(\mathbf{r}) \tag{3.10a}$$

$$F_2(\mathbf{r}) = - - 2\nabla^2\delta\mu(\mathbf{r}) - \frac{C_1^2}{2}\nabla^2\delta\rho(\mathbf{r}) \tag{3.10b}$$

$$F_3(\mathbf{r}) = \frac{C_1^2}{2}\nabla^4\delta\mu(\mathbf{r}) \tag{3.10c}$$

where we have assumed that the quantities $\delta\lambda$, $\delta\mu$ and $\delta\rho$ are square integrable and continuous with continuous fourth order partial derivatives.

On referring to Eq.(3.7) it is seen that if the fluid background and elastic scatterer are independent of frequency (non-viscoelastic) then the transforms $\tilde{F}_j$, $j = 1, 2, 3$ can be separately determined with the result that the functions F_j, $j = 1, 2, 3$ can also be separately reconstructed. The set of Equations (3.10) can then be inverted for the elastic parameters and density of the scatterer. Such a procedure has been developed for the special case of a fluid of non-constant density embedded in a uniform fluid background [20].

4. SUMMARY

We have in this paper presented addressed the problem of determining the elastic parameters and density of an isotropic elastic scatterer embedded in a homogeneous, isotropic background (matrix) from knowledge of the scatterers' compressional and shear wave scattering amplitudes. The theory was based on the T matrix formulation of elastic wave scattering [3] and was developed along lines analogous to those employed in quantum mechanical scattering [2]. The principle finding of the paper was that the elastic wave scattering amplitudes specify the so-called "on-shell" components of the elastic T matrix and that the inverse scattering problem amounts to determining the complete T matrix from these on-shell components.

In the special case of a weak elastic scatterer the components of the T matrix were found to be approximately equal to certain linear combinations of the spatial Fourier transforms of the elastic parameters and density of the scatterer. It was shown that the these relationships could be inverted yielding reconstructions of linear functionals of the elastic parameters and density. Finally, it was shown that for a non-viscoelastic system consisting of an elastic scatterer embedded in a fluid background that separate reconstruction of the two elastic parameters and density is possible if measurements are performed at three or more frequencies.

REFERENCES

1. P.C. Sabatier, "Basic concepts and methods of inverse problems," in *Basic Methods of Tomography and Inverse Problems*, ed. P.C. Sabatier. Philadelphia, Adam Hilger, 1987.

2. J.R. Taylor, *Scattering Theory.* New York, Wiley, 1972.

3. P.C. Waterman, "Matrix theory of elastic wave scattering," *J. Acoust. Soc. Am.* **60**, pp. 567-580, 1976.

4. J.D. Achenbach, *Wave Propagation in Elastic Solids.* Amsterdam, North-Holland, 1973, Chapter 2.

5. A.J. Devaney, "Multiple scattering theory for discrete, elastic, random media," *J. Math. Phys.*

21, pp. 2603-2611, 1980.

6. A. Sommerfeld, *Partial Differential Equations of Physics.* New York, Academic Press, 1967, p. 189.

7. R.G. Newton, "The Marchenko and Gelfand-Levitand methods in the inverse scattering problem in one and three dimensions," in *Conference on Inverse Scattering: Theory and Application.* Philadelphia, SIAM Press, 1983.

8. I. Kay, "The inverse scattering problem," in *Research Report EM-74.* New York, New York University, 1955.

9. V.H. Weston, "On inverse scattering," *J. Math. Phys.* **15**, pp. 209-213, 1974.

10. A.J. Devaney and E. Wolf, "A new perturbation expansion for inverse scattering from three-dimensional finite-range potentials," *Phys. Letts.* **89A**, pp. 269-272, 1982.

11. P.C. Sabatier, "Theoretical considerations for inverse scattering," *Radio Science* **18**, pp.2-18, 1983.

12. V.T. Tatarski, *Wave Propagation in a Turbulent Medium.* New York, McGraw-Hill, 1961.

13. A.J. Devaney and A.B. Weglein, "Inverse scattering within the Heitler approximation," *Inverse Problems* **5**, pp. L49-L52, 1989.

14. R.W. James, *The optical principles of the diffraction of X-rays.* London, Bell, 1948.

15. E. Wolf, "Three-dimensional structure determination of semi-transparent objects from holographic data," *Opt. Commun.* **1**, pp. 153-156, 1969.

16. A.J. Devaney, "Inverse source and scattering problems in ultrasonics," *IEEE Trans. Sonics and Ultrasonics* **SU-30**, pp. 355-364, 1983.

17. K.J. Langenberg, "Applied inverse problems for acoustic, electromagnetic and elastic wave scattering," in *Basic Methods of Tomography and Inverse Problems*, ed. P.C. Sabatier, Philadelphia, Adam Hilger, 1987.

18. A.J. Devaney, "Inversion formula for inverse scattering within the Born approximation," *Optics Letts.* **7**, pp. 11-15, 1982.

19. A.J. Devaney, "A filtered backpropagation algorithm for diffraction tomography," *Ultrasonic Imaging*, Vol. 4, pp. 336-350, 1982.

20. A.J. Devaney, "Variable density acoustic tomography," *J. Acoust. Soc. Am.* **78**, pp. 120-130, 1985.

Elastic Waves and Ultrasonic Nondestructive Evaluation
S.K. Datta, J.D. Achenbach and Y.S. Rajapakse (Editors)

VIBRATION LOCALIZATION BY DISORDER— A VIABLE ALTERNATIVE TO DAMPING?*

Christophe Pierre

Department of Mechanical Engineering and Applied Mechanics
The University of Michigan
Ann Arbor, Michigan 48109-2125

A brief survey of localization phenomena in disordered nearly periodic structural systems is presented. It is shown that *local* periodicity-destroying irregularities have the potential to alter drastically the *global* dynamics of engineering structures, by *localizing* the vibration modes to small geometric regions and *inhibiting* the propagation of incident waves. Remarkably, this is achieved without any dissipation in the structure.

Choosing a *statistical* treatment of the structural irregularities, expressions for the localization factors (the exponential spatial decay rates of the vibration amplitude from the source of excitation) are obtained for generic models of nominally periodic structures. Both wave and modal formulations are introduced. Special attention is paid to the variation of localization effects with frequency. Furthermore, the fundamental distinction between the two phenomena of *weak* and *strong* localization is discussed in terms of system parameters and frequency. It is shown that while weak localization effects are of little interest in most structural dynamics applications, the drastic phenomenon of strong localization is of importance to the structural dynamicist. Finally, the mechanisms of localization and damping are compared and the damping-like effects of irregularities are illustrated.

1. INTRODUCTION

Numerous inherent parameter uncertainties (defects, tolerances, clearances, dissipation estimates, etc.) render the vibrational analysis of complex structures very difficult. Such *local* small-scale variations, commonly disregarded in the modeling and analysis, have the potential to alter drastically the *global* dynamics of engineering structures.

The neglect of local variations is perhaps most critical for *periodic structures*, which consist of repeated identical subsystems along one or more directions; typical examples are truss beams, blade assemblies, and laminated composites. Departure from periodicity always occurs because of manufacturing and material tolerances. Depending on the relative magnitudes of disorder and internal coupling for the nearly periodic structure, these irregularities may localize the free modes to small geometric regions and confine vibrational energy near the source of excitation. This phenomenon is referred to as *mode localization* and was first predicted in solid state physics [1].

Localization has excited the interest of solid state physicists for many years. Though most of the abundant literature is not relevant to our paper,† we must mention two works that impact localization studies in structural dynamics. First, Ishii [3] applied Furstenberg's theorem on the limiting behavior of products of random matrices to study localization in random chains. Modeling each bay, or subsystem, of the periodic system by a transfer matrix, he represented the assembly by a product of random matrices—an approach based on the traveling of *waves* from subsystem to subsystem. Second, Herbert and Jones [4] and Thouless [5] developed a *modal* formulation based upon the spectrum of the disordered chain. These two perspectives form the cornerstone of our study of localization in engineering structures. They differ in that the wave formulation deals with *local* properties (transfer matrices), while the modal method uses *global* information (spectrum).

Hodges [6] was first to recognize that localization can occur in engineering structures and to suggest

* This work is supported by National Science Foundation Grant No. MSM-8700820, Dynamic Systems and Control Program.

† We refer the interested reader to the bibliography by Ziman [2].

that some of the knowledge acquired in physics could be applied to studies of structural dynamics. Using both wave and modal arguments, he discussed localization for chains of coupled pendulums and for beams on randomly spaced supports.

Research on localization in structural dynamics has been limited mostly to *deterministic* analyses of the free modes of disordered one-dimensional (1-D) structures (for example, see [7–11]). Few probabilistic studies of localization have been conducted to date. Hodges and Woodhouse [12] formulated a statistical treatment of the transmission of harmonic forced vibrations from a local source of excitation. They applied the work of Herbert and Jones [4] to calculate localization factors* for a stretched string with irregularly spaced masses attached. Although they did not study localization effects systematically in terms of frequency or system parameters, their work exhibited two types of localization—*weak* and *strong*—depending on the magnitude of internal coupling in the structure. Recently, Kissel [13] chose the wave description of Ishii [3] to calculate localization factors for several infinite 1-D structures. He studied the dependence of the localization factor inside the passbands on the frequency in the limiting case of weak disorder. However, the models he chose did not allow him to vary the relative magnitudes of internal coupling and disorder, thereby restricting his findings to *weak* localization effects. Finally, Pierre [14] performed a statistical investigation of a disordered chain of oscillators in terms of frequency that evidenced both *weak* and *strong* localization. He developed stochastic perturbation methods based upon modal or wave formulations, depending on the localization regime. He conjectured that strong localization effects are most relevant to the structural dynamicist, whereas weak localization, though of concern to solid state physicists, affects most engineering structures very little. Pierre *et al.* [15,16] also calculated localization factors for multi-span beams and assemblies of *multi-mode* subsystems.

To complete this brief review we also recommend the following works to the reader interested in localization: the comprehensive study of continuously disordered one-dimensional systems by Scott [17] and the in-depth review of localization and its relationship to diffusive transport theories such as statistical energy analysis by Hodges and Woodhouse [18].

2. AN EXAMPLE

To illustrate localization, consider the assembly of coupled component beams in Fig. 1; Fig. 2 displays the typical free vibration modes for an ordered assembly of identical beams and a slightly disordered system. The standard deviation of random disorder is 1%, less than typical manufacturing tolerances, and the interbeam coupling stiffness is significant. While the mode of the tuned, or ordered, assembly extends throughout the structure, the mistuned, or disordered, assembly features a mode that is *localized* about a few subsystems. This energy confinement also results in larger amplitudes.

The underlying motivation for studying localization is threefold. First, through the occurrence of localization, parameter uncertainties have the potential to invalidate the predictions of commonly used deterministic modeling and analysis techniques, which reduces the effectiveness of structural models, associated control schemes, and identification techniques. For example, a control strategy based upon the erroneous extended modes of a perfectly periodic structure would be unpredictable. Second, localized vibrations increase amplitudes and stresses locally and may result in severe damage. For instance, single-blade failure in turbomachinery rotors is a plausible result of blade mistuning. Third, when localization occurs, irregularities exhibit a damping-like effect that could be used as a passive control of vibration transmission. This has applications in flexible and lightly damped structures such as truss beams in space.

3. STATISTICAL EXAMINATION OF LOCALIZATION

3.1 Background on Periodic Structures

Periodic structures have very characteristic properties that make much of their dynamics qualitatively the same [19]. This means that useful results for engineering structures can be obtained from elementary models. Periodic structures are made of subsystems, or bays, coupled through one or more coordinates. For example, a multi-span beam on rigid supports is mono-coupled through the bending rotation at the supports. The same beam on elastic supports is bi-coupled, as energy is also transmitted through the bending displacement of the supports. A periodic structure carries as many pairs of (left- and right-traveling) free waves as there are coupling coordinates. Alternating frequency bands in

* The exponential spatial decay rate of the amplitude, defined in §3.1.

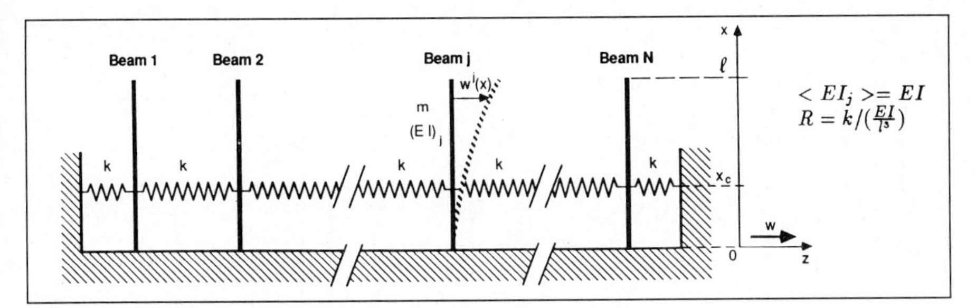

Figure 1. Assembly of mono-coupled, multi-mode component systems (cantilevered beams).

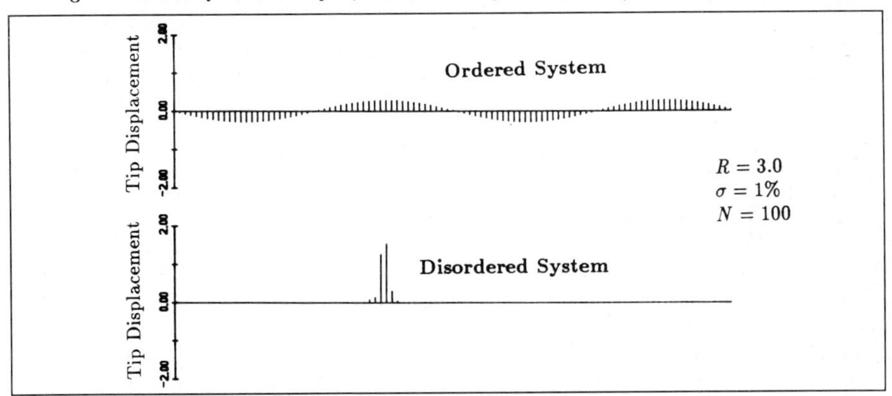

Figure 2. Fourth mode in the third passband (204th mode) for ordered and disordered assemblies of 100 cantilevered Bernoulli-Euler beams. The beams are coupled at the tip, and the coupling strength equals the beams' static stiffness ($R = 3$). The tip displacements are shown. For the disordered system, the standard deviation of stiffness mistuning is 1%. Note the severe change from extended to localized modes.

which the waves either propagate freely (passbands) or are attenuated (stopbands) correspond to each pair of harmonic waves. The number of passbands equals the number of degrees of freedom (DOF) of a subsystem, and energy is only transmitted along the structure through passbands. Dual to the wave properties are the modal properties. The natural frequencies of an N-bay periodic structure lie within the passbands, with N frequencies in each passband. The corresponding mode shapes are extended and feature a nonattenuated shape (see Fig. 2). Mode shapes can be obtained from the free traveling waves of the same frequency by the phase closure principle [20].

A pair of waves is characterized by a *propagation constant*, μ, such that waves propagate according to $e^{\pm\mu}$. The spatial decay of the wave amplitude is thus governed by the real part of the propagation constant, γ, a quantity of special interest in localization. In the passbands of periodic structures, μ is purely imaginary, thus $\gamma = 0$. In the stopbands, attenuation occurs and $\gamma > 0$.

Disruption in periodicity leads to wave attenuation at all frequencies (regardless of dissipation), because traveling waves are partly reflected by the randomized subsystems. This multiple scattering effect over many bays causes localization. The real part of the propagation constant, γ, is then greater than zero even in the passbands of the periodic structure. In this paper we refer to γ as the rate of spatial decay per bay, or the *localization factor*. Because of the wave/mode duality, we expect the mode shapes to become localized as well. Indeed, Matsuda and Ishii [21] argued that localized modes are governed, on the average, by the same exponential envelope as the free waves of same frequency, $e^{-\gamma N}$.* The attenuation of traveling waves also means that energy injected into the structure cannot

* Obviously, this is true for a periodic structure, as $\gamma = 0$ and the modes are extended.

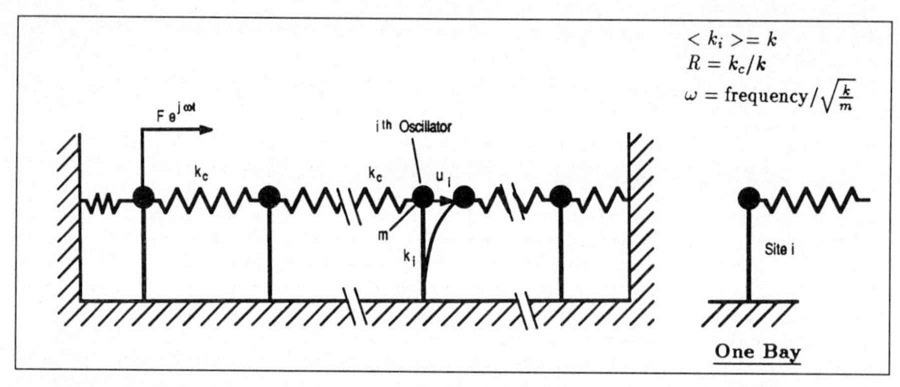

Figure 3. Chain of single-DOF oscillators with random stiffnesses. For the modal formulation we show the left-end excitation. For the wave formulation we show one bay used to form the transfer matrix.

propagate arbitrarily far, but is *confined* near the source.

This suggests that wave attenuation, mode localization, and forced vibration confinement are well described by the localization factor, γ, defined as the average rate of exponential decay for disordered systems. The next section describes techniques to calculate localization factors.

3.2 Calculation of Localization Factors by Stochastic Perturbation Methods

Kissel [13] and Pierre [14] recently made progress in calculating localization factors for simple systems with random disorder. Pierre [14] considered both modal and wave formulations for a chain of single-DOF oscillators with random frequencies, shown in Fig. 3. The *modal* approach is based on the transmission of forced harmonic vibration from the left to the right end of the chain. The localization factor, γ, is expressed in terms of the spectrum of the disordered chain by assuming the amplitude decays exponentially, $|u_N| \sim \exp(-\gamma N)$, for large N [4]:

$$\gamma = -\ln R + \lim_{N\to\infty} \frac{1}{N}\sum_{r=1}^{N} \ln\left|\omega_r^2 - \omega^2\right| \tag{1}$$

where R is the dimensionless coupling between oscillators, ω the (dimensionless) excitation frequency, and ω_r the frequencies of the disordered chain.

The *wave* formulation is based on the propagation of incident waves through a disordered segment. The displacements and forces at the two ends of the ith bay (see Fig. 3) can be related by a random displacement transfer matrix. The wave transfer matrix, $[W_i]$, is obtained from the displacement transfer matrix by a similarity transformation defined by the eigenvectors of the transfer matrix for an ordered site. (Thus, for an ordered site, the wave transfer matrix is diagonal made of the eigenvalues of the displacement transfer matrix.) The wave transfer matrix relates the left- and right-traveling wave amplitudes at site i to those at site $i+1$. Thus, the propagation of a wave incident from the left through an N-bay disordered segment is governed by $\prod_{i=N}^{1}[W_i]$, whose (1,1) term is the inverse of the transmission coefficient τ_N. Hence, the rate of exponential decay of the transmitted amplitude is, asymptotically,

$$\gamma = \lim_{N\to\infty} -\frac{1}{N}\ln|\tau_N| \tag{2}$$

For an ordered assembly, Eqs. (1) and (2) yield $\gamma = 0$ for $\omega^2 \in [1, 1+4R]$, defining the passband. For a disordered system, even though γ is expressed for an infinite chain, it can be viewed as the average decay rate for finite systems of length N and thus is relevant to engineering structures. The limits in Eqs. (1) and (2) cannot be evaluated exactly for disordered chains.

We have developed *stochastic perturbation methods* to obtain approximate *analytical* expressions for the localization factor (see [14] for details). The first scheme treats the (small) random disorder as a perturbation. This *classical* perturbation method gives the approximate localization factor:

$$\gamma \simeq \gamma^{(c)} = \frac{\sigma^2}{2(\omega^2-1)(1+4R-\omega^2)} \qquad \omega^2 \in]1, 1+4R[\qquad O(\frac{\sigma}{R}) < 1 \tag{3}$$

where σ is the standard deviation of disorder and $O(.)$ denotes the order of the argument. Obviously, Eq. (3) holds only for *strong coupling*. The degree of localization depends only on the frequency position in the passband and on the ratio of disorder to coupling, and it increases with this ratio. A typical variation of γ within the passband is depicted in Fig. 4: localization is greatest near the stopbands and least at the midband. Perhaps more important is the localization *effect* in this strong coupling case. For $R = 1$ (the coupling stiffness equals the oscillator stiffness) and $\sigma = 10\%$ (a large disorder), the localization factor at midband is $\gamma^{(c)} = 0.00125$. This means that for an infinite system, the vibration amplitude is governed by $e^{-\gamma^{(c)}N}$, and 555 sites are needed for the amplitude to decay by a factor of two. Even though this effect can be regarded as significant because it results solely from disorder, it is probably unimportant for most engineering structures, which rarely comprise that many subsystems. Consequently, it is termed *weak localization*.

Investigations of the free modes [9,10] show that localization effects are strong for weak internal coupling. Based on this result, we have developed a stochastic *modified* perturbation method to approximate the localization factor in the weak coupling case [14].* This scheme treats the small coupling as a perturbation. At the midband frequency, it yields

$$\gamma^{(m)}_{\text{midband}} = \ln\frac{\sigma}{R} + \ln\sqrt{3} - 1 \qquad (\omega^2 = 1 + 2R) \qquad O(\frac{\sigma}{R}) > 1 \tag{4}$$

Fig. 5 depicts the variation of $\gamma^{(m)}$ versus frequency. Contrary to the strong coupling case (Fig. 4), the localization factor varies little in the passband. A typical value for $R = 0.01$ and $\sigma = 3\%$ is $\gamma^{(m)}_{\text{midband}} = 0.648$, which is several orders of magnitude larger than the weak localization/strong coupling case. A quick calculation reveals that, on the average, the vibration amplitude of the fourth oscillator is $|\frac{u_4}{F}| \sim 0.075$, that is, only $0.075^2 = 0.56\%$ of the energy is transmitted to the fourth oscillator! For the ordered system $\gamma = 0$ and 100% of the energy is transmitted. This drastic effect of disorder in the weak coupling case is termed *strong localization*. We believe it is the most relevant to engineering structures.

The above results seem to indicate that disorder affects only weakly coupled assemblies notably, which would limit the range of our findings. However, we must keep in mind that, in general, subsystems carry many modes and that the modal coupling depends highly on the passband number. Thus, localization effects clearly vary with passband number in assemblies of *multi-mode* (or multi-DOF) subsystems.

Indeed, for an assembly of coupled beams, Cha and Pierre [16] recently showed that in the weak localization regime the localization factor varies approximately according to p^8, where p is the passband number. Thus, the degree of localization is much more sensitive to passband number than to disorder or static coupling strength (γ varies only as $(\frac{\sigma}{R})^2$), implying that the transition from weak to strong localization is very rapid with increasing frequency. Even for strong static coupling and very weak disorder, then, severe localization is unavoidable at high frequencies. This result also suggests that assemblies of beam-like subsystems, such as bladed-disks, become increasingly sensitive to mistuning as frequency increases.

3.3 Localization and Dissipation: Equivalent Damping Factors

Since engineering structures are damped, we must explore how dissipation affects localization. To date, no work on localization in engineering structures has included damping effects. The localization mechanism *confines* energy by disorder: traveling waves are elastically scattered by random irregularities, resulting in the trapping of energy near the excitation source. The damping mechanism, on the other hand, *dissipates* energy as the vibration is transmitted through the subsystems without reflection. While intrinsically distinct, *disorder and dissipation both result in a spatial decay of the amplitude.* This similarity has important practical consequences. For instance, localization could provide an alternative to damping in the passive control of propagating disturbances in large structures,

* For weak coupling, the classical perturbation scheme fails and grossly overpredicts the localization factor [14].

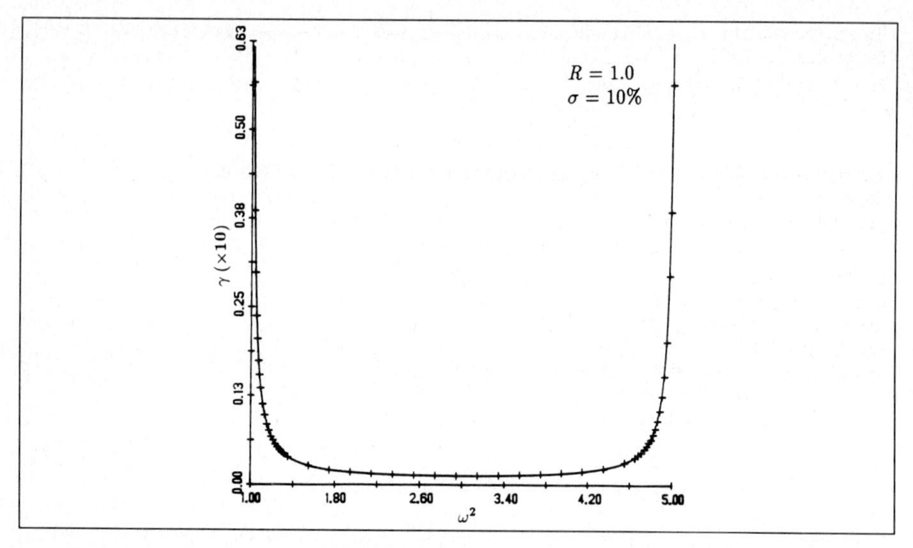

Figure 4. Typical variation of the localization factor in a frequency passband for the weak localization-strong coupling case (for undamped disordered oscillator chains). Classical perturbation results (—) agree very well with Monte Carlo simulations (+).

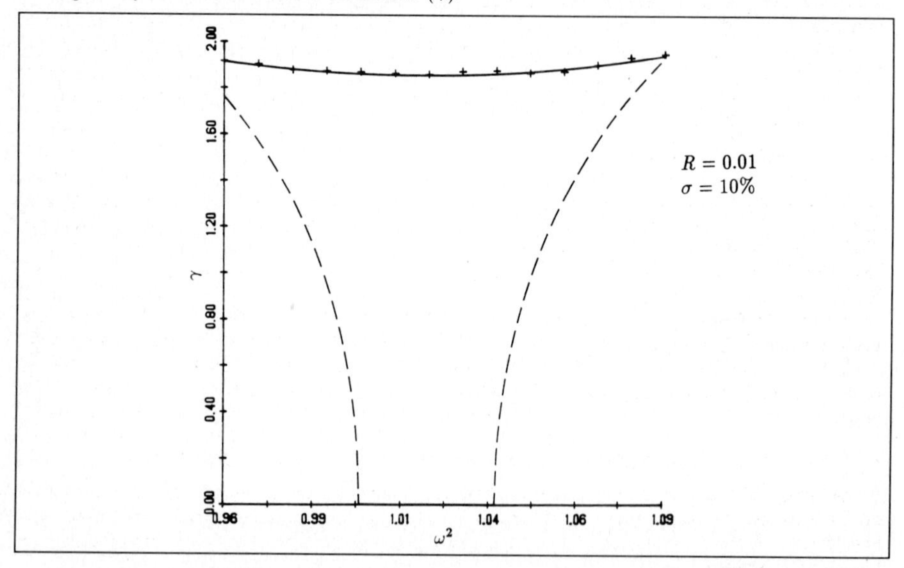

Figure 5. Typical variation of the localization factor in a frequency passband for the strong localization-weak coupling case (for the oscillator chain). The propagation constant of the ordered chain is shown (- - -). Observe the agreement between simulation (+) and modified perturbation results (—).

and in the design of engineering structures, attributing spatial amplitude decay entirely to damping may result in its overestimation. This is important, as under other forcing conditions localization may not occur, leaving the analyst with less damping than estimated.

Our goal is to identify the decay due to damping and that due to disorder in a typical structure.

To achieve this, we explore the damping-like effect of disorder by defining the "equivalent" damping factor due to disorder as the damping factor that results in the same spatial decay, in an ordered system, as disorder does in the undamped system:

$$\gamma^{\text{ordered}}_{\text{damped}}(\delta_{\text{equiv}}, \omega) = \gamma^{\text{disordered}}_{\text{undamped}}(\sigma, \omega) \tag{5}$$

where σ is the disorder in the undamped structure and δ_{equiv} the equivalent damping, which clearly depends on frequency. This factor is useful in evaluating whether localization effects contribute appreciably to dissipation.

We have obtained preliminary results for a chain of structurally damped oscillators* harmonically excited at one end. The decay factor for the damped ordered assembly, $\gamma^{\text{ordered}}_{\text{damped}}$, is simply the real part of the propagation constant [22], and it can be easily calculated. Combining its expression with the localization factors in Eqs. (3) and (4) via definition (5) yields the equivalent structural damping factor that results from a disorder of strength σ. For a midband frequency in the strong (weak localization) and weak (strong localization) coupling cases:

$$\begin{cases} \delta_{\text{equiv}} = \dfrac{\sigma^2}{4R} & O(\dfrac{\sigma}{R}) < 1 \\ \delta_{\text{equiv}} = \sigma\dfrac{\sqrt{3}}{\text{e}} & O(\dfrac{\sigma}{R}) > 1 \end{cases} \tag{6}$$

This gives the damping-like effect of disorder in the undamped structure for an end excitation. Clearly, in the strong coupling case, the equivalent damping is very small: for first-order disorder, it is second-order (probably much smaller than real damping in a typical structure). This reaffirms that weak localization is of little interest to the structural dynamicist. For weak coupling, however, the equivalent damping is proportional to the disorder—a much larger effect, as a disorder of a few percent results in additional "damping" of a few percent. In many structures, we expect the equivalent damping due to disorder to be comparable to or larger than real damping. This could have a profound impact on vibration attenuation at moderate and high frequencies, especially in very lightly damped structures such as those in space.

Nevertheless, there are substantial differences between dissipation and localization, and we must exert care in using equivalent dampings. Eq. (5) compares a deterministic quantity, the decay factor of the damped ordered chain, to the average of a random variable, the localization factor of the undamped disordered chain. Thus, equivalent dampings are *average* quantities and particular realizations of disorder may lead to substantial departure from this average. In particular, in the weak localization case, the dispersion of the localization factor about the small mean is relatively large, and the result of Eq. (6) loses much of its meaning, except for very large structures. Thus, the variances of localization and equivalent damping factors ought to be explored to relate the behavior of a typical system to the average behavior and establish the validity of the average results. In addition, we defined equivalent dampings by examining the amplitude decay due to a single-subsystem excitation. But while damping always reduces vibration amplitudes, the effects of disorder for other excitation conditions are not clear.

4. CONCLUDING REMARKS

¶Small irregularities arising from manufacturing tolerances can alter drastically the dynamics of periodic structures. Typical engineering structures experiencing localization phenomena are truss beams and lattices in space; submarine hulls; turbomachinery rotors; airframes; towed-array sonars; and laminated composites.

¶Weak localization effects are probably unimportant in structural dynamics: beside featuring a very small decay per site, weak localization is hindered by boundary conditions and dissipation.

¶The drastic phenomenon of strong localization occurs in the weak coupling case and is important in many structural dynamics applications. Strong localization can provide substantial amplitude attenuation that may not be available from damping.

* We assumed no dissipation through coupling.

5. REFERENCES

1. Anderson, P. W., "Absence of Diffusion in Certain Random Lattices," *Physical Review*, Vol. 109, 1958, pp. 1492-1505.
2. Ziman, J. M., *Models of Disorder*, Cambridge University Press, Cambridge, England, 1979.
3. Ishii, K., "Localization of Eigenstates and Transport Phenomena in the One-Dimensional Disordered System," *Supplement of the Progress of Theoretical Physics*, No. 53, 1973, pp. 77-138.
4. Herbert, D. C., and Jones, R., "Localized States in Disordered Systems," *Journal of Physics C: Solid State Physics*, Vol. 4, 1971, pp. 1145-1161.
5. Thouless, D. J., "A Relation Between the Density of States and Range of Localization for One Dimensional Random Systems," *Journal of Physics C: Solid State Physics*, Vol. 5, 1972, pp. 77-81.
6. Hodges, C. H., "Confinement of Vibration by Structural Irregularity," *Journal of Sound and Vibration*, Vol. 82, No. 3, 1982, pp. 441-424.
7. Pierre, C., and Dowell, E. H., "Localization of Vibrations by Structural Irregularity," *Journal of Sound and Vibration*, Vol. 114, No. 3, 1987, pp. 549-564.
8. Bendiksen, O. O., "Mode Localization Phenomena in Large Space Structures," *AIAA Journal*, Vol. 25, No. 9, 1987, pp. 1241-1248.
9. Cornwell, P. J., and Bendiksen, O. O., "Localization of Vibrations in Large Space Reflectors," *AIAA Journal*, Vol. 27, No. 2, 1989, pp. 219-226.
10. Pierre, C., and Cha, P. D., "Strong Mode Localization in Nearly Periodic Disordered Structures," *AIAA Journal*, Vol. 27, No. 2, 1989, pp. 227-241.
11. Pierre, C., "Mode Localization and Eigenvalue Loci Veering Phenomena in Disordered Structures," *Journal of Sound and Vibration*, Vol. 126, No. 3, 1988, pp. 485-502.
12. Hodges, C. H., and Woodhouse, J., "Vibration Isolation from Irregularity in a Nearly Periodic Structure: Theory and Measurements," *Journal of the Acoustical Society of America*, Vol. 74, No. 3, 1983, pp. 894-905.
13. Kissel, G. J., "Localization in Disordered Periodic Structures," Ph.D. Dissertation, Massachusetts Institute of Technology, 1988.
14. Pierre, C., "Weak and Strong Vibration Localization in Disordered Structures: A Statistical Investigation," *Journal of Sound and Vibration*, Vol. 138, No. 2, April 1990 (to appear).
15. Bouzit, D., "Vibration Confinement Phenomena in Disordered Multi–Span Beams: A Probabilistic Investigation," Master's Thesis, The University of Michigan, 1988.
16. Cha, P. D., and Pierre, C., "Vibration Localization by Disorder in Assemblies of Mono-Coupled, Multi-Mode Component Systems," submitted for publication to the *ASME Journal of Applied Mechanics*, September 1989.
17. J. F. M. Scott, "The statistics of waves propagating in a one-dimensional random medium," *Proc Roy Soc London*, Vol. A398, 1985, pp. 341-363.
18. Hodges, C. H., and Woodhouse, J., "Theories of Noise and Vibration Transmission in Complex Structures," *Reports in Progress in Physics*, Vol. 49, 1986, pp. 107-170.
19. Brillouin, L., *Wave Propagation in Periodic Structures*, Second Edition, Dover Publications, New York, 1953.
20. Mead, D. J., "Wave Propagation and Natural Modes in Periodic Systems: I. Mono-Coupled Systems," *Journal of Sound and Vibration*, Vol. 40, No. 1, 1975, pp. 1-18.
21. Matsuda, H., and Ishii, K., "Localization of Normal Modes and Energy Transport in the Disordered Harmonic Chain," *Supplement of the Progress of Theoretical Physics*, No. 45, 1970, pp. 56-86.
22. Mead, D. J., "Wave Propagation and Natural Modes in Periodic Systems: II. Multi-Coupled Systems, With and Without Damping" *Journal of Sound and Vibration*, Vol. 40, No. 1, 1975, pp. 19-39.

Elastic Waves and Ultrasonic Nondestructive Evaluation
S.K. Datta, J.D. Achenbach and Y.S. Rajapakse (Editors)

WAVE SCATTERING AND LOCALIZATION IN ANISOTROPIC RANDOM MEDIA

Ping Sheng

Exxon Research and Engineering Co., Rt. 22 East, Clinton Township, Annandale, NJ 08801, U.S.A.

When a wave is multiply scattered in a random medium, its propagation behavior undergoes a transition from coherent, wave-like to incoherent, diffusive transport. However, wave diffusion differs from classical diffusion by the existence of the coherent backscattering effect, which can lead to the "localization" of a wave. This paper delineates the physical basis of the coherent backscattering effect, the sensitivity of the localization phenomenon to spatial dimension of the random system, and the dimensional crossover behavior for anisotropic random media.

I. INTRODUCTION

The concept that a wave can be "localized" through random scatterings was first proposed by P. W. Anderson [1] in 1958 in the context of electron diffusion in random potentials. While the initial concern of the localization theory was mainly in the area of electronic properties of disordered materials, in the past decade there has been an increasing interest in the implications of the localization phenomenon for classical wave transport in random media [2]. In particular, the behavior and parameter dependence of the localized state, the localization-delocalization transition, and the statistical character of multiply-scattered wave field have been the subjects of numerous studies. It is purpose of this paper to delineate the physical basis of localization as derived from these prior studies and to present some new results on the wave scattering and localization behavior in anisotropic random media.

The most common characterization of wave localization is the exponential decay of the wave envelope. In contrast to the evanescent wave (which can result for the electromagnetic wave if the real part of the dielectric constant is negative, such as in metals) or the exponential decay induced by dissipation, here the wave can be locally propagating and the exponential decay is caused purely by random scatterings. Moreover, the decay is statistical in character and therefore can be accompanied by considerable fluctuations. For a one-dimensional randomly-layered medium, the localization and exponential decay of the envelope can be easily seen as a direct result of losing phase coherence. That is, if one denotes by mean-free path s the distances over which the coherence is lost, then by definition of incoherence the total transmission coefficient T through a sample of L is expressible as $T = \prod_{i}^{N} t_i$, where t_i is transmission coefficient over the i th section of thickness s, and $N=L/s$. Alternatively, this can be written as $T=\exp[\langle \ell nt \rangle/s)\cdot L]$, where $\langle \ell nt \rangle = \sum \ell nt_i/N$. The decay length, or the localization length $\ell = -s/\langle \ell nt \rangle$, is seen to be proportional to the mean free path.

Whereas the loss of phase coherence is a sufficient condition for wave localization in one dimensional randomness, the situation is more complicated in higher dimensions. According to the phenomenological scaling theory [3] and the numerical simulation results [4], the spatial dimension two is the marginal dimension for localization, i.e. all waves are localized with infinitesimal amount of randomness in one or two dimensions. However, in two dimensions the localization length is no longer proportional to the mean free path. In three dimensions, there are energy or frequency regions in which the waves are localized, separated from those regions in which the waves are delocalized by so-called mobility edges. Heuristically, the reason that the spatial dimension two is special may be ascribed to the characteristics of diffusion. Since a random walker covers an area L^2=Dt in time t, where D is the diffusion constant, the walker will always return to the neighborhood of the origin in one or two dimensions, but can drift off into infinity in three dimensions. Therefore once the wave transport becomes diffusive due to random scatterings it requires only the additional effect of coherent backscattering, an effect to be described later, to localize all waves in one or two dimensions.

The sensitivity of the localization behavior to the spatial dimension of randomness means that wave scattering and localization in anisotropic systems could potentially be interesting since anisotropic media essentially stradle the different dimensions. The clarification of this question is the subject of the present paper.

II. MODEL DESCRIPTION

For simplicity, we will consider only scalar waves in an anisotropic medium:

$$\nabla^2\phi + \frac{\omega^2}{c^2(\vec{r})}\phi = 0. \tag{1}$$

Here ϕ is the scalar wave amplitude, ω the frequency, and $c(\vec{r})$ denotes the speed of the medium, regarded as a random function of $\vec{r}$. The second term can be alternatively written as $k^2(\vec{r})\phi$, where $k(\vec{r}) = \omega/c(\vec{r})$ is the wave-vector. Equation (1) may be written in the discrete form, let's say on a simple cubic lattice, as

$$\sum_\beta M_{\alpha\beta}\phi_\beta = 0, \tag{2a}$$

$$M_{\alpha\beta} = (k_\alpha^2-2d)\delta_{\alpha\beta}+ \delta_{[\alpha,\beta]}, \tag{2b}$$

where d = 2,3 denotes the spatial dimension, and $\delta_{[\alpha,\beta]} = 1$ for nearest neighbor $[\alpha,\beta]$, and zero otherwise. If we further denote

$$E = \langle k_\alpha^2-2d\rangle, \tag{3a}$$

$$\epsilon_\alpha = (k_\alpha^2-2d) - E, \tag{3b}$$

where < > means averaging, then the wave equation is put in the form of so-called Anderson model[1],

$$H\phi = E\phi, \tag{4a}$$

$$H_{\alpha\beta} = \epsilon_\alpha\delta_{\alpha\beta} + \delta[\alpha,\beta]. \tag{4b}$$

At this level there is no distinction between a quantum-mechanical particle and the classical wave. We see from Eq. (3b) that since k_α is a random number, ϵ_α must also be random.

For our anisotropic model [5], we will specify that ϵ_α consists of two additive random components: one depicts a randomly-layered system and one depicts an isotropic randomness, i.e.

$$\epsilon_\alpha = \theta\eta_{ik} + (1-\theta)\gamma_k, \tag{5}$$

where k denotes the index along the layering direction, i denotes the transverse direction, η and γ are random variables with a distribution function

$$P(\eta_{ik}) = P(\gamma_k) = P(x) = \begin{cases} \frac{1}{W} & , |x| \leq \frac{W}{2} \\ 0 & , \text{otherwise}, \end{cases} \tag{6}$$

and θ is an anisotropy parameter varying between 0 and 1, which interpolates the randomness between a one-dimensional layered (θ=0) and a three-dimensional isotropic (θ=1) system. Physically, this model corresponds to predominantly layered systems, such as the Earth's subsurface, or any randomly layered structures with lateral inhomogeneities that may be either inherent to the system or deliberately introduced.

III. WAVE DIFFUSION AND THE COHERENT BACKSCATTERING EFFECT

Wave transport in disordered materials generally possesses two different characters when viewed on different spatial scales. On a scale less than a mean-free path, s, defined as the probable distance over which the wave suffers a collision, the wave propagates wave-like, just as in homogeneous space. However, on a scale much larger than s in which the wave can suffer many collisions, the transport behavior becomes diffusive. A good example of this transition can be seen in the phonon transport in solids. Multiply-scattered phonons are generally called heat, and they are known to satisfy the diffusion equation.

Mathematically, the character of wave transport in a random medium can be described in ways very analogous to that for a random walker. For example, instead of calculating $\langle\vec{r}\rangle$, the average displacement vector for a random walk, one calculates <G>, the average Green's function (or the average propagator) for the wave. It would be found that just as $\langle\vec{r}\rangle=0$ for a random walker, <G> has a spatial decay length corresponding to the mean free path. However, just as $\langle\vec{r}\rangle=0$ does not imply that a random walker is localized at the origin, the fact that <G> decays over the distance s simply means that beyond the scale of s the transport of wave energy is no longer coherent, or wavelike. To uncover the behavior of incoherent transport, one has to calculate <GG>, which corresponds to $\langle r^2\rangle$ for the random walker. If one writes the exact Green's function as

$$G = \langle G\rangle + \langle G\rangle T\langle G\rangle, \tag{7}$$

where T represents essentially the random local deviation of G from <G>, and <T>=0, then one gets

$$\langle GG\rangle-\langle G\rangle^2 = \langle G\rangle^2\langle TT\rangle\langle G\rangle^2, \tag{8}$$

which shows that $\langle GG\rangle-\langle G\rangle^2$ clearly measures the variance <TT> of the random incoherent scattering. It has been shown that under very general conditions that the behavior of <GG> is diffusive in nature for long travel times and large travel distances [6], just as $\langle r^2\rangle$ for a random walker. The diffusion coefficient formula obtained can also be shown to be identical to that derived from the classical Boltzmann transport theory.

The analogy between wave diffusion and particle diffusion can not be exact, however. In the past decade it has been discovered that random scatterings can not destroy the phase coherence of a wave at the backscattering direction [7-9]. This is due to the fact that the time-reversal invariance of the wave equation defines a mathematical mapping operation for back-scattered waves. Figure 1 illustrates this "coherent back-scattering" effect, sometimes also

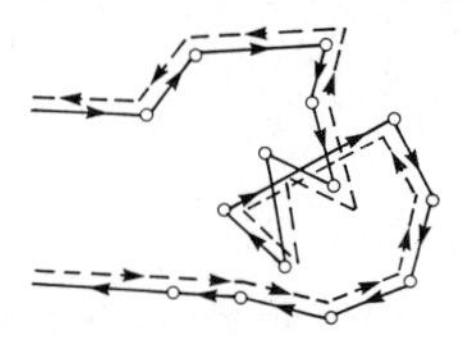

FIGURE 1
Schematic diagram illustrating the phase coherence in the backscattering direction. For any scattering path as indicated by the solid line, there is always a time-reversed path as indicated by the dotted line which has exactly the same phase delay.

denoted as "weak localization" in the physics literature. Two coherent rays, one denoted by the solid line and one by the dashed line, are incident on a random medium. After undergoing random scatterings in the medium, what emerges in the backscattering direction is a sum of many scattering path amplitudes. This is true for both the solid and the dashed rays. In the figure we have shown only one of such scattering paths for illustration. Now, for each scattering path amplitude in the sum for the solid ray, let's say, we can always find an identical time-reversed path in the sum for the dashed ray as shown in Figure 1, and vice versa. In other words, time reversal invariance guarantees that in the backscattering direction everything is coherent, which implies that coherent interference would exactly double the probability density of scattering into the backward direction than into other directions. This effect has been demonstrated experimentally [7-9].

A direct physical consequence of the coherent backscattering effect is to renormalize downward the value of the diffusion constant for wave transport since the effect increases the probability of backward transport instead of forward transport. Mathematically, whereas the calculation of the diffusion constant involves so-called "ladder diagrams" in Eq. (8), the coherent-backscattering effect can be incorporated by including so-called "maximally-crossed diagrams" [10]. Explicit evaluation using these diagrammatic techniques has yielded results in excellent agreement with numerical simulations on the Anderson model [11], including the phenomenon of localization that occurs when the renormalization of the diffusion constant is so strong that it vanishes on a global scale.

IV. WAVE DIFFUSION AND LOCALIZATION IN THE ANISOTROPIC ANDERSON MODEL

Mathematical evaluations of the diffusion coefficient and the coherent backscattering effect have been carried out for the anisotropic Anderson model specified in Section II. Technical details can be found in ref. [5]. Here I describe the essential physical features of the results. First, the results show that the layered model is fairly robust. That is, if the magnitude θ of the isotropic component of the inhomogeneities is small, then the wave scattering behavior can still be cast in the framework of one-dimensional layered system. The effect of the isotropic inhomogeneity is just to modulate the parameters in the theory, such as the localization length. Second, it was found that there is a critical value of θ at which the wave scattering behavior makes a transition and acquires three-dimensional character. When that occurs, for every value of E and θ there is a critical value W_c below which the wave delocalizes. These results are summarized in Fig. 2, where we show both the analytical and numerical simulation values of $W_c(\theta)$ for E=0.

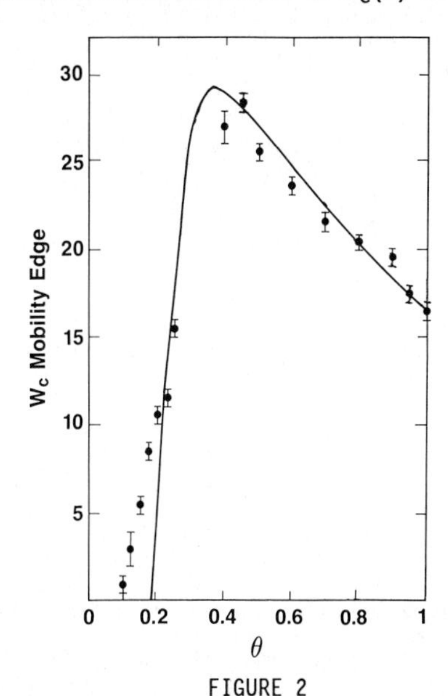

FIGURE 2
Plot of W_c as a function of θ for E=0. The solid line denotes the results calculated from diagrammatical analysis with $\theta_c \simeq 0.19$. Simulation results are denoted by solid circles with extrapolated values of $\theta_c \simeq 0.10$.

The wave behaviors described above should be contrasted with those of the alternative anisotropic hopping model [12] where $\delta_{[i,j]}$ is different for different directions, i.e. in the wave equation instead of having $\nabla^2 = \partial^2/\partial x^2 + \partial^2/\partial y^2 + \partial^2/\partial z^2$, we have $\partial^2/\partial x^2 + k_y\partial^2/\partial y^2 + k_z\partial^2/\partial z^2$, where $k_y \neq k_z \neq 1$. Physically, the anisotropic hopping model corresponds to an isotropic dispersion of anisotropic scatterers. In this case the problem can be simply mapped into an isotropic random one by re-scaling the spatial coordinates, i.e. $y \rightarrow y' = y/\sqrt{k_y}$, $z \rightarrow z' = z/\sqrt{k_z}$. Therefore, the anisotropic hopping model will always exhibit three-dimensional characteristics as soon as the scattering deviates from strict one dimensionality. In that case the interpolation between one and three dimensions is smooth and does not have a critical value of anisotropy.

V. CONCLUDING REMARKS

In this paper I have given a heuristic but physical description of the wave localization phenomenon and its dependence on (and interpolation between) the spatial dimension(s). While much have been learned in the past decade, much more remain to be explored. For example, pulse propagation in a higher dimensional random medium is still a problem to be clarified. Ultimately, it is the hope of this author that the knowledge about wave multiple scattering and localization can lead to statistical inversion routines which would enable us to penetrate the cloudiness barrier imposed by wave multiple scattering. While for layered media this has been successfully carried out [13], for higher-dimensional and anisotropic media the realization of a statistical inversion scheme still seems far away.

REFERENCES AND FOOTNOTES

(a) Work done in collaboration with W. Xue, Z.Q. Zhang, and Q.J. Chu.

[1] Anderson, P.W., Phys. Rev. 109 (1958) 1492.

[2] Sheng, P. (edit.), Scattering and Localization of Classical Waves in Random Media (World Scientific Publishing Co. Inc., 1989).

[3] Abraham, E., Anderson, P.W., Licciardello, D.C. and Ramakrishnan, T.V., Phys. Rev. Lett. 42 (1979) 673.

[4] Mackinnon, A. and Kramer, B., Phys. Rev. Lett. 47 (1981) 1546.

[5] Xue, W., Sheng, P., Chu, Q.J., and Zhang, Z.Q., submitted to Phys. Rev. Lett.

[6] Zhang, Z.Q. and Sheng, P., Wave Diffusion and Localization in Random Composites, in: Scattering and Localization of Classical Waves in Random Media, Sheng, P. (edit.), (World Scientific Publishing Co., Inc., 1989).

[7] Tsang, L. and Ishimaru, A., J. Opt. Soc. Am. A1 (1984) 836.

[8] Van Albada, M.P. and Lagendijk, A., Phys. Rev. Lett. 55 (1985) 2692.

[9] Wolf, P.E. and Maret, G., Phys. Rev. Lett. 55 (1985) 2696.

[10] Vollhardt, D. and Wölfle, P., Phys. Rev. B22 (1980) 4666.

[11] Zdetsis, A.D., et al. Phys. Rev. B32 (1985) 7811.

[12] Apel, W. and Rice, T.M., J. Phys. C16 (1983) L1151.

[13] White, B.S., Sheng, P., Postel, M. and Papanicolaou, G., Phys. Rev. Lett. 63 (1989) 2228.

Elastic Waves and Ultrasonic Nondestructive Evaluation
S.K. Datta, J.D. Achenbach and Y.S. Rajapakse (Editors)
Elsevier Science Publishers B.V. (North-Holland), 1990

THE PECULIARITY OF VIBRATION PROCESS LOCALIZATION IN SEMI-RESTRICTED REGIONS

V.A. Babeshko, I.P. Obratsov, I.I. Vorovich

Kuban State University
Karl Libknekht Street 149
350640 Krasnodar
Soviet Union

1. The spreading waves are known to appear beginning with the critical frequency $\omega^* \geq 0$, which carry over the energy at infinity [1] in semi-restricted linear elastic bodies, such as finite packet of layers, unbounded cylinders under time harmonic loading of their surface. The existence of the unbounded resonance of the massive stamps in the range of frequencies $0<\omega<\omega^*$ [2] is stated for the semi-restricted bodies which have $\omega^*>0$.

The unbounded resonance of the massive stamps was considered impossible [2,3] for such bodies at frequencies $\omega>\omega^*$ and also at any frequencies for the bodies which have $\omega^*=0$. This proposition is being corrected below. It is determined that there are relationships between the characteristics of the contact domains and conditions of loading on the semi-restricted body surface, properties of medium, and frequencies of oscillations for which at the frequency $\omega>\omega^*$, the body has a special regime of dynamic behavior. It is characterized by the absence of energy radiation at infinity. In this regime at certain values of the parameters the system behaves differently than thought possible for such bodies.

Thus, in the case of finite packet of layers during the vibration of stamps on it, for some values of stamp masses their unbounded resonance may take place; the resonance of the weightless finite stamp systems (layer oscillations when stamps are stationary) may occur; for some dimensions of crack-cavity, its unbounded resonance is possible if its surfaces are parallel to the boundary of the domain.

Some necessary and sufficient conditions are obtained for the existence of special regimes of dynamics of the semi-restricted body-stamps-cavities system. These conditions are formulated in the form of analytical formulae called the statical conditions.

2. We shall formulate the statical conditions in the notation of the integral equation solutions of the dynamic contact problems for stamp without friction on the elastic layer or packet of layers. If the domain of contact D admits the splitting into convex domains D_m, then the integral of the dynamic contact problem takes the form [1]:

$$Kq = \iint_D k(x-\xi, y-\eta) q(\xi,\eta) d\xi d\eta = f(x,y), \tag{1}$$

$$x,y \in D;$$

$$k(x,y) = \frac{1}{4\pi^2} \int_{\sigma_1} \int_{\sigma_2} K(u) e^{-i(\alpha x+\beta y)} d\alpha d\beta, \quad u=\sqrt{\alpha^2+\beta^2}, \tag{2}$$

$$D = \bigcup_{m=1}^{N} D_m$$

The description of the kernel properties and integral equations are presented in [1] and the details are omitted here.

With z_n, ζ_n to denote all the positive zeros and poles (their number $p=\infty$) of the even function K(u) which we consider homogeneous we introduce the function

$$K_0(u) = K(u)\prod_{n=1}^{P}(u^2-\zeta_n^2)(u^2-z_n^2)^{-1} > 0 \tag{3}$$

Together with equation (1) we shall consider the integral equation

$$K_0 q_0 = f(x,y),\ \ x,y \in D \tag{4}$$

in the representation of the kernel (2) in which instead of K(u), $K_0(u)$ is introduced.

Owing to its properties the integral equation (4) by virtue of (3) is completely analogous to the equation of static contact problem [4].

Let $P_{[D_m]}$ be projector on domain D_m. We shall construct the system of solutions of the integral equations by one of the methods in [4]. Consider

$$K_0 q_{mn} = P_{[D_m]}\exp\,(iz_n(x\cos\gamma+y\sin\gamma)), x,y \in D, \tag{5}$$

$$n = 1,2,...,p;\ \ m = 1,2,...,N,\ \ 0 \le \gamma < 2\pi$$

If $\mathrm{Re} f(x,y)e^{-i\omega t}$, the amplitude of oscillation of the stamp base in the domain D, then the special regime of layer oscillation will take place under the conditions

$$\iint_D f(x,y)q_{mn}(x,y)dxdy = 0, \tag{6}$$

$$m = 1,2,...,N;\ \ n = 1,2,...,p;\ \ 0 \le \gamma < 2\pi$$

In this regime the elastic system of "semi-restricted body-stamps" doesn't radiate energy at infinity in the absence of the inner absorption, and behaves as a bounded body. It means that the system may possess discrete resonance frequencies. By changing the parameters of the systems without extending this regime, it is possible to find the frequency $\omega > \omega^*$ for resonance, as was done in [2] by changing the stamp mass m at $\omega < \omega^*$. Characteristically, when using the conditions (6) the phenomenon of the vibration process localization is discovered in the zone of stamp location. (Note: The above-mentioned phenomenon may also be traced in other processes such as electrodynamics, diffusion, and thermal conduction in semi-restricted bodies.)

The problems for the stamp, inclusion or crack, which are in the form of thin strips parallel to the layer boundary, are reduced to the search for the integral equations

$$Kq = \int_{-a}^{a} k(x-\xi)q(\xi)d\xi = f(x),\ \ |x| \le a \tag{7}$$

$$k(x) = \frac{1}{2\pi}\int_\sigma K(u)e^{-iux}du$$

Here a is the width of the strip. If only one spreading wave propagates in the elastic layer for that frequency ω, function K(u) is

$$K_0(u) = K(u)(u^2-\zeta^2)(u^2-z^2)^{-1} > 0 \tag{8}$$

$$z > 0,\ \zeta > 0.$$

The statical condition in this case takes the form

$$V(z)K_0^{-1} f=0,\ \ V(z)\phi = \frac{1}{2\pi}\int_{-\infty}^{\infty}\phi(x)e^{izx}dx \tag{9}$$

Here K_0^{-1} is the inverse operator to K_0. The sizes of the stamps a_m and of the cracks a_m^0 for resonance in this case are

$$a_m = [\pi m + \arg K_0^+(z)]z^{-1} + 0(m^{-1}) \tag{10}$$
$$a_m^0 = [\pi(m-0.5) + \arg K_0^+(z)]z^{-1} + 0(m^{-1})$$
$$m \gg 1$$

Function $K_0^+(u)$ is the result of function factorization of $K_0(u)$.

3. Example: Let mass stamp m perform antiplane oscillation on the elastic layer with the rigidly connected lower boundary. Specializing to one-dimensional integral equation, (1) takes the form (2.1) of [5,6], in which for the frequencies $\Omega > \Omega^*$, we use the rule described in [1] of contour location of the kernel K(x). In [1] rule of the contour location in the nucleus representation K(x) is used. Supposing that $\Omega^2 = \rho\omega^2 h^2 \mu^{-1}$ and $\pi < \Omega < 1.5\pi$, i.e. $\Omega > \Omega^* = \pi/2$, we find the values of the dimensions A of the stamp, at which the conditions (6) are satisfied. These values are given in the notations of the work [5] by

$$A_n = \lambda_n^{-1}(n\pi + c)x^{-1}[1 + 0(n^{-1})], \quad n \gg 1, \quad c = \text{const.} \tag{11}$$

Resonance masses m_n have the representation

$$m_n = \frac{Q_n x^2}{(\Omega^2 - 0.25\pi^2)\Omega^2}, \quad Q_n = \int_{-1}^{1} \tau_n(x)dx, \quad x = \sqrt{\Omega^2 - \pi^2}, \tag{12}$$

where $\tau_n(x)$ is the solution of the integral equation (2.1) of the work [5], $\omega=1$, $\lambda=\lambda_n$ with the kernel in which

$$L(\alpha) = \frac{(\alpha^2 - \Omega^2 + 0.25\pi^2)\,\mathrm{th}\sqrt{\alpha^2 - \Omega^2}}{(\alpha^2 - \Omega^2 + \pi^2)\sqrt{\alpha^2 - \Omega^2}} \tag{13}$$

is taken.

REFERENCES

[1] Vorovich, I.I., The Dynamic Mixed Problems of Elasticity Theory for Non-Classical Domains (Nauka, Moscow, 1979).

[2] Vorovich, E.M., Pryakhina, O.D., Tukodova, O.M., PMM 51 (1987), 109.

[3] Babeshko, V.A., Papers of the USSR Academy of Sciences, 295 (1987), 312.

[4] Babeshko, V.A., The Generalized Method of Factorization in the Three-Dimensional Dynamic Mixed Problems of the Elasticity Theory (Nauka, Moscow, 1984).

[5] Vorovich, E.I., Pryakhina, O.D., MTT N3 (1987).

[6] Babeshko, V.A., About Peculiarities of Semi-Restricted Bodies Vibration. Journal of Technical Physics, 14, N8 (1988).

Elastic Waves and Ultrasonic Nondestructive Evaluation
S.K. Datta, J.D. Achenbach and Y.S. Rajapakse (Editors)

ITERATIVE SOLUTION OF INTEGRAL EQUATIONS IN SCATTERING PROBLEMS

P.M. van den Berg* and **R.E. Kleinman†**

*Laboratory of Electromagnetic Research, Department of Electrical Engineering Delft University of Technology, P.O. Box 5031, 2600 GA Delft, The Netherlands

†Center for the Mathematics of Waves, Department of Mathematical Sciences University of Delaware, Newark, DE 19716, U.S.A.

A number of iterative algorithms to solve integral equations arising in scattering problems are discussed. We describe the essential features of the Neumann Series, over-relaxation methods, Krylov subspace methods and the conjugate gradient technique. Relations between all of the methods will be described and numerical performance will be contrasted using a square error criterion.

1. INTRODUCTION

Acoustic scattering problems are often formulated as integral equations and it is this form which serves as the starting point for most numerical solutions. Typically the integral operators which occur are boundary integrals when considering scattering by impenetrable objects and domain integrals for penetrable scatterers. These operators are invariably non-selfadjoint which complicates most numerical approaches. In a large number of cases the integral equations may be put in a Hilbert space frame work. In abstract form the equation is

$$Lu = f \,, \tag{1}$$

where L is a bounded linear operator, not necessarily selfadjoint, which maps a Hilbert space H into itself. The space will be equipped with a norm $\|\cdot\|$ and inner product $\langle \cdot, \cdot \rangle$, linear in the first entry. The right-hand side of (1), f, is an element of H and is known in terms of a prescribed incident field. Throughout we assume that L is bounded, i.e. $\|L\| < \infty$ and (1) is uniquely solvable for every $f \epsilon H$ which means that L^{-1} exists and $\|L^{-1}\| < \infty$. We denote the spectral radius of any operator A by $\sigma(A)$ where

$$\sigma(A) = \lim_{n\to\infty} \|A^n\|^{\frac{1}{n}} \,, \tag{2}$$

In many specific examples the operator L is of the second kind, $L = I - K$, where K is compact.

In this paper we review a class of linear iterative techniques for solving equation (1) based on Krylov subspaces. By this we mean the subspaces spanned by the vectors $\{L^m v;\ m = 0, \cdots, N\}$ for some $v \epsilon H$ to be made explicit below. As will be evident a number known iteration techniques may be included as special cases of the following general method.

2. THE GENERAL ITERATIVE PROCEDURE

We consider the following general iterative procedure for solving (1)

$$\begin{aligned} &u_0 \text{ arbitrary}\,, && r_0 = f - Lu_0\,, \\ &u_n = u_{n-1} + \sum_{m=1}^{n} \alpha_{nm} T r_{m-1}\,, && r_n = f - Lu_n\,. \end{aligned} \tag{3}$$

The residual satisfies the recursion relation

$$r_n = r_{n-1} - \sum_{m=1}^{n} \alpha_{nm} L T r_{m-1}\,. \tag{4}$$

The linear operator T will be taken to be either the identity operator I or the adjoint L^* of L (see Section 8 for comments on other choices of T as preconditioners). Various choices of the constants α_{nm} will result in a number of known iterative procedures. Before illustrating this, we remark that each iterate can be rewritten in the following form:

$$u_n = u_0 + \sum_{m=1}^{n} \beta_{nm} T(LT)^{m-1} r_0 \,, \quad r_n = r_0 - \sum_{m=1}^{n} \beta_{nm} (LT)^m r_0 \,. \tag{5}$$

The expressions of (5) may be established using mathematical induction and (3) - (4) to find β_{nm} in terms of α_{nm}. Thus we see that this general iteration leads to a representation of the residual error at the nth step which is in the Krylov subspace spanned by $\{(LT)^m r_0,\ m = 0, \cdots, n\}$. It is not possible to discuss convergence without putting restrictions on the constants α_{nm} or β_{nm} and this is done in the context of particular choices.

3. NEUMANN ITERATION

The usual Neumann series results from the following specialization of the general method. Let $T = I$; $\alpha_{nm} = 0$, $m < n$; $\alpha_{nn} = 1$. Then (3) and (4) become

$$u_n = u_{n-1} + r_{n-1} = f + (I-L)u_{n-1} \,, \quad r_n = r_{n-1} - Lr_{n-1} = (I-L)r_{n-1} \,, \tag{6}$$

from which we may deduce that

$$u_n = \sum_{m=0}^{n-1} (I-L)^m f + (I-L)^n u_0 \,, \quad r_n = (I-L)^n r_0 \,. \tag{7}$$

It is easy to see that with this choice of α_{nm}, the constants β_{nm} in (5) become

$$\beta_{nm} = (-1)^{m+1} \frac{n!}{m!(n-m)!} \,. \tag{8}$$

It is well known that a condition sufficient to ensure convergence of this iterative process is $\sigma(I-L) < 1$. It should be noted that this is not a necessary condition since there exist examples where the Neumann series converges but $\sigma(I-L) = 1$ [**1**]. Nevertheless for most operators of interest in scattering problems the Neumann series does not converge.

4. STATIONARY OVER-RELAXATION METHOD

The simplest generalization of the Neumann series results from the choice $T = I$; $\alpha_{nm} = 0$, $m < n$; $\alpha_{nn} = \alpha$ in the general method. Then (3) and (4) become

$$u_n = u_{n-1} + \alpha r_{n-1} = \alpha f + (I-\alpha L)u_{n-1} \,, \quad r_n = r_{n-1} - \alpha L r_{n-1} = (I-\alpha L)r_{n-1} \,, \tag{9}$$

from which we may deduce that

$$u_n = \sum_{m=0}^{n-1} (I-\alpha L)^m \alpha f + (I-\alpha L)^n u_0 \,, \quad r_n = (I-\alpha L)^n r_0 \,. \tag{10}$$

In this case the constants β_{nm} in (5) become

$$\beta_{nm} = (-1)^{m+1} \frac{n!\alpha^m}{m!(n-m)!} \,. \tag{11}$$

Convergence of this method, i.e. $\lim_{n\to\infty} r_n = 0$, is assured if $\sigma(I-\alpha L) < 1$. The question of whether it is possible to choose α such that this condition is fulfilled depends on $\sigma(L)$. In [**3**] it was shown that there will exist such an α if $\sigma(L)$ lies in an annular wedge shaped domain in the complex plane with wedge angle less than π excluding a neighborhood of the origin. That this property of $\sigma(L)$ is met depends on the particular operator L. If L is the domain integral operator which occurs in scattering by a penetrable inhomogeneous object then $\sigma(L)$ meets these conditions [**2**]. Of course even if it can be shown that there exist values of α for which $\sigma(I-\alpha L) < 1$ holds, there remains the problem of actually determining such values. It has been proposed that α be chosen to minimize $\|r_1\|$ which results in the explicit value

$$\alpha = \frac{\langle r_0, Lr_0\rangle}{\|Lr_0\|^2} . \tag{12}$$

This choice was shown to be remarkably effective both in domain scattering, where it is known that α exists for which $\sigma(I-\alpha L) < 1$ [**2**], and in rigid body scattering, where the existence of an appropriate α is not yet proven [**3**].

The situation is much different if L is selfadjoint. In that case $\sigma(L)$ is real and since (1) is always uniquely solvable then there always exists an α $(0 < \alpha < 2\|L\|^{-1})$ for which the iteration of (9) converges [**4,5**].

If there exists an α for which the iteration defined by (9) converges, then in fact there will be a domain of α values which ensure convergence and an optimal value which maximizes the rate of convergence. Failure to be close to optimal can have serious numerical consequences as is illustrated in [**3**]. One way to avoid the necessity of estimating the optimal α is found in the next method.

5. SUCCESSIVE OVER-RELAXATION METHOD

The next specialization of the general iterative method results from the choice $T = I$; $\alpha_{nm} = 0$, $m < n$; $\alpha_{nn} = \alpha_n$. That is, we consider the same scheme as before with the exception that instead of choosing α_{nn} to be constant we choose a new value at each step in the iteration. Now (3) and (4) become

$$u_n = u_{n-1} + \alpha_n r_{n-1} = \alpha_n f + (I-\alpha_n L)u_{n-1}\,, \quad r_n = r_{n-1} - \alpha_n L r_{n-1} = (I-\alpha_n L)r_{n-1}\,, \tag{13}$$

from which we may deduce that

$$u_n = \alpha_n f + \sum_{m=1}^{n-1} \prod_{l=m+1}^{n} (I-\alpha_l L)\,\alpha_m f + \prod_{m=1}^{n}(I-\alpha_m L)\,u_0\,, \quad r_n = \prod_{m=1}^{n}(I-\alpha_m L)\,r_0\,. \tag{14}$$

In this equation the constants β_{nm} in (5) may be shown to be given iteratively by

$$\beta_{11} = \alpha_1\,, \quad \beta_{n+1\,m} = \begin{cases} \beta_{n1} + \alpha_{n+1}, & m = 1\,, \\ \beta_{nm} - \alpha_{n+1}\beta_{n\,m-1}\,, & 1 < m \le n\,, \\ -\alpha_{n+1}\beta_{nn}, & m = n+1\,. \end{cases} \tag{15}$$

Now the question of how to choose α_n must be answered before the iteration is fully defined. Here we follow the principle of error minimization which underlies the work of [**6**]. That is we choose α_n to minimize the residual error $\|r_n\|$, at each step. In the stationary method derived previously there was only one constant and this guiding principle led to its determination by minimizing $\|r_1\|$. In the present case we have at our disposal one constant at each step which we find explicitly to be

$$\alpha_n = \frac{\langle r_{n-1}, Lr_{n-1}\rangle}{\|Lr_{n-1}\|^2} . \tag{16}$$

As to convergence of this iterative method of (13) - (16) we have the following result in the general case. If there exists an α such that $\|I-\alpha L\| < 1$ then $\lim_{n\to\infty} r_n = 0$. That is, if in the previous method there exists an α for which not only is $\sigma(I-\alpha L) < 1$ but $\|I-\alpha L\| < 1$ as well, then the present method converges. We remark that even if $\sigma(I-\alpha L) < 1$ it does not follow that $\|I-\alpha L\| < 1$ although it is true that $\sigma(I-\alpha L) < 1 \Rightarrow \|(I-\alpha L)^n\| < 1$ for some $n \ge 1$. But we are unable to show that this weaker condition will guarantee convergence of the present method. The vital difference is that here we do not need to find α in order to implement the process.

If L is selfadjoint and positive ($\langle Lu, u\rangle \geq 0$) then convergence of the iterative procedure given by (13) and (16) is guaranteed with no further conditions on $\sigma(L)$. In fact it may be shown that the rate of convergence is governed by the inequality

$$\|r_n\|^2 \leq \left(1 - \frac{1}{\|L^{-1}\|^2\|L\|^2}\right)^n \|r_0\|^2 . \tag{17}$$

6. KRYLOV SUBSPACE AND CONJUGATE GRADIENT METHOD

Now we consider the full Krylov method by which we mean the general iterative method of (3) - (4) with no a priori assumptions on the coefficients α_{nm} (or β_{nm} in (5)). First we assume $T = I$. Now in keeping with the principle of determining the coefficients to minimize $\|r_n\|$ we see that with (4) the coefficients will be solution of the system of equations [**6,7**]

$$\sum_{m=1}^{n} \alpha_{nm}\langle Lr_{m-1}, Lr_{l-1}\rangle = \langle r_{n-1}, Lr_{l-1}\rangle, \quad l = 1, \cdots, n . \tag{18}$$

Alternatively the same error minimization used with the second equation of (5) leads to a system for the β_{nm}

$$\sum_{m=1}^{n} \beta_{nm}\langle L^m r_0, L^l r_0\rangle = \langle r_0, L^l r_0\rangle, \quad l = 1, \cdots, n . \tag{19}$$

This system is always solvable unless the solution u of (1) is a linear combination of $u_0 \cup \{L^{l-1}r_0,\ l = 1, \cdots, n-1\}$, in which case there exist $\beta_{n-1\,m}$ and $\alpha_{n-1\,m}$ such that $u = u_{n-1}$ and $r_{n-1} = 0$. It may be shown that the general iterative procedure of (3) and (4) may be recast as follows:

$$\begin{aligned} u_n &= u_{n-1} + \alpha_{nn} v_n , \quad r_n = r_{n-1} - \alpha_{nn} L v_n , \quad \alpha_{nn} = \frac{\langle r_{n-1}, Lr_{n-1}\rangle}{\|Lv_n\|^2} , \\ v_n &= r_{n-1} + \sum_{m=1}^{n-1} \gamma_{nm} v_m , \quad \gamma_{nm} = -\frac{\langle Lr_{n-1}, Lv_m\rangle}{\|Lv_m\|^2}, \quad m = 1, \cdots, n-1 , \end{aligned} \tag{20}$$

with error estimate

$$\|r_n\|^2 = \left(1 - \frac{|\langle r_{n-1}, Lr_{n-1}\rangle|^2}{\|Lv_n\|^2\|r_{n-1}\|^2}\right)\|r_{n-1}\|^2 . \tag{21}$$

This method will converge whenever there exists an α such that $\sigma(I-\alpha L) < 1$. This follows from the fact that by choosing β_{nm} to minimize $\|r_n\|$, this value of $\|r_n\|$ is certainly less than or equal to the value of $\|r_n\|$ had we chosen β_{nm} to satisfy (11). But this larger residual error vanishes with n hence in the present case $\lim_{n\to\infty} r_n = 0$ as well. We remark that it was only necessary to know the existence of an α for which $\sigma(I-\alpha L) < 1$ was satisfied. It is not necessary to actually find α nor is the stronger condition, $\|I-\alpha L\| < 1$, required.

When L is selfadjoint but not necessarily positive it is found that

$$\gamma_{nm} = \begin{cases} 0 , & m < n-1 , \\ \frac{\langle r_{n-1}, Lr_{n-1}\rangle}{\langle r_{n-2}, Lr_{n-2}\rangle}, & m = n-1 , \end{cases} \tag{22}$$

and the iteration becomes a conjugate gradient algorithm based on minimizing the residual error in the Hilbert space norm rather the usual conjugate gradient algorithm in which the actual error is minimized in an energy norm [**8**]. If L is also positive then the algorithm will always converge with a rate larger than the one given by (17).

If L is not selfadjoint and we choose $T = L^*$ we end up with the usual conjugate gradient scheme

$$\begin{aligned} u_n &= u_{n-1} + \alpha_{nn} v_n , \quad r_n = r_{n-1} - \alpha_{nn} L v_n , \quad \alpha_{nn} = \frac{\|L^* r_{n-1}\|^2}{\|Lv_n\|^2} , \\ v_n &= L^* r_{n-1} + \gamma_{n\,n-1} v_{n-1} , \quad \gamma_{n\,n-1} = \frac{\|L^* r_{n-1}\|^2}{\|L^* r_{n-2}\|^2} . \end{aligned} \tag{23}$$

This form of the conjugate gradient method will always converge with a rate larger than the one given by (17).

7. NUMERICAL RESULTS FOR THE SCATTERING BY A SLAB

Each method will be demonstrated in the numerical example of the problem of scattering of a time-harmonic plane wave normally incident on a slab of finite width $l = \frac{1}{2}\lambda$, where λ

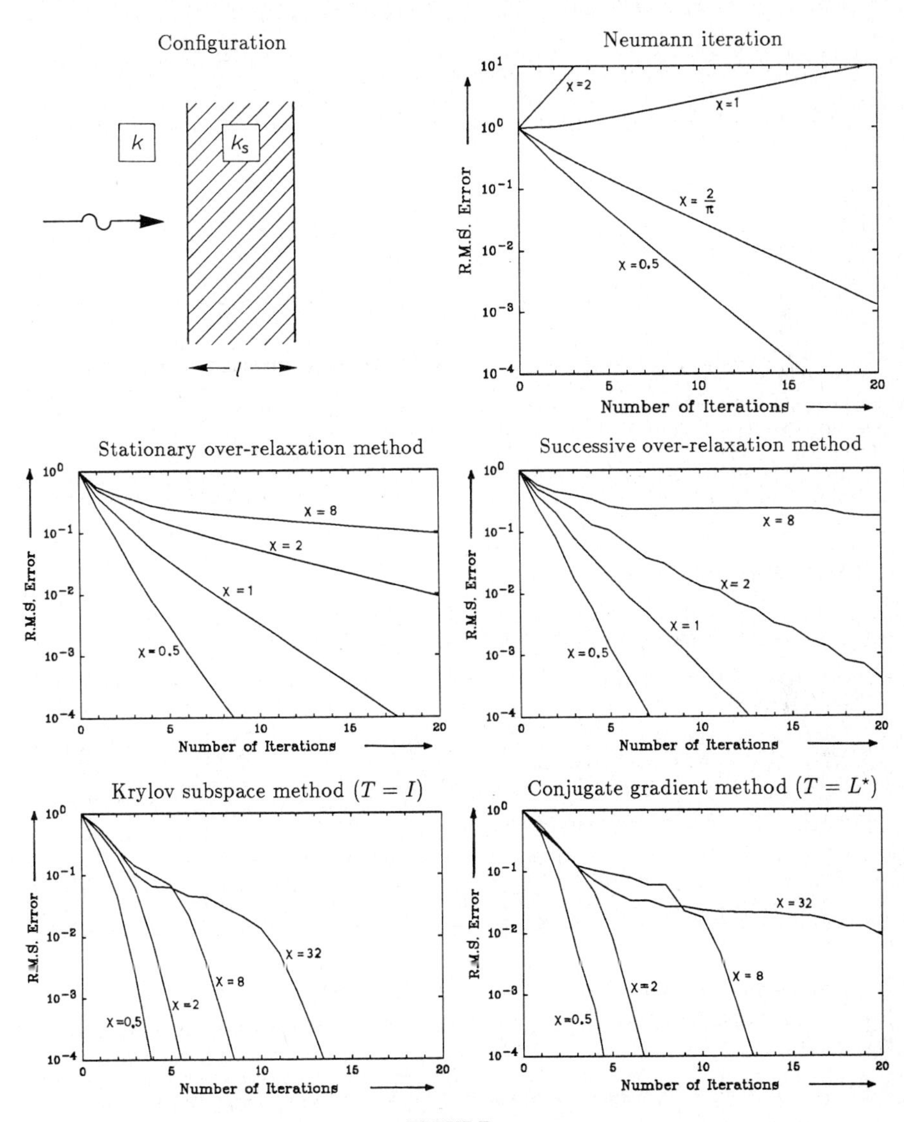

FIGURE 1
Numerical convergence of the various schemes.

is the wavelength. In this case, equation (1) represents a domain integral equation, in which

$$Lu = u(x) - \frac{ik}{2}\int_0^l e^{ik|x-x'|}\,\chi\, u(x')\,dx' \,, \quad f = e^{ikx} \,, \quad x \in (0,l) \,, \tag{24}$$

where $\chi = k_s^2/k^2 - 1$, k_s denotes the wave number of the slab and k denotes the wave number of the surrounding medium. In this example we choose the Hilbert space H to be $L_2(0,l)$. In Figure 1, the numerical convergence of the different schemes are presented by plotting the normalised residual error, $\|r_n\|/\|f\|$, as a function of the number of iterations. It shows that better results come from more sophisticated procedures with the best performance showed by the Krylov subspace method ($T = I$). Note that the occurrence of the operator LL^* in the conjugate gradient method negatively influences the rate of convergence. It also appears that the theoretical orthogonality used in the conjugate gradient scheme is gradually lost due to loss of significant figures in the numerical computations, whereas in the Krylov method it is enforced at each step.

8. CONCLUDING REMARKS

We note that in all of the iterative schemes discussed, the operator T in the general method of (3) - (5) was chosen either as I or L^*. T plays the role of a preconditioner for equation (1) and if a good preconditioner is available then it will accelerate convergence of all of the methods discussed. For example, if $LT = I - L_\varepsilon$ with $\|L_\varepsilon\| < 1$ then there exists an α, namely $\alpha = 1$, for which $\|I - \alpha LT\| < 1$ and all of the iteration schemes based on (3) - (5) using this T will converge. In practical terms one can trade complexity of the iterative method for effective preconditioning. The availability of a sufficiently good preconditioner will enable one to employ the simplest of the iterative methods for computations. Until now most preconditioners are developed from matrix methods which fail to incorporate asymptotic and approximate solutions. Such solutions have been developed through explicit analysis of the particular equation and take into account the physics of the problem which the equation models and it is suggested that future work in preconditioners use these approximate solutions.

ACKNOWLEDGEMENTS

This work was supported under NRL Contract N00014-86-C-2121, NSF Grant No. DMS-8811134, AFOSR Grant-86-0269 and NATO Grant-0230/88.

REFERENCES

[1] Patterson, W.M., Iterative methods for the solution of a linear operator equation in Hilbert space - a survey, Lecture Notes in Mathematics 394 (Springer-Verlag, Berlin, 1974) p. 90.

[2] Kleinman, R.E., Roach, G.F. and Van den Berg, P.M., A convergent Born series for large refractive indices, submitted to J. Opt. Soc. Am., 1989.

[3] Kleinman, R.E., Roach, G.F., Schuetz, L.S., Shirron, J. and Van den Berg, P.M., An over-relaxation method for the iterative solution of integral equations in scattering problems, submitted to Wave Motion, 1988.

[4] Bialy, H., Arch. Rational Mech. Anal. 4, 166-176, 1959.

[5] Kleinman R.E. and Roach, G.F., Proc. Roy. Soc. London A417, 45-57, 1988.

[6] Van den Berg, P.M., IEEE Trans. Antennas Propagat. AP-32, 1063-1071, 1984.

[7] De Hoop, A.T. and Van den Berg, P.M., Integral-equation methods in the radiation and scattering of electromagnetic and acoustic, Series on Acoustic, Electromagnetic and Elastic Wave Scattering, edited by V.K. Varadan and V.V. Varadan, (North-Holland, Amsterdam) in print.

[8] Hayes, R.M., Nat. Bur. Standards Appl. Math. Ser. 39, 71-104, 1954.

Elastic Waves and Ultrasonic Nondestructive Evaluation
S.K. Datta, J.D. Achenbach and Y.S. Rajapakse (Editors)

HOW FINE A SURFACE CRACK CAN YOU SEE IN A SCANNING ACOUSTIC MICROSCOPE?

G.A.D. Briggs

Department of Metallurgy and Science of Materials,
University of Oxford,
England OX1 3PH

The acoustic microscope has great sensitivity to surface cracks. Even when they are much less than a wavelength wide, they can nevertheless show up with strong contrast in an acoustic image. The reason for this is now well known. A dominant role in the contrast is played by Rayleigh waves that are excited in the surface, and their interference with waves specularly reflected from the surface. By defocussing the microscope this interference can be made very sensitive to factors that effect the propagation of the Rayleigh wave, and hence give enhanced contrast from features within a depth of a wavelength or so.

A question that immediately arises is what the limit is to crack detection? Indeed is there any limit at all? There is a fairly strong argument on theoretical grounds that however fine a crack is, it should still scatter Rayleigh waves whether it is filled with water or air, the only limit being the shear wave decay length. The argument seems to be vindicated by observations of fringes of half-Rayleigh-wavelength periodicity. First, these fringes are observed parallel to the crack with almost undiminished strength right up to a point level with the tip of the crack. Second, by adjusting the contrast it is also possible to see semicircular fringes centred on the crack tip, presumably due to diffraction at the tip itself

Radial cracks from indents in a number of different materials have been studied by acoustic microscopy. In every case the acoustic microscope revealed all the cracks that were seen by light microscopy and scanning electron microscopy on the same specimens. More important, the acoustic microscope always gave strong contrast right up to the tip of the crack as observed by both l.m. and s.e.m. So far from the contrast becoming weaker as the crack became finer, Rayleigh waves are scattered no mattter how fine the crack is. The conclusion is that surface cracks very much less than a wavelength wide can be detected with strong contrast in the acoustic microscope, and that if there is a limit it is a very small one indeed.

1. OBSERVATION

It is now well known that the acoustic microscope can reveal surface cracks with very strong contrast [Ilett *et al.* 1984]. This remains true even when the width of the crack at the surface is very much less than the acoustic wavelength in the coupling fluid, and therefore much less than the resolution of the microscope. The origin of the enhanced contrast from surface cracks lies in the strong excitation of Rayleigh waves in the surface of the specimen. These propagate parallel to the surface, and can therefore strike a crack broadside, and be strongly scattered even when it is very fine. The purpose of the results summarized here is to address the question. "How fine can a surface crack be, and still be seen in an acoustic microscope?"

2. THEORY

The contrast in an acoustic microscope is best understood in terms of $V(z)$, the variation of the video signal V with the separation z between the surface of the specimen and the focal plane of the lens, movement towards the lens being taken by convention as negative. A theoretical expression for $V(z)$ can be formulated in terms of the product of the pupil function of the lens and the reflectance function of the specimen, an exponential phase term, and trigonometrical amplitude terms, all summed over the range of angles of waves incident on the specimen [Sheppard and Wilson 1981].

$$V(z) = \int_0^{\pi/2} R(\theta)\, P(\theta)\, e^{ikz\cos\theta} \cos\theta \sin\theta \, d\theta \,. \qquad [1]$$

This form cannot be used when a crack is present, because it implicitly assumes that the scattering of plane waves can be described by a reflectance function in which the angle of incidence and the angle of reflection are equal. In the presence of a crack the integrations over the incident plane waves and the scattered plane waves (specified by the x-components, k_x' and k_x respectively, of their wave-vector) must be performed separately, without assuming that they have the same angle, and with an explicit dependence on the lateral displacement, x, of the lens. In two dimensions [Somekh *et al.* 1985]

$$V(x,z) = \int_{-k_0}^{k_0}\int_{-k_0}^{k_0} \exp[i(k_z'-k_z)z]\, L_1(k_x')\, L_2(k_x)\, S(k_x,k_x')\, \exp[i(k_x'-k_x)x]\, dk_x\, dk_x', \qquad [2]$$

where $k_z' = \omega/v_0 - k_x'$, $k_z = \omega/v_0 - k_x$ are taken as negative and positive respectively, $L_1(k_x')$, $L_2(k_x)$ are the lens functions for transmitted and received waves, and the crack is taken to be at the origin. A two-dimensional scattering function in k-space for light fluid loading may be expressed in the form

$$S(k_x,k_x') = \left[R_0(k_x) + \frac{i4\alpha k_p}{k_x^2 - k_p^2}\right] \delta(k_x - k_x') + \frac{2\alpha}{\pi}\left[\frac{(T-R-1)k_x k_x' + (T+R-1)k_p^2}{(k_x^2 - k_p^2)(k_x'^2 - k_p^2)}\right]. \qquad [3]$$

Here k_x' and k_x are the components of wave vector parallel to the surface of the incident and scattered waves; k_p is the Rayleigh wave pole in the complex k_x plane, and α describes the strength of coupling of the Rayleigh wave to the fluid (in the absence of other attenuation mechanisms in the solid α is equal to the imaginary part of k_p). R_0 is the reflectance function minus the contribution corresponding to Rayleigh wave excitation (*i.e.* the Rayleigh pole [Bertoni 1984]), and R and T are the coefficients of reflection and transmission of Rayleigh waves by the crack. A great advantage of the formulation in this way is that R and T can be taken from independent calculations (albeit usually calculations for the unloaded case [Achenbach 1987]), including calculations for cracks of finite depth (with associated creeping waves and diffraction), crack closure, and different orientations. Line-scans across a crack have been computed using this theoretical model, and they have been compared with line-scans measured using a cylindrical lens with its axis parallel to a crack in a glass specimen [Rowe *et al.* 1986]. A three dimensional theory of the contrast from cracks has since been developed, but it is found that the two-dimensional theory gives a qualitative account of all the most important features that are found in images of cracks and boundaries, such as the enhanced contrast even from very fine cracks, the reversal of contrast with defocus, the half Rayleigh wavelength fringes, and the contrast at grain boundaries and interfaces between different materials.

3. EXPERIMENT

Using the line scan facility on the ELSAM, it is possible to test in more detail how the two-dimensional theory describes the contrast from cracks in an imaging microscope. Figure 1 presents such a comparison. The two upper figures contain theoretical curves of the video signal V as the lens is scanned paralled to the surface at a constant defocus z. The lower figures are pictures of a radial crack from an indent in glass, together with a line scan of the video signal along the straight horizontal line across the middle of the picture. The two figures were chosen to represent cases where the crack appears bright and dark relative to the background respectively. The value of z in the theoretical curves was chosen to optimise the similarity with the experimental curves; the reason for the discrepancy in z is not known, but no doubt the usual uncertainties of thermal drift, frequency, and pupil function all play a role. Neverless the important conclusion from this kind of experiment is that the two-dimensional theory does seem to give some sort of account of what is seen in a three-dimensional microscope with a spherical imaging lens.

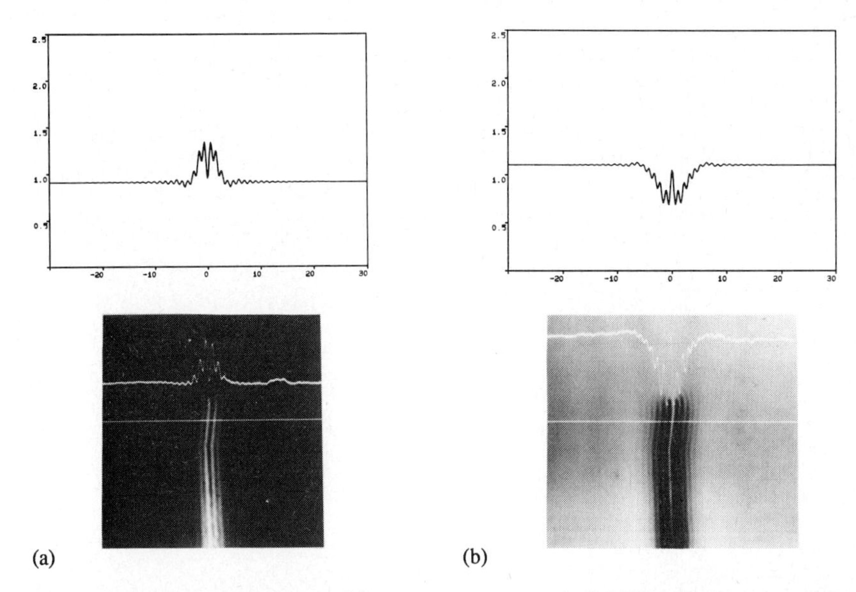

Figure 1. Theoretical and experimental line scans across a crack, 1.5 GHz: (a) theory z = -4.2 μm, experiment z = -3.8 μm; (b) theory z = -6.7 μm, experiment z = -5.2 μm. Each theoretical curve spans ±30 μm either side of the crack, and the experimental figures are reproduced to the same scale.

A Rayleigh wave can exist only because of the ability of a solid to support both longitudinal and shear waves (Kino 1987). Therefore, if a crack locally disrupts the ability to support shear stresses (if it is filled with an incompressible non-viscous fluid), or even the ability to support both shear stresses and longitudinal stresses (if there is no fluid in the crack), then Rayleigh wave propagation will be disrupted, and this will manefest itself in the values of R and T in equation [3]. If the fluid is viscous then there will be a finite decay length for shear stresses. If the viscosity and density of the fluid are η and ρ, then the decay length for shear stress to decay by a factor e at angular frequency ω is

$$d_e = \sqrt{\frac{2\eta}{\rho\omega}}. \qquad [4]$$

At 1 GHz in water at 20°C this gives a value of 18 nm, falling to half that value at 2 GHz and 60°C. In many situations asperity contact would be more important at that separation; this can be modelled by a distribution of springs (Thompson *et al.* 1982, 1983) and hence reflection and transmission coefficients for Rayleigh waves can be calculated (Li and Achenbach 1989).

4. PICTURES

Figure 7 shows acoustic micrographs of the pattern of radial cracks from an indent in glass; the fringes that are present have a periodicity of half a Rayleigh wavelength [Yamanaka and Enomoto 1982]. At zero and positive defocus there are still Rayleigh fringes present because of the reflection of Rayleigh waves even when the crack lies outside the cone of rays from the lens at the Rayleigh angle; these fringes extend a reasonably well defined distance along the crack. The contrast in figures 7c and 7d has been deliberately set to show that in addition to the fringes parallel to the sides of the crack there are also somwhat weaker circular fringes centred at the crack tip, formed by waves reflected, or diffracted, from the tip of the crack itself. These pictures confirm better than any others that the

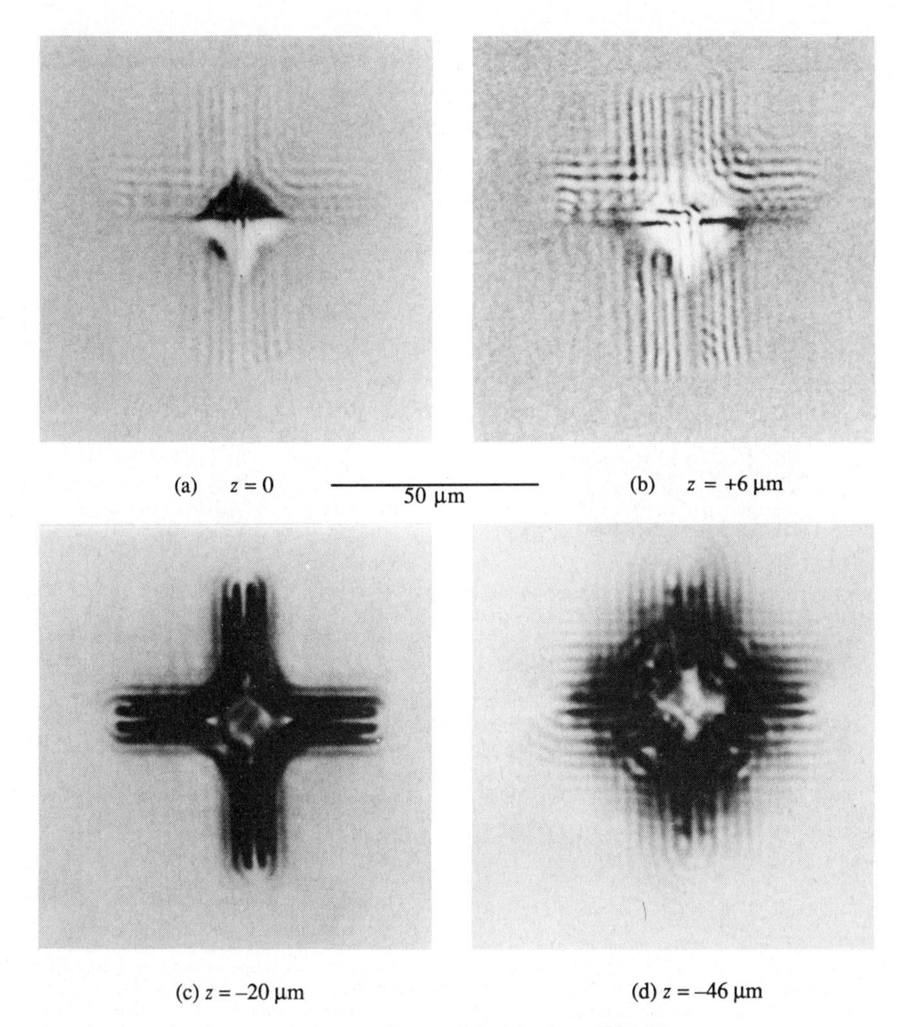

Figure 2. Acoustic micrographs of cracks from an indent in glass, 400 MHz.

contrast from a crack does not gradually become weaker with decreasing crack opening, as might be expected if the transmission of Rayleigh waves across a narrow crack were inversely related to the opening. What does happen is that the fringes continue with full contrast right up to one particular point in the crack, which is itself a strong scatterer. It would seem reasonable to identify this point with the tip of the crack. Thus the crack tip is a well defined feature for the acoustic contrast. The centre of curvature of the fringes could be measured by suitable image analysis techniques with an accuracy considerably better than a wavelength, thus permitting crack length measurement with enhanced resolution. The length of the crack indicated in these acoustic micrographs is in good agreement (within experimental limits) with observations in l.m. and s.e.m. images.

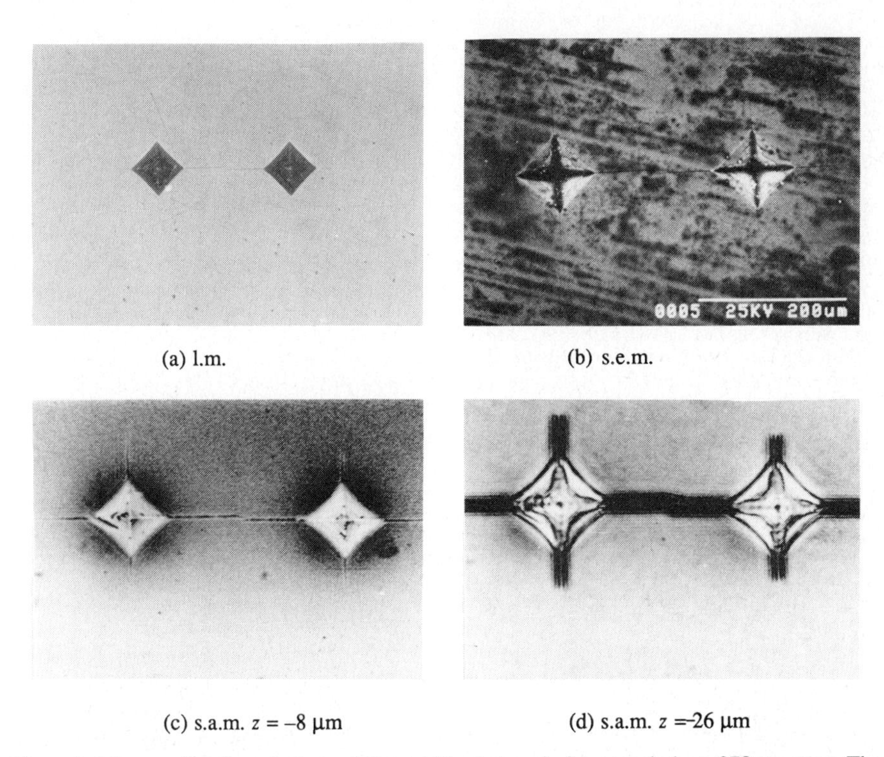

(a) l.m. (b) s.e.m.

(c) s.a.m. $z = -8$ µm (d) s.a.m. $z = -26$ µm

Figure 3. Micrographs of cracks in partially stabilized zirconia from two indents 270 µm apart. The frequency of the scanning acoustic microscope images (s.a.m.) is 400 MHz.

5. APPLICATION

Cracks in ceramics are generally harder to see than cracks in a model material such as glass. Figure 3 shows pictures of cracks from indents in a partially stabilized zirconia (p.s.z.) ceramic. Figure 3a is an l.m. picture; at this magnification it is still just possible to measure the length of the cracks. Figure 3b is an s.e.m. picture. The cracks running top to bottom in figures 3a and 3b seem to be shorter than those running left to right, suggesting that there may be anisotropy in the fracture toughness. Figure 3c is an acoustic image at 400 MHz (at this frequency the Rayleigh wavelength, λ_R, is about 7 µm in zirconia). The cracks running between the two indents show up quite well, and their contrast extends far enough along their length to indicate the overlap that can also be seen in the s.e.m. and l.m. pictures. At this defocus it is not easy to be certain how far the cracks running towards the top and bottom of the images extend. However, by increasing the defocus a strong fringe pattern develops, and this is shown in figure 3d. The fringes stop at a well defined distance along the cracks. Such observations have been extended to indentation cracks in a number of ceramics, including silicon carbide, hot-pressed silicon nitride, reaction bonded silicon nitride, and silicon, and also fatigue cracks in stainless steel. In no case did subsequent examination in l.m. or s.e.m. reveal any cracks or part of a crack that did not give Rayleigh wave contrast in the acoustic microscope; moreover the length of any crack indicated by acoustic microscopy was no shorter than that indicated at the highest magnification in the s.e.m.

6. CONCLUSION

The conclusion from both theory and experiment seems to be that there is no surface crack that is too fine to see provided that it is at least a substantial fraction of a Rayleigh wavelength deep and has a length on the surface of more than a wavelength or so. For a mathematically ideal crack the only limit would seem to be the shear wave decay length, which is about 9 nm at the realistic operating limit of commercial microscopes; in most practical cases other phenomena such as crack closure would occur by that seperation.

7. ACKNOWLEDGEMENTS

A fuller account of the work described here, by G.A.D. Briggs, P.J. Jenkins and M. Hoppe, will be published in *Journal of Microscopy* (1990). The theory will be described in more detail in *Acoustic Microscopy* by G.A.D. Briggs (OUP, 1990). I wish to express thanks to Dr D.S. Spencer for discussions about the decay length of shear waves in a liquid and for plotting the curves in figure 1, to Professor J.D. Achenbach for discussions about the theory of Rayleigh wave scattering by cracks, and to Professor H.L. Bertoni for discussions about the contrast theory of cracks in the acoustic microscope. I am grateful to Professor Achenbach and Professor Datta for funds to enable me to present this paper at the IUTAM conference.

8. REFERENCES

Achenbach, J.D. (1987) Flaw characterisation by ultrasonic scattering methods. In *Solid mechanics research for quantitative non destructive evaluation* (eds J.D. Achenbach and Y. Rajapakse). Nijhoff, Dordrecht, pp 67-81.

Bertoni, H.L. (1984) Ray-optical evaluation of $V(z)$ in the reflection acoustic microscope. IEEE Trans SU-31, 105-116.

Ilett, C., Somekh, M.G. and Briggs, G.A.D. (1984) Acoustic microscopy of elastic discontinuities. *Proc. R. Soc. Lond.* A393, 171-183.

Kino, G.S. (1987) *Acoustic waves: devices, imaging, and analog signal processing*. Prentice-Hall: New Jersey.

Li, Z.L. and Achenbach, J.D. (1989) Reflection and transmission of Rayleigh surface waves by a material interphase (to be published).

Rowe, J.M., Kushibiki, J., Somekh, M.G. and Briggs, G.A.D. (1986) Acoustic microscopy of surface cracks. *Phil. Trans. R. Soc. Lond.* A320, 201-214.

Sheppard, C.J.R. and Wilson, T. (1981) Effects of high angles of convergence on $V(z)$ in the scanning acoustic microscope. Appl. Phys. Lett. 38, 858-859.

Somekh, M.G., Bertoni, H.L., Briggs, G.A.D. and Burton, N.J. (1985) A two dimensional imaging theory of surface discontinuities with the scanning acoustic microscope. *Proc. R. Soc. Lond.* A401, 29-51.

Thompson, R.B., Skillings, B.J., Zachary, L.W., Schmerr, L.W. and Buck, O. (1982) Effects of crack closure on ultrasonic transmission. *Review of Progress in Quantitative NDE2* (eds D.O. Thompson and D.E. Chimenti) Plenum Press.

Thompson, R.B., Fiedler, C.J. and Buck, O. (1983) Inference of fatigue crack closure stresses from ultrasonic transmission measurements. *Nondestructive methods for material property determination*, Plenum Press.

Yamanaka, K. and Enomoto, Y. (1982) Observation of surface cracks with scanning acoustic microscope. *J. Appl. Phys.* 53, 846-850.

Elastic Waves and Ultrasonic Nondestructive Evaluation
S.K. Datta, J.D. Achenbach and Y.S. Rajapakse (Editors)

ACOUSTIC WAVE INTERACTION WITH PARTIALLY CLOSED CRACKS

O. Buck, R. B. Thompson, D. K. Rehbein, and L. VanWyk

Ames Laboratory and Materials Science and Engineering Department
Iowa State University
Ames, IA 50011

A fatigue crack that is able to grow under external load appears acoustically quite different from one that is unable to grow due to the presence of crack closure and crack tip shielding. Relative changes in fatigue crack propagation rates can now be predicted from the ultrasonic response. The technique effectively provides a way of evaluating the load history the crack has experienced.

1. INTRODUCTION

During the growth of a fatigue crack, contact [1,2] between the crack faces is often developed via a variety of mechanisms, including general plastic deformation, sliding of the two faces with respect to one another, or the collection of debris such as oxide particles [3]. Figure 1 schematically sketches such a situation. Compressive stresses, σ_o, are created on either side of the partially contacting crack surfaces. In reaction, opening loads, P_i, arise producing a local stress intensity factor which shields the crack tip from the variations of the externally applied stress intensity factor, as indicated in Fig. 2. This shielding occurs below a stress intensity factor, K_{cl}, at which the first contact during unloading occurs [4]. Thus, a consequence of asperity contact is that the applied stress intensity range, $\Delta K = K_{Imax} - K_{Imin}$, which is usually considered to provide the driving force for fatigue crack propagation, has to be modified [1] to include the effects of crack tip shielding. By using information from acoustic transmission and diffraction experiments [4] we have succeeded in determining the size and density of contacting asperities in the closure region of a fatigue crack, grown under constant ΔK conditions, as well as an estimate of the shielding stress intensity factor, K_{sh}, due to the contacting asperities. The present paper reports on the extension of our earlier work [4,5] in an attempt to determine and quantify the effects of a variable ΔK on fatigue crack propagation by using acoustic experiments, as shown in Fig. 3. As a

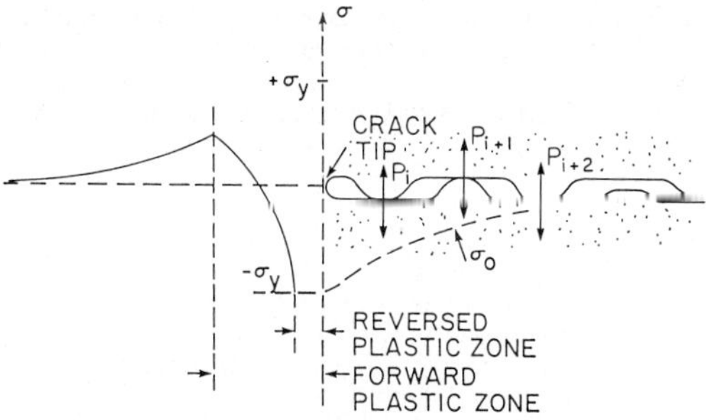

FIGURE 1
Stress field in the vicinity of the crack tip.

result it appears that the acoustic measurements assist in predicting future crack growth rate behavior.

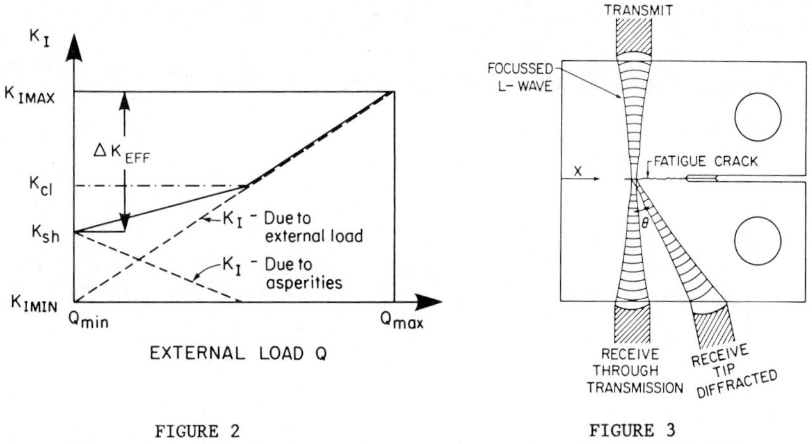

FIGURE 2
Stress intensity factor

FIGURE 3
Acoustic arrangement

2. MODELLING AND EVALUATION OF THE CONTACT TOPOLOGY

The plane of the crack is viewed as consisting of contacting asperities interrupted by crack-like voids which act as scatterers of an elastic wave. The theory of the scattering is based on the electromechanical reciprocity theorem [6] which has been applied to the experimental conditions of normal incidence, shown in Fig. 3. For this case the scattered field is

$$\Gamma \approx \frac{j\omega}{4P} \int_{A^+} \left(2u_i^I - \Delta u_i^T\right) \tau_{ij}^R \, n_i^+ \, dA$$

where P is the electrical power incident on the transmitting transducer, ω is the angular frequency of the signal, u_i^I is the displacement field of the incident acoustic illumination, Δu_i^T is the crack opening displacement due to u_i^I, and τ_{ij}^R is the stress field that would be produced if the receiving transducer illuminated a defect-free material. Integration is performed over the surface A containing the scatterers which has a normal n_i^+. The major problem in evaluating the above equation is in selecting an appropriate description of Δu_i^T. However, Sotiropoulus and Achenbach [7] recently provided the crack opening displacement for crack-like objects which should improve earlier diffraction calculations [4]. For the evaluation of the present results we have used a simpler approach employing an averaged Δu_i^T. The model, which is referred to as the "quasi-static distributed spring model" [8], describes the degree of the contact in terms of a spring constant κ which is primarily a function of the average contact diameter d and separation C. Since $1/\kappa$ is directly proportional to Δu_i^T, evaluation of the above equation leads to a direct prediction of the experimentally determined signal in terms of κ. In high strength aluminum and employing the earlier diffraction calculations [4], we found that the size and density of the contacting asperities in the closure region of a crack, grown at constant ΔK, are as follows: A mean contact separation $\approx 70\mu m \approx$ grain diameter and a mean contact diameter in the vicinity of the crack tip $\approx 35\mu m$, the latter quantity falling off exponentially with distance away from the crack tip.

3. CRACK TIP SHIELDING BY ASPERITIES

Each contact carries its own load P_i, as shown in Fig. 1 and generates its own stress intensity factor. Based on contact pressure calculations [9], residual stresses in the closure region have been calculated [4] with the results agreeing well with those obtained from x-ray diffraction measurements by Welsch et al. [10]. The total shielding stress intensity factor, K_{sh}, is obtained by an integration over all contacts in the closure region, as indicated in Fig. 4. We found, for an unloaded crack, K_{sh} to be about 40% of the maximum stress intensity applied to achieve a constant crack growth rate [4].

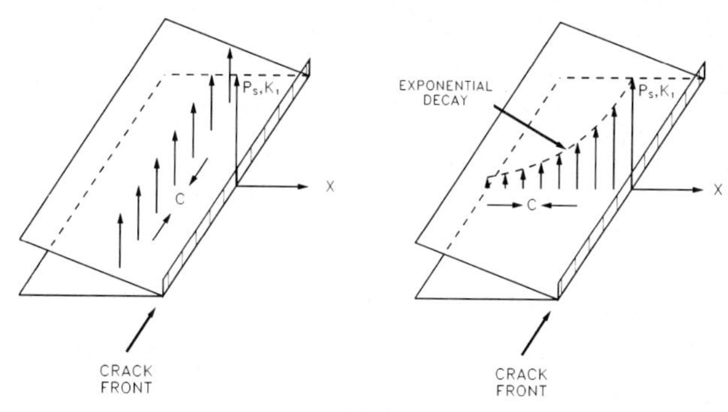

FIGURE 4
Shielding by individual contacts.

4. CRACKS GROWN UNDER VARIABLE ΔK CONDITIONS

We have extended earlier work [4,5,], studying the effects of a variable ΔK on the acoustic transmission signal under two conditions. The first one is indicated schematically on the left hand side of Fig. 5. After a number of

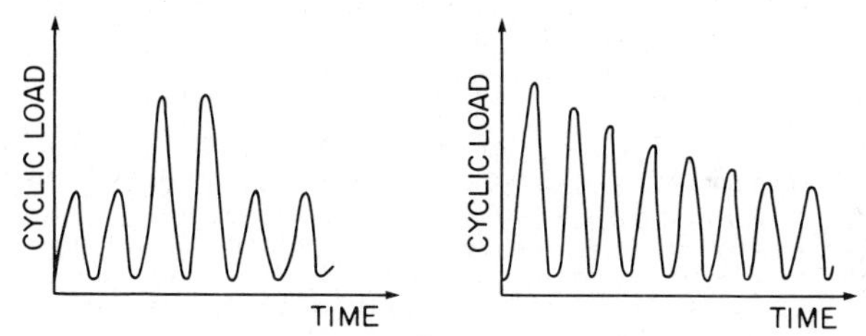

FIGURE 5
Load spectra used in present investigation.

cycles at constant ΔK, the ΔK was raised by a factor of two (overload block) and thereafter was dropped to the previous ΔK. During this experiment, the transmitted signal was determined several times with the specimen under load (or applied K). The second type of experiment is schematically indicated on the right hand side of Fig. 5. In this case the load was dropped continuously such that ΔK decreased with fatigue crack growth. Results of the first experiment on the length of the fatigue crack as a function of fatigue cycles are shown in Fig. 6. The application of the "overload block" at A retarded the crack growth for roughly 120,000 cycles. Transmission data were taken at A, B and C in Fig. 6. Figure 7 presents data at A and B for various K levels applied. Before application of the overload (A) and at the lowest applied K level, the transmitted signals drop off rather slowly with distance away from the crack tip. At the higher applied K levels, the transmitted signals drop off much faster, indicating that the crack is fully open [11]. Immediately after the overload (B), the transmission signals at all K levels indicate that

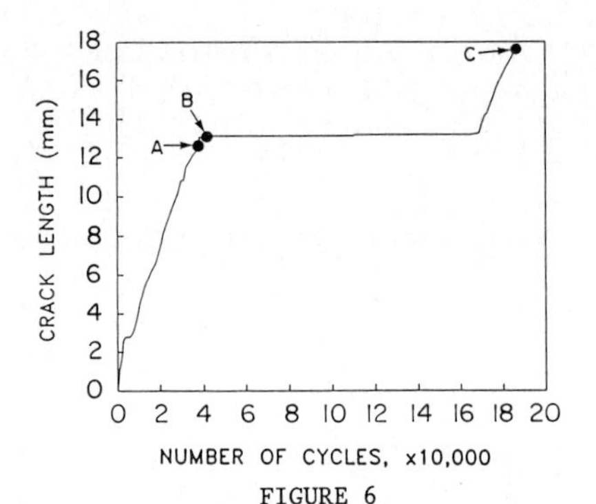

FIGURE 6
Crack length versus number of fatigue cycles (Overload experiment).

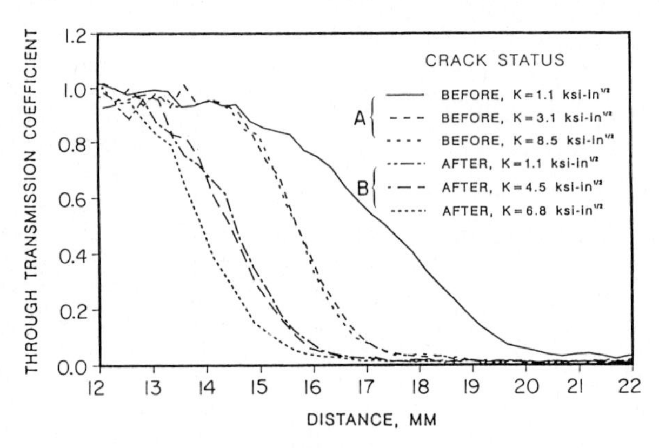

FIGURE 7
Through transmission at 6 MHz before and after overload experiment.

the crack is basically fully open with the apparent crack tip moving slightly to the left as K increases. The data furthermore show that the crack has grown by about 1.5mm during the overload, as confirmed by fractography. In the condition B the crack is retarded, as indicated in Fig. 6, and finally resumes growth. At point C, the transmission signal of the basically unloaded specimen again shows a slow drop off, as before the overload application occurred, with an additional peak in transmission at the original location of the overload application [5]. The resulting κ's at A, B, and C are shown in Fig. 8, together with sketches of the crack closure, consistent with the results of Fig. 7. We have estimated the effects of the peak in κ_3 on K_{sh} [12] and found that the shielding stress intensity factor increases from about 6.8 MPa $m^{1/2}$ at A to 8.0 MPa $m^{1/2}$ at C. The theoretical predictions of the relative change of the crack propagation rate by a modified Paris law $da/dN = \alpha(\Delta K_{eff})^m$, where α and m are materials parameters and ΔK_{eff} is the acoustically determined "driving force" on the crack (see Fig. 2), are in agreement with the experimental observations of a 50% slower propagation rate at point C than at point A . We have not succeeded yet to calculate K_{sh} for the crack immediately after the overload (B). All indications are, however, that there is a short and very tightly closed region (with high K_{sh}) in front of what otherwise appears to be an open crack as indicated in Fig. 8.

Results of the second experiment, in which the applied ΔK was dropped

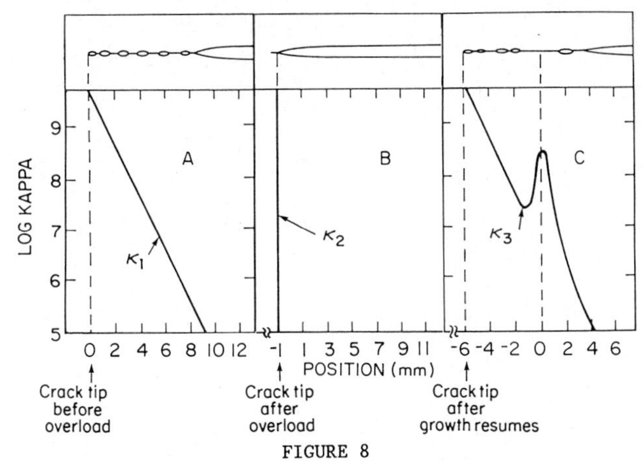

FIGURE 8
Spring constant and asperity contact before and after an overload

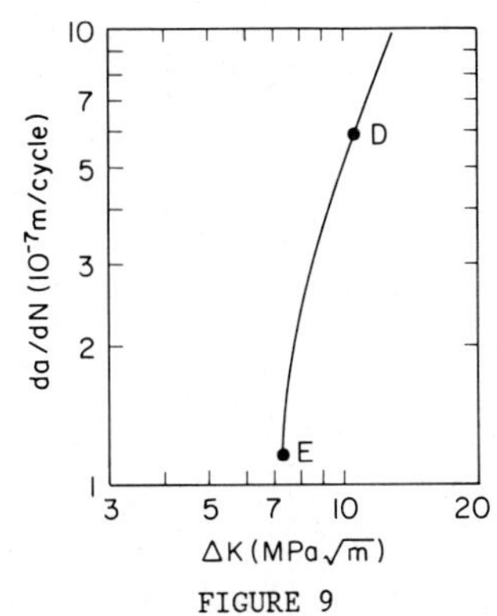

FIGURE 9
Crack growth rate in the "threshold" regime.

continuously, on the resulting crack propagation rate are shown in Fig. 9. The data indicate that the growth regime changed from the Paris to the near-threshold regime. Acoustic transmission data at various acoustic frequencies with no load applied to the specimen were taken in both propagation regimes as shown in Fig. 10. The left hand side of Fig. 10 shows the transmission data in the Paris regime (D) and the right hand side the data in the near-threshold regime (E). As before (A at the lowest K level in Fig. 7), the drop-off in the transmission data indicates an exponentially decaying κ with the closure region as modelled on the left hand side of Fig. 8. Remarkably different is the spatial variation of the transmission data in the near-threshold regime. As seen in the right hand side of Fig. 10, at large distances away from the crack tip the transmission coefficient stays at a relatively high and almost constant level. This indicates to us that the closure region is extended with respect to that in the Paris regime and the resulting K_{sh} is about 45% larger [13]. A conversion of da/dN versus ΔK in Fig. 9 to da/dN versus ΔK_{eff}, defined in Fig. 2, then provides a new correlation for the data shown in Fig. 9 which has a shape close to that in the Paris regime. Thus it appears that the "threshold" behavior, seen in Fig. 9, has been artificially created by the extensive closure, shown on the right hand side of Fig. 10, in agreement with earlier observations [14].

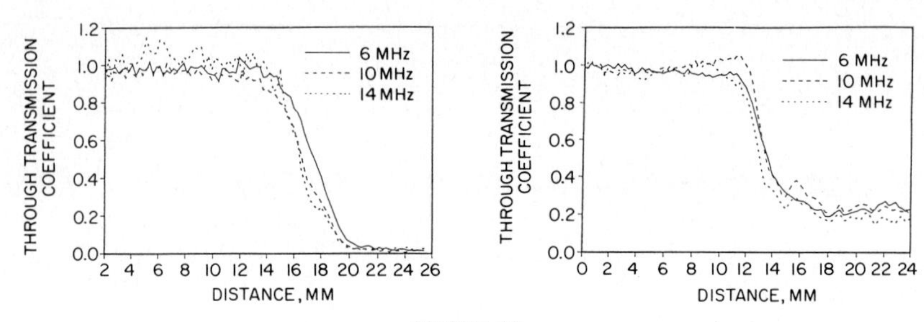

FIGURE 10
Through transmission in the "threshold" regime.

5. CONCLUSIONS

Acoustical measurements on fatigue cracks provide detailed information on the state of the crack due to its past history. Any change in the load spectrum applied to the crack affects the crack face contact and consequently the "shielding" of the crack from the external driving force. Because they can characterize this contact, acoustic measurements thus are able to assist in predicting future crack growth rate changes. Elastodynamic theories of wave scattering play a crucial role in making these interpretations quantitative.

ACKNOWLEDGEMENT

Ames Laboratory is operated for the U.S. Department of Energy by Iowa State University under contract No. W-7405-ENG-82. This work was supported by the Office of Basic Energy Sciences, Division of Materials Sciences.

REFERENCES

[1] Elber, W., in: Damage Tolerance in Aircraft Structures, ASTM STP 486 (Am. Soc. Test. Mat., Philadelphia, 1971) p. 230.
[2] Bowles, C. Q., and Schijve, J., in: Fatigue Mechanisms-Advances in Quantitative Measurement of Physical Damage, ASTM STP 811 (Am. Soc. Test. Mat., Philadelphia, 1983) p. 400.
[3] Suresh, S., and Ritchie, R. O., Scripta Met. 17 (1983) 595.
[4] Buck, O., Rehbein, D. K., and Thompson, R. B., Engr. Fract. Mechanics 28 (1987) 413.
[5] Buck, O., Rehbein, D. K., and Thompson, R. B., in: Liaw, P.K. and Nicholas, T. (eds), Effects of Load and Thermal Histories on Mechanical Behavior of Materials (Met. Soc. of AIME, Warrendale, 1987) p. 49.
[6] Auld, B.A., Wave Motion 1 (1979) 3.
[7] Sotiropoulus, D. A., and Achenbach, J. D., J. Nondestr. Eval. 7 (1988) 123.
[8] Baik, J.-M., and Thompson, R. B., J. Nondestr. Eval. 4 (1984) 177.
[9] Kendall, K. and Tabor, D., Proc. Roy. Soc. London A232 (1971) 231.
[10] Welsch, E., Eifler, D., Scholtes, B., and Macherauch, E., in: Residual Stresses in Science and Technology (DGM Informationsgesellschaft Verlag, Oberursel, 1987) p. 785.
[11] Thompson, R. B. Fiedler, C. J., and Buck, O., in: Ruud, C. O. and Green, R. E., Nondestructive Methods for Material Property Determination (Plenum Press, New York and London, 1984) p. 161.
[12] Buck, O., Thompson, R. B., Rehbein, D. K., Brasche, L. J. H., and Palmer, D. D., in: Advances in Fracture Research (Pergamon Press, Oxford 1989) Vol. 3, p. 3121.
[13] L. M. VanWyk, A Study on Ultrasonic Detection and Characterization of Partialy Closed Fatigue Cracks (Thesis, Iowa State University, 1989).
[14] Ritchie, R. O., Mat. Science Engr. A103 (1988) 15.

Elastic Waves and Ultrasonic Nondestructive Evaluation
S.K. Datta, J.D. Achenbach and Y.S. Rajapakse (Editors)

INVERSE SCATTERING IN GENERAL ANISOTROPIC ELASTIC MEDIA

David E. Budreck
Dept. of Mathematics, and
Dept. of Engineering Science and Mechanics
Iowa State University
Ames, IA. 50011

James H. Rose
Ames Laboratory
Iowa State University
Ames, IA. 50011

1. INTRODUCTION

Rigorous inverse scattering methods for either isotropic or anisotropic media are virtually nonexistent. Some approximate inverse scattering methods are available for isotropic elastic solids, however even the simplest approximate inversion algorithms have not been extended to general anisotropic elastic media.

In this paper we present and discuss an exact result for inverse scattering theory in the context of anisotropic elastodynamics. Specifically, we can reduce the solution of the inverse scattering problem to the solution of an integral equation, which itself is driven by near-field scattering data. The scatterer here is taken to be a localized region in which both the elastic constant tensor and mass density are spatially varying, and where the latter is taken to be unknown. The scatterer is set against an otherwise uniform anisotropic solid, and the scattering data consists of impulse response functions obtained from point emitters and point receivers. These point emitters and point receivers are taken to lie on some bounding surface S (see Fig. 1a) which is in the neighborhood of and includes the scatterer. We use data for one transmitter and for receivers at all points on S, and measured for all time.

The integral equation discussed above has been derived, in a more general form and together with other results, by the authors of this article in [1]. This derivation represents an extension of the method of Rose and Cheney [2] for the scalar wave equation to that of the fully anisotropic vector wave equation. Their method is itself linked to that of Newton [3] for the Schrödinger equation. It is our intent in this article to focus solely on the integral equation, which we feel is the key result of [1].

This manuscript is structured as follows. In section 2 we present the formalism for forward scattering in the anisotropic elastic media. In section 3 we present and discuss the key result of [1], which we call a near-field version of the Newton-Marchenko equation for anisotropic elastodynamics. The extraction of the spatially varying density function from the solution to the integral equation will complete the presentation of the inverse scattering result. In section 4 we end with a brief concluding section.

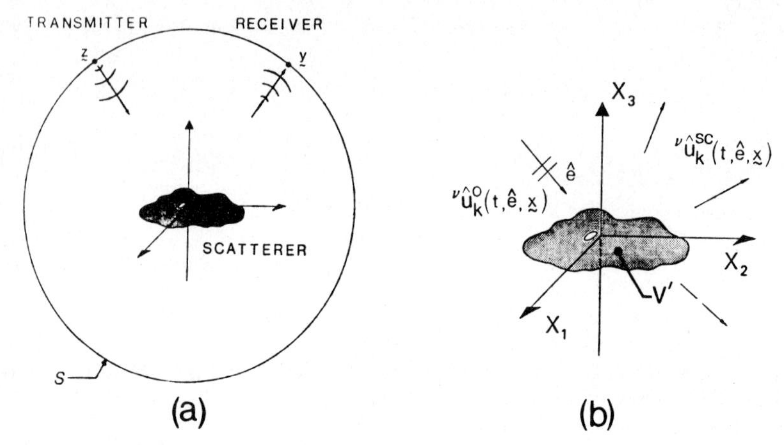

Fig. 1. Scattering Geometries for (a) Inverse problem, and (b) Forward problem..

2. Elastic Wave Forward Scattering Theory

In this section we present the theory for the scattering of waves in a general anisotropic elastic medium, where the scatterer is a region of inhomogeneity with compact spatial support. This material can be found in [1], and deals exclusively with the forward scattering problem. Thus this section is a necessary precursor to the inverse scattering section which follows.

Consider first a homogeneous, anisotropic, linear elastic, and infinitely extended domain, $V \equiv \mathbf{R}^3$, with Cartesian elastic constant tensor $C^0_{ijk\ell}$ and mass density ρ^0. The field equation governing time-dependent wave motion for $\underset{\sim}{x} \in V$ is

$$\hat{L}^0_{ik}(t,\underset{\sim}{x})\left[{}^{\nu}\hat{u}^0_k(t,\hat{e},\underset{\sim}{x})\right] = 0 \quad , \tag{2.1}$$

where ${}^{\nu}\hat{u}^0_k$ is the kth component of the vector displacement wavefield, $\hat{e}$ is a unit vector pointing in the direction of propagation, and

$$\hat{L}^0_{ik}(t,\underset{\sim}{x}) \equiv C^0_{ijk\ell}\partial_{x_j}\partial_{x_\ell} - \rho^0\delta_{ik}\partial_t^2 \quad . \tag{2.2}$$

Here ∂_x and ∂_t denote spatial and time derivatives, respectively, and δ_{ik} is the Kronecker delta. Also, all such "L" operators in this paper act on an appropriate weighted L^2 space. In Eq. (2.1) $\nu = 1-3$, which serves to indicate the three modes which propagate freely (no scattering) through the homogeneous domain. The "o" superscripts in Eqs. (2.1) and (2.2) emphasize homogeneity, and hence free propagation of the wavefields. The "^" superscript denotes the time-domain Fourier transform of both operator and function in Eq. (2.1). The Fourier transform convention is defined by

$$\hat{f}(t) = (2\pi)^{-1}\int d\omega\, e^{-i\omega t}f(\omega), \qquad f(\omega) = \int dt\ e^{i\omega t}\hat{f}(t) \quad . \tag{2.3a-b}$$

The convention taken throughout this article is that unspecified limits of integration denote integration over the variable's entire domain of definition. Thus the domain of integration in each integral above is $(-\infty,\infty)$.

Solutions of Eq. (2.1) are well known:

$$ {}^{\nu}\hat{u}_k^o(t,\hat{e},\underset{\sim}{x}) = \overset{\nu}{\phi}_k(\hat{e})\delta(c_\nu(\hat{e})t - \hat{e}\cdot\underset{\sim}{x}) \quad , \tag{2.4} $$

where again $\nu = 1-3$, and $\overset{\nu}{\phi}_k$, the polarization vectors, are the eigenfunctions of the Christoffel equation

$$ \left[C^o_{ijk\ell}\hat{e}_j\hat{e}_\ell - \rho^o\delta_{ik}c^2_\nu(\hat{e})\right]\overset{\nu}{\phi}_k(\hat{e}) = 0 \quad , \tag{2.5} $$

with c_ν the corresponding eigenvalues. Thus the solutions to Eq. (2.1) are plane-wave delta-functions propagating in the $\hat{e}$-direction, with phase velocity $c_\nu(\hat{e})$, and polarization $\overset{\nu}{\phi}_k(\hat{e})$.

Consider next the interaction of the freely propagating wavefields (2.4) with a bounded region of inhomogeneity, $V' \in V$, in which the elastic constant tensor and mass density are spatially varying (see Fig. 1b):

$$ C_{ijk\ell}(\underset{\sim}{x}) \equiv C^o_{ijk\ell} + C'_{ijk\ell}(\underset{\sim}{x}), \quad \underset{\sim}{x} \in V' \quad , \tag{2.6} $$

$$ \rho(\underset{\sim}{x}) \equiv \rho^o + \rho'(\underset{\sim}{x}), \quad \underset{\sim}{x} \in V' \quad . \tag{2.7} $$

Here we demand that

$$ C'_{ijk\ell}(\underset{\sim}{x}) = \rho'(\underset{\sim}{x}) \equiv 0, \quad \underset{\sim}{x} \in V - V' \quad . \tag{2.8} $$

Thus the scatterer, contained in the region V', can be thought of as an inhomogeneity in an otherwise uniform host medium that has a density ρ^o and elastic constants $C^o_{ijk\ell}$. Throughout this paper we assume that the density and elastic constant functions are real, non-negative, and scalar-valued.

The scatterer gives rise to a scattered displacement wavefield, ${}^{\nu}\hat{u}_k^{sc\pm}(t,\hat{e},\underset{\sim}{x})$, and the total wavefield everywhere in V is given by

$$ {}^{\nu}\hat{u}_k^{\pm}(t,\hat{e},\underset{\sim}{x}) \equiv {}^{\nu}\hat{u}_k^{sc\pm}(t,\hat{e},\underset{\sim}{x}) + {}^{\nu}\hat{u}_k^{o}(t,\hat{e},\underset{\sim}{x}) \quad . \tag{2.9} $$

The "ν" and "$\pm$" superscripts of the scattered, and thus total, wavefields derive their meaning from the incident field ${}^{\nu}\hat{u}_k^o$ which gives rise to them. The "ν" in ${}^{\nu}\hat{u}_k^{sc}$ indicates a scattered field due to an incident field with $\nu = 1-3$, and the "$\pm$" indicates one of the following initial (final) conditions:

$$ {}^{\nu}\hat{u}_k^{\pm}(t,\hat{e},\underset{\sim}{x}) \equiv {}^{\nu}\hat{u}_k^{0}(t,\hat{e},\underset{\sim}{x}), \quad t \to \mp\infty \quad , \tag{2.10$^\pm$} $$

where we take ($2.10^\pm$) to be satisfied over an arbitrarily small (but nonzero) interval in t. Also, the $\hat{e}$-dependence is carried through to the scattered and total wavefields as a reminder of the incident direction that gave rise to them.

The total wavefield (2.9) must satisfy the full wave equation through all of V:

$$ \hat{L}_{ik}(t,\underset{\sim}{x})[{}^{\nu}\hat{u}_k^{\pm}(t,\hat{e},\underset{\sim}{x})] = 0 \ , \tag{2.11} $$

where

$$\hat{L}_{ik}(t,\underset{\sim}{x}) \equiv \partial_{x_j} C_{ijk\ell}(\underset{\sim}{x})\partial_{x_\ell} - \rho(\underset{\sim}{x})\delta_{ik}\partial_t^2 \ . \tag{2.12}$$

We define that portion of the operator $\hat{L}_{ik}$ which contains the effect of the inhomogeneity as

$$\hat{L}'_{ik}(t,\underset{\sim}{x}) \equiv \hat{L}_{ik}(t,\underset{\sim}{x}) - \hat{L}^{\circ}_{ik}(t,\underset{\sim}{x}) \ , \tag{2.13}$$

where Eqs. (2.12), (2.13), (2.2) together with Eqs. (2.6) and (2.7) give

$$\hat{L}'_{ik}(t,\underset{\sim}{x}) \equiv \partial_{x_j} C'_{ijk\ell}(\underset{\sim}{x})\partial_{x_\ell} - \rho'(\underset{\sim}{x})\delta_{ik}\partial_t^2 \ . \tag{2.14}$$

We will need the Green's functions corresponding to the operators $\hat{L}^0_{ik}$ and $\hat{L}_{ik}$ given by Eqs. (2.2) and (2.12), respectively. We define the free-propagating Green's function as that which satisfies

$$\hat{L}^0_{ik}(t,\underset{\sim}{x}) \left[{}^0\hat{G}^{m\pm}_k(t-t',\underset{\sim}{x},\underset{\sim}{x}')\right] \equiv -\delta_{im}\delta(t-t')\delta(\underset{\sim}{x}-\underset{\sim}{x}') \quad , \tag{2.15}$$

where $\underset{\sim}{x},\underset{\sim}{x}' \in V$, and $+(-)$ indicates radiation and incoming boundary conditions, respectively. We note that causality implies that

$$ {}^0\hat{G}^{m\pm}_k(t-t',\underset{\sim}{x},\underset{\sim}{x}') \equiv 0, \quad \pm(t-t') < 0 \ . \tag{2.16$^\pm$}$$

In the notation used above the lower (upper) Latin index corresponds to the field point $\underset{\sim}{x}$ (source point $\underset{\sim}{x}'$). Physically, ${}^0\hat{G}^{m+}_k(t-t',\underset{\sim}{x},\underset{\sim}{x}')$ is the displacement in a uniform host medium at $\underset{\sim}{x}$ in the k-direction at time t due to a point load of unit amplitude applied at $\underset{\sim}{x}'$ in the m-direction at time t'.

We further define the fully-interacting Green's function as that which satisfies

$$\hat{L}_{ik}(t,\underset{\sim}{x}) \left[\hat{G}^{m\pm}_k(t-t',\underset{\sim}{x},\underset{\sim}{x}')\right] \equiv -\delta_{im}\delta(t-t')\delta(\underset{\sim}{x}-\underset{\sim}{x}') \quad , \tag{2.17}$$

where again $\underset{\sim}{x},\underset{\sim}{x}' \in V$, and $+(-)$ indicates radiation (incoming) boundary conditions. Again causality implies

$$\hat{G}^{m\pm}_k(t-t',\underset{\sim}{x},\underset{\sim}{x}') \equiv 0, \quad \pm(t-t') < 0 \ . \tag{2.18b$^\pm$}$$

This fully-interacting Green's function has the same physical interpretation as its free-propagating counterpart, except here the effect of the inhomogeneity (the scatterer) is taken into account.

Elastic Wave Inverse Scattering Results

In this section we present and discuss exact results for inverse scattering in an anisotropic elastic media, see [1]. The scattering geometry remains as described in Fig. 1a and Sec. 2, i.e. an isolated inhomogeneity in an otherwise uniform anisotropic space. Here the scatterer is contained by a smooth, simply-connected, closed surface, S. It is assumed that scattering measurements can be made from any point on the surface S, which will also be called the observation surface. The scattering data are obtained as follows. A time-dependent delta-function point load of unit strength is applied at time $t = t'$ in the m-direction at a point $\underset{\sim}{z}$ on S, and the resulting displacement in the i-direction is measured at all points $\underset{\sim}{y}$ on S and for all time t. That is, the data are just the Green's functions $\hat{G}_i^{m+}(t-t', \underset{\sim}{y}, \underset{\sim}{z})$ for $\underset{\sim}{y}, \underset{\sim}{z} \in S$ (see Eq. (2.17)).

We now cite the major result to be presented in this article, what we call a near-field version of the Newton-Marchenko equation for anisotropic elastodynamics, or elastic wave N-M equation for short. We will then indicate how this equation may be interpreted as a linear integral equation from which the spatially varying density function $\rho(\underset{\sim}{x})$ may be ascertained, assuming the spatially varying elastic constant tensor $C_{ijk\ell}(\underset{\sim}{x})$ is known. The elastic wave N-M equation is

$$\begin{aligned} &\hat{G}_p^{m+}(t, \underset{\sim}{x}, \underset{\sim}{y}) = 0 \qquad\qquad t < 0 \;, \\ &\hat{G}_p^{m+}(t, \underset{\sim}{x}, \underset{\sim}{y}) = C^o_{ijk\ell} \int_{s^2} d^2\hat{z}|\underset{\sim}{z}|^2 \int d\,t' \\ &\Big\{ \hat{G}_p^{i-}(t', \underset{\sim}{x}, \underset{\sim}{z}) \hat{G}_m^{k,\ell+}(t-t', \underset{\sim}{y}, \underset{\sim}{z}) \\ &\qquad - \hat{G}_p^{k,\ell-}(t', \underset{\sim}{x}, \underset{\sim}{z}) \hat{G}_m^{i+}(t-t', \underset{\sim}{y}, \underset{\sim}{z}) \Big\} n_j(\hat{z}), \quad t > 0 \;. \end{aligned} \tag{3.1}$$

Here $\partial_{z_k} \hat{G}_j^i(t, \underset{\sim}{y}, \underset{\sim}{z}) \equiv \hat{G}_j^{i,k}(t, \underset{\sim}{y}, \underset{\sim}{z})$, and the first integral is over the surface of a unit ball. Equation (3.1) is a simplified version of a more general result derived in [1]. Key to that derivation is the exploitation of the causal nature of the Green's functions, in the sense that $\hat{G}_j^{i+}$ and $\hat{G}_j^{i-}$ have disjoint supports in time (see Eqs. (2.18$^\pm$)). The equivalent result of Eq. (3.1) which appears in [1] is more general in having used a stronger causality condition than that of Eqs. (2.18$^\pm$), i.e. one which takes into account the finite speed of propagation of the medium.

In Eq. (3.1) both $\underset{\sim}{y}$ and $\underset{\sim}{z}$ are chosen to lie on the observation surface S, while $\underset{\sim}{x}$ is a point in the inaccessible region that is interior to S (see Fig. 1a). Thus Eq. (3.1) is a linear integral equation for the Green's function $\hat{G}_j^{i-}(t, \underset{\sim}{x}, \underset{\sim}{z})$ in terms of the data $\hat{G}_j^{i+}(t-t', \underset{\sim}{y}, \underset{\sim}{z})$. Let us assume the integral equation can be solved, and let us further assume that the spatially varying elastic constant tensor (Eq. (2.6)) is known. Then the spatially varying portion of the density function, $\rho'(\underset{\sim}{x})$, can be extracted from the low frequency asymptotics of the Green's function $G_j^{i-}(w, \underset{\sim}{x}, \underset{\sim}{z})$, which is the frequency-domain version of the solution to the integral equation (3.1). This is evidenced by the second major result cited from [1], which is

$$\rho'(\underset{\sim}{x}) = \frac{-\hat{p}_i \hat{q}_n [\partial x_j C_{ijk\ell}(\underset{\sim}{x}) \partial x_\ell Q_k^n(\underset{\sim}{x}, \underset{\sim}{z})]}{\hat{p}_i \hat{q}_n G_i^{n-}(\omega = 0, \underset{\sim}{x}, \underset{\sim}{z})} \;. \tag{3.2}$$

In Eq. (3.2) above the $\hat{p}_i$ and $\hat{q}_n$ are components of any two unit vectors in $\mathbf{R}^3$, and $\underset{\sim}{z}$ is any vector $\underset{\sim}{z} \in S$. The Q_k^n also follows from the $\omega \to 0$ limit of $G_j^{i-}(w, \underset{\sim}{x}, \underset{\sim}{z})$ via

$$Q_k^n(\underset{\sim}{x}, \underset{\sim}{z}) \equiv \lim_{\omega \to 0} \omega^{-2} \left[G_k^{n\pm}(\omega, \underset{\sim}{x}, \underset{\sim}{z}) - H_k^{n\pm}(\omega, \underset{\sim}{x}, \underset{\sim}{z}) \right] \;, \tag{3.3}$$

where $H_k^{n\pm}$ is a Green's function which satisfies the operator equation

$$[\partial_{x_j} C_{ijk\ell}(\underset{\sim}{x})\partial_{x_\ell} + \rho^o\omega^2\delta_{ik}]H_k^{m\pm}(\omega, \underset{\sim}{x}, \underset{\sim}{z}) = -\delta_{im}\delta(\underset{\sim}{x} - \underset{\sim}{z}) \ , \tag{3.4}$$

together with the same initial conditions as $G_k^{m\pm}$ and thus also those of ${}^oG_k^{m\pm}$. This Green's function is "intermediate" between ${}^oG_k^{m\pm}$ and $G_k^{m\pm}$ in that it uses knowledge of the exact $C_{ijk\ell}$ everywhere, but it uses only knowledge of ρ^o (the background density). Thus $H_k^{m\pm}(\omega, \underset{\sim}{x}, \underset{\sim}{z})$ is known, since we have assumed $C_{ijk\ell}(\underset{\sim}{x})$ is known for all $\underset{\sim}{x} \in \mathbf{R}^3$.

Little is known about the use of Eq. (3.1) to determine $\hat{G}_j^{i-}(t, \underset{\sim}{x}, \underset{\sim}{z})$ from scattering data. On the positive side Eq. (3.1) was derived in analogy with the three-dimensional Marchenko equation that Newton successfully solved for the far-field inverse problem for Schrödinger's equation. However, the high-frequency asymptotics of Schrödinger's equation are particularly simple and these asymptotics played a key role in Newton's approach. The elastodynamic equation on the other hand has rather complicated high-frequency asymptotics. This arises since the local velocity is spatially varying. Consequently the wave fronts (characteristic surfaces) can be curved, and even bifurcated. Attempts to solve Eq. (3.1) directly for $\hat{G}_j^{i-}$ will almost surely have to confront this issue regarding the high-frequency asymptotics directly. We have seen, however, that Eq. (3.2) is sufficient to solve our inverse problem provided $G_i^{n-}(\omega, \underset{\sim}{x}, \underset{\sim}{z})$ is known where $\underset{\sim}{x}$ ranges over the region of the inhomogeneity and $\underset{\sim}{z}$ is a point on the observation surface S. We remind the reader that the purpose of the integral equation (3.1) is to provide the $G_i^{n-}(\omega, \underset{\sim}{x}, \underset{\sim}{z})$ needed by this approach.

4. Summary and Conclusions

An exact result for inverse scattering in an anisotropic elastic media has been presented and discussed. The result is a generalization of the Newton-Marchenko equation, which itself led to a solution of the far-field inverse scattering problem for the Schrödinger equation. The elastic wave N-M equation presented herein is an integral equation driven by near-field data. The spatially varying density function was shown to follow from the low- frequency asymptotics of the solution to this integral equation, assuming the spatially varying elastic constant tensor is known.

References

1. Budreck, D. E. and Rose, J. H., "Inverse Scattering for Anisotropic Elastodynamics," submitted for publication to *Inverse Problems.*

2. Rose, J. H. and Cheney, M. (1987), "Self-consistent equations for variable-velocity three-dimensional inverse scattering," *Phys. Rev. Lett.*, **59**, 954.

3. Newton, R. G. (1980), "Inverse scattering II. Three dimensions," *J. Math. Phys.* **21**, 1698–1715.

Elastic Waves and Ultrasonic Nondestructive Evaluation
S.K. Datta, J.D. Achenbach and Y.S. Rajapakse (Editors)

NUMERICAL TECHNIQUES FOR ELASTIC WAVE PROPAGATION AND SCATTERING

F. Fellinger, K.J. Langenberg

Dept. Electrical Eng., FB 16
University of Kassel
3500 Kassel, FRG

Based on an integral formulation of the elastodynamic equations of motion — Newton's and Hooke's laws — we present a method of discretization resulting in a numerical scheme to solve these equations within a rectangular grid. The technique is called EFIT for *E*lastodynamic *F*inite *I*ntegration *T*echnique and compares well with Finite Difference and Finite Element methods. Its accuracy is checked against a reference solution for a twodimensional test example, i.e. a piston transducer of finite size radiating into a solid halfspace.

1. INTRODUCTION

System modeling for nondestructive testing with ultrasound requires powerful numerical schemes to predict signal amplitudes and time histories of elastic wave scattering in widely arbitrary environments. Up to now, essentially Finite Difference and Finite Element Techniques [2,1] have been proposed to meet this challenge. Here, we present a new numerical scheme competing with the above techniques, which is based on Yee's method to solve Maxwell's equations [4] and which has been further evaluated by Weiland [3] being available now as a user oriented software package [5] for a variety of applications relying on electromagnetic waves. Weiland calls the method EMFIT for *E*lectro*M*agnetic *F*inite *I*ntegration Technique, hence, we adopt the acronym EFIT to match the method to elastic wave propagation. The basic idea of EMFIT is to use Maxwell's equations in integral form and to apply them to a set of elementary cubic cells called Yee's lattice. This way, a number of physical properties of the electromagnetic field are sustained in the numerical scheme, like, for instance, the divergence relation of the electric field strength in source-free regions; this prevents the occurence of spurius modes as they are observed in the Finite Difference Technique.

2. INTEGRAL FORMULATION OF THE ELASTODYNAMIC EQUATIONS OF MOTION

Newton's equation of motion as written down for a volume element V of a solid naturally comes in integral form:

$$\iint_{S_g} \mathbf{t}\, dS + \iiint_V \mathbf{f}\, dV = \iiint_V \rho \ddot{\mathbf{u}}\, dV \tag{1}$$

Here, S_g is the closed surface of the volume V, and equation 1 expresses the condition of equilibrium between the inertial forces given by the density ϱ of the solid times the second time derivative $\ddot{\mathbf{u}}$ of the (differential) displacement $\mathbf{u}$, and the prescribed volume force density $\mathbf{f}$ together with the surface traction $\mathbf{t}$. Defining the stress tensor $\mathbf{T}$ with the outer surface normal $\mathbf{n}$ and the surface element $\mathbf{dS} = \mathbf{n} dS$

$$\mathbf{t} = \mathbf{n} \cdot \mathbf{T} \tag{2}$$

we obtain

$$\iint_{S_g} \mathbf{dS} \cdot \mathbf{T} + \iiint_V \mathbf{f}\, dV = \iiint_V \rho \ddot{\mathbf{u}}\, dV \tag{3}$$

Introducing the velocity instead of the displacement through $\mathbf{v} = \dot{\mathbf{u}}$ and disregarding volume forces, Newton's equation of motion is obtained in integral form:

$$\iiint_V \rho \dot{\mathbf{v}}\, dV = \iint_{S_g} \mathbf{n} \cdot \mathbf{T}\, dS \tag{4}$$

Obviously, another equation is needed relating the stress and the displacement, and in linear elastodynamics, it is provided by Hooke's law of elasticity:

$$\mathbf{T} = \mathbf{c} : \mathbf{S} \tag{5}$$

i.e. relating the stress tensor $\mathbf{T}$ linearly to the strain tensor $\mathbf{S}$ via the stiffness tensor $\mathbf{c}$, a tensor of rank four, whence the double contraction as indicated by the double dot. The strain can be expressed by the displacement through

$$\begin{aligned} \mathbf{S} &= \nabla_s \mathbf{u} \\ &= \frac{1}{2}\left[\nabla\mathbf{u} + (\nabla\mathbf{u})^{21}\right] \end{aligned} \tag{6}$$

where ∇_s indicates the symmetric part of the displacement gradient dyadic, and the upper index 21 accounts for the indicial exchange after computation of the dyadic product $\nabla\mathbf{u}$.

To derive an integral version of Hooke's law, we integrate equ. 5 over the volume V and obey equ. 6

$$\iiint_V \mathbf{T}\, dV = \frac{1}{2}\mathbf{c} : \iiint_V \left[\nabla\mathbf{u} + (\nabla\mathbf{u})^{21}\right] dV \tag{7}$$

For simplicity we have assumed constant stiffness within the volume V. Applying a dyadic Gaussian theorem the result

$$\iiint_V \mathbf{T}\, dV = \frac{1}{2}\mathbf{c} : \iint_{S_g} (\mathbf{nu} + \mathbf{un})\, dS \tag{8}$$

is obtained, where $\mathbf{nu}$ and $\mathbf{un}$ denote dyadic products. Restricting for the moment to the case of an isotropic solid, we can introduce Lamé's constants λ and μ, and, using the itentity dyadic operator $\mathbf{I}$ we get

$$\iiint_V \mathbf{T}\, dV = \lambda\mathbf{I} \iint_{S_g} \mathbf{n} \cdot \mathbf{u}\, dS + \mu \iint_{S_g} (\mathbf{nu} + \mathbf{un})\, dS \tag{9}$$

which is the desired integral form of Hooke's law. Differentiating with respect to time, the resulting equation

$$\iiint_V \dot{\mathbf{T}}\, dV = \lambda\mathbf{I} \iint_{S_g} \mathbf{n} \cdot \mathbf{v}\, dS + \mu \iint_{S_g} (\mathbf{nv} + \mathbf{vn})\, dS \tag{10}$$

is, together with equ. 4, the building block of our numerical scheme to be described below. In contrast to Maxwell's equations, where three components of the electric field strength are related to three components of the magnetic field strength, we observe here, that the coupled system of elastodynamic equations relates three velocity components to six components of the symmetric stress tensor, but, as it is true for Maxwell's equations, only first time derivatives are involved.

3. DISCRETIZATION METHOD: EFIT

We discretize equ.'s 4 and 10 via approximation of the volume and surface integrals over a cubic cell of volume Δ^3. Fig. 1 shows such a cell with edges paralell to the 1,2,3-axes of a cartesian coordinate system; the pertinent orthonormal unit vectors are denoted by $\mathbf{e}_i$, the components of the velocity vector by v_i, the latter ones being dislocated from the local coordinate origin by $\Delta/2$. The computation of the surface integral in Newton's equation requires the knowlegde of

$$\mathbf{n} \cdot \mathbf{T} = \begin{cases} \pm\mathbf{e}_1 \cdot \mathbf{T} = \left(T_{11}^{r,l}, T_{12}^{r,l}, T_{13}^{r,l}\right) \\ \pm\mathbf{e}_2 \cdot \mathbf{T} = \left(T_{12}^{f,b}, T_{22}^{f,b}, T_{23}^{f,b}\right) \\ \pm\mathbf{e}_3 \cdot \mathbf{T} = \left(T_{13}^{u,d}, T_{23}^{u,d}, T_{33}^{u,d}\right) \end{cases} \tag{11}$$

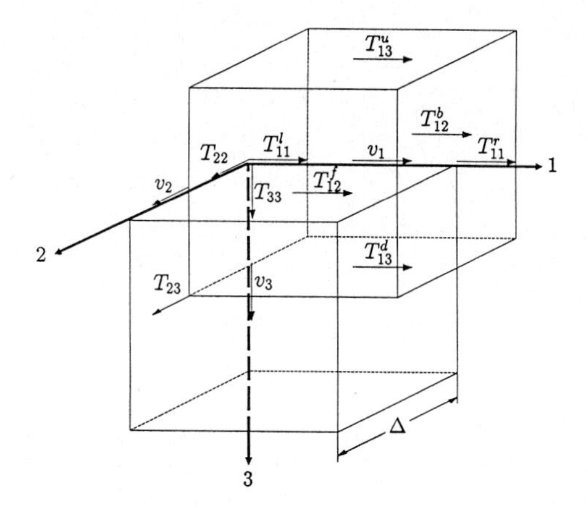

Figure 1: Elementary cubic cell

where we refer to the right and left side, front and back side, and up and down side of the cube via r, l, f, b, u, d. In Fig. 1 we indicated that the off-diagonal elements of the stress tensor are also dislocated to the midpoints of the cell surfaces. This dislocation is equally appropriate for Newton's as well as Hooke's law. We obtain as discretized version of equ. 4

$$\begin{aligned}
\rho\dot{v}_1\Delta^3 &= \left(T^r_{11} - T^l_{11} + T^f_{12} - T^b_{12} + T^d_{13} - T^u_{13}\right)\Delta^2 \\
\rho\dot{v}_2\Delta^3 &= \left(T^r_{12} - T^l_{12} + T^f_{22} - T^b_{22} + T^d_{23} - T^u_{23}\right)\Delta^2 \\
\rho\dot{v}_3\Delta^3 &= \left(T^r_{13} - T^l_{13} + T^f_{23} - T^b_{23} + T^d_{33} - T^u_{33}\right)\Delta^2
\end{aligned} \tag{12}$$

For the diagonal elements the discretized Hooke's law reads

$$\begin{aligned}
\dot{T}_{11}\Delta^3 &= \lambda\left(v^r_1 - v^l_1 + v^f_2 - v^b_2 + v^d_3 - v^u_3\right)\Delta^2 + 2\mu\left(v^r_1 - v^l_1\right)\Delta^2 \\
\dot{T}_{22}\Delta^3 &= \lambda\left(v^r_1 - v^l_1 + v^f_2 - v^b_2 + v^d_3 - v^u_3\right)\Delta^2 + 2\mu\left(v^f_2 - v^b_2\right)\Delta^2 \\
\dot{T}_{33}\Delta^3 &= \lambda\left(v^r_1 - v^l_1 + v^f_2 - v^b_2 + v^d_3 - v^u_3\right)\Delta^2 + 2\mu\left(v^d_3 - v^u_3\right)\Delta^2
\end{aligned} \tag{13}$$

and the off-diagonal elements result in the following equations

$$\begin{aligned}
\dot{T}_{12}\Delta^3 &= \mu\left(v^f_1 - v^b_1 + v^r_2 - v^l_2\right)\Delta^2 \\
\dot{T}_{13}\Delta^3 &= \mu\left(v^d_1 - v^u_1 + v^r_3 - v^l_3\right)\Delta^2 \\
\dot{T}_{23}\Delta^3 &= \mu\left(v^d_2 - v^u_2 + v^f_3 - v^b_3\right)\Delta^2
\end{aligned} \tag{14}$$

Discretization of a whole test specimen requires a threedimensional mesh, where we number the gridpoints through

$$\begin{aligned}
&\text{1-direction:} \quad i = 1, 2, \ldots, I \\
&\text{2-direction:} \quad j = 1, 2, \ldots, J \\
&\text{3-direction:} \quad k = 1, 2, \ldots, K
\end{aligned} \tag{15}$$

An arbitrary gridpoint is then given by $n(i,j,k)$

$$n(i,j,k) = 1 + (i-1) + (j-1)I + (k-1)IJ \tag{16}$$

when we move through the grid in $(1 \to 2 \to 3)$-direction. We then define the following vectors as solutions of our discrete elastodynamic equations of motion

$$\begin{aligned} v_\alpha &= \left[v_\alpha^{(1)}, \ldots, v_\alpha^{(n)}, \ldots, v_\alpha^{(IJK)}\right] && (17) \\ T_{\alpha\beta} &= \left[T_{\alpha\beta}^{(1)}, \ldots, T_{\alpha\beta}^{(n)}, \ldots, T_{\alpha\beta}^{(IJK)}\right] && (18) \\ &\text{with } \alpha, \beta = 1,2,3 \\ &\text{and } T_{\alpha\beta} = T_{\beta\alpha} \end{aligned}$$

Discretization of time is performed in full and half time steps Δt according to

$$\begin{aligned} v_\alpha(t) &\Longrightarrow v_\alpha(l\Delta t) \text{ with } l \in \mathcal{Z} && (19) \\ T_{\alpha\beta}(t) &\Longrightarrow T_{\alpha\beta}\left(l\Delta t - \frac{1}{2}\Delta t\right) && (20) \end{aligned}$$

or — in shorthand notation —

$$\begin{aligned} v_\alpha(l\Delta t) &= v_\alpha^l && (21) \\ T_{\alpha\beta}\left(l\Delta t - \frac{1}{2}\Delta t\right) &= T_{\alpha\beta}^{l-\frac{1}{2}} && (22) \end{aligned}$$

Causality then requires

$$v_\alpha^l = 0, \quad T_{\alpha\beta}^l = 0 \text{ for } l \le 0 \tag{23}$$

The time derivative is appropriately approximated by central differences

$$\begin{aligned} v_\alpha^l &= v_\alpha^{l-1} + \dot{v}_\alpha^{l-\frac{1}{2}}\Delta t && (24) \\ T_{\alpha\beta}^{l+\frac{1}{2}} &= T_{\alpha\beta}^{l-\frac{1}{2}} + \dot{T}_{\alpha\beta}^l \Delta t && (25) \end{aligned}$$

In equ. 24 v_α^{l-1} is known from the preceding time step, and $\dot{v}_\alpha^{l-\frac{1}{2}}$ can be computed via the discretized Newton equation from $T_{\alpha\beta}^{l-\frac{1}{2}}$; for $l = 1$ $T_{\alpha\beta}^{\frac{1}{2}}$ is prescribed as initial condition. In equ. 25 $T_{\alpha\beta}^{l-\frac{1}{2}}$ is also known from the preceding time step — for $l = 1$ the initial condition is inserted —, and $\dot{T}_{\alpha\beta}^l$ enters from Hooke's law via v_α^l.

4. A TWODIMENSIONAL TEST EXAMPLE

Twodimensional problems are computationally simpler than threedimensional problems; this specialization is readily obtained from our above equations omitting all 3-components. Then, a twodimensional test example is defined, where an analytical reference solution permits an accuracy check of EFIT. The geometry is sketched in Fig. 2: a strip-like piston transducer of width 20 mm specifies a T_{22}-component proportional to

$$f(t) = \begin{cases} (1 - \cos \pi f_c t) \cos 2\pi f_c t & \text{for } 0 < t < T_D \\ 0 & \text{otherwise} \end{cases} \tag{26}$$

with $f_c = 2/T_D = 1.88$ MHz and $T_D = 64\, t_0$ with $t_0 = 17$ ns on a "specimen" with otherwise stress-free boundary conditions; it radiates into a solid with pressure wave speed $c_P = 5900$ m/s and shear wave speed $c_S = 3200$ m/s. The specimen has dimensions of 100 mm × 75 mm requiring 1026 × 771 nodes and $\Delta t = 0.5\, t_0$ for EFIT to produce numerically stable results; 2000 time steps yield a 3 h computer run on an apollo domain DN 10000 workstation, or a 25 s run on a Cray-YMP with 8 processors. In Fig. 2, an arrow indicates that point on the surface to which the displayed

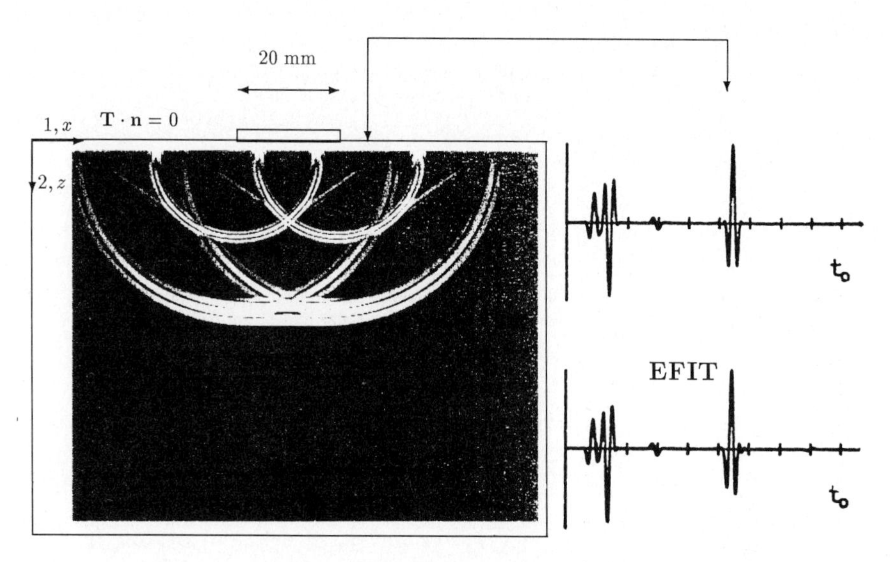

Figure 2: Twodimensional test example

v_1-component-signal refers to. Typically, the pressure wave pulse from the adjacent corner of the radiating aperture is immediately followed by a Rayleigh-pulse; the pressure pulse from the remote corner is only small due to radiation damping but the pertinent Rayleigh-pulse is of equal amplitude with regard to the first one because in two dimensions there is no amplitude decay with increasing distance.

Fig. 2 also displays the EFIT-computation for the well-known $|\mathbf{v}|$-snap-shot of wavefronts, which can be mathematically as well as physically identified [6].

In addition to the EFIT-result, a reference signal is given in Fig. 2, which has been computed quite differently. Starting with a Huygens-type solution of the differential equation for the displacement, the following spectral wave decomposition of, say, $u_x(x, z, \omega)$ with ω denoting circular frequency, can be derived — we assume $e^{-j\omega t}$ time dependence —

$$u_x(x,z,\omega) = -\frac{j}{2\pi\mu} F(\omega) \int_{-\infty}^{+\infty} \frac{\sin(ak_P\alpha)}{k_P\alpha} \alpha \frac{(\kappa^2 - 2\alpha^2)e^{jk_P z \alpha_{Pz}} - 2\alpha_{Pz}\alpha_{Sz} e^{jk_P z \alpha_{Sz}}}{R(\alpha)} e^{jk_P x \alpha}\, d\alpha \quad (27)$$

with the Rayleigh-function

$$R(\alpha) = (\kappa^2 - 2\alpha^2)^2 + 4\alpha^2 \alpha_{Pz}\alpha_{Sz} \quad (28)$$

and

$$\alpha_{Pz} = \sqrt{1-\alpha^2} \quad (29)$$

$$\alpha_{Sz} = \sqrt{\kappa^2-\alpha^2} \quad (30)$$

The velocity relation is given by $\kappa = c_P/c_S$, and $k_P = \omega/c_P$ is the pressure wave number; $F(\omega)$ denotes the frequency spectrum of $f(t)$, and a is the half width of the aperture. The inverse Fourier integral 27 with regard to α can now be numerically evaluated the following way: the Rayleigh-pole contribution is treated analytically, and the remainder is done by FFT's for a set of freqencies ω, which are needed to cover the bandwidth of $F(\omega)$. Both parts are then combined in the x-domain and inverted with regard to ω by an FFT again to end up in the time domain. The result, which compares with our EFIT-example is displayed in Fig. 2 (top) as a reference; obviously, the agreement between both methods is excellent.

The mathematical details of the methods described in this paper and more examples can be found in [7].

References

[1] R. Ludwig, W. Lord: A finite-element formulation for the study of ultrasonic NDT systems. IEEE Trans. Ultrasonics, Ferroel., and Frequ. Contr. 35 (1988) 809

[2] L.J. Bond, M. Punjani, N. Saffari: Ultrasonic wave propagation and scattering using explicit finite difference methods. In: *Mathematical Modelling in Nondestructive Testing* (Eds.: M. Blakemore, G.A. Georgiou), Clarendon Press, Oxford 1988

[3] T. Weiland: Numerical solution of Maxwell's equations for static, resonant and transient fields. Proc. Int. URSI Symposium on Electromagnetic Theory, Budapest, August 1986

[4] K.S. Yee: Numerical solution of initial boundary value problems involving Maxwell's equations in isotropic media. IEEE Trans. Ant. Propagat. AP-14 (1966) 302

[5] MAFIA User Guide, The MAFIA Collaboration: DESY, Los Alamos National Laboratory, KFA-Jülich, February 1987

[6] K.J. Langenberg, U. Aulenbacher, G. Bollig, P. Fellinger, H. Morbitzer, G. Weinfurter, P. Zanger, V. Schmitz: Numerical modelling of ultrasonic scattering. In: *Mathematical Modelling in Nondestructive Testing* (Eds.: M. Blakemore, G.A. Georgiou), Clarendon Press, Oxford 1988

[7] P. Fellinger: Ein Verfahren zur numerischen Lösung elastischer Wellenausbreitungsprobleme im Zeitbereich durch direkte Diskretisierung der elastodynamischen Grundgleichungen. Ph.D. Thesis, Dept. Electrical Eng., University of Kassel, FRG Kassel 1989

Elastic Waves and Ultrasonic Nondestructive Evaluation
S.K. Datta, J.D. Achenbach and Y.S. Rajapakse (Editors)

RECIPROCITY, DISCRETIZATION, AND THE NUMERICAL SOLUTION OF ELASTODYNAMIC PROPAGATION AND SCATTERING PROBLEMS

Adrianus T. de Hoop

Laboratory of Electromagnetic Research,
Department of Electrical Engineering,
Delft University of Technology,
P.O. Box 5031, 2600 GA Delft, The Netherlands

The larger part of wave propagation and scattering problems in elastodynamics has to be modeled with the aid of numerical methods. Many of these methods can be envisaged as being discretized versions of appropriate 'weak' formulations of the pertinent operator (differential or integral) equations. For the relevant problems as formulated in the time Laplace-transform domain it is shown that the Betti-Rayleigh reciprocity theorem encompasses all known weak formulations, while its discretization leads to the discretized form of the corresponding operator equations. The role of properly selected error criteria in the discretization procedures is emphasized.

1. INTRODUCTION

Except for the canonical problems whose solution can be expressed in terms of analytic functions of a not too complicated nature (examples of which can be found in Pao and Mow [**1**]) and for analytic approximation techniques (usually of an asymptotic nature), both in the long and the short wavelength regimes, that can be applied to a wider variety of cases, (see, for example, Achenbach, Gautesen and McMaken [**2**]) wave propagation and scattering problems in elastodynamics have to be addressed with the aid of numerical methods. Many of these methods can be envisaged as being discretized versions of appropriate 'weak' formulations of the pertinent operator (differential or integral) equations. Specifically, the finite-element method, based on the method of weighted residuals (see, for example, Zienkiewicz and Taylor [**3**]), applied to the elastodynamic partial differential equations, and the collocation method (method of point matching) applied to the source type elastodynamic integral relations (see, for example, Herman [**4**]) can be grouped in this category. The present paper is an attempt to systematize the different approaches and to bring, to a certain extent, consistency and coherence in the procedures.

To this end, the time Laplace-transform domain (s-domain) elastodynamic Betti-Rayleigh reciprocity theorem for time-invariant configurations is taken as point of departure. By taking the transform parameter s to be positive and real, as is done in the Cagniard method for calculating impulsive waves in stratified media (see, for example, Van der Hijden [**5**]), or complex in the right half $\mathrm{Re}(s) > 0$ of the complex s-plane, the causality of the wave motion is ensured by requiring the time Laplace-transform domain wave field quantities to be bounded functions of position in space, especially at large distances from the sources (of bounded extent) that generate the wave field. Also, arbitrarily inhomogeneous and anisotropic materials with arbitrary relaxation behavior can in this way be incorporated in the analysis in an easy manner. From the s-domain results, the time-domain counterparts follow upon using some standard rules of the one-sided Laplace transformation, while the results for the limiting case of sinusoidally in time varying field quantities follow upon replacing s by $j\omega$, where j is the imaginary unit and ω the angular frequency of the oscillations, on the condition that imaginary values of s are approached via the right half of the complex s-plane.

Taking one of the two states in the reciprocity theorem to be the elastodynamic state to be actually computed and the other to be an 'auxiliary' or a 'computational' one, it is shown that the reciprocity theorem encompasses all known 'weak' formulations of the elastodynamic differential equations and source-type integral relations. For this reason, the Betti-Rayleigh reciprocity relation is considered to best serve as the point of departure for the computational modeling of elastodynamic wave fields. (Note that this theorem is global rather than local in nature.)

Turning to numerics, first a geometical discretization procedure is applied, with the tetrahedron (simplex in three-dimensional space) as the basic building block. The mesh size of the discretization (maximum diameter of the tetrahedra) should be consistent with both the geometry of the configuration and the (inhomogeneous) distribution of matter in it, in the sense that it guarantees that the discretized configuration and the actual one differ relatively not more than a given fractional number according to some agreed-upon relative error criterion (for example, the relative root-mean-square error). Subsequently, the wave field to be computed is expanded in terms of a base in an appropriately chosen linear function space, and a weighting procedure with an appropriately chosen computational state is carried out. This procedure leads to a system of linear algebraic equations in the expansion coefficients. To solve this system of equations, again, an error criterion is needed to define what the 'best approximation' to the solution is (cf. Van den Berg [**6**]). In this manner, several versions of the finite-element method can be understood, as well as certain discretized versions of the integral equations describing scattering phenomena.

Finally, it is remarked that both forward (direct) and inverse modeling are natural consequences of the reciprocity theorem (cf. De Hoop [**7**]). Through an analysis of the type presented here, it is, at least theoretically, feasible that the whole operation of arriving at numerical results of a predetermined accuracy can be computer controlled.

2. THE ELASTODYNAMIC WAVE FIELD IN THE CONFIGURATION

The configuration in which the elastodynamic wave field is present consists of a solid material that occupies three-dimensional space R^3. In the bounded subdomain $D^s \subset R^3$ the solid is, in general, inhomogeneous and anisotropic. The boundary surface ∂D^s of D^s is assumed to be piecewise smooth. In the unbounded domain D_0 that is the complement of $D^s \cup \partial D^s$ in R^3, the solid is homogeneous and isotropic.

The position of observation in the configuration is specified by the coordinates $\{x_1, x_2, x_3\}$ with respect to a fixed, orthogonal, Cartesian reference frame with origin O and the three mutually perpendicular base vectors $\{\mathbf{i}_1, \mathbf{i}_2, \mathbf{i}_3\}$ of unit length each. In the indicated order, the base vectors form a right-handed system. The subscript notation for Cartesian vectors and tensors is used and the summation convention applies. The corresponding lowercase Latin subscripts are to be assigned the values $\{1, 2, 3\}$. Whenever appropriate, the position vector will be denoted by $\mathbf{x} = x_m \mathbf{i}_m$. The time coordinate is denoted by t. Partial differentiation is denoted by ∂; ∂_m denotes differentiation with respect to x_m, ∂_t is a reserved symbol for differentiation with respect to t.

For any causal space-time function $u = u(\mathbf{x}, t)$ the one-sided Laplace transform is introduced as

$$\hat{u}(\mathbf{x}, s) = \int_{t=0}^{\infty} \exp(-st) u(\mathbf{x}, t) dt, \tag{1}$$

where the instant $t = 0$ marks the onset of the events. Obviously, for bounded $|u(\mathbf{x}, t)|$, $\hat{u}(\mathbf{x}, s)$ is an analytic function of the complex transform parameter s in the right half $\mathrm{Re}(s) > 0$ of the complex s-plane. For ease of notation the circumflex over a symbol denoting its s-domain counterpart will be omitted in the remainder of the paper. The elastodynamic wave motion is started from a configuration at rest; then, under the one-sided Laplace transformation the operator ∂_t is replaced by an algebraic factor s.

In each subdomain of the configuration where the solid's acoustic properties vary continuously with position, the elastodynamic wave field quantities are continuously differentiable and satisfy the s-domain equations

$$-\Delta_{k,m,p,q} \partial_m \tau_{p,q} + s\rho_{k,r} v_r = f_k, \tag{2}$$

$$\Delta_{i,j,m,r} \partial_m v_r - s S_{i,j,p,q} \tau_{p,q} = h_{i,j}, \tag{3}$$

in which $\tau_{p,q}$ is the stress, v_r is the particle velocity, $\rho_{k,r}$ is the volume density of mass, $S_{i,j,p,q}$ is the compliance, f_k is the volume source density of force, and $h_{i,j}$ is the volume source density of rate of deformation. Full anisotropy in the inertia and (visco-)elastic properties of the solid is taken into account. For lossy solids (i.e., solids with relaxation), $\rho_{k,r}$ and $S_{i,j,p,q}$ are s-dependent, subject to the condition of causality which entails analyticity in $\mathrm{Re}(s) > 0$. Across interfaces between different kinds of solids (which are assumed to be in rigid contact), v_r and $\Delta_{k,m,p,q}, \nu_m \tau_{p,q}$ are continuous, where ν_m is the unit vector to the normal along such an interface. The symmetrical unit tensor of rank four

$$\Delta_{k,m,p,q} = (1/2)(\delta_{k,p}\delta_{m,q} + \delta_{k,q}\delta_{m,p}), \tag{4}$$

where $\delta_{k,p}$ is the symmetrical unit tensor of rank two (Kronecker tensor), is characteristic for elastodynamics and automatically selects from any tensor of rank two with which it is contracted, the symmetrical part. In the configuration, objects can be present that are impenetrable to the elastodynamic wave motion. On the boundary surface of such an object explicit boundary conditions hold; admissible ones are: either $\Delta_{k,m,p,q}\nu_m\tau_{p,q} \to 0$ (void) or $v_r \to 0$ (immovable rigid object).

In the domain D_0 we have $\rho_{k,r} = \rho_0\delta_{k,r}$ and $S_{i,j,p,q} = \Lambda_0\delta_{i,j}\delta_{p,q} + 2M_0\Delta_{i,j,p,q}$, in which ρ_0, Λ_0 and M_0 are position independent. The elastodynamic wave field quantities that are generated by known sources in such a solid are analytically known (cf. Achenbach [**8**], De Hoop [**9**]). In deriving the relevant representations causality plays, again, an essential role.

3. THE RECIPROCITY RELATION

For our further analysis, the s-domain reciprocity relation that is associated with Eqs.(2) and (3) will serve as point of departure. A general wave field reciprocity theorem interrelates, in a specific manner, the quantities that characterize two different physical states that could occur in one and the same domain in space-time. For time-invariant configurations the application of the one-sided Laplace transformation of Eq.(1) to the convolution-type reciprocity theorem leads to an equivalent s-domain result. For this to be applicable to the configuration under investigation, the solids in the two states should be present in one and the same time-invariant domain D. The two states will be distinguished by the superscripts A and B, respectively.

First, the local reciprocity theorem will be derived. From it, the global reciprocity theorem for the entire configuration (or a subdomain of it) will be obtained. The local reciprocity relation follows upon considering the interaction quantity $\Delta_{k,m,p,q}\partial_m(\tau^A_{p,q}v^B_k - \tau^B_{p,q}v^A_k)$ and evaluating this quantity with the use of Eqs.(2) and (3) for the States A and B, respectively. The result is

$$\begin{aligned}\Delta_{k,m,p,q}\partial_m(\tau^A_{p,q}v^B_k - \tau^B_{p,q}v^A_k) &= s(\rho^A_{k,r} - \rho^B_{r,k})v^A_r v^B_k - s(S^A_{i,j,p,q} - S^B_{p,q,i,j})\tau^A_{p,q}\tau^B_{i,j} \\ &\quad -f^A_k v^B_k + f^B_r v^A_r + h^B_{p,q}\tau^A_{p,q} - h^A_{i,j}\tau^B_{i,j}. \end{aligned} \tag{5}$$

Equation (5) holds at any point of D in the neighborhood of which the properties of the solids in the States A and B vary continuously with position. As far as the right-hand side of Eq.(5) is concerned, the terms fall into two categories. In the first set of terms, the medium properties of the solids in the States A and B occur. These terms terms vanish if $\rho^A_{k,r} = \rho^B_{r,k}$ and $S^A_{i,j,p,q} = S^B_{p,q,i,j}$. If these properties hold, the solid present in State B is denoted as the adjoint of the solid present in State A. In case these properties hold for one and the same solid, this solid is denoted as self-adjoint or reciprocal. The second set of terms at the right-hand side is associated with the source distributions in the States A and B. Obviously, these terms vanish in a sourcefree subdomain.

The global reciprocity relation that holds for some domain D in the configuration is obtained by integrating Eq.(5) over the domain D and applying Gauss' divergence theorem to the resulting left-hand side. With this, the following relation is obtained:

$$\begin{aligned}&\int_{\partial D} \Delta_{k,m,p,q}\nu_m(\tau^A_{p,q}v^B_k - \tau^B_{p,q}v^A_k)dA \\ &= \int_D [s(\rho^A_{k,r} - \rho^B_{r,k})v^A_r v^B_k - s(S^A_{i,j,p,q} - S^B_{p,q,i,j})\tau^A_{p,q}\tau^B_{i,j} \\ &\quad -f^A_k v^B_k + f^B_r v^A_r + h^B_{p,q}\tau^A_{p,q} - h^A_{i,j}\tau^B_{i,j}]dV, \end{aligned} \tag{6}$$

where the contributions from interfaces have canceled in view of the pertaining boundary conditions of the continuity type, while the contributions from the boundary surfaces of impenetrable objects have vanished in view of the pertaining boundary conditons of the explicit type. Equation (6) is the global reciprocity relation that will be used in the considerations that follow. In it, ∂D is the boundary surface of the domain D and ν_m is the unit vector along its normal, pointing away from D.

In some of the applications, D will be the entire three-dimensional space. To address this situation, Eq. (6) is first applied to the domain interior to the sphere S_Δ of radius Δ and with center at the origin of the chosen reference frame, after which the limit $\Delta \to \infty$

is taken. From some Δ onward, S_Δ will be entirely situated in a homogeneous, isotropic, perfectly elastic solid. On S_Δ, the far-field representations for the elastodynamic wave field quantities can, for sufficiently large values of Δ, be used, from which the contribution from S_Δ can be shown to vanish in the limit $\Delta \to \infty$ (cf. De Hoop [**10**]).

4. DISCRETIZATION PROCEDURE AND ERROR CRITERIA

Each quantity $Q = Q(\mathbf{x}, s)$ occurring in the elastodynamic wave problem (be it a scalar, a vector, or a tensor of arbitrary rank) and defined on some domain D has, after discretization, a discretrized counterpart $[Q] = [Q](\mathbf{x}, s)$ defined on the discretized version $[D]$ of D. The local error δQ in Q is defined as

$$\delta Q(\mathbf{x}, s) = [Q](\mathbf{x}, s) - Q(\mathbf{x}, s) \quad \text{for } \mathbf{x} \in [D] \cap D. \tag{7}$$

The global error $ERR_\Omega(Q)$ in Q over some domain Ω is given by

$$ERR_\Omega(Q) = ERROP_\Omega(\delta Q), \tag{8}$$

where $ERROP$ is some positive definite operator acting on δQ over the domain Ω. For example, the root-mean-square error is given by

$$ERR_\Omega(Q) = RMS_\Omega(\delta Q), \tag{9}$$

where

$$RMS_\Omega(\delta Q) = \left[\int_{\mathbf{x}\in\Omega} \|\delta Q(\mathbf{x}, s)\|^2 dV\right]^{1/2}, \tag{10}$$

in which $\|\cdot\|$ denotes a suitably defined norm over Ω. Equation (8) defines the absolute global error in the quantity Q over the domain Ω. The relative global error $err_\Omega(Q)$ is taken as

$$err_\Omega(Q) = ERROP_\Omega(\delta Q)/ERROP_\Omega(Q). \tag{11}$$

Obviously, $err_\Omega(Q) = 0$ if $\delta Q = 0$ for all $\mathbf{x} \in \Omega$, and $err_\Omega(Q) = 1$ if $\delta Q = -Q$ for all $\mathbf{x} \in \Omega$.

Discretization of the computational domain

The actual machine computations are finite in number and can therefore only be carried out for some bounded computational domain $D \in R^3$. In the scheme to be presented, the inhomogeneous part D^s of the configuration has to be entirely incorporated in D, so we take D^s to be a proper subset of D. Note that the elastodynamic wave motion is defined in the entire R^3, which implies that D also contains some part of D_0, where the solid is homogeneous and isotropic. Without loss of generality we can therefore take the discretized version of the domain of computation identical to the actual one, i.e. $[D] = D$. Owing to this, the error analysis can be carried out over D, since both the exact and the discretized wave field quantities are then defined over D.

We now discretize D by taking it to be the union of a finite number of tetrahedra (simplices in R^3) that all have vertices, edges and faces in common. For any bounded domain, such a discretization can be carried out up to an error $O(h)$ as $h \to 0$, where h is the supremum of the maximum diameters of the tetrahedra (see, Naber [**11**]). The vertices of the tetrahedra will also be denoted as the nodes of the (geometrical) mesh and h will be denoted as the mesh size. Each quantity $Q = Q(\mathbf{x}, s)$ occurring in the elastodynamic wave field problem (be it a scalar, a vector, or a tensor of arbitrary rank) will, in the interior of each tetrahedron, be approximated by the linear interpolation of its values at the vertices. Let $\{\mathbf{x}(0), \mathbf{x}(1), \mathbf{x}(2), \mathbf{x}(3)\}$ denote the position vectors of the vertices of a particular tetrahedron (simplex) $SMPLX$, then the corresponding linear interpolation is given by

$$[Q](\mathbf{x}, s) = \sum_{IV=0}^{3} Q(IV, s)\lambda(IV; \mathbf{x}) \quad \text{for } \mathbf{x} \in SMPLX, \tag{12}$$

in which

$$Q(IV, s) = [Q](\mathbf{x}(IV), s), \quad \text{for } IV = 0, 1, 2, 3, \tag{13}$$

is the value of Q at the vertex with ordinal number IV, and $\{\lambda(IV; \mathbf{x}); IV = 0, 1, 2, 3\}$ are the barycentric coordinates of $\mathbf{x}$ in $SMPLX$. The latter have the property

$$\lambda(IV, \mathbf{x}(JV)) = \{1, 0\} \quad \text{if } \{IV = JV, IV \neq JV\}, \tag{14}$$

and are expressed in terms of the vectorial areas of the faces of $SMPLX$ through

$$\lambda(IV;\mathbf{x}) = 1/4 - (1/3V)(x_m - b_m)A_m(IV) \quad \text{with } IV = 0,1,2,3, \tag{15}$$

where $A_m(IV)$ is the outwardly oriented vectorial area of the face opposite the vertex with ordinal number IV and $b_m = (x_m(0) + x_m(1) + x_m(2) + x_m(3))/4$ is the position vector of the barycenter of $SMPLX$.

Discretization of the medium parameters

In the discretization of the medium parameters a distinction must be made between subdomains in which these parameters vary continuously with position and subdomains in which surfacess of discontinuity in these parameters occur. It is assumed that across such surfaces of discontinuity in medium parameters, the parameter values jump by finite amounts. Especially in applications where accurate values of the wave field quantities up to these surfaces are needed (such as, for example, in the modeling of borehole measurement situations in exploration geophysics, and for applications in the non-destructive evaluation of mechanical structures), special measures have to be taken to model accurately the behavior of these quantities. In principle, the medium properties are allowed to jump across any face of any tetrahedron of the discretized geometry. To accomodate this feature, all nodes of the geometrical mesh are considered as multiple nodes, where the multiplicity of each node is equal to the number of tetrahedra that meet at that node. The values of the constitutive parameters at the vertices that are needed in the local expansion Eq.(12), follow either from user-supplied input expressions that are spatially sampled in the interior of each tetrahedron close to each of its vertices (as is the case in direct or forward profiling problems) or from computationally derived values (as is the case in inverse profiling problems). Out of the thus constructed local expansions of the medium parameters, their global expansions over the domain of computation are composed by combining the local expansions. If in the latter procedure, at a particular node no discontinuity turns up, the multiple node is changed into a simple one, with an associated single value of the relevant constitutive parameter. For the non-scalar constitutive parameters the components with respect to the background Cartesian reference frame are used in the entire discretization procedure.

Discretization of the volume source densities

For the discretization of the volume source densities the same procedure as for the discretization of the medium parameters is followed.

Discretization of the wave field quantities

In the discretization of the elastodynamic wave field quantities the situation is more complicated. Here, some components are by necessity continuous across an interface of discontinuity in material properties, while other components show a finite jump across such a discontinuity surface. To preserve accuracy in the computational results, it is necessary, both in the modeling of direct or forward problems and in the modeling of inverse problems, to take computational measures that enforce the continuity conditions across an interface (in machine precision) and leave the non-continuous components free to jump by finite amounts. For this purpose, local expansions of the type of Eq.(12) have been developed where a non-scalar quantity at a vertex is expressed in terms of its components that are continuous across an interface of discontinuity in material properties. These components are, in general, not the components with respect to the background Cartesian reference frame. For the elastodynamic wave field quantities the relevant difficulty does not arise in connection with the particle velocity: all components of this quantity are continuous across the interfaces of discontinuity in material properties, since we have assumed the two media at either side of each interface to be in rigid contact. For the stress, however, the components normal to an interface (that together form the traction) are continuous across the interface, while the remaining components show a jump discontinuity. To guarantee the continuity of the traction across each face of adjoining tetrahedra, we consider each node as a multiple node and construct at each vertex the stress out of the three tractions at the three faces that meet at that vertex and use the relevant values in the local expansion Eq.(12). The relevant elements are denoted as face elements. Out of the thus constructed local expansions, the global expansions over the domain of computation are composed by combining the local expansions. If in this procedure simple nodes are met, the stress is just expressed in terms of its components in the background Cartesian reference frame. The face-element representation for the global expansion of the stress could also be used at simple nodes, but this leads to an unnecessarily large number of expansion coefficients to be computed (without yielding an increased accuracy) since the number of tetrahedral faces that meet at a particular node is larger than three.

5. FINITE-ELEMENT MODELING

To construct the system of equations that results from the finite-element modeling of elastodynamic wave problems, we apply the global Betti-Rayleigh reciprocity relation Eq.(6) to the discretized computational domain $[D]$, substitute the proper expansions for State A in it, put the constitutive coefficients for State B equal to zero, and successively take for the wave field in State B one of the global expansion functions of State A, while at the boundary surface $\partial[D]$ of the domain of computation an 'absorbing' boundary condition is invoked.

6. INTEGRAL-EQUATION MODELING

To construct the system of equations that results from the integral-equation modeling of elastodynamic wave problems, we apply the global Betti-Rayleigh reciprocity relation Eq.(6) to the discretized computational domain $[D]$, substitute the proper expansions for State A into it, put the constitutive coefficients for State B equal to the ones of the embedding D_0, and successively take for the sources in State B one of the global expansions of State A. (In this way we construct, in fact, the discretized Green's functions of the embedding.)

ACKNOWLEDGEMENTS

The research reported in this paper has been financially supported through Research Grants from the Stichting Fund for Science, Technology and Research (a companion organization to the Schlumberger Foundation in the U.S.A., from Schlumberger-Doll Research, Ridgefield, CT., U.S.A., and from Etudes et Productions Schlumberger, Clamart, France. This support is gratefully acknowledged.

REFERENCES

[1] Pao, Y.H. and C.C. Mow, *Diffraction of elastic waves and dynamic stress concentrations*, Crane, Russak, New York, 1973.

[2] Achenbach, J.D., A.K. Gautesen, and H. McMaken, *Ray methods for waves in elastic solids*, Pitman, Boston, 1982.

[3] Zienkiewicz, O.C. and R.L. Taylor, *The finite element method*, McGraw-Hill, London, 4th. ed., 1989.

[4] Herman, G.C., *Scattering of transient acoustic waves in fluids and solids*, Report Nr. 1981-13, Laboratory of Electromagnetic Research, Department of Electrical Engineering, Delft University of Technology, Delft, the Netherlands, 1981, 183 pp.

[5] Van der Hijden, J.H.M.T., *Propagation of transient elastic waves in stratified anisotropic media*, North-Holland, Amsterdam, 1987.

[6] Van den Berg, P.M., "Iterative computational techniques based upon the integrated square error criterion", IEEE Transactions on Antennas and Propagation, Vol. AP-32, No. 10, October 1984, pp. 1063-1071.

[7] De Hoop, A.T., "Time-domain reciprocity theorems for elastodynamic wave fields in solids with relaxation and their application to inverse problems", Wave Motion **10** (1988) 479-489.

[8] Achenbach, J.D., *Wave propagation in elastic solids*, North-Holland, Amsterdam, 1973, p. 103.

[9] De Hoop, A.T. *Representation theorems for the displacement in an elastic solid and their application to elastodynamic diffraction theory*, Thesis, Department of Electrical Engineering, Delft University of Technology, Delft, the Netherlands, 1958, 84 pp.

[10] De Hoop, A.T., "A time-domain energy theorem for the scattering of plane elastic waves", Wave Motion **7** (1985) 569-577.

[11] Naber, *Topological methods in Euclidean space*, Cambridge University Press, Cambridge, 1980.

Elastic Waves and Ultrasonic Nondestructive Evaluation
S.K. Datta, J.D. Achenbach and Y.S. Rajapakse (Editors)

INVERSE SCATTERING AND IMAGING

K.J. Langenberg, T. Kreutter, K. Mayer, P. Fellinger

Dept. Electrical Eng., FB 16
University of Kassel
3500 Kassel, FRG

Following the guidelines of a recently developed unified theory of scalar inverse scattering within the linearizing Born or Kirchhoff approximations — culminating in a 3D imaging system for ultrasonic NDT applications — we extend its basic ideas to elastic wave inverse scattering, especially for the case of a scatterer with stress-free boundaries being illuminated by a plane pressure or shear wave. Two inversion schemes, both yielding the same result, i.e. the "visible" part of the singular function of the scattering surface, are proposed: far-field inversion, provided the data are collected in the remote region of the defect, or, far-field inversion combined with a near-field far-field transformation of the measured data for the case of defects closed to a measurement surface. The latter is computationally very effective in terms of Fourier transforms provided the mesurement surface is planar or circular cylindrical. In addition, we point out why the powerful idea of backpropagation resulting in the scalar Porter-Bojarski integral equation is not very appropriate for the elastodynamic case. Our algorithms are checked against simulations and discussed as elastodynamic extensions of the scalar SAFT-algorithm in terms of a mode-matched SAFT; their use in advanced imaging systems is presently only limited due to the requirement of tangential as well as normal components of the displacement vector to be measured on the material surface.

1. INTRODUCTION

If the differential equation for the displcement vector $\mathbf{u}(\mathbf{R},t)$ as function of position $\mathbf{R}$ and time t

$$(\lambda+2\mu)\nabla\nabla\cdot\mathbf{u}(\mathbf{R},t)-\mu\nabla\times\nabla\times\mathbf{u}(\mathbf{R},t)-\varrho\frac{\partial^2}{\partial t^2}\mathbf{u}(\mathbf{R},t)=0 \tag{1}$$

— λ and μ denote Lamé's constants of the homogeneous and isotropic bulk material of density ϱ — is "scalarized" in terms of the potentials

$$\mathbf{u}(\mathbf{R},t)=\nabla\Phi(\mathbf{R},t)+\nabla\times\Psi(\mathbf{R},t)\text{ with }\nabla\cdot\Psi(\mathbf{R},t)=0 \tag{2}$$

by concentrating only on the scalar potential $\Phi(\mathbf{R},t)$, which is approximately correct if we just consider an incident pressure wave and observe only in the scattered pressure wave mode, we can formulate a unified and concise theory of scalar inverse scattering for typical canonical scattering defects, i.e. penetrable defects like material inclusions, and perfectly scattering defects like voids, which are characterized by soft or rigid boundary conditions [1,2]. This theory could be called a generalized diffraction tomography, because, first, it applies only if the Born or Kirchhoff approximations hold, which linearize the inverse scattering problem, and second, it reduces to conventional diffraction tomography in terms of the Fourier Diffraction Slice Theorem if planar measurement surfaces are under concern for monofrequent multiple angle plane wave excitation. But, as a matter of fact, it also holds for the case of frequency diversity, either for pitch-catch or pulse-echo experiments, and, interesting enough, it allows the evaluation of explicit time domain backpropagation schemes, thus providing a rigorous diffraction theory of the otherwise heuristically proposed SAFT-algorithm [1], which has found wide application in ultrasonic nondestructive testing [3]. That way, Fourier transform alternatives to SAFT-processing of time domain synthetic aperture data have been developed for planar apertures [4,1,5] and circular cylindrical apertures [6], which even allow 3D ultrasonic imaging [7]; an example is given in Fig. 1, where the 3D image of an arificially induced defect in a diffusion bond is displayed (details of the system can be found in [7]).

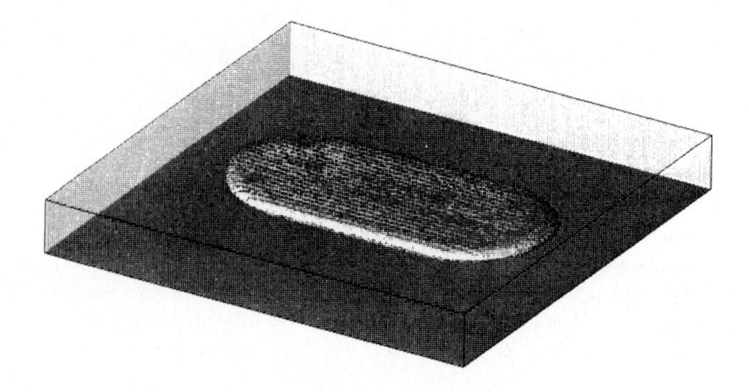

Figure 1: 3D iso-contour plot image of an artifical defect in a diffusion bond as obtained with the GhK-TET 3D ultrasonic imaging system based on FT-SAFT

The above-mentioned theory allows three approaches:

- Writing down the solution for $\Phi(\mathbf{R},\omega)$ — ω denoting circular frequency — in the far-field of the scatterer, i.e. assuming far-field measurements on a closed measurement surface S_M, the far-field scattering amplitude can be made the basis of a spatial Fourier space inversion scheme, if the spatial Fourier vector $\mathbf{K}$ is defined in terms of experimental parameters according to

$$\mathbf{K} = k_P(\hat{\mathbf{R}} - \hat{\mathbf{k}}_i) \tag{3}$$

  where k_P is the pressure wave number, and $\hat{\mathbf{k}}_i$ is the unit vector of propagation of the incident plane wave, whereas $\hat{\mathbf{R}}$ denotes the unit vector to the point of observation.

- For S_M in the near-field of the scatterer, the scattering amplitude can be defined alternatively in terms of a Huygens-type integral extending over S_M [8,9,2], which provides a near-field far-field transformation; then it is used as above. This approach is particularly effective, if S_M is planar or circular cylindrical, allowing the computation of the Huygens integral via Fourier transforms.

- Instead of defining the scattering amplitude in terms of a Huygens integral *outside* the measurement surface, a generalized holographic field can be defined *inside* that surface applying the principle of backpropagagtion through insertion of the complex conjugate of Green's function. Manipulation of this representation yields the so-called Porter-Bojarski integral equation relating the equivalent sources defining the "pressure wave" defect [10,11] and the holographic field, to which one has access via measurements. This integral equation gives rise to the above-mentioned unified theory of linearized inverse scattering if it is integrated with regard to ω or with regard to $\hat{\mathbf{k}}_i$.

In fact, all three approaches are equivalent with regard to the results, they only differ in algorithmic effectivity, and in computational usefulness to verify relationships between different algorithms. Hence, in order to obtain the full elastodynamic counterpart of this theory, we are free to choose that version, which is most appropriate. It turns out, that the near-field far-field transformation together with the far-field inversion gives rise to the easiest access, whereas elastodynamic backpropagation seems somewhat cumbersome.

2. ELASTODYNAMIC FAR-FIELD INVERSION FOR SCATTERERS WITH STRESS-FREE BOUNDARIES

A Huygens-type solution of equ. 1 for the displacement $\mathbf{u}_s$ scattered by a defect with a stress-free surface S_c is given by

$$\mathbf{u}_s(\mathbf{R},\omega) = \int\!\!\int_{S_c} \mathbf{n}'\mathbf{u}(\mathbf{R}',\omega) : \mathbf{\Sigma}(\mathbf{R}-\mathbf{R}',\omega)\, dS' \tag{4}$$

where $\mathbf{n}'$ denotes the outer normal on S_c [12]; we do not give the third rank Green's tensor $\mathbf{\Sigma}$ explicitly, because we only need its far-field approximation

$$\begin{aligned}\mathbf{\Sigma}^{far}(\mathbf{R}-\mathbf{R}',\omega) &= -\frac{j\mu k_S^3}{\varrho\omega^2}\left[\hat{\mathbf{R}}\mathbf{I} + (\hat{\mathbf{R}}\mathbf{I})^{213} - 2\hat{\mathbf{R}}\hat{\mathbf{R}}\hat{\mathbf{R}}\right] G_S^{far}(\mathbf{R}-\mathbf{R}',\omega)\\ &\quad -\frac{jk_P^3}{\varrho\omega^2}(\lambda\mathbf{I}\hat{\mathbf{R}} + 2\mu\hat{\mathbf{R}}\hat{\mathbf{R}}\hat{\mathbf{R}})G_P^{far}(\mathbf{R}-\mathbf{R}',\omega)\\ &= \frac{e^{jk_SR}}{4\pi R}e^{-jk_S\hat{\mathbf{R}}\cdot\mathbf{R}'}\mathbf{\Sigma}_S(\hat{\mathbf{R}},\omega) + \frac{e^{jk_PR}}{4\pi R}e^{-jk_P\hat{\mathbf{R}}\cdot\mathbf{R}'}\mathbf{\Sigma}_P(\hat{\mathbf{R}},\omega)\end{aligned} \tag{5}$$

with

$$G_{P,S}^{far}(\mathbf{R}-\mathbf{R}',\omega) = \frac{e^{jk_{P,S}R}}{4\pi R}e^{-jk_{P,S}\hat{\mathbf{R}}\cdot\mathbf{R}'} \tag{6}$$

Here, S is the "shear wave index", and $\mathbf{I}$ is the dyadic identity operator. We obtain the far-field approximation of equ. 4 through

$$\begin{aligned}\mathbf{u}_s^{far}(\mathbf{R},\omega) &= \frac{e^{jk_SR}}{4\pi R}\int\!\!\int\!\!\int_{V_c} \mathbf{U}_c(\mathbf{R}',\omega) : \mathbf{\Sigma}_S(\hat{\mathbf{R}},\omega)e^{-jk_S\hat{\mathbf{R}}\cdot\mathbf{R}'}\, d^3\mathbf{R}' +\\ &\quad \frac{e^{jk_PR}}{4\pi R}\int\!\!\int\!\!\int_{V_c} \mathbf{U}_c(\mathbf{R}',\omega) : \mathbf{\Sigma}_P(\hat{\mathbf{R}},\omega)e^{-jk_P\hat{\mathbf{R}}\cdot\mathbf{R}'}\, d^3\mathbf{R}'\\ &= \frac{e^{jk_SR}}{R}\mathbf{C}_S(\hat{\mathbf{R}},\omega) + \frac{e^{jk_PR}}{R}\mathbf{C}_P(\hat{\mathbf{R}},\omega)\\ &= \mathbf{u}_{sS}^{far}(\mathbf{R},\omega) + \mathbf{u}_{sP}^{far}(\mathbf{R},\omega)\end{aligned} \tag{7}$$

where we have introduced dyadic equivalent surface sources $\mathbf{U}_c(\mathbf{R}',\omega)$ through

$$\mathbf{U}_c(\mathbf{R}',\omega) = \gamma(\mathbf{R}')\mathbf{n}'\mathbf{u}(\mathbf{R}',\omega) \tag{8}$$

with the singular function $\gamma(\mathbf{R})$ of S_c [13]; the introduction of the latter allows the extension of the surface integral in equ. 4 to a volume integral over the volume V_c of the scatterer. The quantities $\mathbf{C}_\beta(\hat{\mathbf{R}},\omega)$ with $\beta = P, S$ play the role of vectorial scattering amplitudes in the β-mode, and it is easily verified that the following properties hold

$$\mathbf{C}_P(\hat{\mathbf{R}},\omega) \times \hat{\mathbf{R}} = 0 \tag{9}$$

$$\mathbf{C}_S(\hat{\mathbf{R}},\omega) \cdot \hat{\mathbf{R}} = 0 \tag{10}$$

This permits computational selection of a special mode in the far-field

$$\mathbf{u}_{s\beta}^{far}(\mathbf{R},\omega) = \frac{e^{jk_\beta R}}{R}\mathbf{C}_\beta(\hat{\mathbf{R}},\omega) \tag{11}$$

to be considered separately for inversion. We linearize in terms of physical elastodynamics (PE) [14]

$$\begin{aligned}\mathbf{U}_c(\mathbf{R}',\omega) &\Rightarrow \mathbf{U}_{c\alpha}^{PE}(\mathbf{R}',\omega)\gamma_u(\mathbf{R}')\\ &= \mathbf{U}_{0\alpha}(\mathbf{n}',\hat{\mathbf{k}}_i,\omega)\gamma_u(\mathbf{R}')e^{jk_\alpha\hat{\mathbf{k}}_i\cdot\mathbf{R}'}\end{aligned} \tag{12}$$

with

$$\mathbf{U}_{0\alpha}(\mathbf{n}',\hat{\mathbf{k}}_i,\omega) = \mathbf{u}_{0\alpha}(\mathbf{n}',\hat{\mathbf{k}}_i,\omega)\mathbf{n}' \tag{13}$$

where k_α is the wave number of the incident plane wave in the mode $\alpha = P, S$; $\mathbf{u}_{0\alpha}(\mathbf{n}', \hat{\mathbf{k}}_i, \omega)$ denotes the surface displacement "induced" by the incident wave as computed by PE, whence the occurrence of only *that* part of the singular function, which is illuminated by that wave, i.e. $\gamma_u(\mathbf{R}')$, the index u indicating multiplication of $\gamma(\mathbf{R}')$ with a step-function. We obtain

$$\mathbf{u}_{s\alpha\beta}^{far}(\mathbf{R},\omega) = \frac{e^{jk_\beta R}}{4\pi R} \int\int\int_{V_c} \gamma_u(\mathbf{R}')\mathbf{U}_{0\alpha}(\mathbf{n}', \hat{\mathbf{k}}_i, \omega) : \mathbf{\Sigma}_\beta(\hat{\mathbf{R}}, \omega) e^{-j(k_\beta \hat{\mathbf{R}} - k_\alpha \hat{\mathbf{k}}_i)\cdot \mathbf{R}'} \, d^3\mathbf{R}' \tag{14}$$

Computing a "stationary phase normal" $\mathbf{n}' = \mathbf{n}_{\alpha\beta}$ [2] according to

$$\mathbf{n}_{\alpha\beta} = \frac{k_\beta \hat{\mathbf{R}} - k_\alpha \hat{\mathbf{k}}_i}{|k_\beta \hat{\mathbf{R}} - k_\alpha \hat{\mathbf{k}}_i|} \tag{15}$$

we get rid of the $\hat{\mathbf{R}}$-dependence of the source term in the integral 14 replacing $\mathbf{n}'$ by $\mathbf{n}_{\alpha\beta} = \mathbf{n}_{\alpha\beta}(\hat{\mathbf{R}}, \hat{\mathbf{k}}_i)$ to end up with

$$\mathbf{u}_{s\alpha\beta}^{far}(\mathbf{R},\omega) = \frac{e^{jk_\beta R}}{4\pi R} \mathbf{U}_{0\alpha\beta}(\hat{\mathbf{R}}, \hat{\mathbf{k}}_i, \omega) : \mathbf{\Sigma}_\beta(\hat{\mathbf{R}}, \omega) \int\int\int_{V_c} \gamma_u(\mathbf{R}') e^{-j\mathbf{K}\cdot\mathbf{R}'} \, d^3\mathbf{R}' \tag{16}$$

with the Fourier vector

$$\mathbf{K} = k_\beta \hat{\mathbf{R}} - k_\alpha \hat{\mathbf{k}}_i \tag{17}$$

and

$$\mathbf{U}_{0\alpha\beta}(\hat{\mathbf{R}}, \hat{\mathbf{k}}_i, \omega) = \mathbf{u}_{0\alpha}(\mathbf{n}_{\alpha\beta}, \hat{\mathbf{k}}_i, \omega)\mathbf{n}_{\alpha\beta} \tag{18}$$

The double contraction of the dyad $\mathbf{U}_0$ with the triad $\mathbf{\Sigma}_\beta$ yields a vector

$$\mathbf{U}_{0\alpha\beta} : \mathbf{\Sigma}_\beta = \mathbf{C}_{0\alpha\beta}(\hat{\mathbf{R}}, \hat{\mathbf{k}}_i, \omega) \tag{19}$$

Hence, in shorthand notation equ. 16 reads

$$\mathbf{u}_{s\alpha\beta}^{far}(\mathbf{R},\omega) = \frac{e^{jk_\beta R}}{4\pi R} \mathbf{C}_{0\alpha\beta}(\hat{\mathbf{R}}, \hat{\mathbf{k}}_i, \omega)\, \tilde{\gamma}_u(\mathbf{K}) \tag{20}$$

where $\tilde{\gamma}_u(\mathbf{K})$ denotes the threedimensional Fourier transform of $\gamma_u(\mathbf{R})$. Equ. 20 is our desired far-field inversion scheme, because, due to the knowlegde of $\mathbf{C}_{0\alpha\beta}(\hat{\mathbf{R}}, \hat{\mathbf{k}}_i, \omega)$ from experiments, we can easily compute the Fourier transform of the "visible" singular function from

$$4\pi R e^{-jk_\beta R} \frac{\mathbf{C}_{0\alpha\beta} \cdot \mathbf{u}_{s\alpha\beta}^{far}}{|\mathbf{C}_{0\alpha\beta}|^2} = \tilde{\gamma}_u(\mathbf{K}) \tag{21}$$

and, therefore, a threedimensional Fourier inversion will get us back to the spatial domain. Obviously, the result of the far-field inversion scheme 21 is neither dependent on the incoming mode nor dependent on the observed wave mode, hence, it might be called a mode-matched inversion scheme. From the scalar theory, we know that the time domain counterpart of 21 is a backprojection algorithm [1], as it is the case for far-field SAFT, so the name mode-matched FT-SAFT for equ. 21 seems appropriate, where "FT" indicates that spatial Fourier transforms are involved.

In case we do not account for the compensation of the elastodynmaic nature of the experiment by taking explicitly care of $\mathbf{C}_{0\alpha\beta}(\hat{\mathbf{R}}, \hat{\mathbf{k}}_i, \omega)$ we could instead compute

$$\hat{\mathbf{R}} \cdot \mathbf{u}_{s\alpha P}^{far} = \frac{e^{jk_P R}}{4\pi R} \hat{\mathbf{R}} \cdot \mathbf{C}_{0\alpha P}(\hat{\mathbf{R}}, \hat{\mathbf{k}}_i, \omega)\, \tilde{\gamma}_u(\mathbf{K}) \tag{22}$$

$$\hat{\mathbf{R}} \times \mathbf{u}_{s\alpha S}^{far} = \frac{e^{jk_S R}}{4\pi R} \hat{\mathbf{R}} \times \mathbf{C}_{0\alpha S}(\hat{\mathbf{R}}, \hat{\mathbf{k}}_i, \omega)\, \tilde{\gamma}_u(\mathbf{K}) \tag{23}$$

which would, via Fourier inversion, result in a weighted singular function depending on the selection of the outgoing mode, which, in a real experiment, could be realized for instance by time gating. At least the inversion according to equ. 22 is strictly scalar, and, in fact, it is nothing but a "mode-dependent" P-wave FT-SAFT, i.e. a "scalarized" FT-SAFT, if, for example, α is chosen as P. Fig. 2

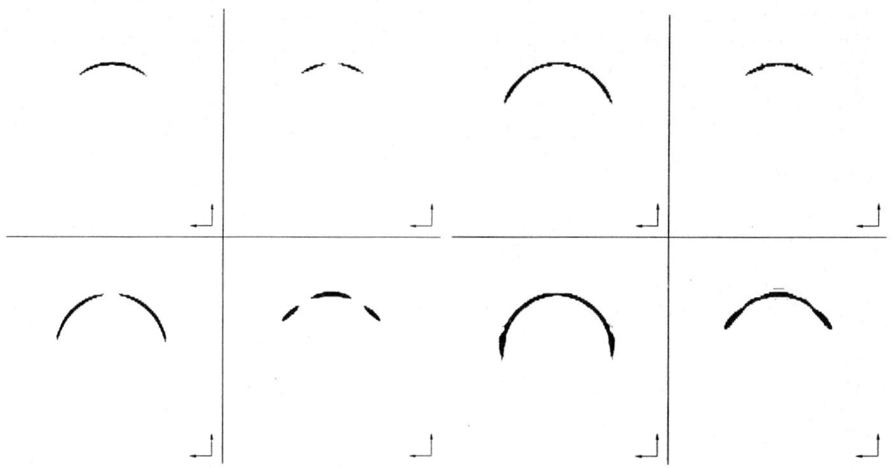

Figure 2: Left: Result of the mode-dependent far-field FT-SAFT inversion as applied to a circular cylinder with stress-free boundary for an incident plane pressure or shear wave coming from the top; top: result of $\hat{\mathbf{R}} \cdot \mathbf{u}_{sPP}^{far}$ and $\hat{\mathbf{R}} \cdot \mathbf{u}_{sSP}^{far}$; bottom: result of $\hat{\mathbf{R}} \times \mathbf{u}_{sPS}^{far}$ and $\hat{\mathbf{R}} \times \mathbf{u}_{sSS}$. Right: Result of the mode-matched far-field FT-SAFT inversion as applied to a circular cylinder with stress-free boundary for an incudent plane pressure or shear wave coming from the top; clockwise: $\alpha\beta = PP, SP, PS, SS$.

shows computer simulations of the output of either inversion scheme 21 or inversion schemes 22, and 23, respectively; the twodimensional example of a circular cylinder with stress-free boundary condition was chosen to compute the scattered field for every $\alpha\beta$-combination, and then, both inversion algorithms have been applied. Obviously, in both cases, the visible part of the singular function is imaged, but, as predicted, the mode-matched algorithm compensates for the angular weighting of the induced surface displacement.

3. ELASTODYNAMIC NEAR-FIELD FAR-FIELD TRANSFORMATION

For measurement surfaces S_M in the near-field of the scatterer we can apply the complete version of the elastodynamic Huygens principle, which, in contrast to equ. 4, also involves the stress tensor $\mathbf{T}$ and a dyadic Green's function $\mathbf{G}$, because no particular boundary conditions are imposed:

$$\mathbf{u}_s(\mathbf{R},\omega) = \iint_{S_c} \{\mathbf{u}(\mathbf{R}',\omega) \cdot [\mathbf{n}' \cdot \mathbf{\Sigma}(\mathbf{R}-\mathbf{R}',\omega)] - [\mathbf{T}(\mathbf{R}',\omega) \cdot \mathbf{n}'] \cdot \mathbf{G}(\mathbf{R}-\mathbf{R}',\omega)\}\, dS' \tag{24}$$

For observation points $\mathbf{R}$ in the far-feld of S_M we compute the β-mode scattering amplitude from equ. 24 as

$$\mathbf{C}_\beta(\hat{\mathbf{R}},\omega) = \frac{1}{4\pi} \iint_{S_M} [\mathbf{u}\mathbf{n}' : \mathbf{\Sigma}_\beta - (\mathbf{T} \cdot \mathbf{n}') \cdot \mathbf{G}_\beta]\, e^{-jk_\beta \hat{\mathbf{R}} \cdot \mathbf{R}'}\, dS' \tag{25}$$

with

$$\mathbf{\Sigma}_P(\hat{\mathbf{R}},\omega) = -\frac{jk_P^3}{\varrho\omega^2}(\lambda \mathbf{I}\hat{\mathbf{R}} + 2\mu \hat{\mathbf{R}}\hat{\mathbf{R}}\hat{\mathbf{R}}) \tag{26}$$

$$\mathbf{\Sigma}_S(\hat{\mathbf{R}},\omega) = -\frac{j\mu k_S^3}{\varrho\omega^2}\left[\hat{\mathbf{R}}\mathbf{I} + (\hat{\mathbf{R}}\mathbf{I})^{213} - 2\hat{\mathbf{R}}\hat{\mathbf{R}}\hat{\mathbf{R}}\right] \tag{27}$$

$$\mathbf{G}_P(\hat{\mathbf{R}},\omega) = \frac{k_P^2}{\varrho\omega^2}\hat{\mathbf{R}}\hat{\mathbf{R}} \tag{28}$$

$$\mathbf{G}_S(\hat{\mathbf{R}},\omega) = \frac{k_S^2}{\varrho\omega^2}(\mathbf{I} - \hat{\mathbf{R}}\hat{\mathbf{R}}) \tag{29}$$

If the data have been produced with an α-mode, we can explicitly account for that adding that index, which yields

$$\mathbf{u}_{s\alpha\beta}^{far} = \frac{e^{jk_\beta R}}{R}\mathbf{C}_{\alpha\beta}(\hat{\mathbf{R}},\omega) \tag{30}$$

where

$$\mathbf{C}_{\alpha\beta}(\hat{\mathbf{R}},\omega) = \frac{1}{4\pi}\iint_{S_M}\left[\mathbf{u}_\alpha\mathbf{n}' : \mathbf{\Sigma}_\beta - (\mathbf{T}_\alpha\cdot\mathbf{n}')\cdot\mathbf{G}_\beta\right] e^{-jk_\beta\hat{\mathbf{R}}\cdot\mathbf{R}'}\,dS' \tag{31}$$

Equ. 30 can then be used as input into the inversion scheme 21.

4. ELASTODYNAMIC BACKPROPAGATION

Similar to the scalar case, we can define a displacement vector holographic field in terms of the elastodynamic Huygens principle introducing complex conjugate Green's tensors, thus accounting for elastodynamic backpropagation. The resulting Porter-Bojarski equation can be integrated with regard to frequency, provided the approximation of physical elastodynamics is introduced, but the subsequent dyadic inversion turns out to be very cumbersome; hence, we prefer the above inversion schemes.

References

[1] G.T. Herman, H.K. Tuy, K.J. Langenberg, P. Sabatier. *Basic Methods of Tomography and Inverse Problems*. Adam Hilger, Bristol 1987

[2] K.J. Langenberg. Wave Motion 11 (1989) 99

[3] V. Schmitz, W. Müller, G. Schäfer. In: *Review of Progress in QNDE*, Plenum Press, New York 1986

[4] K.J. Langenberg, M. Berger, T. Kreutter, K. Mayer, V. Schmitz. NDT International 19 (1986) 177

[5] K. Nagai. Proc. IEEE 72 (1984) 748

[6] T. Kreutter, K.J. Langenberg. In: *Review of Progress in QNDE*, Plenum Press, New York 1990

[7] K. Mayer, R. Marklein, K.J. Langenberg, T. Kreutter. Ultrasonics (1989) (to appear)

[8] A.J. Devaney, G. Beylkin. Ultrasonic Imaging 6 (1984) 181

[9] M. Cheney, J.H. Rose. Inverse Problems 4 (1988) 435

[10] K.J. Langenberg. Proc. 12th World Conf. NDT, Elsevier, Amsterdam 1989

[11] K. Mayer, R. Marklein, K.J. Langenberg, T. Kreutter. In: *Review of Progress in QNDE*, Plenum Press, New York 1990

[12] Y.H. Pao, V, Varatharajulu. J. Acoust. Soc. Am. 59 (1976) 1361

[13] N. Bleistein. Wave Motion 11 (1989) 113

[14] J.D. Achenbach. *Wave Propagation in Elastic Solids*, North-Holland, Amsterdam 1973

Elastic Waves and Ultrasonic Nondestructive Evaluation
S.K. Datta, J.D. Achenbach and Y.S. Rajapakse (Editors)

Scattering of elastic waves by rough cracks

Patricia A. Lewis
Mathematics Department,
University of Manchester,
Manchester,
M13 9PL,
England

1 Introduction

In the theory of the ultrasonic evaluation of metallurgical defects in solid components there arises two important questions:

1. Given a crack of prescribed overall size, roughness and range, what is the probability of detection at a given angle of incidence?

2. What is the relation between the edge of the ultrasonic signal and the geometrical edge of the crack?

In order to begin to answer these questions, it is necessary to provide a rational theory for scattering by rough cracks in elasticity theory.

The most common approach to date has been Kirchhoff Theory (or the Tangent Plane Method). However this method does not correctly model diffraction by the edge of the crack and also neglects shadowing and multiple reflections. As roughness increases these latter effects become increasingly important and so we now consider an exact approach. The theory is an extension of that given in [1] for the problem of an incident SH wave. We show how this may be developed for the more general elastic problem.

2 Statement of the problem

We consider an infinite homogeneous isotropic elastic solid containing a crack ∂D. A small time harmonic disturbance of frequency ω is established and is diffracted by the defect. The displacement vector $\mathbf{u}(P)$ satisfies the following system of equations

$$-\mu\underline{\nabla}\times\underline{\nabla}\times\mathbf{u} \quad + \quad (\lambda+2\mu)\underline{\nabla}(\underline{\nabla}.\mathbf{u}) = -\rho_o\omega^2\mathbf{u}, \tag{1}$$

$$\overset{\nu}{\mathbf{T}} \quad = \quad \underline{0}, \quad P\in\partial D \tag{2}$$

where λ and μ are the Lamé coefficients, $\overset{\nu}{\mathbf{T}}$ is the stress vector on the surface whose normal is ν and ρ_o is the density of the elastic solid in the undisturbed state (see figure 1). The radiation condition, assuming a time factor $e^{-i\omega t}$, is the Sommerfeld condition, namely

$$r^{\frac{1}{2}}\left(\frac{\partial\phi}{\partial r}-iK\phi\right) \quad \rightarrow \quad 0 \qquad r^{\frac{1}{2}}\phi = O(1) \tag{3}$$

$$r^{\frac{1}{2}}\left(\frac{\partial\psi}{\partial r}-ik\psi\right) \quad \rightarrow \quad 0 \qquad r^{\frac{1}{2}}\psi = O(1) \tag{4}$$

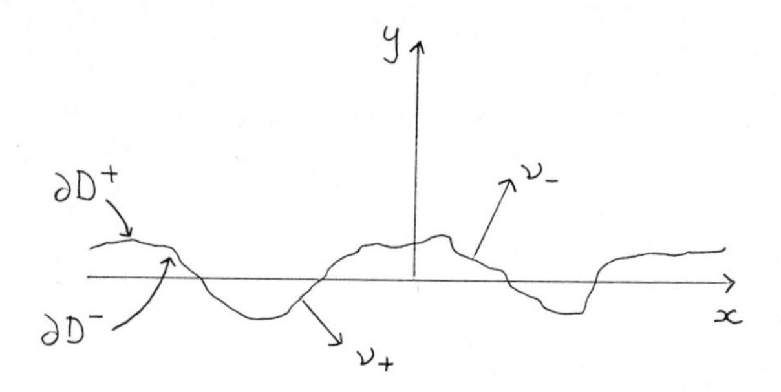

Figure 1: Geometry of the problem

as $r \to \infty$ uniformly in angle θ, where (r, θ) are plane polar coordinates and

$$u_i = \phi_{,i} + \varepsilon_{ij3}\psi_{,j}, \tag{5}$$

and K and k are the wave numbers of the compression and shear waves respectively. Finally we also impose an edge condition i.e. there exists $M > 0$ such that

$$|\mathbf{u}| < M \tag{6}$$

in every neighbourhood of the crack edge. This condition is sufficient to ensure that the edges themselves are not sources of energy, but does not preclude singular behaviour there. If we now write

$$\mathbf{u}(P,t) = [\mathbf{u}^{(o)}(P) + \mathbf{u}^{(s)}(P)]e^{-i\omega t}, \tag{7}$$

where $\mathbf{u}^{(o)}$ is the disturbance which would exist in an unbounded uncracked solid and $\mathbf{u}^{(s)}$ is the field scattered by the defect, the Reciprocal Theorem of Elastodynamics gives

$$u_m^{(s)}(P) = -\int_{\partial D} u_i(q)\Sigma_m^i(P,q)d|q|, \tag{8}$$

where Σ_m^i is calculated using Hooke's Law, namely

$$\Sigma_m^i = \lambda\nu_m U_{k,i}^i + \mu\nu_k(U_{m,k}^i + U_{k,m}^i), \tag{9}$$

and $U_m^i(P,q)$ is our fundamental solution given by

$$U_m^i(P,q) = \delta_{im}\Psi + (\Phi - \Psi)_{,m,i}, \tag{10}$$

where

$$\Phi = -\frac{i}{4}H_o^{(1)}(kR), \qquad \Psi = -\frac{i}{4}H_o^{(1)}(KR), \tag{11}$$

$H_o^{(1)}(kR)$ is the Hankel function of the first kind and $R = |P - q|$.

3 The Crack Green Function

We consider a Green Function of the form

$$H_m^{(l)}(q,P) = -\int_{\partial D} \rho_i^{(l)}(t,q)\Sigma_i^m(t,P)d|t|, \quad q \in \partial D, \quad t \in \partial D. \tag{12}$$

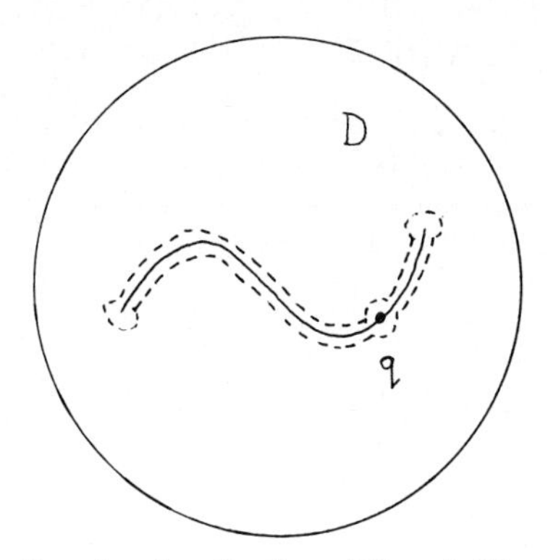

Figure 2: Domain of application of Green's Theorem

We assume that ρ has the special properties which ensure that $H_m^{(l)}$ is a fundamental solution of the elastic wave equation (1) which is *discontinuous across* ∂D, satisfies the edge and radiation conditions (3),(4) and (6), and *corresponds to equal and opposite line forces at* $P = \pm q$. A full discussion of ways of choosing $\rho(t,q)$ for the SH or acoustic problem is given in [1]. It turns out that in the problems considered here it is only necessary to choose ρ_i^m to be proportional to

$$\rho^*(t,q) = \ln |\, x_t - x_q \,| - \ln |\, 1 - x_t x_q + \sqrt{1-x_t^2}\sqrt{1-x_q^2}\,|, \tag{13}$$

where x_t and x_q denote distance measured along a "mean plane" of the crack. The rationale for this choice will be published elsewhere.

4 Derivation of the Integral equation

We apply Green's Theorem to $u_i(p)$ and $H_i^{(l)}(p,q)$ in the domain D exterior to the crack as shown in figure 2. This yields, upon using the boundary and radiation conditions

$$[u_l(q)] \quad - \quad \int_{\partial D} \nu_j k_{ji}^{(l)}(p,q)[u_i(p)]d|p| = \int_{\partial D} \nu_j [H_i^{(l)}(p,q)]\sigma_{ij}^o(p)d|p| \tag{14}$$

for l=1,2 and

$$[u_3(q)] \quad - \quad \int_{\partial D} K(p,q)[u_3(p)]d|p| = \int_{\partial D} \rho(p,q)\frac{\partial u^o}{\partial \nu_p}d|p| \tag{15}$$

where the square parentheses denote jumps across the crack. Equations (14) and (15) are a system of coupled Fredholm Integral equations of the second kind for the in-plane motion (u_1,u_2), and a single Fredholm equation for the anti-plane motion respectively, where

$$\sigma_{ij}^o \quad = \quad \lambda\delta_{ij}u_{k,k}^o + \mu(u_{i,j}^0 + u_{j,i}^0), \tag{16}$$

and

$$K(p,q) \quad = \quad \frac{\partial}{\partial \nu_p}\int_{\partial D} \rho(t,q)\frac{\partial}{\partial \nu_t}H_o^{(1)}(kR)d|t|. \tag{17}$$

Finally in this section, we note that the choice of $\rho_i^{(l)}(t,q)$ is not unique. Any continuous function $\sigma(t,q)$ may be added to $\rho(t,q)$ and $[\mathbf{u}]$ will satisfy the resulting Fredholm equation. It turns out that the simplest choice $\rho_i^{(l)}(t,q)$ is

$$\rho_i^{(l)}(t,q) = -\delta_{il}\frac{2(1-\nu)\mu}{\pi}\rho^*(t,q) \tag{18}$$

5 Model calculation

We consider, as an illustration of the method, an SH wave incident on a class of cracks, whose profiles, $y = f(x)$, have a 1-1 mapping on to the mean plane $y = 0$. Thus we suppose that the gradient of the surface is everywhere finite but not necessarily small. The kernel, after we have mapped our surface integral down onto the mean plane, is

$$\begin{aligned} K(p,q) &= \frac{ik^2}{4}\int_{-1}^{1}\rho(x_t,x_p)\times \\ & \left\{\left[f'(x_t)f'(x_p)+1-\frac{(x+f'(x_p)s)(x+f'(x_t)s)}{x^2+s^2}\right]H_o^{(1)}(k\sqrt{x^2+s^2})\right. \\ &- \left[1+f'(x_t)f'(x_p)-2\frac{(x+f'(x_p)s)(x+f'(x_t)s)}{x^2+s^2}\right]\frac{H_1^{(1)}(k\sqrt{x^2+s^2})}{k\sqrt{x^2+s^2}} \\ &+ \left.\frac{2i}{\pi k^2 x^2}\right\} dx_t \end{aligned} \tag{19}$$

where $x = x_p - x_t$ and $s = f(x_p) - f(x_t)$ and we have scaled our coordinates so that the edges of the crack lie at $(\pm 1, 0)$.

Next we write the Hankel function in the following form

$$H_o^{(1)}(kR) = \ln(kR)\sum_{n=0}^{\infty} a_n(kR)^n + \sum_{n=0}^{\infty} b_n(kR)^n. \tag{20}$$

Thus

$$\begin{aligned} K(p,q) &= a_o(x_p)\int_{-1}^{1}\rho(x_t,x_q)\ln\mid x_t-x_p\mid dx_t \\ &+ a_1(x_p)\int_{-1}^{1}\rho(x_t,x_q)(x_t-x_p)\ln\mid x_t-x_p\mid dx_t \\ &+ a_2(x_p)\int_{-1}^{1}\rho(x_t,x_q)(x_t-x_p)^2\ln\mid x_t-x_p\mid dx_t \\ &+ \int_{-1}^{1}\rho(x_t,x_p)F_o(x_t,x_p)dx_t. \end{aligned} \tag{21}$$

The first three terms can be evaluated explicitly and so it is only necessary to find a way of evaluating the remainder.

The remainder term $F_o(x_t, x_p)$ is a twice continuously differentiable function. This allows us to use various interpolation formulae, including cubic-spline approximations, B-spline interpolants, or Chebychev polynomial approximations. This last method is the highest order approximation and the most efficient, as it allows us to exploit the behaviour of $\rho(x_t, x_q)$ and evaluate the remainder analytically. We write

$$F_o(x_t,x_p) = \sum_{n=o}^{N}\alpha_n(x_p)T_n(x_t), \tag{22}$$

and express the Chebychev polynomials of the first kind, $T_n(x_t)$, in terms of the corresponding polynomials of the second kind, $U_n(x_t)$. It is therefore necessary to evaluate integrals of the form

$$\begin{aligned} I_n(x_q) &= \int_{-1}^{1}\rho(x_t,x_q)U_n(x_t)dx_t \\ &= -\pi\sqrt{1-x_q^2}U_{n+1}(x_q). \end{aligned} \tag{23}$$

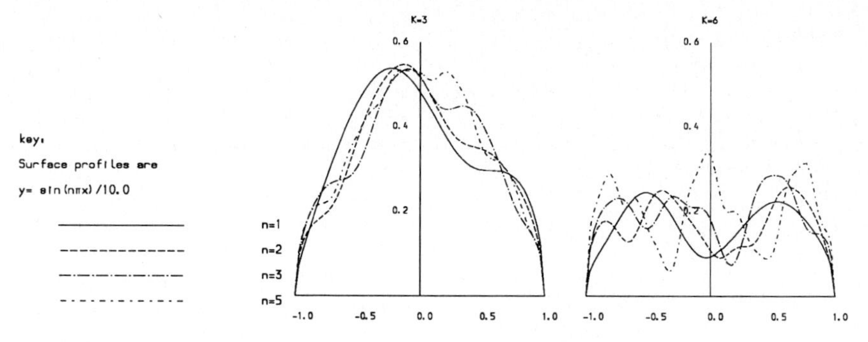

Figure 3: C.O.Ds for sinusoidal surfaces

Finally we introduce

$$[u](x) = \sqrt{1-x^2}\mu(x) \tag{24}$$

so that we can model correctly the behaviour of the C.O.D. at the crack tips. This leads to an equation for μ, which is

$$\sqrt{1-x_q^2}\mu(x_q) - \int_{-1}^{1}\sqrt{1-x_p^2}K(x_p,x_q)dx_p = \int_{-1}^{1}\rho(x_t,x_q)\frac{\partial u^o}{\partial n_t}ds_t \tag{25}$$

The importance of the factor $\sqrt{1-x_q^2}$ in equation (23) now becomes apparent. The kernel in the above integral is non-singular and so therefore all the standard methods of solution are available to us.

6 Results

The results presented here use the simplest method of solution possible, namely the Nystrom Method, whereby we replace the integral by a Gaussian quadrature rule. The first graphs are plots of the C.O.D for normally incident plane waves (k=3.0, 6.0) for surfaces of the form $y = \sin(n\pi x)/10.0$ (n=1,2,3,5). The case k=6.0 clearly shows a periodicity in the C.O.D. corresponding to the periodicity in the surface. The k=3.0 graph shows a different feature of the C.O.D., namely a self-similarity property, whereby features of the n=1 solution repeat themselves, on an increasingly smaller scale, in the n=2,3,5 solutions.

To conclude, we now show an example of an ultrasonic scan of a beam across the surface $y = \sin(3\pi x)/10.0$. This situation corresponds to a crack of width 5mm embedded in a material at a depth of 25mm, with incident wavenumber of k=15. The scattered field from the rough surface is approximately a third that of the corresponding result for a flat plate, the energy being scattered diffusely over a larger area, as would be expected. The horizontal axis here represents distance in the plane of the crack measured in crack lengths.

7 Conclusion

The advantages of the Crack Green Function Method are

1. The theory is exact and therefore, in principle, there are no restrictions on k and the surface heights and gradients.

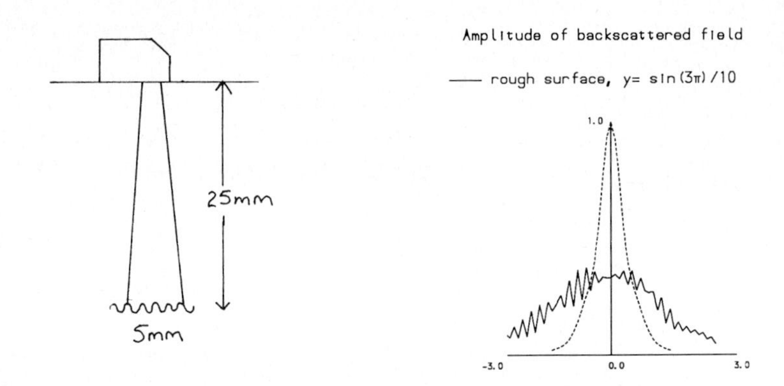

Figure 4: A simulation of an ultrasonic scan across a rough surface

2. The method explicitly displays the known singularity in the C.O.D. Other methods, such as the finite element or boundary element methods, require the use of special elements to correctly model this singular behaviour.

3.There is an abundance of suitable numerical schemes for solving Fredholm Integral Equations of the form (25)

4. The integral equation provides an interpolation formula, thus it is possible to calculate the C.O.D. at any point, irrespective of the solution method.

In conclusion, we have an exact formulation of the scattering problem. This allows us to begin to answer the questions posed in the introduction in two ways:

1. We may sample the scattering from an ensemble of rough surfaces by numerically solving our integral equation for each realization and thereby gain information about the statistics of the scattered field, or

2. We may try to gain statistical knowledge of the scattered field directly, by deriving integral equations for certain moments of the solution (25). But this approach leads to restrictions on the r.m.s. heights and gradients of the types of surface we may examine.

References

[1] G.R. Wickham *Integral equations for boundary value problems exterior to open arcs and surfaces.* in Treatment of Integral Equations by Numerical Methods, Ed C. T. H. Baker et. al., Academic Press (1982)

Elastic Waves and Ultrasonic Nondestructive Evaluation
S.K. Datta, J.D. Achenbach and Y.S. Rajapakse (Editors)
Elsevier Science Publishers B.V. (North-Holland), 1990

A MODEL FOR THE EFFECTS OF DEFECT SURFACE ROUGHNESS ON ULTRASONIC DETECTION AND SIZING

J. A. Ogilvy
Theoretical Physics Division,
Harwell Laboratory,
Didcot,
Oxon, OX11 0RA, U.K.

A model for studying the effects of defect surface roughness on ultrasonic signals is described. Results from the model are presented, showing how roughness can enhance or reduce detectability, depending on the level of defect mis-orientation.

1 INTRODUCTION

When an ultrasonic wave impinges onto a rough defect the distribution of scattered and diffracted energy is affected by the roughness. The narrowly scattered coherent field is reduced in strength and energy is scattered into a more widely spread diffuse field. This redistribution of scattered energy affects ultrasonic inspection techniques, where the capability of any ultrasonic technique to detect and size defects is influenced by the nature of the wave-defect interactions.

A theoretical model has been developed for studying the effects of defect surface roughness on ultrasonic inspection. The model is currently acoustic, so that mode-conversion effects are not included, and is based on Kirchhoff theory, thus limiting the levels of defect roughness which may be considered. However, the model has been used to predict changes in defect detectability which result from roughness on the faces of defects, and comparison with experiment shows good general agreement. The model is briefly described here and some predictions are presented.

2 MODELLING DEFECT ROUGHNESS

Many real defects of interest in ultrasonic Non-Destructive Testing are rough. The form of this roughness depends on the growth mechanism for the defect and the material in which it is situated. In general, however, the roughness is stochastic in nature such that the defect profiles may be predicted only in a statistical sense, with parameters such as r.m.s. height and correlation length being used to describe classes of defects. Individual defect surfaces within these classes are termed *realisations* and represent the many different defect profiles which may arise within a set of defects such as, for example, fatigue cracks. The model therefore uses statistical techniques to generate, numerically, sets of random but correlated data taken to represent defect surface profiles. The surface is discretised in two perpendicular directions, and surface points are represented by the locations $x_i = i\Delta x$ and $y_j = j\Delta y$, where Δx and Δy are the discretisation intervals in the two directions. Each surface is described in terms of its deviation from a smooth, planar surface. If this deviation, termed the 'height', is represented as a discrete random variable h_{ij}, where this represents the surface height at the point (x_i, y_j) then methods of time-series analysis [1] show that this may be represented by:

$$h_{ij} = \sum_{pq} w_{i+p,j+q} u_{pq} \tag{1}$$

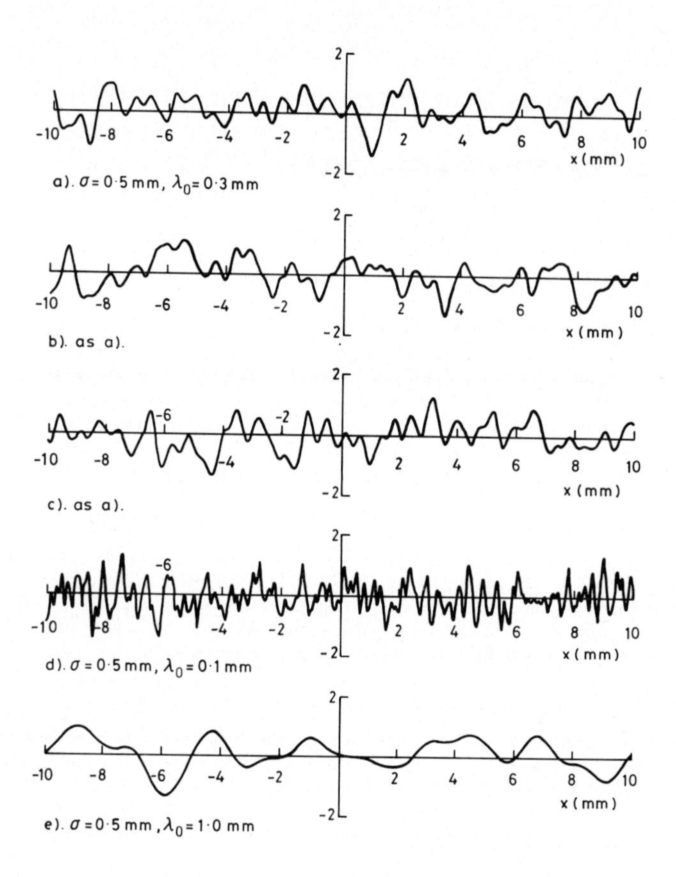

Figure 1: Profiles of numerically generated surfaces, showing the random but correlated nature of the surfaces.

where the u_{pq} are random, *uncorrelated* data (i.e. white noise) and the w_{lm} are weights chosen to smooth, or correlate, the initial data. The choice of weights determines the correlation function for the data. In this model the initial data and weights are chosen such that the generated surfaces possess Gaussian height distributions and correlation functions [2,3]. Evidence that real defects possess these statistics is conflicting, and depends on the types of defects considered [4]. In general this model is realistic for defects with roughness formed by a large number of independent processes.

Figure 1 shows a set of surface realisations generated using this process. The surfaces are randomly rough, in the sense that the surface profile cannot be predicted, but are correlated to give some length scale to the height variation.

3 KIRCHHOFF THEORY

Acoustic Kirchhoff theory is used to calculate the field scattered by the numerically generated randomly rough surfaces. This theory gives the unknown field *on* the scattering surface, relating this to the known incident field through local surface properties such as height and gradient. Each surface point is taken to scatter as though it were part of an infinite plane surface, parallel to the

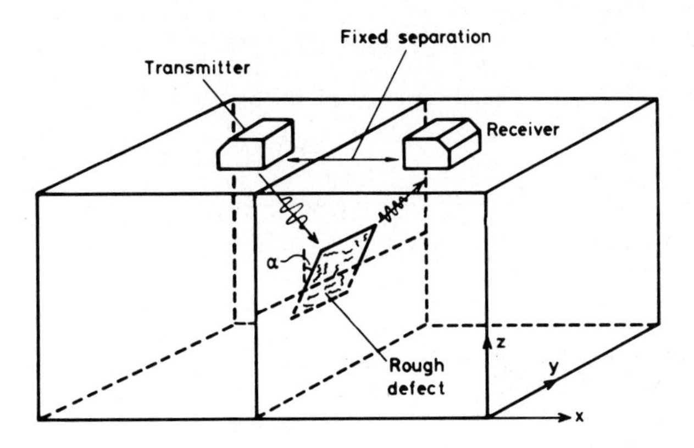

Figure 2: Three-dimensional model geometry.

local surface tangent at that point [5]. The approximation is therefore exact for infinite smooth surfaces but remains good even for quite rough surfaces. The Helmholtz scattering formula is then used, in conjunction with Kirchhoff theory, to give the far field scattered wave in terms of an integral over the rough surface [2,3,6].

4 THE MODEL GEOMETRY

The model geometry is shown in figure 2. A rectangular defect of arbitrary orientation is at some depth below the surface of an inspection block. The defect surface is statistically rough, in the sense described above. The transmitter and receiver positions are specified, as are details of the transducers, such as crystal size, beam centre angle and pulse shape along the beam axis. The transducers are modelled as simple piston sources, with the far field amplitude given by a Bessel function profile [3]. For each transmitter and receiver position the scattered pulse is calculated, as outlined in the next section. By specifying single locations of the transducers, or a scan direction and an increment, A, B or C scans may be modelled.

5 CALCULATING THE SCATTERED SIGNAL

For each transmitter and receiver location the scattered field is calculated by first dividing the defect face into small, planar, rectangular integration regions, or patches. Each patch is taken to be parallel to the local surface tangent at the mid-point of the patch. If the patches are sufficiently small that that the incident field amplitude may be assumed constant across each patch, analytical integration gives the contribution to the scattered field of each small area. By combining information on the transmitter profile, the receptivity of the receiver and the incident pulse then the total scattered field is determined by summation of the contributions from all the patches. The details of the process can be found in reference [3].

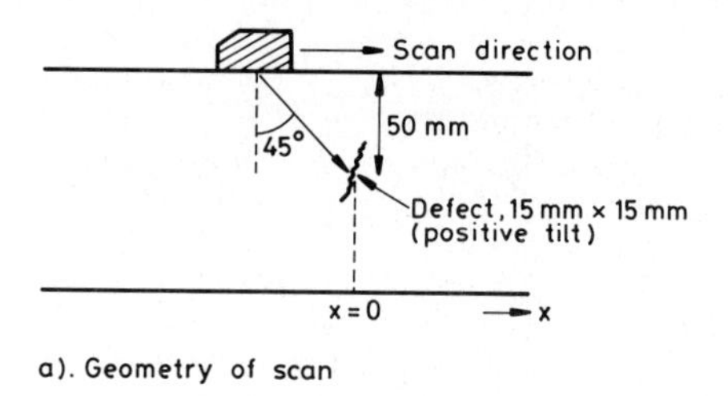

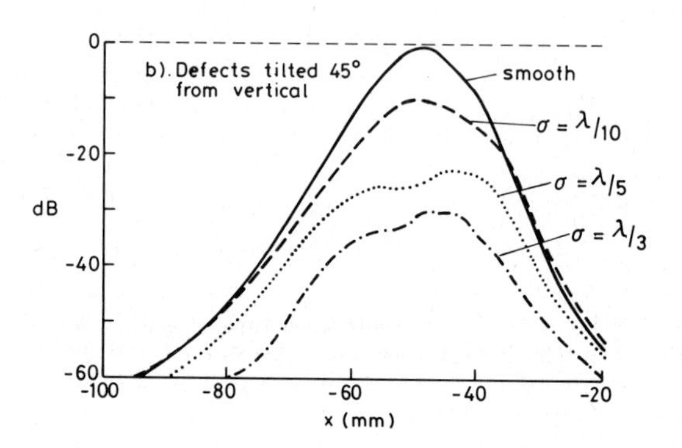

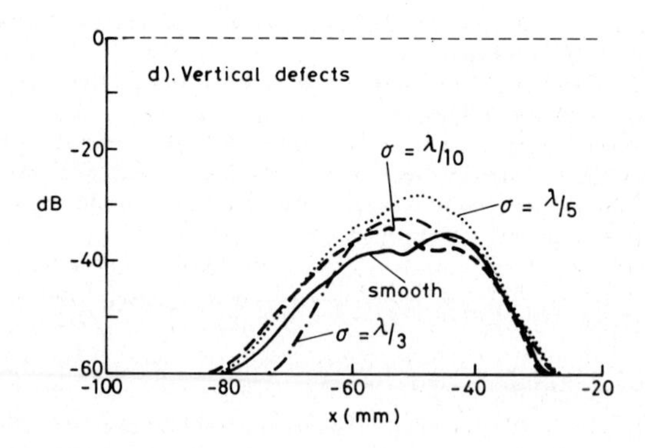

Figure 3: Predicted scan patterns for Tandem inspection of defects of various orientations and levels of roughness. The effects of roughness depend on the effective level of defect mis-orientation.

6 PREDICTIONS OF THE MODEL

The model gives scattered pulse shapes or, using the peak amplitude of these pulses, amplitude curves for specified scans. A range of inspection geometries may be modelled, including Time-of-Flight, pulse-echo or Tandem geometries. Sample results for a pulse-echo geometry are shown in figure 3. Peak signal amplitudes are shown as a function of scan position, for three defect orientations and four levels of roughness, for each orientation. These results indicate that the effects of roughness depend on the inspection geometry: for well-oriented defects (i.e. those oriented to give a specular signal from smooth defects) roughness decreases detectability through loss of the strong specular signal. For defects of significant mis-orientation (i.e. those for which detection of smooth defects is through off-specular signals) roughness can enhance detection, because of the presence of the widely spread diffuse field. For defects of intermediate orientation, roughness usually has little effect on detectability.

These results can be summarised by examining the effects of a given level of roughness on defect detectability, for a range of defect mis-orientations. Such information is presented in figure 4. Roughness enhances detectability, beyond some level of mis-orientation, dependent on the defect size, level of roughness and inspection geometry. For the level of roughness shown here (r.m.s. height $\sim \lambda/5$) mis-orientation beyond around $10° - 20°$ leads to increased signal amplitudes. One of the important parameters here is the defect size: the larger the defect the narrower the spread of the specular signal from smooth defects and hence the smaller the degree of mis-orientation beyond which detection is outside the strong signal. The diffuse field from rough defects therefore overtakes the signal from smooth defects at smaller mis-orientations, the larger the defect.

7 CONCLUSIONS

A model simulating many aspects of ultrasonic inspection of buried, rough defects has been outlined here. The model relies on numerical simulation techniques for calculating signal amplitudes and, as such, is computationally intensive, especially when time-dependent calculations are required. The model uses acoustic Kirchhoff theory and cannot therefore consider mode-conversion effects or very rough defects. However, the model has been shown to give good agreement with experimental data [3], and can be used in a range of situations to predict the effects of roughness on ultrasonic inspection techniques. In general roughness is shown to affect detectability to an extent strongly dependent on the inspection geometry and defect size. For well-oriented defects signal amplitudes can be reduced by roughness, through around $20 - 30$dB. Conversely, roughness on the faces of defects mis-oriented by angles greater than around $10 - 20°$ may lead to signal amplitudes which are enhanced by around $10 - 20$dB.

These conclusions, however, are dependent on parameters such as level of roughness, inspection geometry, defect depth and size and, perhaps most importantly, inspection *frequency*. This last parameter arises because it is the effective roughness, relative to the incident wavelength, which is crucial in determining whether a defect behaves as a rough or smooth scatterer. The higher the inspection frequency the smaller the incident wavelength and hence the greater the effective defect roughness. For inspection systems where roughness is thought to be advantageous (such as where defects are known to be mis-oriented and it is required to enhance the detected signal amplitude) then increasing the inspection frequency may be helpful (provided other effects such as increased attenuation do not negate the beneficial effects). Conversely, if roughness is reducing detection signals through loss of the strong specular signal then a reduction in inspection frequency may be helpful, to reduce the effective roughness and hence increase the proportion of energy which is scattered into and around the specular direction.

ACKNOWLEDGEMENT

This work has been funded by the UK Fast Reactor Programme.

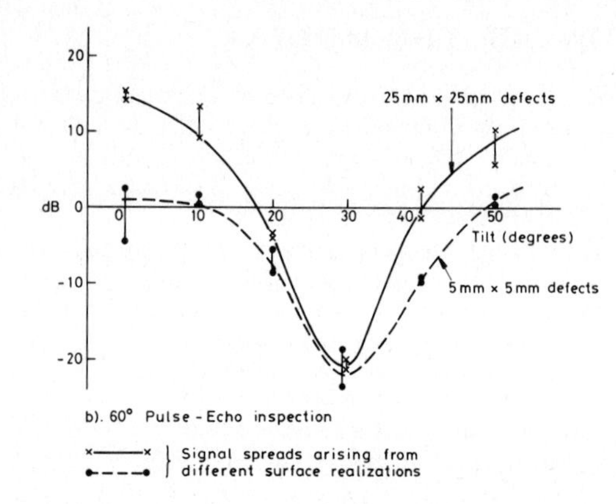

Figure 4: Effects of defect mis-orientation on rough defect detectability, for pulse-echo inspection. Beyond around 10° – 20° of mis-orientation roughness can enhance detectability. This conclusion, however, depends on the defect size, the level of roughness and the inspection geometry.

REFERENCES

[1] Priestley, M.B. *Spectral analysis and time series, vol I: univariate series.* (Academic, London, 1981.)

[2] Ogilvy, J.A. Computer simulation of acoustic wave scattering from rough surfaces. *J. Phys. D: Applied Physics*, 1988, **21**, 260–277.

[3] Ogilvy, J.A. Model for the ultrasonic inspection of rough defects. *Ultrasonics*, 1989, **27** 69–79.

[4] Thomas, T.R. (Ed) *Rough surfaces.* (Longman, London, 1982.)

[5] Beckmann P and Spizzichino A. *The scattering of electromagnetic waves from rough surfaces.* (Pergamon, Oxford, 1963.)

[6] Pao Y-H and Mow C-C. *Diffraction of elastic waves and dynamic stress concentrations.* (Hilger, London, 1973.)

Elastic Waves and Ultrasonic Nondestructive Evaluation
S.K. Datta, J.D. Achenbach and Y.S. Rajapakse (Editors)

THE NULL FIELD AND T MATRIX METHODS FOR ELASTODYNAMIC SCATTERING PROBLEMS: CRACKS AND OTHER TYPES OF FLAWS

Peter OLSSON and Anders BOSTRÖM

Division of Mechanics
Chalmers University of Technology
Göteborg, Sweden

The null field approach (by which we mean an approach involving explicit use of the extinction theorem), which originated in the work on electromagnetics by Waterman, and the T matrix method (by which we mean a certain building block method for solving multiple scattering problems) are two methods which have evolved together, but which really are logically distinct methods. They have been extended by many researchers to handle a number of different scattering problems also in elastodynamics and acoustics. In the present paper we review some of the results in elastodynamics, with particular emphasis on recent results on scattering by planar and nonplanar cracks with various conditions of contact.

1. INTRODUCTION

It is a curious fact of history that the theory of elastodynamics owes much of its mathematical development during the nineteenth century to the interest in the theory of propagation of light, i.e. electromagnetics. More recently, elastodynamics has benefitted from investigations into electromagnetic theory in another way, in that methods originally developed for solving electromagnetic scattering problems have been imported successfully into the domain of elastic wave scattering. An example of this is provided by the case of the null field approach (in the following often abbreviated by NFA), a method originally developed by Waterman for solving time harmonic electromagnetic scattering problems [1]. Soon after the invention, it was realized by Waterman [2,3] and others [4] that the NFA could be extended to cover also acoustic and elastodynamic scattering problems. But it also turned out that the NFA could be used for other purposes than solving scattering problems. This includes problems like that of determining the dispersion relations for surface waves on corrugated half spaces [5] or along noncircular bore holes [6], but also purely elastostatic problems [7]. Another type of application is radition of water waves [8].

In parallel with the development of the NFA, and in close contact with the NFA, another method has evolved which allows an effective treatment of many multiple scattering problems. This is the T matrix method (in the following often abbreviated TMM), of which some original works in elastodynamics can be found in Refs. [9,10]. The reader should immediately be warned that in the litterature the names NFA and TMM are used essentially interchangeably, and that the distinction between the two designations which is argued in the present paper is as yet nonstandard. (In this context we should perhaps point out that other names have been given to the NFA. Watermans own suggestion, "the extended boundary condition" [11], is one such perhaps not very fortunate choice of designation.) The reason for the terminological distinction which we make here is that we are in fact dealing with two logically distinct methods. In the following sections we will describe this distinction by looking at how the methods apply to the canonical elastodynamic single and multiple scattering problems, but let us first state the use of the designations NFA and TMM which we think is the most useful:

By the *null field approach* we mean

> A method for solving various wave propagation problems (including but not limited to the problem of finding the transition matrix of a single scatterer) which makes explicit use of the extinction part of the integral representation theorem and utilizes global rather than local basis functions.

By the *T matrix method* we mean

> An approach to multiple scattering problems where the problems are solved in terms of the transition matrices of the individual scatterers (and in terms of reflection and transmission coefficients of planar surfaces, if any such are present, etc.). It is a "building block" approach, where the individual T matrices could have been obtained by any method, and are assumed to be known.

The aim of the present paper is twofold. The first aim is to advocate this terminological distinction. The second is to review some recent results on how the two methods can be applied to elastodynamic scattering by a class of flaws which is of great importance for applications in ultrasonic nondestructive testing. These results, due to ourselves and some other researchers, are on scattering by cracklike flaws, i.e. flaws which are essentially two dimensional, in the sense that they do not extend over an appreciable volume. *Bona fide* open cracks, both pennyshaped and nonplanar are among the ones that have been treated, but also perfectly lubricated fluidfilled ones, as well as flaws with spring contact and dissipative forces between the faces of the flaw.

2. THE NULL FIELD APPROACH

Let us to begin by considering the canonical single scattering problem in elastodynamics. We thus consider an inhomogeneity (of finite volume) bounded by a closed surface S residing within a linear, homogeneous, and isotropic elastic solid. The outward pointing unit normal of S we denote by $\hat{n}$. By subscripts + and –, we denote limits taken from the positive and negative side of S, as defined by the direction of the outward pointing normal, respectively. On S some boundary conditions are assumed to hold, but for the purposes of the present discussion we will not specify them. The nature of the inhomogeneity is likewise left unspecified for the time being.

The direct scattering problem is the following. Given some incident, prescribed displacement field $\mathbf{u}^i$ of circular frequency ω (the time factor $\exp(-i\omega t)$ is suppressed throughout), we want to compute the scattered field $\mathbf{u}^s$ caused by the presence of the inhomogeneity. Choosing an origin O within the volume bounded by S, such that the sphere centered at O and just touching S has a nonvanishing radius, we expand the incident and scattered fields in terms of regular and outgoing spherical vector waves with respect to this origin as

$$\mathbf{u}^i = \sum_n a_n \operatorname{Re}\psi_n \tag{1}$$

$$\mathbf{u}^s = \sum_n f_n \psi_n \tag{2}$$

For a linear problem, there of course exists a linear operator which maps the incident field into the scattered, and an explicit matrix represtation of this linear operator in the particular choice of bases in (1) and (2) above is the transition matrix, or T matrix, $T = \{ T_{nn'} \}$ which thus satisfies

$$f = T\, a \tag{3}$$

Here $f = \{f_n\}$ and $a = \{a_n\}$ are column vectors formed of the expansion coefficients. The transition matrix is related to the scattering matrix, the S matrix, in a simple manner, viz. $S = 1 + 2\,T$.

For the present problem, the null field approach can be utilized to compute the transition matrix in the following manner. Starting from the outer integral representation [4]

$$\mathbf{u}^i + \frac{k_s}{\mu} \int_S [\, \mathbf{u}_+ \cdot (\hat{n} \cdot \overline{\Sigma}\,) - \mathbf{t}_+ \cdot \overleftrightarrow{\mathbf{G}}\,] dS = \begin{cases} \mathbf{u} & \text{outside of S} \\ 0 & \text{inside of S} \end{cases} \tag{4}$$

and inserting expansions of all field quantities except the surface fields, we obtain expressions for the f_n and the a_n which give these coefficients as surface integrals over S of the unknown surface fields in the limit from the outside of the scatterer.

It should be noted that the expression for the f_n is derived from the first line of the integral representation, and that it is unproblematic to got to the limit where the volume inside S vanishes as far as this is concerned. On the other hand the expression for the a_n is derived from the second line of (4), called the extinction theorem, and this derivation rests crucially on the assumption that an inscribed sphere of nonvanishing radius exists. This is the main problem when trying to use the null field approach in solving scattering problems involving cracks and cracklike flaws. Below we discuss how to overcome this difficulty.

For a scatterer of finite volume, the way to obtain the transition matrix from the expressions for f_n and a_n is relatively straightforward. Information about the scatterer (in the form of, e.g., an equation of motion or an inner integral representation) is used together with the boundary conditions, and the appropriate surface fields are approximated in a complete set of functions on the surface of the scatterer. The relations obtained from this information and from the expressions for $\{f_n\}$ and $\{a_n\}$ are then used to obtain matrix equations. These are solved for $\{f_n\}$ in terms of $\{a_n\}$, and the coefficients of the linear relation obtained are thus the elements of the transition matrix. In practice of course this can only be done up to a finite truncation. The mathematical questions of convergence, error bounds, etc. for this procedure are difficult. Some results are given in Refs. [11–16].

Examples of items in the library of transition matrices which have been obtained by essentially this approach include a cavity [3], a welded elastic inclusion [3], a "smooth" elastic inclusion [17], an elastic inclusion surrounded by a thin interface layer [18], an incompressible elastic inclusion [19], a welded rigid immovable inclusion [20], and a "smooth" rigid immovable inclusion [21].

The limitations on the above procedure are mainly the following. The surface of the inclusion should not be too "unspherical" in the sense that there should not be "too many" points of inflection on the surface, and that the ratio between the maximum and minimum radii should not be greater than roughly 3. There are exceptions to the latter limitation, see below. In numerical applications only rotationally symmetric scatterers have been treated in elastodynamics, but in acoustics and electromagnetics examples of nonrotationally symmetric scatterers have been published [22,23]. Of course, the scattererers can easily by placed in nonrotationally symmetric configurations [10], and the incident waves do of course not have to be incident along the symmetry axis of the scatterer. The more important limitation is the one in frequency: The method in practice only converges for low to intermediate frequencies.

In general the above method is efficient and attractive to use in the cases where it works. It is therefore of interest to see if it is possible to circumvent the difficulty with the inscribed sphere of nonvanishing radius, or, in practice, the limitation on the ratio between the maximum and minimum radii. To see how this can be done, let us consider the case where the scatterer does not occupy any volume, but rather resides on an oriented open surface S_c with edge C. The boundary conditions we take to be those of a spring contact model, i.e.

$$\mathbf{t}_+ = \mathbf{t}_- \qquad \text{on } S_c \qquad (5)$$

$$\mathbf{u}_+ - \mathbf{u}_- = \frac{1}{\mu k_s}(\alpha\,\hat{n}\hat{n}\cdot\mathbf{t}_- - \beta\,\hat{n}\times(\hat{n}\times\mathbf{t}_-)) \qquad \text{on } S_c \qquad (6)$$

The model incorporates dissipative effects, since α and β can have imaginary parts. In general α and β can be frequency dependent. The important limiting cases of an open crack ($|\alpha| \to \infty$, $|\beta| \to \infty$) and a fluidfilled, perfectly lubricated crack ($\alpha = 0$, $|\beta| \to \infty$) are also included.

Since S_c is open and does not bound any volume, we cannot directly use the NFA. But if we form a closed surface by adjoining to S_c a fictitious surface S_w (subscript w for "welded") which does not represent any material boundary, and over which welded contact boundary conditions apply, we can form a closed surface $S_c + S_w$, bounding a scatterer of finite volume. In the cases of open or perfectly lubricated cracks, the edge conditions must be used to form surface fields which are preferably continuous (at the very least bounded) [24,25] to get convergence, but for the generic spring contact case this is not necessary [26]. In Table 1 the importance of using the edge conditions in farfield computations for an open crack is demonstrated. There l_{max} is the truncation in the crucial index, determining the matrix sizes. What the table shows is that if the surface fields are not constructed so that they are at least bounded, we have no numerical convergence. For bounded but discontinuous surface fields, the convergence of the total scattering cross section is essentially as one over the truncation squared, and for continuous surface fields it is as one over the third power of the truncation [24].

Surface fields:	Continuous	Bounded	Unbounded
l_{max}	$l_{max}^3 \times \lvert\Delta\sigma\rvert$	$l_{max}^2 \times \lvert\Delta\sigma\rvert$	$10^3 \times \lvert\Delta\sigma\rvert$
20	2.4	2.5	1.9
22	2.1	2.4	1.9
24	2.4	2.3	1.9
26	2.3	2.3	1.8
28	2.4	2.3	1.7
30	2.3	2.3	1.6

Table 1. Numerical rate of convergence of the total cross section σ at $k_s a = 8$ for hemispherical capshaped crack. P wave incident along axis of symmetry. $\Delta\sigma = \sigma_{l_{max}} - \sigma_{l_{max}-2}$.

The effects of the crack not being planar are demonstrated clearly in the computations of Ref. [24], and in Ref. [25] the possibility of distinguishing between an open and a perfectly lubricated crack is demonstrated in time domain computations. Another application is presented in [27], where the dynamical stress intensity factors for nonplanar cracks, both open and fluidfilled are computed. The question of how weak the interfacial forces in the spring contact model should be to give results which do not differ appreciably from those of an open crack is discussed in [26].

Finally we should say something about the exception to the requirement on the ratio between radii stated above. The exception is the case of an ellipsoid. For an oblate spheroidal cavity it is possible to go down to an oblateness where the quotient between the greater and the smaller axis is 5 (at $k_s R_{max} = 5$, say). In Ref. [28] Peterson discusses how one can, by deleting some of the numerically troublesome parts which integrate analytically to zero in the computation of the relevant matrices, go down even further in oblateness. The approach is essentially the same as that used by Burden [6]. Peterson is able to get convergence even for the case where $R_{max} / R_{min} = 1000$ for the oblate spheroidal cavity. In fact, it should be possible to handle the case of an elliptic crack by a similar, analytical computation, cf. the results for the acoustic and electromagnetic cases [22,23].

3. THE T MATRIX METHOD

As stressed in the introduction, by the T matrix method we mean a building block method for multiple scattering problems. Here each inhomogeneity is completely characterized by its T matrix and the distances between the inhomogeneities enter via the translational properties of the wave functions that the T matrices are referred to.

Consider first the canonical case with two inhomogeneities. The total T matrix is then given by [9]

$$T = R(\mathbf{d}_1)\, T^1\, [1 - S(\mathbf{d})\, T^2\, S(-\mathbf{d})\, T^1]^{-1}\, [R(-\mathbf{d}_1) + S(\mathbf{d})\, T^2\, R(-\mathbf{d}_2)]$$
$$+ R(\mathbf{d}_2)\, T^2\, [1 - S(-\mathbf{d})\, T^1\, S(\mathbf{d})\, T^2]^{-1}\, [R(-\mathbf{d}_2) + S(-\mathbf{d})\, T^1\, R(-\mathbf{d}_1)] \tag{7}$$

where T^1 and T^2 are the individual T matrices of the two inhomogeneities. T is here referred to an arbitrary master origin, $\mathbf{d}_1$ and $\mathbf{d}_2$ are the vectors to the origin inside the inhomogeneities that T^1 and T^2 are referred to, and $\mathbf{d} = \mathbf{d}_2 - \mathbf{d}_1$. R and S are matrices describing the translational properties of the regular and outgoing spherical waves, respectively.

From a systematic point of view the form of T just given is excellent. But computationally this form is rather awkward as the truncations needed in a numerical computation are mainly dependent on the radius of the circumscribing spheres which might be much bigger for T than for T^1 and T^2. It is then better to use an approach where no master origin is used [28]. Instead the incident wave is expanded around both the origins inside the scatterers (as in Eq. (1)) and the scattered field is likewise expanded as

$$\mathbf{u}^s = \sum_n f_n^1\, \psi_n(\mathbf{r}_1) + \sum_n f_n^2\, \psi_n(\mathbf{r}_2) \tag{8}$$

where the expansion coefficients for each origin is

$$f^1 = T^1\, [1 - S(\mathbf{d})\, T^2\, S(-\mathbf{d})\, T^1]^{-1}\, [a^1 + S(\mathbf{d})\, T^2\, a^2] \tag{9}$$

$$f^2 = T^2\, [1 - S(-\mathbf{d})\, T^1\, S(\mathbf{d})\, T^2]^{-1}\, [a^2 + S(-\mathbf{d})\, T^1\, a^1] \tag{10}$$

In this way the needed truncations are only dictated by the ones needed to accurately give T^1 and T^2 but is independent of the separation of the two inhomogeneities.

To consider more than two inhomogeneities is a straightforward generalization of the above [9,29]. Another type of generalization is to look at a bounded inhomogeneity in a half-space or layered structure [10,30]. The bounded inhomogeneity is still characterized by its T matrix, whereas the interfaces are characterized by their reflection and transmission operators.

The T matrices for individual inhomogeneities that are needed within the T matrix method have usually been computed by the null field approach. For spheres separation-of-variables has of course been used. Other methods are also possible and an example is given by Peterson [28], who has computed the T matrix of the penny-shaped crack using an integral equation method. The inhomogeneities are generally rotationally symmetric but by applying a rotation matrix it is possible to consider an arbitrary orientation of the inhomogeneity, e.g., in a half-space [10].

REFERENCES

[1] Waterman, P.C., Proc. IEEE **53** (1965) 805.
[2] Waterman, P.C., J. Acoust. Soc. Am. **45** (1969) 1417.
[3] Waterman, P.C., J. Acoust. Soc. Am. **60** (1976) 567.
[4] Varatharajulu, V. and Pao, Y.-H., J. Acoust. Soc. Am. **60** (1976) 560.
[5] Boström, A., J. Acoust. Soc. Am. **85** (1989) 1549.
[6] Burden, A.D., Wave Motion **7** (1985) 153.
[7] Olsson, P., Applied Scientific Research **42** (1985) 131.
[8] Martin, P.A., J. Fluid Mech. **113** (1981) 315.
[9] Boström, A., J. Acoust. Soc. Am. **67** (1980) 399.
[10] Boström, A. and Kristensson, G., Wave Motion **2** (1980) 335.
[11] Waterman, P.C., Wave Motion **5** (1983) 273.
[12] Wall, D.J.N., Methods of overcoming numerical instabilities associated with the T-matrix method, in: Varadan, V.K. and V.V. Varadan, V.V., (eds.), *Acoustic, Electromagnetic and Elastic Wave Scattering – Focus on the T-Matrix Approach*, (Pergamon Press, New York, 1980) pp. 269–286.
[13] Martin, P.A., Wave Motion **4** (1981) 391.
[14] Colton, D. and Kress, R., Q. Journal Mech. Appl. Math. **36** (1983) 87.
[15] Ramm, A.G., J. Math. Phys. **23** (1982) 1123.
[16] Kristensson, G., Ramm, A.G. and Ström, S., J. Math, Phys. **24** (1983) 2619.
[17] Boström, A., J. Acoust. Soc. Am. **67** (1980) 1904.
[18] Olsson, P., Datta, S.K. and Boström, A., Elastodynamic scattering from inclusions surrounded by thin interface layers, in print.
[19] Olsson, P., Appl. Sci. Res. **44** (1987) 313.
[20] Olsson, P., Wave Motion **7** (1985) 421.
[21] Olsson, P., J. Acoust. Soc. Am. **79** (1986) 1237.
[22] Kristensson, G., J. Sound Vib. **103** (1985) 487.
[23] Björkberg, J. and Kristensson, G., Can. J. Phys. **65** (1987) 723.
[24] Boström, A. and Olsson, P., Wave Motion **9** (1987) 61.
[25] Olsson, P., J. Nondestructive Eval. **5** (1986) 161.
[26] Olsson, P., Elastodynamic scattering from pennyshaped and nonplanar cracklike flaws with dissipation and restoring forces, in print.
[27] Olsson, P. and Boström, A., Dynamic stress intensity factors for 3D non-planar cracks, in: *Elastic Wave Propagation*, McCarthy, M.F. and Hayes, M.A., (eds.), (North-Holland, Amsterdam, 1989) pp. 399–404 .
[28] Peterson, L., thesis, Institute of Theoretical Physics, Göteborg, Sweden (1989).
[29] Peterson, B. and Ström, S., Phys. Rev. D **8** (1973) 3661.
[30] Boström, A. and Karlsson, A., Geophys. J. R. astr. Soc. **89** (1987) 527.

Elastic Waves and Ultrasonic Nondestructive Evaluation
S.K. Datta, J.D. Achenbach and Y.S. Rajapakse (Editors)

SOLUTION OF THE INVERSE PROBLEM IN ACOUSTIC SCATTERING BY USE OF AN ARTIFICIAL INTELLIGENCE METHOD

Gérard QUENTIN and Alain CAND

Groupe de Physique des solides
Université Paris 7 - Tour 23 - 2, place Jussieu
75251 Paris Cedex 05 - France

We present , in this paper, the application of an artificial intelligence method, the so-called A* algorithm, to solve the inverse problem of scattering by elastic homogeneous cylinders. The high frequency domain case is studied using a rays theory. The low frequency domain case is introduced and an experimental method for identification of the resonances of cylindrical scatterers is presented.

1. INTRODUCTION

The past decades have seen the development of many technics for inverse problem solving [1]. The majority of them uses mathematical inversion . Few others use Trial and Error Methods. The main advantage of these latters are that the mathematics needed are only the ones used for solve the forward problem. Nethertheless some drawbacks arise when these methods are used : at first, the difficulty to assess the unicity of the solution and secondly, the limits of computing time induced by the great amount of forward problems to solve.

The method presented here belongs to the class of trial and error methods. It is derived from a general basic artificial intelligence method, the A* algorithm, applied to the acoustical scattering by cylindrical elastic bodies immersed in a fluid inverse problem.

2. THE A* ALGORITHM

The A* algorithm was first introduced by Hart, Nilsson and Raphael [2,3], ensuring the Heuristic Determination of Minimum Cost Paths. It allows the convergence of the search to a solution if this latter exists. Mathematics and games were its first and main application fields as for the exemple presented in section 2.3 .

The search takes place in a graph or simply in a tree which is made of nodes and branches. The nodes are the different configurations of the system that may occur and the branches are the paths which connect the nodes to some others, to their nearest neighbours.The successors are the nodes directly generated from parent nodes by generation rules. The optimization of the search of a solution achieved by this procedure brings a cure to the second major drawback, the limit due to the computing time.

2.1. The Heuristic Evaluation Function of Graph Nodes

This heuristic search procedure as for the best-first-search procedures [4] uses an heuristic evaluation function of nodes , **F**. This function drives the searching process and at each step it provides the best node value. Then it selects the more promising node (and the more promising branch) which will be tes-ted at first. Its extremization ,minimization or maximization according to the considered problem, provides the solution. For instance, **F** is the number of misplaced or wellplaced pieces in a puzzle problem or the distance covered by the Commercial Traveller in the connected problem. For the A* procedure, **F** is splited into two parts :

$$\mathbf{F}(n) = \mathbf{G}(n) + \mathbf{H}(n)$$

The **G** function measures the cost of the path from the initial node to the current one, n. **H** estimates the additionnal cost of path to reach a goal-node or solution from n.

2.2. The Nodes Sets

This procedure uses three kinds of nodes. The first ones are the nodes which were not evaluated because they were not generated. The two others are classified into the opened nodes set, **O** , and the closed nodes set , **C**. **O** contains evaluated nodes which seemed not more promising nodes at the first examination. Consequently, their successors were not generated. **C** contains evaluated nodes which seemed more promising nodes at the first examination but their successors appeared as less promi-sing and the exploration of the corresponding branches was temporarily stopped.

When the termination criterion of the search is true for the current node, the search is ended.

2.3. A* Search for 8-Pieces-Puzzle Problem

We present here from N. Nilsson, a famous puzzle problem created by Sam Loyd. The aim of this problem is to find the path to reach the solution from the initial configuration N_0 using only vertical and horizontal shifts of the pieces.

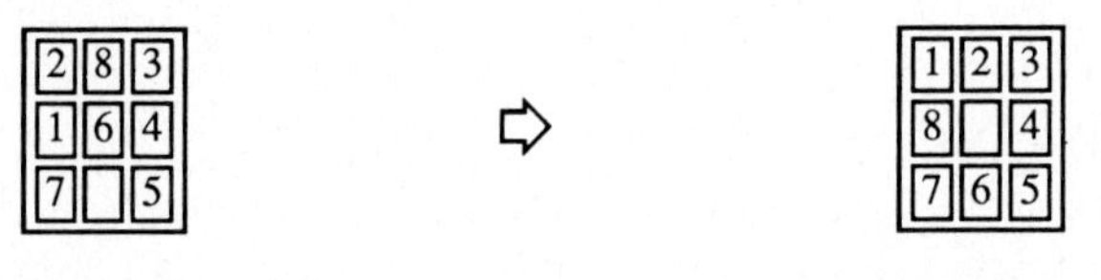

Initial Configuration N_0 Solution Configuration N_{Goal}

Figure 1
The aim of the search for a 8-pieces puzzle

Here, **G**(n) is the number of moves from N_0 to the current node n. **H**(n) is the number of misplaced pieces. At each depth level in the graph, the A* algorithm selects at first the best situation for which F is minimum.

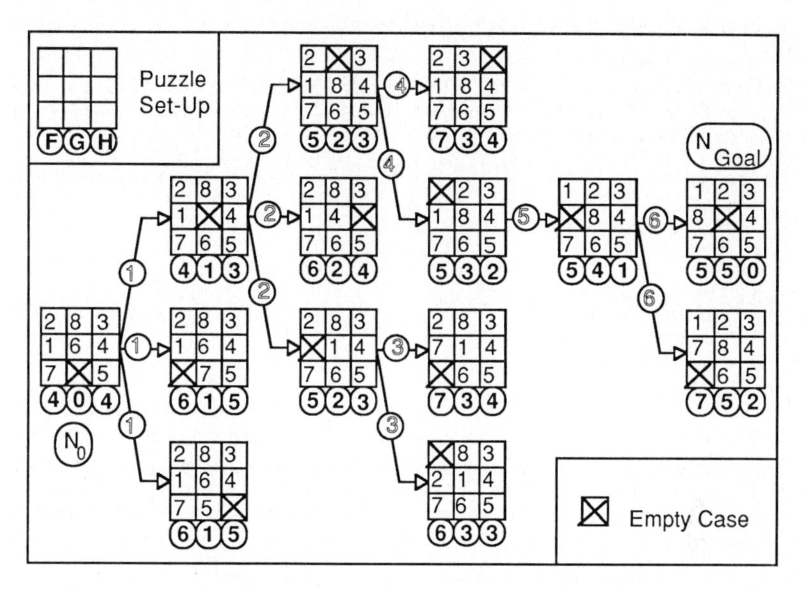

Figure 2
A* Search Graph for the 8-Pieces-puzzle problem

3. SET-UP AND BRIEF REVIEW OF THE FORWARD PROBLEM

3.1. Set-Up

We consider an elastic homogeneous cylinder immersed in water which is irradiated by short ultrasonic pulses (≈ 200 nanoseconds). These pulses are transmitted by piezoelectric transducers immersed in the fluid at a distance r from the axis of the scatterer. The densities of the fluid and the cylinder are respectively ρ_F and ρ_C. The velocities are c_F, c_S and c_L, respectively for bulk waves in the fluid, bulk shear and longitudinal waves in the solid.All quantities related to the scattering phenomenon are studied versus the dimensionless wavenumber ka (k is the wavenumber in the surrounding fluid medium and a the radius of the cylinder).

3.2. The High ka Domain : Use of Rays Theory

For large ka (ka > 200), we just have to use a monostatic configuration of probe-target geometry in a backscattering study. The probe's axis is converging with the cylinder's one and perpendicular to it. Then the transducer works in pulse-echo mode. We use target with radii of the order of 1 centimeter. Since we consider large ka waves, the scattering phenomena are well described by a geometrical theory of acoustics [5,6,7]. Then it is more relevant to analyze the scattering signal in the time domain.

The scattered echoes are defined by their arrival time either after emission, $T_{(n,m)}$, or after receiving the specular echo (0,0), $t_{(n,m)}$, and by their maximum relative amplitudes $a_{(n,m)}$. These (n,m) echoes are the part of the scattered signal corresponding to acoustical rays generated with incidence α_i upon the fluid-cylinder boundary ($-\alpha_F \leq \alpha_i \leq +\alpha_F$) for which the path inside a cross-section of the culinder is a set of n rays among which m have a shear polarization.

$$t_{(n,m)} = t_{(n,m)}\,(c_F, r, a, c_L, c_S, \alpha_i)$$
$$a_{(n,m)} = a_{(n,m)}\,(c_F, r, a, c_L, c_S, \alpha_i, \rho_F, \rho_C)$$
$$r = a + c_F \frac{T^M_{(0,0)}}{2}$$

3.3. The Low ka Domain : Use of the P.R.I.M.

In this case (ka < 30), we have to use a bistatic configuration where the transmitter is steady and the receiver moves around the elastical cylindrical scatterer's axis. Here it is necessary to use frequency analysis of the scattered signals. Effectively, informations about scatterer are mainly carried out by its acoustical resonances which are characterized by the numbers (n,ℓ), n being the modal number and ℓ being related to the type of surface wave. The frequencies and the widths of the resonances are experimentally measured. Using broadband pulses experiments lead to observation of resonance spectra [8]. Resonance frequencies and widths are directly measurable on it. To get enough informations to solve the inverse problem, measurements of the mode number n for each resonance is necessary. We have have developped a new method called the Pulsed Resonance Identification Method or P.R.I.M. [9] which leads to a rapid identification of the modal number n of all well - resolved resonances contained in the probes bandwidths. It uses only one angular scanning of the receiver around the cylindrical scatterer. The P.R.I.M. allows to identify the order n of a resonance by the study of the shape of the scattering angular diagram.(Fig.3). At a resonance frequency we count the number of lobes which is equal to 2 times the number n. These patterns are obtained by plotting on a polar diagram the amplitude of the resonance in the spectrum versus the sampled angular position of the receiver, θ_i, around the scatterer. A similar method has been developped simultaneously by another french team [10].

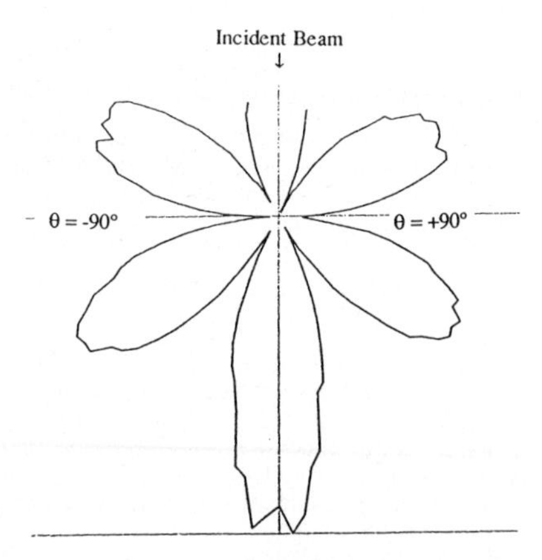

Figure 3
Resonance (3,1) angular diagram of a copper wire (Ø 1.13 mm)
at $k_1a^* = 5.45$ with an angulart sampling interval of 2.5°.

4. APPLICATION OF A* ALGORITHM TO SOLVE THE INVERSE PROBLEM OF THE ACOUSTICAL SCATTERING BY ELASTIC CYLINDERS AT LARGE KA

4.1. The Tree of the Search

It is a sampling of the continuous 3-dimensionnal space $[a, c_L, c_S]$ where a, c_L, c_S are the searched parameters of the search. The sampling steps are Δa, Δc_L, Δc_S. The samples are the nodes n_i which are connected with their six nearest neigbours ($n_i = (a^{[i]}, c_L^{[i]}, c_S^{[i]})$). The moves allowed by the generation rules of nodes are only one step move along one of the 3 co-ordinates axis. The border of the space of the search are:

$$\begin{cases} a \in [9\text{ mm} - 11\text{ mm}] \text{ with } \Delta a = 0.1\text{ mm} \\ c_L \in [2.7\text{ km/s} - 10.5\text{ km/s}] \text{ with } \Delta c_L = 0.05\text{ km/s} \\ c_S \in [1.5\text{ km/s} - 3.6\text{ km/s}] \text{ with } \Delta c_S = 0.05\text{ km/s} \\ \frac{3}{2} c_S \leq c_L \leq 3\, c_S + 2\text{km/s} \end{cases}$$

where stay the great majority of elastic solids bulk velocities.

4.2. The Heuristic Evaluation Function

The **G** function is equal to the sum of a simple move cost Δ, fixed to 1 in our simulations, from the initial node, N_0, to the current one n_i :

$$G(n_i) = \sum_{n=N_0}^{n_i} \Delta$$

The **H** function is a mathematical norm built upon the differences between calculated $t_{(n,m)}{}^C$ and measured $t_{(n,m)}{}^M$ arrival times of some relevant echoes (in microseconds) with a normalisation time fixed to $\Delta t = 1$ µs. We have chosen the (2,0), (3,0), (3,1) and (3,3) echoes detected in backscattering and easily identifiable due to their large relative amplitudes.

$$\mathbf{H}(n_i) = \text{INT}\left(\frac{\sqrt{\sum_{p=1}^{4} \left| t_p^C(n_i) - t_p^M(n_i) \right|^2}}{\Delta t} \right)$$

where the index p represents the (n,m) echoes used. When **H** is less than a limiting value defining the zero level for **H**, the search stops and the current node is the solution. This is the termination criterion.

In addition, for each possible solution the method uses a scanning on the value of the density of the cylinder which compares the calculated amplitudes of echoes and their measured values. If the difference between them is less than a limiting value the solution is kept, else the search is continued. We scan the range 2.6 10^3 kg/m^3 $\leq \rho_C \leq$ 10.6 10^3 kg/m^3 by steps $\Delta\rho_C$ = 50 kg/m^3 .

4.3. Some Results of the Search

We present results for two samples : a duraluminum cylinder of 20.48 mm diameter and a steel cylinder of 19.95 mm. For these results the precision of this method is better then 1%.

	Duraluminum Cylinder			Steel Cylinder		
	Measured Values	Calculated Values	N_0	Measured Values	Calculated Values	N_0
r (mm)	160.0	160.1		160.0	160.0	
a (mm)	10.24	10.3	11.0	9.98	10.0	10.0
c_L (km/s)	6.37	6.40	6.00	5.85	5.85	6.6
c_T (km/s)	3.13	3.10	3.50	3.30	3.30	2.60

5. DISCUSSION AND CONCLUSION

This method, derived from the A* algorithm, provides a relevant solution if the initial node is not "too far" from the exact solution and if the zero level for H is well chosen because in fact , due to the sampling of the continuous physical space, the exact solution stays exceptionnally on a node of the tree. The choice of the heuristic evaluation function is very important. However, if these conditions are fulfilled, this method provides rapidly and with a good accuracy the best approximate solution.

ACKNOWLEDGEMENTS

This work is supported by the French Direction des Etudes et Recherches under contract 88/093.

REFERENCES

[1] Tarantola, A., Inverse Problem Theory (Elsevier, 1987)
[2] Hart , P.E., Nilsson, N.J., and Raphael, B., IEEE Transactions on SSC, vol.4 (1968).
[3] Hart , P.E., Nilsson, N.J., and Raphael, B., SIGART Newsletter, vol.37 (1972).
[4] Rich, E., Artificial Intelligence (McGraw-Hill, 1983).
[5] Welton, P.J., de Billy, M., Hayman, A., and Quentin, G., J.Acoust.Soc.Am. 67 (1980) 470.
[6] Quentin, G., de Billy, M., and Hayman, A., J.Acoust.Soc.Am. 70 (1981) 870.
[7] Quentin, G., and Fekih, M., Rev. CETHEDEC- Ondes et Signal 72 (1982) 91.
[8] De Billy, M., J.Acoust.Soc.Am. 79 (1986) 219.
[9] Quentin, G., and Cand, A., Electron.Lett. 25 (1989) 353.
[10] Pareige, P., Rembert, P., Izbicki, J.L., Maze, G., and Ripoche,J., Phys.Lett. A 135 (1989) 143.

Elastic Waves and Ultrasonic Nondestructive Evaluation
S.K. Datta, J.D. Achenbach and Y.S. Rajapakse (Editors)

SCATTERING OF ACOUSTIC AND ELASTIC WAVES FROM CRACKS USING HYPERSINGULAR BOUNDARY INTEGRAL EQUATIONS

Thomas J. Rudolphi, Frank J. Rizzo and Guna Krishnasamy

Department of Engineering Science and Mechanics
Iowa State University
Ames, Iowa 50011, USA

1. INTRODUCTION

Locating and sizing both surface-breaking and subsurface cracks is one of the fundamental problems in the field of NDE (Nondestructive Evaluation). Ultrasonic measurements, whereby acoustic or elastic waves are transmitted into the host medium and subsequently received and interpreted after being scattered by the crack, are one of the primary means to locate and characterize cracks. The echo signature of these impressed fields, possibly of a pulsed nature or as continuous harmonic waveforms, provides information regarding the location, orientation and size of scatterers such as cracks. Analytical methods to predict the response of scatterers, except for the most simplified models and input waveforms, are hopelessly complex. Numerical approaches are able to address some problems with practical realism and, in this regard, the boundary element method has proven particularly effective.

Crack problems pose an especially interesting problem to boundary element methods in that the usual boundary integral equation becomes degenerate for infinitely thin, internal boundaries as those used for modeling cracks. The derivative boundary integral equation overcomes this degeneracy, but it is of a more singular character, which itself poses a challenge for computation. In the present work, we present a formulation in which the higher order singularities of the so-called hypersingular integral equation are systematically reduced to regular, weakly singular or Cauchy principal value integrals. Given is the formulation for the general, time-harmonic field produced by a crack of arbitrary shape embedded in an infinitely large elastic host medium. Specialization is subsequently made for the case of an arbitrarily shaped, flat crack under static load.

2. DEVELOPMENT

For an unbounded elastic medium with an infinitely thin interior crack with surface S, reference unit normal n_i and subjected to a time-harmonic incident wave u_i^i (explicit frequency variation $e^{-i\omega t}$ suppressed), the scattered displacement field can be expressed by the representation integral [1,2]

$$u_m^s(\xi) = C_{ijkl} \int_S \left[G_{im}^D \Delta u_{k,l} n_j - G_{km,l}^D \Delta u_i n_j \right] dS \quad (1)$$

where C_{ijkl} are the material constants, G_{im}^D are the components of the dynamic (time-harmonic) fundamental solution [2], Δu_i are the crack surface displacement discontinuities and the total displacement field is the sum of the incident and scattered, i.e., $u_i = u_i^i + u_i^s$. If the crack surface is self-equilibrated, then the first term of the integral above vanishes to leave

$$u_m^s(\underline{\xi}) = -C_{ijkl} \int_S G_{km,l}^D(\underline{\xi},\underline{x}) \Delta u_i(\underline{x}) n_j(\underline{x}) dS(\underline{x}) \quad (2)$$

In this expression for u_m^s the arguments for each function involved have been given to clarify the point dependence and the derivatives denoted by the subscript-comma notation are with respect to the coordinates at the integration point $\underline{x}$.

If the field point $\underline{\xi}$ is taken to either crack surface in Eq. (2), say $\underline{\xi} \to \underline{\xi}_o$, as the boundary integral equation is normally acquired, the resulting limit expression on the right side of Eq. (2) is interpretable as a Cauchy principal value integral and a coefficient times $\Delta u_i(\underline{\xi}_o)$. Then the equation will involve both the crack surface displacement and the displacement discontinuity. It is thus incapable of resolving either. An alternate relation which does not involve both unknowns is possible through the stress field, which requires the displacement gradients at $\underline{\xi}$ and the constitutive equations. From Eq. (2) the gradient equation is

$$\frac{\partial u_m^s}{\partial \xi_n}(\underline{\xi}) = -C_{ijkl} \frac{\partial}{\partial \xi_n} \int_S G_{km,l}^D \Delta u_i n_j ds \quad (3)$$

With these gradients, the constitutive equation $\sigma_{ij} = C_{ijkl}\varepsilon_{kl} = C_{ijkl}u_{k,l}$ and Cauchy's relation $t_i = \sigma_{ij}n_j$, we have

$$t_i^s(\underline{\xi}) = -C_{ijkl} n_j(\underline{\xi}) C_{mnop} \frac{\partial}{\partial \xi_l} \int_s G_{ok,p}^D \Delta u_m n_n ds$$

$$= -C_{ijkl} n_j(\underline{\xi}) \frac{\partial}{\partial \xi_l} \int_s T_{mk}^D \Delta u_m ds \quad (4)$$

where $T^D_{mk}(\underline{\xi},\underline{x}) = C_{mnop}n_n(\underline{x})G_{ok,p}(\underline{\xi},x)$.

It is observed that the kernels $G^D_{im,l}$ of Eq. (2) are $O(1/r^2)$ and if we let $\underline{\xi}\to\underline{\xi}_o \in S$, the resulting integrals are singular and produce, in the limit, a principal part or jump terms of $C_{ki}(\underline{\xi}_o)\Delta u_i(\underline{\xi}_o)$, where C_{ki} are constants, and a Cauchy principal value integral. The subsequent differentiation to determine the displacement gradients produces kernels with stronger $O(1/r^3)$ singularities. Thus, in Eq. (4) for the tractions, if we take the limit $\underline{\xi}\to\underline{\xi}_o \in S$ before the differentiation, then the derivative is that of a Cauchy principal value integral. This type of derivative of a Cauchy principal value integral is interpretable [3,4,5] as a finite part integral, which we write as

$$t^s_i(\underline{\xi}_o) = C_{ijkl}n_j(\underline{\xi}_o) \oint_S T^D_{mk,l}\Delta u_m ds = \oint_S S^D_{mij}\Delta u_m ds \; n_j(\underline{\xi}_o) \tag{5}$$

where we have utilized the result $\frac{\partial T_{mk}}{\partial \xi_l} = -\frac{\partial T_{mk}}{\partial x_l} = -T_{mk,l}$ and have defined $S^D_{mij} = C_{ijkl}T_{mk,l}$.

Quadrature formulas for integration of the $O(1/r^3)$ singularities to numerically determine the finite part integrals have been devised by Kutt [6], but calculations based upon the finite part approach for the static problem [7] have presented considerable difficulties.

An alternate and perhaps more intuitive approach to determine the limit and differentiation required in Eq. (4) to determine the crack surface scattered traction is to first perform the differentiation under the integral when the point $\underline{\xi}$ is not on the boundary, so the differentiation and integration interchange is permissible. Then before the limit $\underline{\xi}\to\underline{\xi}_o \in S$ is taken, the integral can be regularized and converted to ordinary integrals through the use of Stokes' theorem. This approach has been taken by Krishnasamy, et al. [8] for three-dimensional problems and by Rudolphi, et al. [9] for static, two-dimensional problems.

For the current problem, an intermediate result showing part of the regularization is

$$t^s_i(\underline{\xi}) = C_{ijkl}n_j(\underline{\xi})\left\{ \int_S \left[T^D_{mk,l} - T_{mk,l}\right]\Delta u_m dS + \int_S T_{mk,l}\Delta u_m dS \right\} \tag{6}$$

where T_{mk} is the static or zero frequency limit of the dynamic tensor T^D_{mk}. The first term on the right of Eq. (6) is weakly singular, so a subsequent limit as $\underline{\xi} \rightarrow \underline{\xi}_o \in S$ is trivial.

Before proceeding to this limit, the last term of Eq. (6) is further divided into a regular term and singular terms. Presuming Δu_m is Hölder continuous at the limit point $\underline{\xi}_o$, we write

$$\int_S T_{mk,l} \Delta u_m ds = \int T_{mk,l} \left[\Delta u_m - \Delta u_m(\underline{\xi}_o) - \frac{\partial \Delta u_m}{\partial x_p}(\underline{\xi}_o)(x_p - \xi^o_p) \right] ds$$

$$+ \int_S T_{mk,l} ds \Delta u_m(\underline{\xi}_o) + \int_S T_{mk,l}(x_p - \xi^o_p) ds \frac{d\Delta u_m}{dx_p}(\underline{\xi}_o) \qquad (7)$$

where ξ^o_p denotes the coordinates at $\underline{\xi}_o$. Effectively, the strong $(O(1/r^3)$ singularity is now isolated in the second term on the right of Eq. (7); the last term is $O(1/r^2)$. These last two terms can be integrated, through the use of Stokes' theorem [8] and put back into Eq. (7) and Eq. (6) to give

$$t^s_p(\underline{\xi}_o) = - \int_S \left[S^D_{ipq} - S_{ipq} \right] \Delta u_i dS \, n_q(\underline{\xi}_o)$$

$$- \int_S S_{ipq} \left[\Delta u_i - \Delta u_i(\underline{\xi}_o) - \Delta u_{i,n}(\underline{\xi}_o)(x_n - \xi^o_n) \right] dS \, n_q(\underline{\xi}_o)$$

$$+ C_{pqmr} C_{ijkl} n_q(\underline{\xi}_o) \left\{ \epsilon_{jrq} \oint_C G_{km,l} dx_q \Delta u_i(\underline{\xi}_o) \right.$$

$$+ \epsilon_{rlq} \oint_C G_{km} dx_k \frac{\partial \Delta u_i}{\partial \xi_j}(\underline{\xi}_o)$$

$$\left. + \epsilon_{jrq} \oint_C G_{km,l}(x_p - \xi^o_p) dx_q \frac{\partial \Delta u_i}{\partial \xi_p}(\underline{\xi}_o) \right\}$$

$$- \frac{1}{8\pi} C_{pqmr} n_q(\underline{\xi}_o) \oint_C \left[\epsilon_{mji} R,_{pp} + \frac{1}{1-\nu} \epsilon_{jpi} R,_{pm} \right] dx_i \frac{\partial \Delta u_j}{d\xi_r}(\underline{\xi}_o) \qquad (8)$$

where C is the boundary curve of the crack surface S and $R = |\underline{x} - \underline{\xi}_o|$. All terms of Eq. (8) have thus been reduced to regular or weakly singular integrals.

If the crack surface is stress-free, then $t_i = t_i^i + t_i^s = 0$ so $t_i^i = -t_i^s$, and with a prescribed incident wave, one can put that field into the left side of Eq. (8) and then solve the integral equation for the crack opening displacement. Solution methods for numerically solving that equation for general, curved crack geometries are being developed. Some simplified versions of the general expression have been solved and one is presented below. The counterpart of this equation for acoustic scattering has been solved for the case of a circular, plane crack with normal [8] and oblique incidence.

We now consider the specialization of Eq. (8) to the very particular case of normal loading of a plane crack in the zero frequency limit, or elastostatic case. The plane crack is oriented so that it lies in the x_1, x_2-plane with reference normal components $n_1 = n_2 = 0$, $n_3 = -1$ and the medium is loaded so that $\sigma_{33} = p_o$ in the remote field. To solve the problem of a stress-free crack, the scattered (or perturbed) tractions on the crack become

$$t_p^s(\underline{\xi}_o) = -\delta_{3p} p_o \tag{9}$$

and $\Delta u_1(\underline{\xi}_o) = \Delta u_2(\underline{\xi}_o) = 0$ (10)

For the zero frequency problem, the first term of Eq. (8) is zero and the remainder simplifies, after some manipulation, to

$$\begin{aligned} 4\pi p_o\left(\frac{1-\nu}{\mu}\right) &= \int_S \frac{1}{r^3}\left[\Delta u_3(\underline{x}) - \Delta u_3(\underline{\xi}_o) - \Delta u_{3,n}(\underline{\xi}_o)(x_n - \xi_n^o)\right] dS \\ &+ \oint_C \left(\frac{1}{r^2}\right)\left[r,_1 dx_2 - r,_2 dx_1\right] \Delta u_3(\underline{\xi}_o) \\ &+ \oint_C \frac{1}{r}\left[\Delta u_{3,1}(\underline{\xi}_o) dx_2 - \Delta u_{3,2}(\underline{\xi}_o) dx_1\right] \end{aligned} \tag{11}$$

With the exception of the coefficient on the left side of this equation, it is identical to the zero frequency acoustic, or potential problem with a potential jump of p_o across the crack surfaces.

3. DISCUSSION AND CONCLUSION

An integral equation formulation of the time-harmonic wave scattering from arbitrarily shaped cracks has been given, both in terms of a finite part integral and a regularization of that integral which contains only weakly singular integrals and integrals over the contour of the crack. The latter approach is well suited for computation and both the static and the acoustic counterparts of the integral equation have been numerically solved.

ACKNOWLEDGEMENTS

This work was sponsored in part by the Office of Naval Research, Contract Nos. N00014-86-K-0551 and N00014-89-K-0109, under Y. Rajapakse.

REFERENCES

[1] Achenbach, J. D., Gautesen, A. K. and McMaken, H., Ray Methods for Waves in Elastic Solids (Pitman, New York, 1982).
[2] Rizzo, F. J., Shippy, D. J. and Rezayat, M., A Boundary Integral Equation Method for Radiation and Scattering of Elastic Waves in Three Dimensions, Int. J. Num. Methods in Engr., 21 (1985), pp. 115-129.
[3] Mangler, K. W., Improper Integrals in Theoretical Aerodynamics, Royal Aircraft Est., Ministry of Supply, Aero. Res. Council, Current Papers No. 94 (London, 1952).
[4] Brandão, M. P., Improper Integrals in Theoretical Aerodynamics: The Problem Revisited, AIAA Journal, 25, No. 9 (1987), pp. 1258-1260.
[5] Kaya, A. C. and Erdogan, F., On the Solution of Integral Equations with Strongly Singular Integrals, Q. Appl. Math, XLV, No. 1 (1987), pp. 105-122.
[6] Kutt, H. R., The Numerical Evaluation of Principal Value Integrals by Finite Part Integration, Num. Math., 24 (1975), pp. 200-210.
[7] Polch, E. Z., Cruse, T. A. and Huang, C. J., Traction BIE Solutions for Flat Cracks, Comp. Mech., 2 (1987), pp. 253-267.
[8] Krishnasamy, G., Schmerr, L. W., Rudolphi, T. J. and Rizzo, F. J., Hypersingular Boundary Integral Equations: Some Applications in Acoustics and Elastic Wave Scattering, accepted in J. Appl. Mech.
[9] Rudolphi, T. J., Krishnasamy, G., Schmerr, L. W. and Rizzo, F. J., On the Use of Strongly Singular Integral Equations for Crack Problems, Proc. 10th Int. Conf. Boundary Elements, Southampton, Vol. 3 (1988), pp. 249-263.

Elastic Waves and Ultrasonic Nondestructive Evaluation
S.K. Datta, J.D. Achenbach and Y.S. Rajapakse (Editors)

Theoretical aspects of image formation in the scanning acoustic microscope

M G Somekh
Dept. of Electrical and Electronic Engineering
University of Nottingham
Nottingham NG2 7RD
England

INTRODUCTION

It has been appreciated in recent years that one of the principal advantages of the scanning acoustic microscope lies in its ability to access quantitative information about the mechanical properties of the sample with good spatial resolution. The acoustic microscope is commonly used in one of two modes: non-scanning V(z) mode or imaging mode. In the former mode the entire purpose is usually determination of the surface wave velocity and attenuation with high accuracy [1] . For imaging applications useful work may be performed without recourse to detailed quantitative interpretation, but it is gradually being realised that for material science applications, at least, the features observable with the SAM are for the most part observable with other microscopies although not always with as good contrast or so easily. The greatest challenge for the use of the SAM on hard materials thus lies in the area of quantitative image acquisition and interpretation.

The purpose of this paper is to discuss some of the important issues that need to be addressed in order to provide quantitative interpretation of SAM images. The length of the paper precludes detailed review of all aspects of SAM image formation but a brief overview will be included in the next section in order to guide the reader to appropriate references in the literature. Section 3 summarises the properties of image formation in the microscope whilst section 4 illustrates some of the unique difficulties encountered in image interpretation due to the predominant role of surface wave excitation.

2. OVERVIEW

The initial theoretical study on image formation in the acoustic microscope was developed by Lemons in his doctoral thesis and subsequent publications [2]. This work appreciated the confocal (discussed in section 3) nature of image formation and gave a good account of the step response and resolution of the instrument for an 'ideal' object with no surface wave excitation.

2.1 Laterally homogeneous surfaces

The role of the excitation of surface waves was first noticed by Weglein and Wilson [3] who explained the observed periodicity of the output voltage of the lens response as the lens/sample separation was changed. This was explained in using a simple ray optical model which took account of the interference between two principal contributions arising from specular reflection and surface wave excitation [4]. This effect was then called the acoustic materials signature (although it is now more usually and less colorfully called the V(z) response). This approach was subsequently applied by Parmon and Bertoni [5] to give a more consistent and corrected expression.

Around the same time Atalar [6] developed a Fourier optical model describing the V(z) response. This work closely followed the approach used by Goodman [7] for optical system using a paraxial approximation. Sheppard and Wilson [8] subsequently applied the Fourier optical method to increase the range of validity to high angles of convergence and derived one of the more convenient expressions for the V(z) response:

$$V(z) = \int_{-\infty}^{\infty} P(\theta)R(\theta)\exp(2jkz\cos\theta)\cos(\theta)\sin(\theta)d\theta \qquad (1)$$

where θ is the angle of incidence and $P(\theta)$ is the pupil function and $R(\theta)$ is the reflection coefficient as a function of angle.

The ray-optical approach was further extended by Bertoni [9] to predict the full V(z) curve (as opposed to just the periodicity of the materials signature) in which he gave explicit expressions for the geometrical and Rayleigh wave contributions. Although the approach does not give the most accurate quantitative predictions it does give a clear physical picture of contrast mechanisms in the microscope, but more importantly provides convenient lens design criteria in order to tailor the microscope response.

Application of the Fourier optical formulation to predict contrast from anisotropic materials was subsequently developed by Somekh et al. [10]. Where as work on layered samples has been carried our by several authors such as Bray [11].

2.2 Laterally inhomogeneous samples

The theory for laterally inhomogeneous sample naturally lagged behind the that of homogeneous samples. The effects of cracks and discontinuities in the image of the reflection microscope were studied using and approximate Greens function approach by Somekh et al. [12], and by Cox and Addison [13] using similar but more computationally based approach.

Recently a more exact Green's function approach has been applied by Roberts [14] which should ultimately allow very accurate predictions of the forward problem for a wide range of material structures.

2.3 Inverse problems

It was realised from early on that the V(z) response of the microscope could be inverted by means of a Fourier transform (see equation [1] to retrieve $R(\theta)$. Liang et al. [15] have demonstrated the effectiveness of this approach by measuring both the amplitude and phase and performing a Fourier inversion. The technique works well close to the Rayleigh wave critical angle but the reconstruction away from this part of the angular spectrum appears to be rather sensitive to noise. Fright et al. [16] have applied phase retrieval inversion techniques to invert modulus only V(z) measurements. Once again the reconstruction away from the surface wave criticality was not generally satisfactory .

Ray based methods of reconstruction using *a priori* knowledge of the principal contributions to the V(z) response have been applied satisfactorily to the cylindrical lens [1] and to either cylindrical or spherical lenses [17].

At the present time inversion methods even for homogeneous samples have met with only limited success, no efforts have yet been made on inhomogeneous samples (except for one highly specialized case [18]). such inversions remain an unresolved problem in the microscope. Section 4 will discuss some relevant considerations if such problems are to be seriously attempted.

3. IMAGING CHARACTERISTICS OF THE ACOUSTIC MICROSCOPE FOR 'PERFECT' SAMPLES

This section will describe the image formation in the acoustic microscope drawing where possible an analogy with scanning optical microscopy. It is assumed that the reader is familiar with the hardware configuration of the acoustic lens, for those who are not there are few better descriptions than the review by Lemons and Quate [2]. We first consider the response of the microscope to a point object at the focal plane. We consider, here, only a perfect object in which no surface waves are excited. The focal plane distribution on the object surface is limited by diffraction even for a simple spherical lens due to the large impedance mismatch between the coupling fluid and the lens material [2].

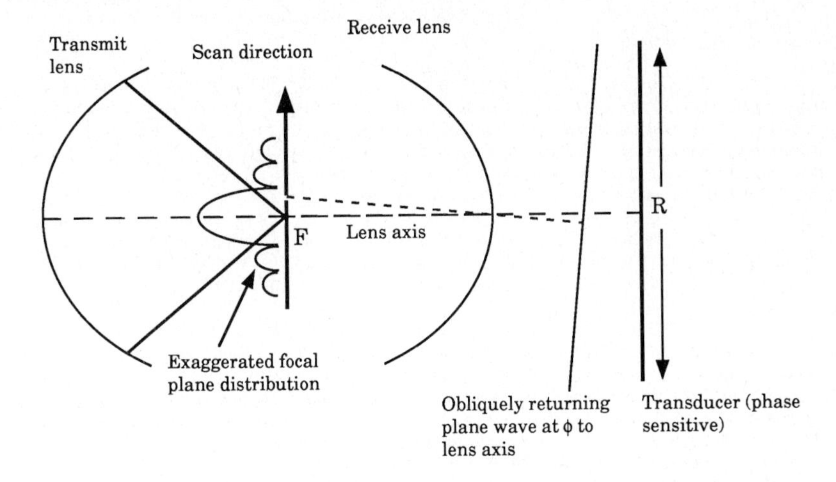

F is the geometrical focus, focal length is f and the object position is x

Figure 1. Simplified schematic of formation of point image the SAM

Figure 1 shows a simplified one dimensional representation of the formation of a point image in the SAM. The system is shown in transmission for clarity although the argument applies equally to a reflection microscope by folding the system over around the focal plane. The receiving lens illuminates the sample with a diffraction limited focal spot (in one dimension given by a sinc function). The receiving lens thus images a point source onto the transducer, the amplitude of this source will be proportional to the field strength in the focal plane. The lens will refocus the point source into an approximate plane wave propagating at an angle of $\phi=\tan^{-1}(x/f)$ with the lens axis. Since the transducer is a phase sensitive detector the signal from the transducer is proportional to the complex field amplitude integrated over the area of the transducer. For a one dimensional system and assuming complete overlap between the returning beam and the transducer this gives:

$$\text{Amplitude of point source} \propto \frac{\sin(kRNA)}{NA}$$

$$\text{Image amplitude} \propto \text{Amplitude of source} \times \frac{\sin(kRNA)}{NA}$$

$$\therefore \text{Image amplitude of point source} \propto \{\frac{\sin(kRNA)}{NA}\}^2$$

where NA is the numerical aperture of the lens.

The output amplitude distribution from a focal plane will thus be proportional to sinc^2 and the intensity distribution will be proportional to sinc^4. It may be seen therefore that the microscope behaves in a manner analogous to an optical interference microscope which detects only the interference term (in this sense the response is exactly similar to the AC term extracted from a scanning heterodyne interferometer). The phase sensitive transducer thus has a role analogous to the reference beam in an optical interferometer.

The presence of the phase detector can be seen to sharpen up the point response of the microscope expected from a conventional incoherent microscope. The SAM thus exhibits the $\sqrt{2}$

improvement in point response expected from a confocal microscope. The analogy between the imaging performance of scanning interference and conventional confocal optical microscopes is examined in reference [19].

The corollary of improved lateral resolution arising from confocal operation is the improved depth resolution bestowed by coherent detection. As the sample is defocused the wavefront of the radiation returning to the transducer will be curved. A Fourier optics analysis of the sample response as a function of defocus for a 'perfect' sample with unity reflection coefficient for a spherical transducer with a $\sin\theta$ pupil function is given by [15]: $V(z)=C\exp[-j\pi z(1+\cos(\alpha)]\mathrm{sinc}(2z(1-\cos\alpha)$, where α is the semiangle of the lens.

Bertoni [9] has analysed the effect of defocus using a ray-optical approach. The main conclusion as far as reflection from a perfect reflector is concerned is that the $|V(z)|$ response is not symmetrical as predicted when $P(\theta)$ is real in equation [1].

To summarise image formation in the SAM is precisely analogous to the scanning optical heterodyne interferometer and as such the methods applicable to such optical systems may be applied, provided similar simplifying assumptions about the nature of the sample can be applied. This is broadly the case for subsurface imaging applications, not involving surface wave excitation. Surface wave excitation, however, complicates the situation so that familiar concepts such as resolution and point spread function need to be redefined. These issues will be addressed in the next section.

4. SAMPLE RELATED CHARACTERISTICS OF SAM IMAGING

We now briefly look at the effects of surface wave propagation on the imaging response of the microscope with a view to examining how concepts such as point spread function, lateral resolution etc. need to be modified. In order to do this the sample can be approximated as an elastic membrane which allows the excitation and reradiation of surface waves. The model makes no pretence at being an accurate representation of the situation actually applying to the real sample, but more appropriately, for our purposes, it does illustrate the manner in which image formation must be modified when the sample supports surface waves.

4.1 The membrane model

This is probably the simplest model that can take account of surface wave excitation. The governing equations and boundary conditions governing the situation are:
For the fluid:

$$(\nabla^2 + k^2)\phi = 0 \qquad (2)$$

where ϕ is the displacement potential, k is the wave vector in the fluid.

For the membrane:

$$\mu u_{tt} - T(u_{xx} + u_{yy}) = p \qquad (3)$$

where the subscripts denote partial differentiation, u is the displacement in the membrane, $\mu(x,y)$ is the surface density of the membrane as a function of spatial position. T is the line tension in the membrane. The boundary conditions are (a) continuity of pressure (b) continuity of pressure at the interface.

Substituting a sinusoidal variation into $\mu(x,y)$ allows one to calculate the microscope response to a a sinusoidal grating. Non-periodic samples can also be modelled by substitution of the appropriate form of $\mu(x,y)$ where the differential equation may be transformed to form a Fredholm integral equation [20].

4.2 Imaging with surface waves.

In section 3 we discussed how the microscope can be modelled by analogy to a scanning optical microscope the obvious question to ask is therefore: can one define the imaging performance in terms of the concepts used so fruitfully in coherent optical systems? From the point of view of the hardware alone there is no reason to expect that the transfer function approach could not be applied to the SAM. The problem, however, rests with the nature of the sample where surface wave excitation means that energy is transferred laterally.

4.2.1 Criterion for image quality when examining inhomogeneous samples

The microscope is often operated in a defocused condition in order to enhance the contrast. The first question that arises is what criterion one uses to assess the quality of an image. The microscope does not image any simple sample property since different features are imaged at different object defocuses. The most reasonable criterion appears to be how the V(z) of a sample with lateral variation differs from that obtained from a homogeneous sample with the same properties as those on the sample axis. This has been discussed in some detail in reference [20] but the overall conclusion is that all the diffracted orders making up the image should be compared with the equivalent components contributing to an image with same magnitude of density variation but with periodicity many times larger than the illuminated area.

Although this criterion is somewhat cumbersome and rather difficult to apply experimentally without access to a uniform sample of the same material, it does avoid some of the difficulties which arise when scattering is enhanced by resonant effects on the sample surface, this will be discussed later in this sub section.

We illustrate the problem with a specific example. Figure 2 shows a sinusoidal sample with a peak to peak density variation of 5%, the defocus position is chosen so that the V(z) is at a minimum for the mean value of surface density. The results show that because of the choice of defocus position (close to the minimum of the V(z) response) even the *amplitude* (as opposed to intensity) response in the image gives contrast at twice the spatial frequency of the property variation. We also note that the magnitude of the response to the large spatial frequency (sample wavelength=1.2 water wavelengths) is much larger than expected due to resonant Bragg scattering of waves propagating along the sample surface.

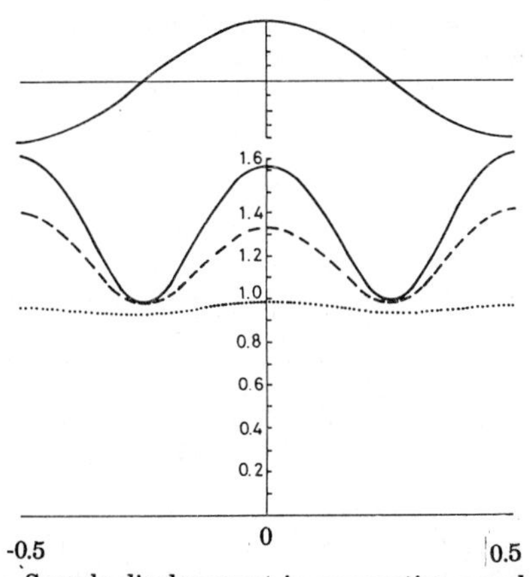

Sample displacement in corrugation wavelengths

Figure 2. Theoretical microscope response close to minimum of V(z) response, as spherical lens is scanned over 5% density variation, shown on the top trace: solid line corresponds to period of density variation equal to 1000 water wavelengths (ie very long compared to spot size), dashed 10 water wavelengths and dotted 1.2 water wavelengths.

Clearly a rigorous analysis of the scanning optical microscope requires that effects such as lateral shifting of energy and vector wave scattering be taken into account, nevertheless, optical systems can be usefully characterised by treating the sample as one in which the diffraction efficiency is uniform for all angles of incidence and spatial frequencies. This allows transfer functions to be effectively used to predict the major properties of the optical microscope. These simplifications also allow for the possibility of inverting the response to access the sample properties. In the acoustic microscope the membrane model described above shows that (a) the diffraction efficiency is greater by typically a factor of between 30 and 100 for waves incident close to the surface wave critical angle and (b) resonant scattering occurs when the spatial frequency of the corrugations is close to half the surface wavelength. The first point means that even a 'weak' object with density variation of less than 5% will diffract energy into high diffracted orders so that the response of the microscope is not linear in terms of sample properties. The second point means that the diffraction efficiency is highly dependent on the spatial frequency of the object. Both these factors make the transfer function approach more difficult to apply since the image response cannot be regarded as the linear sum of the image responses from harmonic or point objects. The concepts of modulation transfer function and point response are thus of much more limited value than the equivalent optical concepts.

To conclude solution to forward problems in the SAM seems to in a satisfactory state providing good predictions of image contrast from known features provided the lens pupil function is well characterised. On the other hand, the Rayleigh wave contrast mechanism means that progress on inversion problems on laterally inhomogeneous samples will require considerable effort before useable methods can be employed without using *a priori* assumptions.

5.0 REFERENCES

1. J Kushibiki and N Chubachi, IEEE Sonics and Ultason. vol. SU-32, pp.189-212, 1985
2. R A Lemons and C F Quate, Ch 1, *Physical Acoustics* Ed. W P Mason and R P Thurston, VolXIV, pp.1-92, 1979
3. R D Weglein and R G Wilson, Elect. Lett., vol.14, pp.352-354, 1978
4. R D Weglein, Appl. Phys. Lett. Vol.34(3), pp.179-181, 1979
5. W Parmon and H L Bertoni, Elect. Lett., vol.15(21), pp.684-686, 1979.
6. A Atalar, J. Appl. Phys., vol. 50, pp.664-672, 1979.
7.J W Goodman, *Introduction to Fourier Optics,* McGraw-Hill, New York, 1968.
8. C J R Sheppard and T Wilson, Appl. Phys. Lett.,vol. 38(12), pp.858-859, 1981.
9. H L Bertoni, IEEE Trans. on Sonics. and Ultrasonics., vol. SU-31, pp.105-116, 1984.
10. M G Somekh, G A D Briggs and C Ilett., Phil. Mag. vol.49A(2) pp.179-204, 1984.
11.R C Bray, *Acoustic and Photoacoustic Microscopy,* G L Report No. 3243, Stanford University, 1981.
12.M G Somekh, H L Bertoni, G A D Briggs and N J Burton, Proc. Roy. Soc. Lond. vol. A401, pp. 29-51.
13. B N Cox and R Addison, In *Review of quantitative NDE,* (Ed. D O Thompson D E Chimenti) vol.3B, Plenum, 1984.
14. R Roberts, J. A. S A. vol. 85, pp.861-566, 1989.
15. K K Liang, G S Kino and B T Khuri-Yakub, IEEE Trans. on Sonics. and Ultrasonics. Vol. SU-32(2), pp.213-224, 1985.
16. W R Fright, R H T Bates, J M Rowe, D S Spencer, M G Somekh and G A D Briggs, J. Micros. vol.153(1), pp.103-117, 1989
17. J M Rowe and M G Somekh, IEEE Ultrasonics Symp. Proceedings, pp.909-913, 1988.
18. K K Liang, S D Bennett, B T Khuri-Yakub and G S Kino,IEEE Trans. on Sonics. and Ultrasonics. Vol. SU-32(2), pp.266-273, 1985.
19. C J R Sheppard and T Wilson, Phil Trans Roy Soc., vol. 295, pp.513-536, 1980
20. M G Somekh, IEE Proc. Pt. A, vol.134(3), pp.290-300, 1987.

Elastic Waves and Ultrasonic Nondestructive Evaluation
S.K. Datta, J.D. Achenbach and Y.S. Rajapakse (Editors)

EXPERIMENTAL FLAW DETECTION BY SCATTERING OF PLATE WAVES

Hartmut SPETZLER*

Cooperative Institute for Research in Environmental Sciences
and Department of Geological Sciences
University of Colorado and NOAA
Campus Box 216
Boulder, Colorado 80309

Subhendu DATTA*

Cooperative Institute for Research in Environmental Sciences
and Department of Mechanical Engineering
University of Colorado and NOAA
Campus Box 216
Boulder, Colorado 80309

The recent development of a capacitive transducer with a wide and flat frequency response made it possible to obtain absolute surface displacements for elastic waves scattered from flaws. Here we report the experimental results from the scattering of Rayleigh-Lamb waves from a surface breaking crack in a two dimensional plate. Theoretical results are being compared with experimental data.

1. INTRODUCTION

The study of guided plate waves is of interest in a variety of scientific and engineering fields, e.g. seismology, electronics, ultrasonics and nondestructive testing. Here we present an application of scattered Rayleigh-Lamb waves to the detection of a surface breaking flaw in a two dimensional plate. The development of a new non-contacting detector [1] with a wide and flat frequency response allowed the measurement of absolute normal surface displacements.

In this paper we concentrate on the experimental aspects of a combined theoretical and experimental investigation which has been described elsewhere [2] in greater detail.

*The work was funded by grants from the Office of Naval Research (#N00014-86-K- 0280; Scientific Office: Dr.Y Rajapakse), The National Science Foundation (#MSM-86-09813) and NASA (#NAGW-1388).

2. EXPERIMENTAL ARRANGEMENT

Figure 1 shows the experimental arrangement which was used for the propagation of Rayleigh-Lamb waves in a two dimensional plate. The plate is 2280 mm long, 25.4mm high and 6.4mm wide. While we have used a variety of signal sources, here we report results obtained by dropping a steel ball bearing on the glass surface. A dropping ball provides a simple source which can be readily modelled theoretically. The ball dropper consists of a carefully honed brass tube which allows laminar flow of air past the ball as it descends through the tube. This assures a high degree of repeatability of the source which is very important if received signals are to be compared quantitatively. The steel ball bearing is being held in place by a magnet which upon removal initiates the dropping of the ball. A common time base is obtained from a piezoelectric transducer suitably placed near the ball bearing impact. The capacitive transducer rests on the rails of the aluminum support structure and can be moved along this rail to receive signals anywhere on the top surface. An inverted support system has been used in some cases to be able to get signals directly opposite to the impact site.

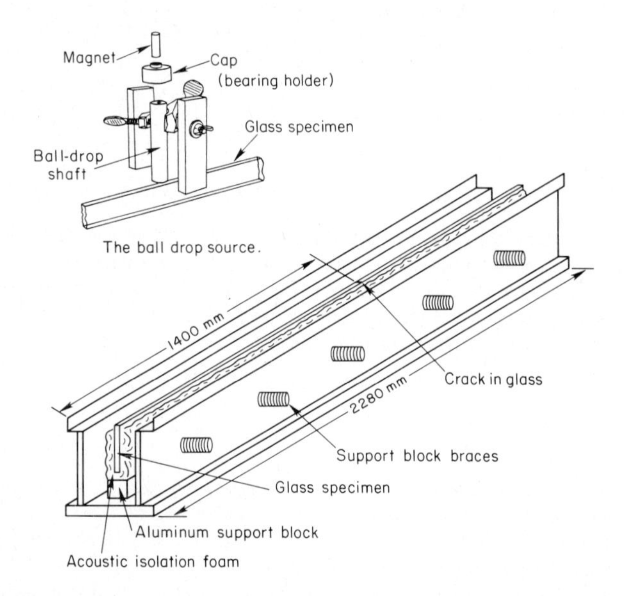

Figure 1
The specimen and support structure with the source. The structure supports the specimen but is mechanically decoupled by a high attenuation, impedance mismatch urethane foam. This prevents loss of signal to the support structure and isolates the specimen from external vibrations. Aluminum coated microscope slides are glued to the sample surface to provide the electrically conductive and smooth surface required by the sensor.

The capacitive transducer (CT) (see Figure 2) developed in our laboratory [1] forms the heart of the detection system. A flame polished platinum needle with a diameter of approximately .3mm is held about 100nm above the conductive surface of the specimen. The conductive surface in the present experiments consists of microscope cover slides with vacuum deposited aluminum on their top surfaces. The change in capacitance C between the needle and the conductive surface is measured by the transducer. The electronic circuitry is a modified version of that developed by RCA (Radio Corporation of America) for their VideoDisk player. A 930 MHz resonant circuit is modulated by C. Demodulation of this high frequency signal results in the detected signal. For a given constant gap size between the needle and the conductive surface the output of the transducer is a DC signal. In the present version of our transducer we utilize a low frequency quasi DC signal as input to a servo system which assures that the gap size is maintained at a constant value. In practice this involves a proportional controller (similar to furnace controllers with adjustable set points) which supplies current to a small heater. The heater controls the temperature of a hollow aluminum cylinder to which the platinum needle is attached. By operating the heater above ambient temperature, thermal expansion and contraction of the cylinder is used to maintain a constant gap at low frequencies.

Rapid changes in the gap size are not compensated for by the servo control and represent the desirable acoustic signal. To within 7 dB the output of the transducer is flat from below 10kHz to above 6 MHz. In Figure 3 we give the sensitivity of the transducer for the experimental range of importance for the present experiments. At .44 V/nm and a noise level of 4mV the sensitivity of this transducer approaches that of piezoelectric transducers. At a one to one signal to noise level the detectability limit is therefore .01nm. The dynamic range extends to about 50 nm thus covering more than three orders of magnitude.

Data received with the capacitance transducer is recorded on digital oscilloscopes and then transferred to a work station class computer for analysis.

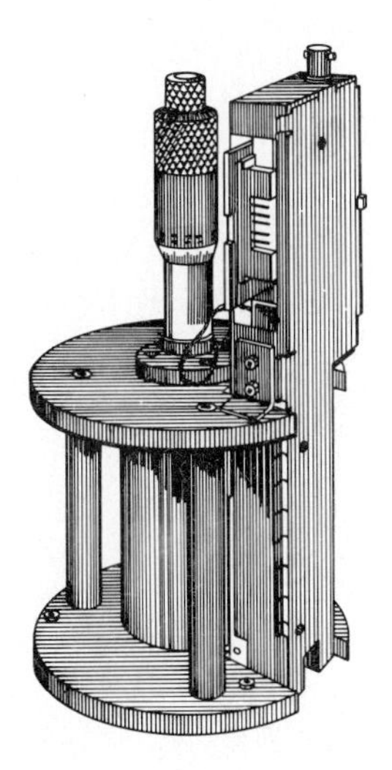

Figure 2
This drawing of the capacitive transducer (C.T.) shows the rectangular box containing the electronic circuits. The micrometer is used to bring the platinum needle (not shown) to within a few micrometers of the sample. The heater for fine positioning is housed in the central cylinder.

Figure 3
Sensitivity of the CT versus frequency. Surface displacements were generated by a resonating rod excited by a piezoelectric transducer. The displacements were calibrated using a Michaelson interferometer. The CT's response to the resonating rod was then measured to obtain a calibration of the CT.

3. EXPERIMENTAL RESULTS

The experiments had a dual purpose. In support of the theoretical work, we needed to perform experiments which were amenable to theoretical treatment. In search of crack detection we were free to exploit any experimental arrangement. Here we report first the experimental results in support of the theoretical work.

The theoretical calculations were carried out for a two dimensional plate. In order to check the validity of this assumption we moved the impact points across the width of the plate and detected the signals on the bottom of the plate. Figure 4 shows the experimental arrangements and the results.

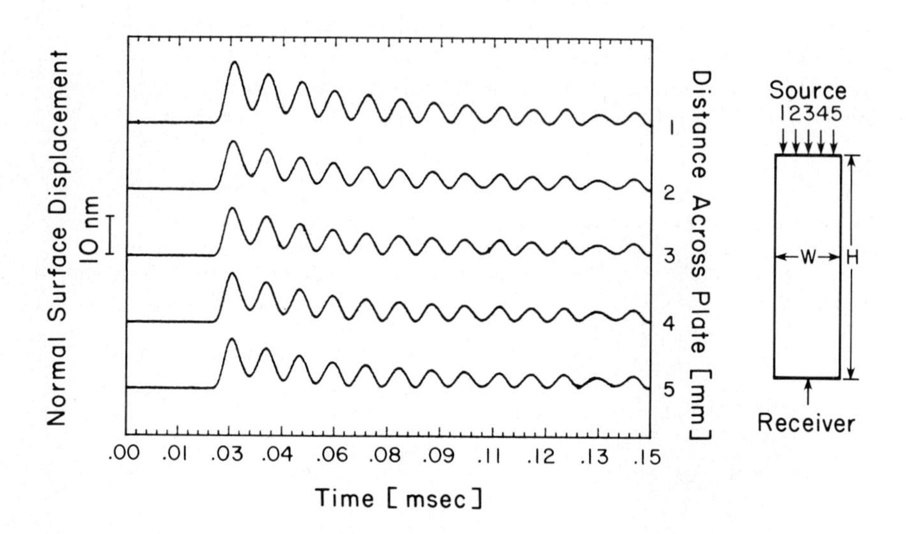

Figure 4
Confirmation of two dimensional plate assumption. The impact point of the steel bearings is moved across the 6.4 mm sample in 1 mm increments. No appreciable difference was observed unless the balls were glancing off the edge of the sample.

Theory was compared with experiment by using experimentally recorded signals in the far field and calculating a source function from them. This source function was then used to calculate further signals at different distances in the far field. Experimentally obtained signals were compared with theoretically derived signals in the time domain. Figure 5 shows examples of these comparisons.

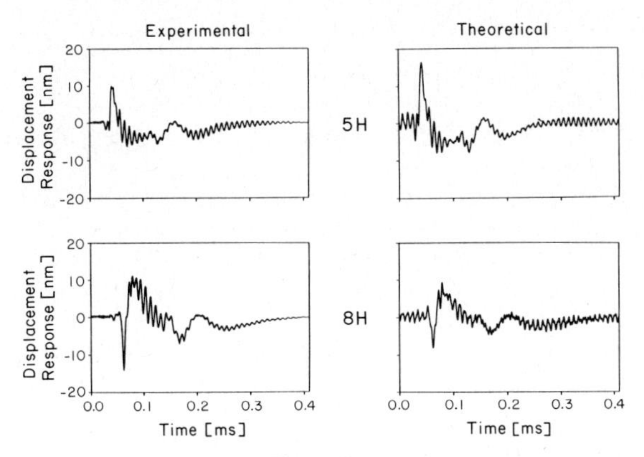

Figure 5
Experimental data and theoretical simulation of the surface displacement at 5H and 8H without a crack. The experimental traces are multiplied with a cosine window to ease theoretical computation. The experimental surface response at 5H was used with the Green's function to compute the source function. This source function was then used for modeling the surface displacement at 8H. The analogous technique was used to calculate the surface displacement at 5H.

Finite elements were used to model the crack in the far field and to calculate power spectra with the crack present and with it absent. Comparison with the experimental results in the frequency domain were accomplished by calculating power spectra from the experimental signals. Figure 6 shows the experimental arrangement and the finite element grid which was used in the calculations. Calculated and experimental spectra are compared in Figure 7.

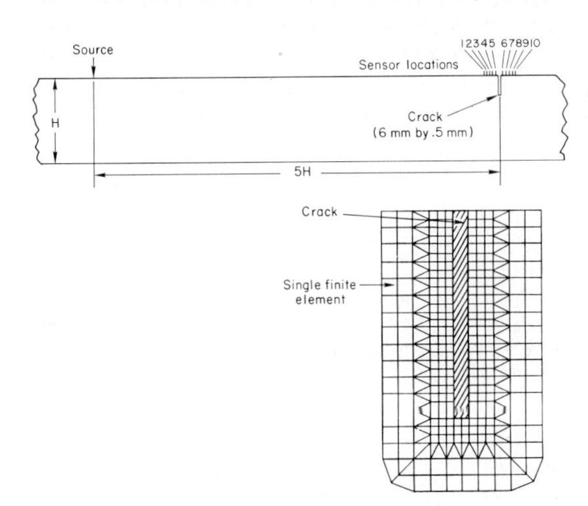

Figure 6
(Above) Experimental set-up to study the scattering in the vicinity of the crack. Receiver locations are positioned at 1 mm increments (H=24.5 mm). Theoretical calculations in the vicinity of the crack were made using a finite element grid. An expanded view of the grid is shown at right.

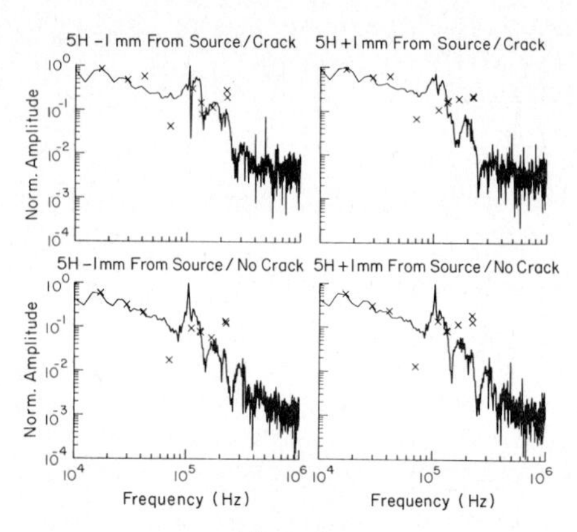

Figure 7
Amplitude spectra of normal surface displacements 1 mm on either side of 5H from the source with and without a crack. 5H is the distance between the source and the crack. Movement of the receiver location by 2 mm changes the amplitude spectrum very little without a crack (bottom figures) and an appreciable amount with a crack (top figures). The solid lines are experimental data, the crosses are the theoretical results.

Experimental detection of the crack in the near field was accomplished with the set up shown in Figure 8. The transducer was kept stationary and the source was moved across the crack. Again the situation with and without the crack are compared. The comparisons are shown in Figure 9. Note that between 0.2 and 0.4 ms on traces 2 and 3, and to a lesser extend on trace 4, there is much less high frequency energy in the no crack case than there is in the case where a crack is present. Apparently destructive interference occurs in this region if no crack is present. The presence of the crack prevents this destructive interference. Also note the opposite sense of the first arrival in trace 4 for the two cases.

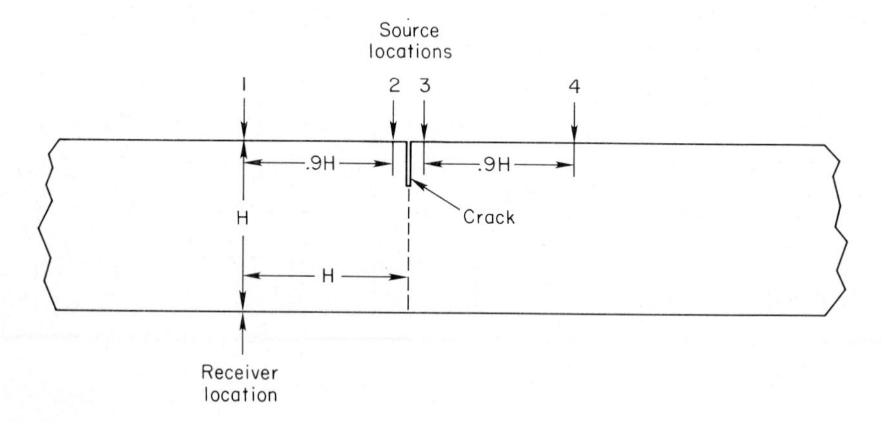

Figure 8
Experimental configuration used to study the effect of the crack. Normal surface displacements on the bottom of the sample are measured in response to ball bearing impacts at locations 1,2,3 and 4 above the sample. The results are illustrated in Figure 9.

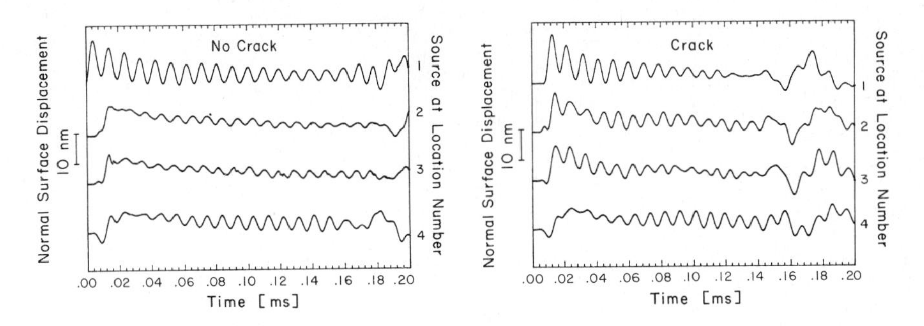

Figure 9
Normal surface displacements on the opposing side of the sample from the source. The source is placed at one of the four positions shown in Figure 8. The receiver is stationary and on the opposite side of the sample. Times after 0.12 ms are subject to reflections from the ends, and should be ignored.

4. CONCLUSIONS

The development of a transducer with a small probe (in comparison to a wave length of the acoustic waves in question) and a flat frequency response enabled us to directly compare experimentally obtained signals with theoretical calculations both in the time and the frequency domain. Generally good agreement is found, the agreement being better in the time domain for waves propagating down an infinite two dimensional plate than for the frequency domain for scattering from a crack. The crack is easily detectable in the far field when the spacings of the receiver locations are small. It is also easily detected in the near field. The latter arrangement lends itself readily to the design of a compact tool for crack detection. Source and receiver could easily be housed in the same physical instrument. The results from the present study suggest further development of the instrumentation and the computational technique toward a practical flaw detection instrument. The theoretical considerations will guide the instrument design and data analysis in terms of choosing design parameters such as frequency band and source to receiver spacing.

ACKNOWLEDGEMENTS

We wish to thank our collaborators in this project J. Paffenholz, J.W. Fox, X.Gu, G.S. Jewett, T. Chakraborty, H. M. Smith, I. C. Getting, M. Bouden, and E. Field. The experimental work was carried out in the Geophysics laboratory in CIRES (Cooperative Institute for Research in Environmental Sciences) at the University of Colorado in Boulder. The theoretical work was carried out in the Mechanical Engineering research facilities at the University of Colorado, Boulder. Numerical computations were carried out at the National Center for Supercomputing Applications at the University of Illinois, Urbana.

REFERENCES

[1] Boler, F.M., Spetzler, H.A., and Getting, I.C., Rev. Sci. Instrum. 55 (1984).
[2] Paffenholz, J., Fox, J.W., Gu, X., Jewett, G.S., Datta, S.K., and Spetzler, H.A., in press.

Elastic Waves and Ultrasonic Nondestructive Evaluation
S.K. Datta, J.D. Achenbach and Y.S. Rajapakse (Editors)
Elsevier Science Publishers B.V. (North-Holland), 1990

NUMERICAL TECHNIQUES FOR WAVE PROPAGATION AND SCATTERING IN INHOMOGENEOUS ANISOTROPIC MATERIALS

Andrew Temple and Jill Ogilvy

Solid State and Materials Modelling Group
Theoretical Physics Division
United Kingdom Atomic Energy Authority
Harwell Laboratory
Didcot, Oxon OX11 ORA, United Kingdom

The welds in austenitic stainless steel, used in the construction of fast breeder reactors, are inhomogeneous and anisotropic on the scale of a few millimetres. It is highly desirable, from economic and safety considerations, to be able to inspect the welds using ultrasound. We apply both ray-tracing and finite difference methods for predicting the behaviour of ultrasound in typical manual-metal-arc weldments. We demonstrate that the two approaches are consistent and complementary then conclude that inspection may not be possible with ultrasonic compression waves at 2MHz nor with shear waves at higher frequencies.

1. INTRODUCTION

As part of the UK work on the development of fast breeder reactors, we are investigating the inspectability of different types of welds in austenitic steel. Ultrasonic inspection is the preferred technique since it is capable of providing the through-wall extent of any crack-like defects and hence can, in principle, yield information necessary for fracture mechanics assessments of defects. Our aim here is to illustrate both ray-tracing and finite difference methods applied to one particular type of weld structure. In general, the microstructure of a weld is dependent on both the welding process and the process variables employed. For a review of the microstructure of manual-metal-arc (MMA), submerged arc (SA), metal inert gas (MIG) and flux cored arc (FCA) welds, see Worrall and Hudgell [1]. We have chosen manual-metal-arc welds to illustrate our approaches. We discuss the ray-tracing and finite difference methods in the next section with applications to MMA welds in section 3.

2. NUMERICAL TECHNIQUES

A typical MMA weld considered in [1] has the form shown in the central portion of figure 1, resembling a region tesselated with scallop shells. In our model idealisation of this, the regions are bounded by semi-circles. In practice the boundary of each weld bead will be somewhat less regular than this. Within each bead, columnar grains form on cooling with the orientation of their long axes approximately as shown schematically. Unlike other types of weld, such as FCA or SA, there is very little epitaxial growth of the grains through the boundary of each weld bead. The orientation of the material thus exhibits sharp discontinuities at the boundaries between adjacent beads. The weld metal formed is inhomogeneous on the scale of the weld bead, typically a few millimetres. Because the columnar grains are also relatively large, the intrinsic anisotropy of the material controls the propagation of elastic waves whose wavelengths are a few millimetres at typical MHz frequencies.

To determine whether a given weld is inspectable, it is necessary to understand the propagation of elastic waves through inhomogeneous, anisotropic materials where the scale of the inhomogeneities is the same as the wavelength. This problem is generally intractable analytically and must be solved numerically.

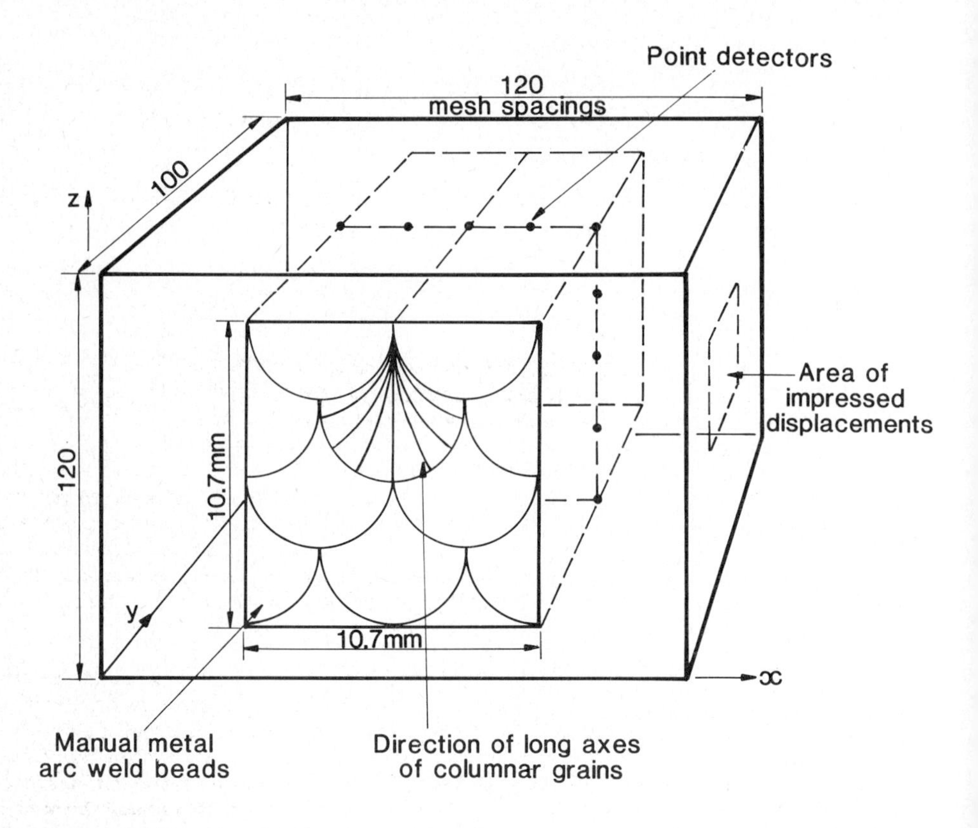

FIGURE 1
Schematic of MMA weld structure as used in the ray tracing and finite difference models.

2.1 Ray Tracing

One approach is to calculate the local elastic constants based on knowledge of the direction of a given crystal symmetry axis. The particular axis chosen is the [001] direction of a transversely isotropic material. It is assumed that this direction corresponds to the visible markings on a macrograph. Neutron scattering or local texture measurements using a Rayleigh wave goniometer could be used to prove this assumption but, as far as we know, this has not been done.

Within the region of smoothly varying elastic constants, curved ray paths are mapped out a small step at a time [2,3]. At each step along the group velocity direction there is a change in material properties. Changes in velocity, ray direction and ray amplitude can be calculated by assuming a fictitious boundary between the two ends of the step. At this boundary, the full elastic wave conditions of continuity of stress and displacement can be solved to give the refracted wave amplitude and direction. The results of repeating this process with a small step are curved ray paths and amplitude changes which represent the energy changes due to refraction. The transit time of the ray is also calculated so that wavefront roughening effects can be predicted by plotting markers on the rays at equal time intervals.

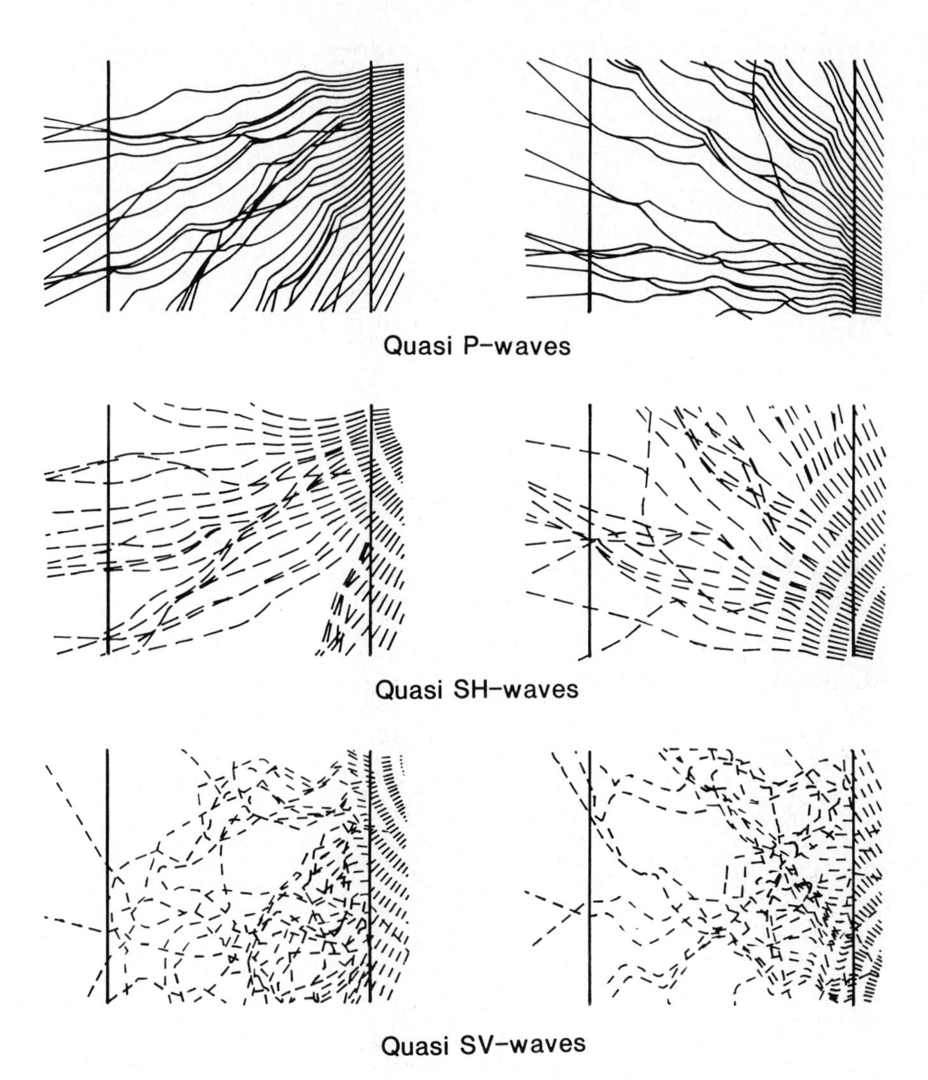

FIGURE 2
Ray paths traversed through an MMA weld by waves of different polarizations diverging from a point source.

The ray tracing model has been used [4] to demonstrate that even welds with a small volume of inhomogeneous material, such as narrow gap tungsten inert gas welds, can cause considerable perturbation to an ultrasonic beam.

Some results of ray paths for P–, SH– and SV-waves traversing the model MMA weld at 45° are shown in figure 2. The modes are strictly mixtures but are labelled by their predominant character. Waves originating from a point source and propagating down through the weld at 45° are shown on the left, whilst rays propagating up at 45° are shown on the right. Solid lines have been used for quasi-P-waves and dotted lines for quasi-shear waves. From the figure we conclude that P- and SH-waves are less affected by the structure than SV-waves.

2.2 Finite Difference Solution

Whereas the ray tracing solution to the propagation problem is essentially that of a particle with given initial velocity in a force field, the finite difference technique discretizes the governing wave equation in both space and time. Other techniques, such as finite element methods, will also do this and may be better for dealing with irregular boundaries. We chose the finite difference method for ease of computer programming. Complete details of the model and computer program were given in [5]. Since both the ray tracing and finite difference approaches are solving the same problem, we would expect them to give similar results. Ray tracing is, however, a high frequency approximation which performs surprisingly well at the frequencies of interest. Also, the two approaches calculate different things: position (and amplitude) of particles after a specified time since they were released with a given initial velocity compared to displacements everywhere at the later time following forcing, as if by a transducer, for some initial number of steps.

The elastic constants used in both models are the same and are those appropriate to 308 austenitic weld metal. They are given in table 1. The full geometry of the finite difference model is shown in figure 1. Displacements are applied over a small area of the ZY plane at x=0 (not as shown on the figure where the area is shown as the x=120 plane for ease of illustration), representing a plane wave with given direction and polarization incident on a rigid baffle with a hole cut in it. These displacements are specified with the correct time dependence to induce a single sinusoidal cycle of frequency 2MHz. To obtain a reasonably accurate solution the regular cubic mesh spacing is 0.13mm which allows a maximum timestep of 0.0125 μs for stability.

Detectors can be located anywhere in the mesh. Some examples are shown in figure 1. These record the three components of displacement, the divergence and the curl of these displacements, at every timestep. Taking the divergence and curl produces components which are essentially the compression and shear-wave parts. Plots of the resulting traces as a function of time are called A-scans. Examples from three detectors, located along a line parallel to the x-axis through the centre of the weld, are shown in figure 3. Results for all three incident modes are shown together with the incident signal detected soon after generation at the baffle. Notice how the single cycle input acquires a more complicated shape as it propagates through the weld beads and how the peak compression-wave amplitude decreases. Based on the peak amplitudes shown, we predict a decrease of the P-wave signal by about 6 dB over 13mm of weld metal. For real welds, where the ultrasonic pulse-echo pathlength in weld metal can easily be 50-100mm, this would represent signal to noise difficulties for non-destructive inspection which might render them uninspectable. From the A-scans we can estimate the group velocity of the different waves. Figure 3a gives a compression wave speed of 5.5 mm/μs, figure 3b gives an SH-wave speed of 3.1 mm/μs and figure 3c gives an SV-wave speed of 3.0 mmμs. These values were taken from the times of the zero-crossing between the two largest peaks and could be compared with arrival times predicted by RAYTRAIM.

As well as A-scans, the finite difference model plots snapshots of the compression or shear-wave components of the wavefield at specified times on arbitrary planes. Examples were given in [6]. Defects with stress-free surfaces, representing cracks or voids, can also be included in the model very simply [6].

3. CONCLUSIONS

The models show visually how different wave modes propagate. As well as this qualitative understanding they can yield a great deal of quantitative information. The calculations shown here are not for identical cases. RAYTRAIM results at $\pm 45^o$ for infinite frequency are used as illustrations whereas the finite difference results are for waves running horizontally across the structure. The finite difference results are for 2MHz waves with 26 nodes per compression wavelength. The finite difference solution is for an incident beam which spreads in all three directions and is truly three dimensional and suggests that the structure would not be readily inspectable with P-waves at 2MHz. Results for the shear waves are less conclusive since they have been allowed to propagate only half as far through the structure as the P-waves, i.e. they have covered just one weld bead. Both have been run on a CRAY-2 supercomputer, although the ray-tracing program is also available on a VAX. Run times for the problems illustrated here are a few minutes for the ray tracing and a few hours for the finite difference solution (due partly to the large amounts of graphical output requested but not included here).

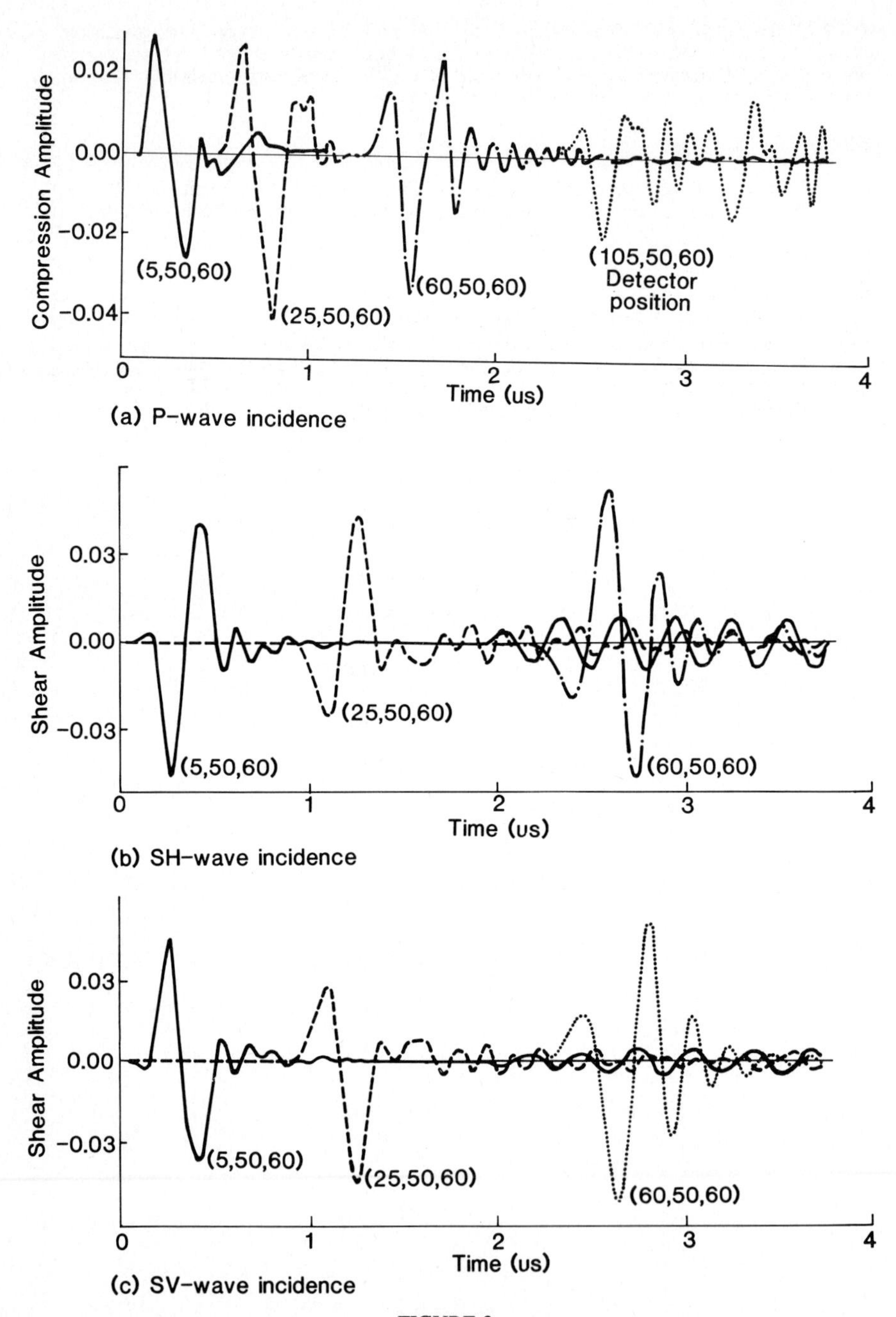

FIGURE 3
Plots of the compression and shear-wave components at various locations outside and inside the weld.

ACKNOWLEDGEMENTS

We are grateful to the Fast Reactor project of the UKAEA for supporting this work.

REFERENCES

[1] Worrall, G.M. and Hudgell, R.J., Ultrasonic Characterisation of Austenitic Welds, in Brook C. and Hanstead P.D., (eds.), Reliability in Non-Destructive Testing (Pergamon Press, Oxford, 1989) pp 187-198.
[2] Ogilvy, J.A., NDT Int. 18 (1985) 67-77.
[3] Ogilvy, J.A., Ultrasonics 24 (1986) 337-347.
[4] Ogilvy, J.A., J.Brit.Inst. NDT 29 (1987) 147-156.
[5] Temple, J.A.G., J.Phys.D:Appl.Phys. 21 (1988) 859-874.
[6] Temple, J.A.G., Optimisation for Ultrasonic Inspection of Austenitic Welds and Components by Three Dimensional Modelling of Elastic Wave Propagation and Scattering in Inhomogeneous and Anisotropic Media, in Brook C and Hanstead P.D., (eds.) Reliability in Non-Destructive Testing (Pergamon Press, Oxford, 1989) pp 199-207.

Table 1 Elastic constants of 308 austenitic weld metal

Density	8.12×10^3 kg m^{-3}
$C_{11} = C_{22}$	$= 262.7 \times 10^9$ Nm^{-2}
C_{12}	$= 98.2 \times 10^9$ Nm^{-2}
$C_{13} = C_{23}$	$= 145.0 \times 10^9$ Nm^{-2}
C_{33}	$= 216.0 \times 10^9$ Nm^{-2}
$C_{44} = C_{55}$	$= 129.0 \times 10^9$ Nm^{-2}
$C_{66} = 1/2\,(C_{11} - C_{12})$	

Elastic Waves and Ultrasonic Nondestructive Evaluation
S.K. Datta, J.D. Achenbach and Y.S. Rajapakse (Editors)

A UNIFIED APPROACH TOWARDS INVERSE PROBLEMS IN ULTRASONIC NDT

L.F. van der Wal [1], M. Lorenz [1] and A.J. Berkhout [2]

[1] TNO Institute of Applied Physics, P.O. Box 155, 2600 AD Delft, The Netherlands

[2] Delft University of Technology, Laboratory of Applied Physics, Group of Seismics and Acoustics, P.O. Box 5046, 2600 GA Delft, The Netherlands

At the end of 1988 a research project was started to develop an applicable, high-resolution imaging method for ultrasonic NDT, based on the synthetic aperture focusing technique. This project is initiated by the TNO Institute of Applied Physics and two major companies in The Netherlands. In this paper our proposed approach and the underlying acoustical theory are presented.

1. INTRODUCTION

There is an increasing interest in the industry for the development of NDT-methods capable of producing high-resolution images of various types of defects. The ultrasonic Synthetic Aperture Focusing Technique (SAFT), amongst others, proved to be a suitable tool to achieve this. As can be concluded from the results of projects regarding the use of SAFT in defect-imaging, the achieved resolution is reasonably good (Seydel [1]). However, the imaging capabilities strongly depend on defect orientation and the way data-acquisition is carried out (Doctor et al. [2]). The methods are generally using zero-offset data and are limited to the focusing of direct energy only. We propose a method, based on SAFT, that makes full use of the information contained in the direct (diffracted and reflected) and indirect (reflected via medium boundaries) energy, together with all apriori knowledge on the object under investigation. We are currently investigating the contributions of all types of waves and wavepaths to the resulting image. To concur with practice, the underlying acoustical theory is presented in this paper in a discretized way and in the form of matrices (Berkhout [3]).

2. THE DATAMODEL

In practice ultrasonic measurements are always discrete in time and space. In linear wave theory and time-invariant media the imaging problem may be described in the temporal frequency domain. In addition, as the recording length is finite (T), the frequency domain may be discrete ($\Delta\omega=2\pi/T$). Therefore our approach is presented as a discrete model in the space/temporal frequency domain, i.e. the (x-ω)-domain.

At the acquisition plane $z=z_0$ a measurement can be carried out by transmitting an ultrasonic signal into the medium under investigation, and by recording the response to that signal at a number of receiver positions. For such an experiment the vector $\vec{P}(z_0)$ describes the monochromatic response at $z=z_0$ as a function of the lateral coordinate x. For a number of physical experiments the data vectors $\vec{P}(z_0)$ may be combined in a data matrix $\mathbf{P}(z_0)$.

3. THE FORWARD PROBLEM

The vector notation presented above can be used to formulate the forward problem. $\vec{P}^+(z_0)$ represents the downgoing source wavefield at the surface. Using wavefield extrapolation we may obtain the downgoing wavefield $\vec{P}^+$ at depthlevel z_m by applying the propagation operator $\mathbf{W}^+$. Each column of this matrix represents the response at depthlevel z_m due to a dipole source at the surface. At depthlevel z_m reflection occurs, which is described by the matrix operator $\mathbf{R}^+$. By applying $\mathbf{R}^+$, the reflected wavefield $\vec{P}^-$ is obtained at depthlevel z_m. Finally, by applying the upward propagation operator $\mathbf{W}^-$ we obtain the reflected wavefield $\vec{P}^-$ at the surface. Each column of the matrix $\mathbf{W}^-$ represents the response at the surface due to a dipole source at depthlevel z_m. Note, that in this formulation vector notation is not only used for the data, but also for the operators, which are presented in the form of matrices.

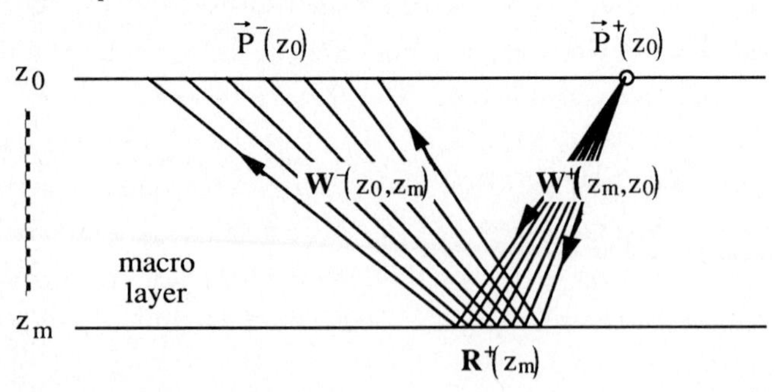

FIGURE 1: A model of the forward problem.

The $\mathbf{W}^+$ and $\mathbf{W}^-$ operators may refer to different wavetypes and wavepaths. In a homogeneous halfspace we only deal with direct wavepaths. In a bounded monolayer we have to include indirect wavepaths, and even refraction effects have to be included in a two-media layer (see figure 2). The propagation operators may be calculated by using either finite difference modeling (Rietveld, [4]) or (Gaussian beam) ray tracing (Kinneging [5]).

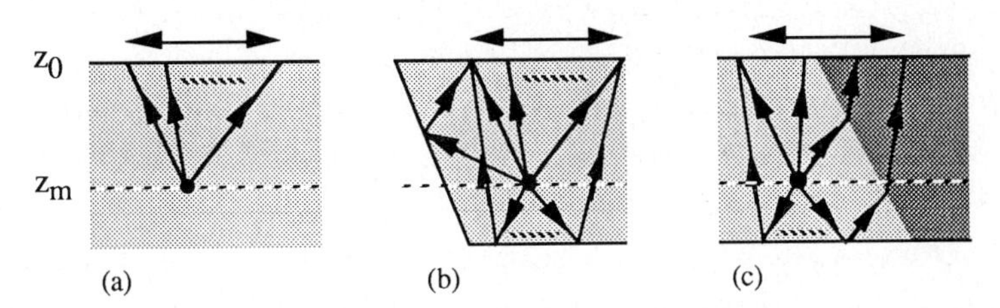

FIGURE 2: Wavepaths in a homogeneous halfspace (a), a bounded monolayer (b) and in a two-media layer (c).

The forward model may be combined into a single matrix equation, including all depthlevels:

$$\vec{P}^-(z_0) = \left[\sum_{m=0}^{\infty} \mathbf{W}^-(z_0,z_m)\, \mathbf{R}^+(z_m)\, \mathbf{W}^+(z_m,z_0) \right] \vec{P}^+(z_0) \ . \tag{1}$$

Finally, the relations between the induced source function and the downgoing wavefield (matrix $\mathbf{D}^+$), and between the measured detector signals and the upgoing wavefield (matrix $\mathbf{D}^-$), have to be added to make the description complete (see figure 3).

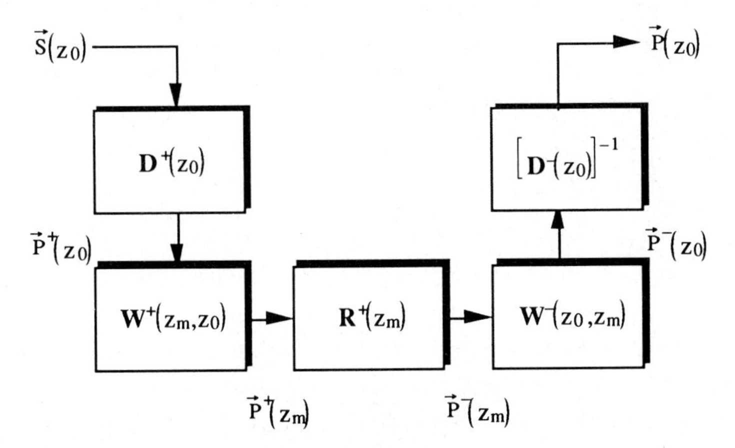

FIGURE 3: The forward cycle for a defect at depth level $z = z_m$.

4. THE INVERSE PROBLEM

To be able to image reflectivity, the reflectivity information has to be recovered from $\mathbf{R}^+$. Given the measured data $\vec{P}(z_0)$, the reflectivity matrix $\mathbf{R}^+$ may be retrieved as a function of depth. Looking at figure 3, this implies that the source and receiver transfer matrices, as well as the downward and upward propagation matrices, have to be removed.

To achieve this, the inverse problem is subdivided in three steps:

1. Surface-related pre-processing, which involves decomposition into one-way waves, and possibly multiple elimination (for instance when using wedges) (Wapenaar & Berkhout [6]). Here discrete signal theory is used.
2. Elimination of propagation effects, which involves construction of the inverse operators, applying them to recover and image reflectivity. Here elastic wave theory is used.
3. Target-related pre-processing, which involves elastic inversion, and possibly defect oriented inversion. Here parametric inversion techniques may be used, together with (non acoustic) apriori information.

The two most important aspects are the decomposition into one-way waves, and the construction of the inverse operators. Two focusing operators have to be designed: focusing operator $\mathbf{F}^+$ to compensate for the upward propagation, and focusing operator $\mathbf{F}^-$ to compensate for the downward propagation:

$$\mathbf{F}^+(z_m,z_0) = [\mathbf{W}^-(z_0,z_m)]^{-1} \approx [\mathbf{W}^+(z_m,z_0)]^* , \tag{2a}$$

$$\mathbf{F}^-(z_0,z_m) = [\mathbf{W}^+(z_m,z_0)]^{-1} \approx [\mathbf{W}^-(z_0,z_m)]^* . \tag{2b}$$

Here the "matched-filter" approach is proposed, suppressing evanescent waves and assuming lossless media, resulting in a stable solution.

5. THE APPROACH

To be able to image the reflectivity accurately, making characterization of defects from the imaging results possible, we will concentrate on the elimination of propagation effects first. To achieve this in practice, the inverse wavefield extrapolation will be carried out for a specific region (volume) of interest only, taking different wavetypes and different wavepaths into account. Also the effects of mode-conversion will be included. Referring to practice, the use of pre-calculated operators (table-driven inversion, Blacquière [7]) is proposed, making efficient monitoring of certain regions of interest easier. The calculation of these operators should preferably be based on (elastic) wave theory (appropriate finite difference or ray tracing schemes) and optimum use should be made of the well-known macro-properties of the medium.

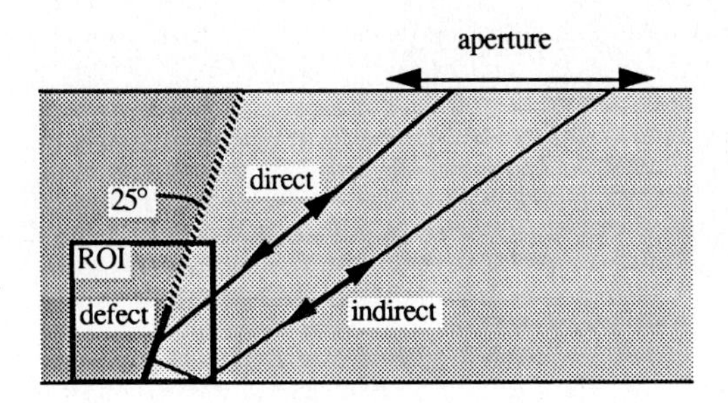

FIGURE 4: The model of an inclined (weld-) defect; the ROI indicates the part imaged in figure 5.

6. AN EXAMPLE

To illustrate the role of indirect wavepaths in SAFT imaging an example is presented here. Figure 4 shows the model of an inclined (weld-) defect of 5 mm length in a 20 mm steel plate, and figure 5 shows the two images obtained by focusing the direct and indirect energy. Zero-offset data-acquisition and shear waves were used, and 70° cosine-weighted angled-beams. The input signal was a pulse with 3 MHz center frequency and 3 MHz (-3 dB) bandwidth. Clearly the orientation, shape and dimensions of the defect are better imaged by focusing the indirect energy. Note from figure 5b that the direct energy causes no distortions of the image, due to the differences in arrival time. To image the indirect energy correctly, use is made of the knowledge of the object-parameters (such as thickness and wave velocity): the macro-properties of the medium.

7. DISCUSSION

The example shown above is simple, but it illustrates elegantly the value of the information contained in the indirectly reflected energy. We are working on a method that will use all the relevant information on the defect(s) contained in all ultrasonic data measured in practice. By combining different images, resulting from focusing different wavepaths/types, the resolution of the final (combined) image may be improved and characterization of defects may be possible from the imaged reflectivity.

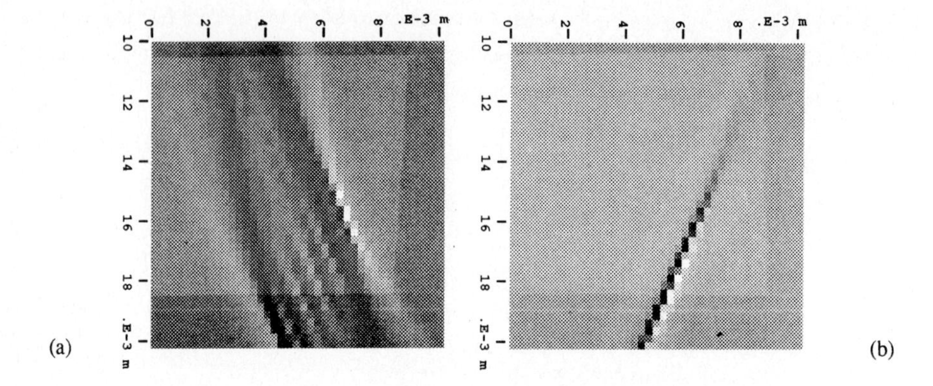

FIGURE 5: The imaging results of zero-offset data-acquisition focusing direct (a) and indirect (b) energy from the (weld-) defect of figure 4.

ACKNOWLEDGEMENT

This research project is currently supported by KEMA Arnhem, Shell Research Amsterdam, and Gasunie Groningen, all in The Netherlands.

REFERENCES

[1] Seydel, J., *Ultrasonic synthetic aperture focusing techniques in NDT*, in: Sharpe, R.S. (Ed.), *Research techniques in nondestructive testing, vol. VI*, Academic Press, 1982, pp.1-47.

[2] Doctor, S.R., Hall, T.E., and Reid, L.D., *SAFT - The evolution of a signal processing technology for ultrasonic testing*, NDT International, vol. 19 no. 3, June 1986.

[3] Berkhout, A.J., *Applied seismic wave theory*, Elsevier Science Publishers Co., Inc., 1987.

[4] Rietveld, W.E.A., *Full elastic modeling for nondestructive evaluation; a finite difference approach*, internal report TNO Institute of Applied Physics, Delft, 1988.

[5] Kinneging, N.A., *Three-dimensional redatuming of seismic shot records*, PhD-thesis, Delft University of Technology, 1989.

[6] Wapenaar, C.P.A., and Berkhout, A.J., *Elastic wave field extrapolation*, Elsevier Science Publishers Co., Inc., to be published in 1989.

[7] Blacquière, G., *3D Wave field extrapolation in seismic depth migration*, PhD-thesis, Delft University of Technology, 1989.

B

GUIDED WAVES IN MULTILAYERED AND COMPOSITE MEDIA

Elastic Waves and Ultrasonic Nondestructive Evaluation
S.K. Datta, J.D. Achenbach and Y.S. Rajapakse (Editors)

WAVE PROPAGATION IN COMPOSITE MEDIA AND MATERIAL CHARACTERIZATION

Subhendu K. Datta[a*], A.H. Shah[b†] and W. Karunasena[b†]

[a]Department of Mechanical Engineering and CIRES
Center for Space Construction
University of Colorado
Boulder, CO 80309-0427

[b]Department of Civil Engineering
University of Manitoba
Winnipeg, Canada R3T 2N2

Characteristics of wave propagation in an undamaged composite medium are influenced by many factors, the most important of which are: microstructure, constituent properties, interfaces, residual stress fields, and ply lay-ups. Measurements of wave velocities, attenuation, and dispersion provide a powerful tool for nondestructive evaluation of these properties. In this paper we review recent developments in modeling of ultrasonic wave propagation in fiber and particle reinforced composite media. Additionally, we discuss some modeling studies of the effects of interfaces and layering on attenuation and dispersion. These studies indicate possible ways of characterizing material properties by ultrasonic means.

1. INTRODUCTION

Ultrasonic waves provide an efficient means of characterizing the effective mechanical properties of a nonhomogeneous material. Several theoretical studies show that for long wavelengths one can model the effective wave speeds and attenuation of plane longitudinal and shear waves propagating through a medium containing a distribution of inclusions or fibers. It is possible also to model the changes in speeds and attenuation of waves propagating in the presence of microcracks or voids. At long wavelengths the wave speeds predicted by these models are nondispersive and hence provide the values for the static elastic moduli of the bulk material.

Speeds of propagation of elastic waves in the presence of a distribution of inclusions have been modeled [1-9]. References to other works can be found in those cited and in the review articles [10,11]. Although most of the early works dealt with spherical inclusions, effect of inclusion shape and orientation has been modeled in [4,6-9]. Anisotropic wave speeds caused by oriented ellipsoidal inclusions were modeled and compared with experiments [9].

As a special case of propagation through a particle-reinforced composite one can also obtain the results for a medium permeated by cracks or voids (pores). There are numerous works that have dealt with the problem of speed and attenuation of waves in such a medium. References to these

*The work reported here was supported in part by grants from the National Science Foundation (#INT-8521422 and #INT-8610487), by a grant from the Office of Naval Research (#N00014-86-K-0280; Scientific Officer: Dr. Y. Rajapakse), and by NASA (#NAGW-1388).
†Supported by a grant from the Natural Science and Engineering Research Council of Canada (#OGP007988).

can be found in [12-15]. Effect of void shape on phase velocities has been reported in [10] and [15].

Most of the works dealing with particle reinforced composites have assumed perfect bonding between the inclusions and the matrix material. Effect of interface (interphase) layers has received some attention [3,16-20]. It has been found that diffuse interface has very weak influence on wave propagation characteristics. However, effect of an interface layer is quite pronounced. Such a layer increases attenuation. Both the density and stiffness of the layer, and the curvature of the interface affect the attenuation.

The literature on wave propagation in a fiber-reinforced composite material is vast. There are numerous models that have been proposed to predict dispersion and attenuation of elastic waves propagating perpendicular to the aligned continuous fibers. References to many of the works can be found in [21]. Most of the reported works deal with isotropic fibers. Anisotropic fibers were considered in [21,22]. In [22] model predictions in conjunction with experiment were used to inversely determine the graphite fiber transverse isotropic elastic constants.

Because most structural composites are laminated or layered, wave propagation in laminated composite plates and shells has received considerable attention. Several papers in this volume deal with this problem and the reader is referred to those for additional references.

In this paper we will focus our attention on guided wave propagation in a composite plate. Two problems will be considered: (1) Guided waves in a cross-ply periodically laminated plate, and (2) dispersion of waves in a bonded plate with isotropic homogeneous layers. These two problems are chosen to illustrate the effect of layering and the interface (bond) layer properties on the dispersion characteristics of guided waves.

2. PROBLEM FORMULATION AND SOLUTION

In this section we will first consider a periodically laminated composite plate where each lamina is made up of a continuous fiber reinforced material. Then in the second part of the section we will discuss guided waves in a plate made up of two isotropic homogeneous layers bonded together by a thin layer of bond material.

2.1 Guided waves in a cross-ply laminated plate

Consider a cross-ply laminated plate, which is composed of alternate layers of continuous fiber reinforced materials of equal thickness. It will be assumed that fibers are oriented at 90° to one another in adjacent layers and that the configuration is symmetric in the plate. Thus the top and bottom layers have fibers oriented in the same direction. A global Cartesian coordinate system with origin on the mid-plane of the middle layer will be chosen. x-axis is chosen parallel to the direction of propagating guided waves which is either parallel or perpendicular to the fibers in the middle layers. y-axis lies in the middle plane and z-axis perpendicular to the plane. The thickness of the plate is taken to be H so that the thickness of each lamina is $h=H/n$, where n is the number of laminae.

If the wavelengths of the propagating waves are much larger than the fiber diameters and spacings then, as has been shown before [21,22], each lamina can be modeled as transversely isotropic with the symmetry axis parallel to the fibers. Thus the problem reduces to that of wave propagation in a plate with layers of transversely isotropic material, where the axes of symmetry in adjacent layers are perpendicular to one another. Our object here is to analyze the effect of the number of layers on dispersion of guided waves propagating either along the x-axis or along the y-axis.

Since we will be considering a large and varying number of layers it will be convenient to resort to a numerical technique in which the number and properties of layers can be altered arbitrarily without substantially changing the solution procedure. Such a technique was proposed earlier by us [23] and also by others [24-26]. In [24-26] authors present a stiffness method in which the thickness variations of the displacements are approximated by quadratic functions of the thickness variable. The generalized coordinates in this prepresentation are the displacements at the top, middle, and bottom of each layer. In [23] an alternative higher order polynomial representation was proposed where generalized coordinates were the displacements and *tractions* at the top and bottom of each layer. This was found to give better results at high frequencies. However, because both displacements and tractions were involved, it entailed much more cumbersome algebra than the scheme used in [24-26]. To avoid this algebraic complexity we will use the quadratic interpolation functions used in [24-26].

Since we consider waves propagating either in the symmetry direction or perpendicular to it in each layer, the problem separates into two uncoupled ones: plane strain in which the displacement components are u_x, 0, u_z, and SH or antiplane strain when the only non-zero displacement is u_y. In this paper we will consider the plane strain problem only.

In order to achieve numerical accuracy each lamina is divided into several sublayers. A local coordinate system ($x^{(k)}$, 0, $z^{(k)}$) is chosen in each sublayer with the origin in the mid-plane. The strain-displacement relations in each sublayer are, for non-vanishing strain components,

$$\varepsilon_{xx}^{(k)} = u_{x,x}^{(k)}\,,\; \varepsilon_{zz}^{(k)} = u_{z,z}^{(k)}\,,\; \varepsilon_{xz}^{(k)} = \frac{1}{2}\gamma_{xz}^{(k)} = \frac{1}{2}(u_{x,z}^{(k)} + u_{z,x}^{(k)}) \tag{1}$$

where comma denotes differentiation. The stress-strain relation in this sublayer is

$$\{\sigma\} = [c^{(k)}]\,\{\varepsilon\} \tag{2}$$

where

$$\{\sigma\}^T = [\sigma_{xx}, \sigma_{zz}, \sigma_{xz}] \tag{3a}$$

$$\{\varepsilon\}^T = [\varepsilon_{xx}, \varepsilon_{zz}, \gamma_{zx}] \tag{3b}$$

$$[c^{(k)}] = \begin{bmatrix} c_{11}^{(k)} & c_{13}^{(k)} & 0 \\ c_{13}^{(k)} & c_{33}^{(k)} & 0 \\ 0 & 0 & c_{55}^{(k)} \end{bmatrix} \tag{3c}$$

For convenience the superscript (k) on u, σ, and ε has been dropped above and in the subsequent development. Using the interpolation polynomials in the z-direction, the displacement components are approximated as,

$$\{U\} = [N]\,\{q\} \tag{4}$$

where

$$\{U\}^T = [u_x, u_z] \tag{5a}$$

$$\{q\}^T = [u_x^b, u_z^b, u_x^m, u_z^m, u_x^f, u_z^f] \tag{5b}$$

$$[N] = \begin{bmatrix} n_1 & 0 & n_2 & 0 & n_3 & 0 \\ 0 & n_1 & 0 & n_2 & 0 & n_3 \end{bmatrix} \tag{5c}$$

In equation (5) the generalized displacements $u_x^b(x,t)$, $u_z^b(x,t)$, $u_x^m(x,t)$, $u_z^m(x,t)$, $u_x^f(x,t)$, and $u_z^f(x,t)$ are taken at the back, middle, and front (top) nodal surfaces of the sublayer. The interpolation polynomials n_i are quadratic functions given by

$$n_1 = -\hat{z}+2\hat{z}^2,\; n_2 = 1-4\hat{z}^2,\; n_3 = 2\hat{z}^2+\hat{z} \tag{6}$$

where $\hat{z}=z^{(k)}/h^{(k)}$, $h^{(k)}$ being the thickness of the sublayer.

Using Hamilton's principle the governing equation for the entire plate is found [26] to be

$$[K_1]\{Q\}'' + [K_2^*]\{Q\}' - [K_3]\{Q\} - [M]\{\ddot{Q}\} = 0 \tag{7}$$

Here $[K_1]$, $[K_3]$, and $[M]$ are symmetric and $[K_2^*]$ is skew symmetric. Primes and dots denote differentiation with respect to x and t, respectively. $\{Q\}$is the vector of all the nodal displacement components. We will consider propagating waves in the x-direction. Thus $\{Q\}$ is assumed of the form

$$\{Q\} = \{Q_o\}e^{i(kx-\omega t)} \tag{8}$$

Substituting (8) in (7) we get the eigenvalue problem

$$[-K_1k^2 + K_2^* ik - K_3 + M\omega^2]\ \{Q_o\} = 0 \tag{9}$$

Equation (3) can be solved to find ω for a given k or to find k for given ω. Some results are discussed in section 3.

2.2 Rayleigh-Lamb Waves in a Bonded Plate

The characterization of bond quality using ultrasonic techniques has long been a subject of study by many researchers. Ultrasonic methods provide a powerful tool for detecting debonding or weakening of bond strength. Their success in measuring bond strength largely depends on the understanding of the nature of the changes in the wave propagation characteristics due to the changes in the material properties of the bond layer.

A study of guided waves in thin layers can be found in [27]. For thin bonded layers there have been several studies [28-34]. In most of these the case of normal incidence is considered. Oblique incidence has been considered in [35]. In these studies the bond layer is approximated as a massless spring or a fluid layer that allows jump in the displacement keeping tractions continuous. Attempts at using this spring or slip model to detect weak bonding has shown its applicability to a class of bonds, although there are indications that adjustable coefficients may be needed to explain the behavior at high frequencies [35-37]. In [20,36,37] a shell model that combines the effects of density (inertia) and stiffness has been developed to study scattering from inclusions with thin interface (bond) layers. It is found that the effect of density dominates in the cases considered. Baik and Thompson [38] have also considered a shell model to analyze dispersive behavior and a density model has been used by Nayfeh and Nassar [39].

In an effort to assess the feasibility of characterizing bond properties by ultrasonic means we [40] have made a parametric study of the exact spring and density models for the thin interface layer in a bonded plate. A summary of the results is presented in the following.

We consider a sandwich plate of two outer layers and an interface thin bond layer of materials that are isotropic and homogeneous. A global Cartesian coordinate system (x,y,z) with origin at the free surface of the top layer, and x,y-axis parallel and z-axis perpendicular to the surface, is considered. For convenience our attention will be focused on the two-dimensional problem. Thus it is assumed that the displacement components $(u_x,0,u_z)$ uncouples from the antiplane strain motion $(0,u_y,0)$. Here we present some results for the plane strain case.

In plane strain deformation the nonzero displacements and tractions at an interface z=constant form a four vector

$$\{S\} = [u_x,\ u_z,\ \tau_{xz},\ \tau_{zz}]^T \tag{10}$$

Let $\{S^-\}$ and $\{S^+\}$ denote the values of the four vector $\{S\}$ as $z \to h^-$ and h^+, respectively. Here z=h is the depth of an interface from the top free surface. If the bond is perfect at the interface then

$$\{S^-\} = \{S^+\} \tag{11}$$

Now consider two layers occupying $0 \leq z \leq h_1$ and $h_2 \leq z \leq h_3$ are bonded together by a thin layer of thickness $h_2 - h_1 = h_0$. If h_0 is very small compared to h_1, $h_3 - h_2$, and the wavelength of the propagating wave, then the thin layer is often approximated as one of vanishing thickness and the interface conditions are applied at $z = h_1 \approx h_2$. In the shell model it is assumed that

$$\{S^+ - S^-\} = \frac{1}{2}\begin{bmatrix} 0 & F \\ D & 0 \end{bmatrix}\{S^- + S^+\} \tag{12}$$

where

$$[F] = \text{Diag}[a\frac{h_o}{\mu_o} \quad a\frac{h_o}{\lambda_o + 2\mu_o}] \tag{13}$$

$$[D] = \text{Diag}[-b\rho_o\omega^2 h_o - b\rho_o\omega^2 h_o] \tag{14}$$

where a,b are some parameters, and λ_o, μ_o, and ρ_o are the Lamé constants and density, respectively, of the layer. In writing (12) a harmonic time dependence of the form $e^{-i\omega t}$ has been assumed, ω being the circular frequency. Note that when a=0 one obtains the density model and for b=0 the spring model holds.

Equation (12) is often an adequate approximation for the thin layer, the regimes of validity of which have not been systematically investigated. In [40] a parametric study has been performed for a typical plate. Some of these results are discussed in the next section.

3. NUMERICAL RESULTS AND DISCUSSION

In this section we present some selected numerical results showing the effect of the number of layers and bond properties on the dispersion of guided waves.

3.1 Dispersion of Waves in a Laminated Plate

In order to understand the effect of the number of laminae we consider a graphite fiber-reinforced laminate. The properties of 0° and 90° laminae are given in Table 1. Here 0° signifies fibers aligned with the wave propagation direction (x-axis) and 90° signifies fibers aligned with the y-axis.

Table 1. Properties of the laminae
(All stiffnesses are in units of 10" N/m^2)

Layer	ρ(g/cm^3)	c_{11}	c_{33}	c_{13}	c_{44}	c_{55}
0° lamina	1.2	1.6073	0.1392	0.0644	0.0350	0.0707
90° lamina	1.2	0.1391	0.1392	0.0350	0.0707	0.0350

Figures 1(a) and 2(a) show the variation of phase velocities of different modes with frequency in three-layered (0°/90°/0°) and 39-layered (...0°/90°/0°...) plates. Corresponding results for propagation in differently oriented plates (90°/0°/90°;...90°/0°/90°...) are presented in Figures 1(b) and 2(b). It is seen that dispersion characteristics depicted by Figs. 1(a) and 1(b) are vastly different. However, Figs. 2(a) and 2(b) are remarkably similar. In fact, these results agree quite well with predictions of an effective modulus theory (for details the reader is referred to [41]). Thus it appears that for a sufficiently large number of layers the plate behaves isotropically in its plane. This is a remarkable result and clearly the number of layers necessary to show in-plane isotropic behavior must depend on the material properties and stacking sequence of the laminae. We hope to pursue this further in the future.

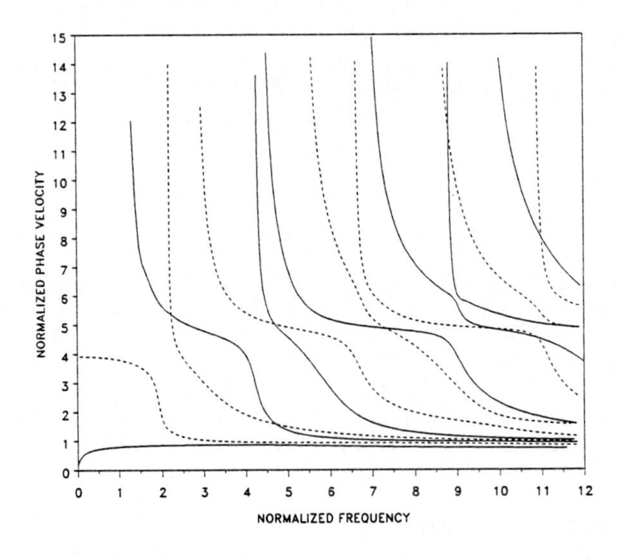

(a)

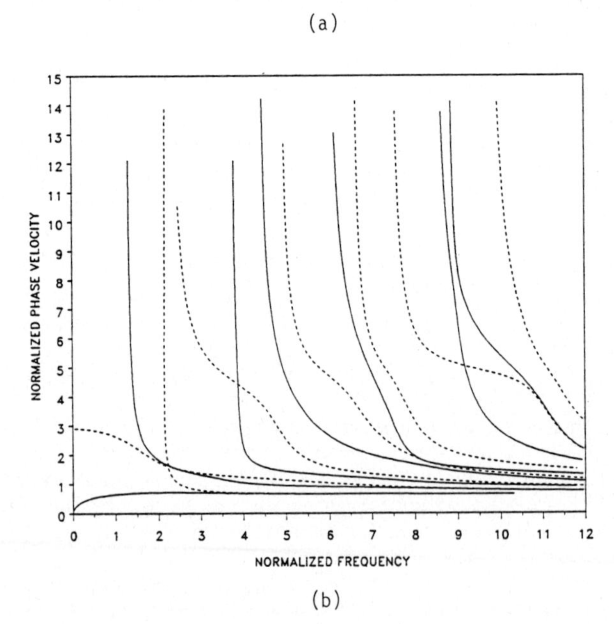

(b)

FIGURE 1
Guided waves in a (0°/90°/0°) three-layered plate.
(a) Propagation in the 0° direction. (b) Propagation in the 90° direction.

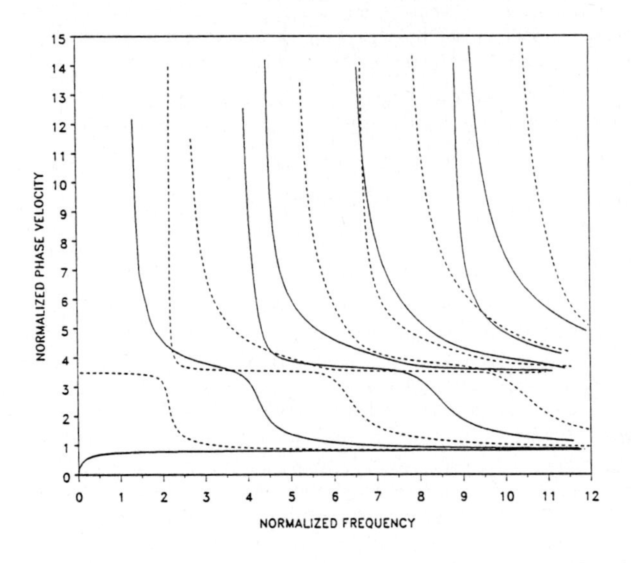

(a)

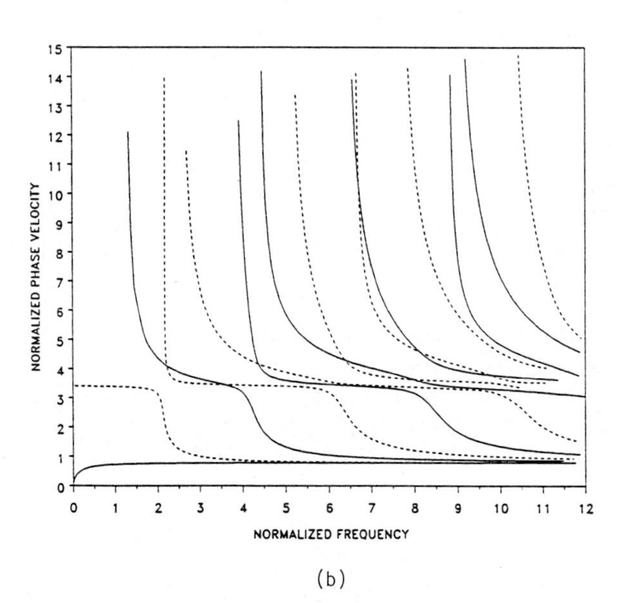

(b)

FIGURE 2
Same as Figure 1 for a 39-layered plate.

3.2 Interface Layer Effects on Dispersion

To exhibit the implications of different approximations of thin interface layers on the dispersion of guided waves we consider a plate made up of an outer layer of gold on a nickel-iron substrate. Properties are given in Table 2. Properties of the interface layer are varied.

Table 2. Properties of a Bonded Plate
(Longitudinal wave speed=c_p, Shear wave speed=c_s)

Material	ρ (g/cm^3)	c_p (mm/μs)	c_s (mm/μs)	Thickness (mm)
Gold (au)	19.32	3.24	1.22	0.45
Bond	--	--	--	0.5
Fe - 42% Ni	8.10	4.86	2.60	10.0

Figures 3 and 4 show the dispersion curves predicted by the actual model and those obtained by the approximate spring and mass models. In these figures M is the ratio of the shear moduli of the bond layer and the gold layer, and D is the ratio of the densities. It is seen from Fig. 3 that when the bond layer has both high modulus and high density then the spring model predictions are higher than the actual. In this case the density model predictions are very close to the actual for all the modes except the third, when the results agree only at low frequencies. The reason for this anomalous behavior is not clear. Figure 4 shows the results for the case when the bond layer is very weak (low density and low modulus). It is seen that in this case the spring model agrees very well with the exact, whereas the density model predictions are higher. It is interesting to note that in both cases the two model predictions become close to the exact at high frequencies for all the modes except the first and the third. Other features (not shown) of interest are:

(a) both the spring and density model predictions agree with the exact when the bond is high modulus and low density;

(b) none of the models is very good for all the modes when the bond is low modulus and high density.

4. CONCLUSION

It has been demonstrated that ultrasonic velocity techniques provide sensitive means of characterizing the properties of bulk composite materials and interfaces between the different phases. It is also shown that dispersion characteristics of waves in a plate are very sensitive to the following parameters:

1. Number of layers and stacking sequence in a laminated plate.
2. Properties of the interface bond layer.

Thus the ultrasonic waves may be used to characterize these properties. Furthermore, it appears that a laminated plate with a sufficiently large number of layers can be modeled as effectively quasi-isotropic. This feature may have very important significance for ultrasonic characterization of thick composites.

ACKNOWLEDGEMENTS

The authors are grateful to Dr. P.-C. Xu for carrying out the parametric study of a bonded plate.

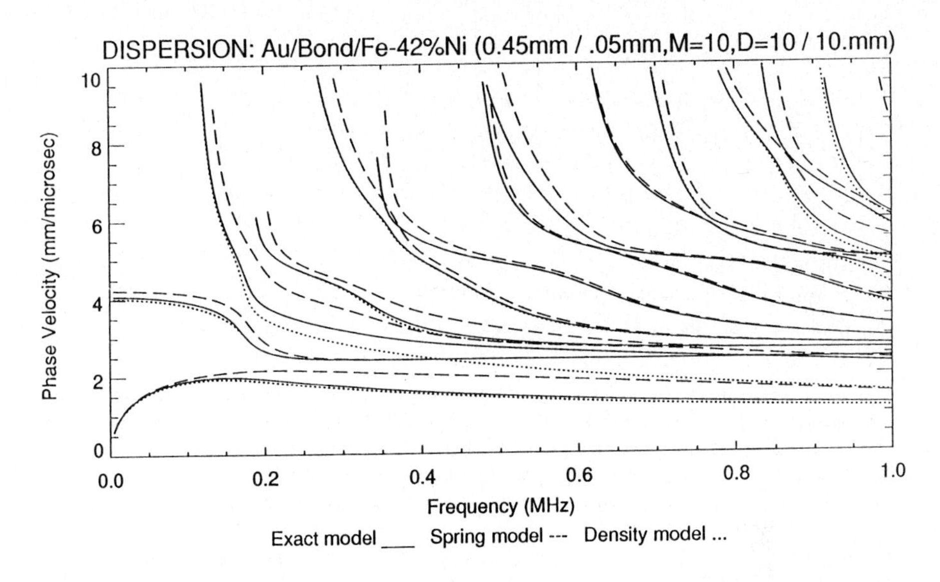

FIGURE 3
Effect of a heavy stiff bond layer on guided waves.

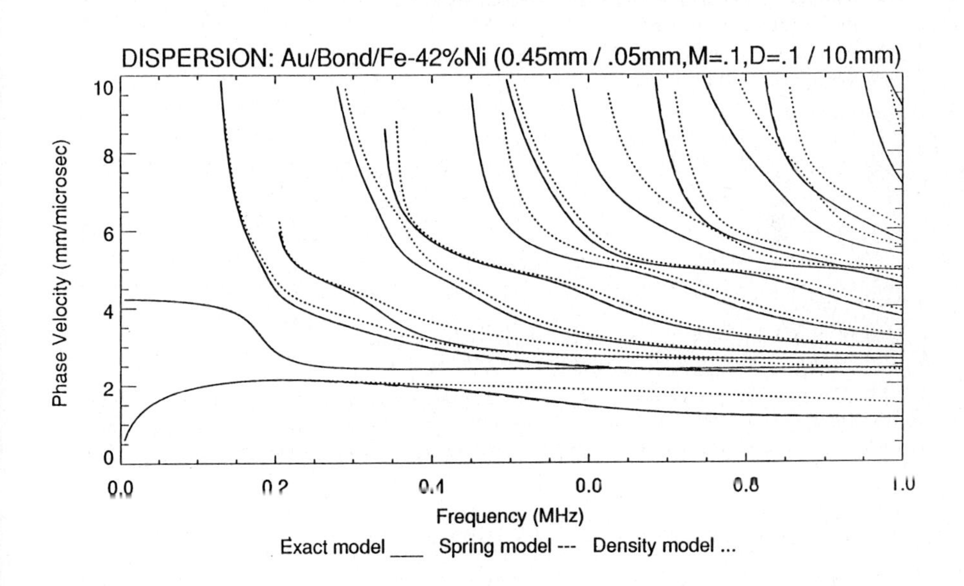

FIGURE 4
Effect of a light soft bond layer on guided waves.

REFERENCES

[1] Fikioris, J.G. and Waterman, P.C., J. Math Phys. 5 (1964) 1413.

[2] Mal, A.K. and Knopoff, L., J. Inst. Math. Appl. 3 (1967) 376.

[3] Mal, A.K. and Bose, S.K., Proc. Camb. Philos. Soc. 76 (1974) 587.

[4] Datta, S.K., J. Appl. Mech. 44 (1977) 657.

[5] Devaney, A.J., J. Math. Phys. 21 (1980) 2603.

[6] Berryman, J.G., J. Acoust. Soc. Am. 68 (1980) 1809.

[7] Willis, J.R., J. Mech. Phys. Solids 28 (1980) 307.

[8] Varadan, V.K., Ma., Y.C., and Varadan, V.V., J. Acoust. Soc. Am. 77 (1985) 375.

[9] Ledbetter, H.M. and Datta, S.K., J. Acoust. Soc. Am. 79 (1986) 239.

[10] Datta, S.K. and Ledbetter, H.M., in: Johnson, G.C. (ed.), Wave Propagation in Homogeneous Media and Ultrasonic Nondestructive Evaluation, AMD-Vol. 62 (American Society of Mechanical Engineers, New York, 1984) pp. 141-153.

[11] Datta, S.K. and Ledbetter, H.M., in: Lamb, J.P. (ed.), Proceedings of the Tenth U.S. National Congress of Applied Mechanics (American Society of Mechanical Engineers, New York, 1987) pp. 377-387.

[12] Chatterjee, A.K., Mal, A.K., Knopoff, L., and Hudson, J.A., Math. Proc. Camb. Phil. Soc. 88 (1980) 547.

[13] Hudson, J.A., Geophys. J. R. Astr. Soc. 64 (1981) 133.

[14] Sayers, C.M., Ultrasonics 26 (1988) 73.

[15] Ledbetter, H.M., Lei, M., and Datta, S.K., in: Holbrook, J. and Bussière, J., Proceedings of the Symposium on Nondestructive Monitoring of Materials Properties (Materials Research Society, Pittsburgh, PA, 1989).

[16] Sayers, C.M., Wave Motion 7 (1985) 95.

[17] Datta, S.K. and Ledbetter, H.M., in : Selvadurai, A.P.S. and Voyiadjis, G.Z. (eds.), Mechanics of Materials Interfaces (Elsevier, Amsterdam, 1986) pp. 131-141.

[18] Datta, S.K., Ledbetter, H.M., Shindo, Y., and Shah, A.H., Wave Motion 10 (1988) 171.

[19] Paskaramoorthy, R., Datta, S.K., and Shah, A.H., J. Appl. Mech. (1988) 871.

[20] Olsson, P., Datta, S.K., and Boström, A., Elastodynamics Scattering from Inclusions Surrounded by Thin Interface Layers, to be published.

[21] Datta, S.K., Ledbetter, H.M., and Kriz, R.D., Int. J. Solids Str. 20 (1984) 429.

[22] Ledbetter, H.M., Datta, S.K., and Kyono, T., J. Appl. Phys. 65 (1989) 3411.

[23] Datta, S.K., Shah, A.H., Bratton, R.L., and Chakraborty, T., J. Acoust. Soc. Am. 83 (1988) 2020.

[24] Dong, S.B. and Nelson, R.B., J. Appl. Mech. 39 (1972) 739.

[25] Dong, S.B. and Pauley, K.E., J. Eng. Mech. ASCE, 104 (1978) 802.

[26] Dong, S.B. and Hwang, K.H., J. Appl. Mech. 52 (1985) 433.

[27] Farnell, G.W., and Adler, E.L., in: Mason, W.P. and Thurston, R.N. (eds.), Physical Acoustics, Vol. 9 (Academic Press, New York, 1972) pp. 35-127.

[28] Jones, J.P. and Whittier, J.S., J. Appl. Mech. 34 (1967) 905.

[29] Alers, G.A. and Graham, L.J., in: de Klerk, J. (ed.), Proceedings of the IEEE Ultrasonics Symposium (IEEE, New York, 1975) 579.

[30] Schoenberg, M., J. Acoust. Soc. Am. 68 (1980) 1516.

[31] Rokhlin, S., Hefets, M., and Rosen, M., J. Appl. Phys., 51 (1980) 3579.

[32] Rokhlin, S., Hefets, M., and Rosen, M., J. Appl. Phys., 52 (1980) 2847.

[33] Rokhlin, S.I. and Marom, D., J. Acoust. Soc. Am. 80 (1986) 585.

[34] Tsukahara, Y. and Ohira, K., Ultrasonics 27 (1989) 3.

[35] Mal, A.K. and Xu, P.-C., in: McCarthy, M.F. and Hayes, M.A. (eds.), Elastic Wave Propagation (North-Holland, Amsterdam, 1979) pp. 67-73.

[36] Datta, S.K., Olsson, P., Boström, A., in: Ting, T.C.T. and Mal, A.K. (eds.), Wave Propagation in Structural Composites, AMD-Vol. 90 (American Society of Mechanical Engineers, New York, 1988) pp. 109-116.

[37] Olsson, P., Datta, S.K., and Boström, A., in: McCarthy, M.F. and Hayes, M.A., (eds.), Elastic Wave Propagation (North-Holland, Amsterdam, 1989) pp. 381-386.

[38] Baik, J.M. and Thompson, R.B., J. Nondestr. Eval. 4 (1984) 177.

[39] Nayfeh, A.H. and Nassar, E.A., J. Appl. Mech. 45 (1978) 822.

[40] Xu, P.-C. and Datta, S.K., Guided Waves in a Bonded Plate: A Parametric Study, to be published.

[41] Karunasena, W., Datta., S.K., and Shah, A.H., Wave Propagation in a Multi-layered Laminated Cross-ply Composite Plate, to be published.

Elastic Waves and Ultrasonic Nondestructive Evaluation
S.K. Datta, J.D. Achenbach and Y.S. Rajapakse (Editors)

FREE WAVES AT A FLUID/LAYERED-COMPOSITE INTERFACE

Arthur M. B. BRAGA and George HERRMANN

Division of Applied Mechanics
Stanford University
Stanford, California 94305-4040

The propagation of free harmonic waves along the interface between an acoustic fluid and an anisotropic laminated composite is investigated. The dispersion equation is written in terms of the impedance of the fluid and the surface impedance tensor of the layered composite. An algorithm for the numerical evaluation of the surface impedance tensor of anisotropic laminated plates is presented. Attention is focused on the subsonic Scholte-Gogoladze-like wave.

1. INTRODUCTION

The propagation of free waves at a fluid/solid interface finds important applications in ultrasonic non-destructive evaluation and in structural acoustics. In a homogeneous half-space, two types of such waves are known to exist [1]. The first, with decaying amplitude as it propagates along the interface, and which is equivalent to the Rayleigh wave in the solid as the fluid density becomes negligible, is called a *leaky* or *generalized Rayleigh* wave. The other, named here the *Scholte-Gogoladze* wave after the two researchers that first, independently, uncovered its existence in 1948 [2,3], propagates unattenuated parallel to the boundary, while decaying exponentially in both directions away from the interface.

In the present contribution, the propagation of free harmonic waves along the interface between an acoustic fluid and an anisotropic laminated composite is investigated. This problem has attracted the attention of many researchers in recent years. Most of the efforts have focused on devising efficient techniques to model the composite. Matrix methods based on the concept of a propagator matrix, or on the Thompson-Haskell matrices, are the most widely used [4-6]. It is well-known, however, that this approach leads to numerical instabilities at high frequencies [6], and special schemes (see, for instance, [7]) have to be employed in order to avoid such problems. In the case of composites made of anisotropic layers, when all the stress and displacement components are coupled, the use of these schemes becomes very difficult. An efficient method, the Global Matrix technique, has been devised recently by Mal [8]. In this paper we use an alternative approach, based on the work of Hager and Rostamian [9] for isotropic layered media, and present a numerically stable algorithm for the evaluation of the surface impedance tensor of anisotropic laminated composites. As will become clear, this rank two tensor is a fundamental concept in the type of fluid/solid interaction problems discussed here.

2. THE IMPEDANCE TENSOR

Let $\boldsymbol{u}$ and $\boldsymbol{\sigma}$ represent respectively the displacement and Cauchy stress tensor in a inhomogeneous elastic solid. The traction vector acting across planes normal to a fixed unit vector $\boldsymbol{n}$ is given by

$$\boldsymbol{t} = \boldsymbol{\sigma}\boldsymbol{n} \ . \tag{1}$$

In the present investigation we are concerned with plane harmonic motions. Therefore, the following assumption is made

$$\boldsymbol{u}(x,z,t) = \bar{\boldsymbol{u}}(z;\omega,k_x)e^{i(\omega t - k_x x)} \quad \text{and} \quad \boldsymbol{t}(x,z,t) = \bar{\boldsymbol{t}}(z;\omega,k_x)e^{i(\omega t - k_x x)}, \tag{2}$$

Furthermore, we let the unit vector $\boldsymbol{n}$ be parallel to the Cartesian axis z. The impedance tensor $\boldsymbol{G}(z;\omega,k_x)$ is defined by the expression*

$$\bar{\boldsymbol{t}}(z) = i\omega \boldsymbol{G}(z)\,\bar{\boldsymbol{u}}(z)\,. \tag{3}$$

3. FLUID/SOLID INTERFACE

Let an acoustic fluid of density ρ_f and speed of sound c_f occupy the semi-infinite region $z > 0$. This fluid half-space is in contact at $z = 0$ with the plane surface of an inhomogeneous elastic solid. We consider plane harmonic motions of the type

$$\boldsymbol{v}_f(x,z,t) = \bar{\boldsymbol{v}}_f(z)e^{i(\omega t - k_x x)} \quad \text{and} \quad p_f(x,z,t) = \bar{p}_f(z)e^{i(\omega t - k_x x)}, \tag{4}$$

where $\boldsymbol{v}_f$ and p_f are, respectively, the particle velocity and pressure fields in the fluid. The solid surface is characterized by the *surface impedance tensor* $\boldsymbol{G}$, which is the value of the impedance tensor of this inhomogeneous solid at the plane $z = 0$. The unit vector $\boldsymbol{n}$, parallel to the Cartesian z-axis, is normal to the plane interface. In the absence of interior sources, the motion in the fluid half-space is described, in addition to (4), by equations

$$\bar{\boldsymbol{v}}_f(z)\cdot\boldsymbol{n} = e^{-i\varphi z}\,\bar{\boldsymbol{v}}_f(0)\cdot\boldsymbol{n} \quad \text{and} \quad \bar{p}_f(z) = \mathcal{Z}_f\,\bar{\boldsymbol{v}}_f(z)\cdot\boldsymbol{n}\,, \tag{5}$$

where the phase $\varphi = \hat{\varphi}(\omega,k_x)$ is chosen as

$$\hat{\varphi}(k_x,\omega) = \begin{cases} -i\left(k_x^2 - \omega^2/c_f^2\right)^{\frac{1}{2}}, & \omega/c_f < k_x\,; \\ \left(\omega^2/c_f^2 - k_x^2\right)^{\frac{1}{2}}, & \omega/c_f \geq k_x\,; \end{cases} \tag{6}$$

in order to satisfy the radiation condition. In equation $(5)_2$, $\mathcal{Z}_f = \hat{\mathcal{Z}}_f(\omega,k_x)$ is the fluid impedance given by

$$\hat{\mathcal{Z}}_f(k_x,w) = \rho_f\omega/\hat{\varphi}(k_x,\omega)\,. \tag{7}$$

Along the fluid/solid interface the following continuity conditions must be obeyed:

$$\bar{\boldsymbol{v}}_f(0)\cdot\boldsymbol{n} = i\omega\,\bar{\boldsymbol{u}}\cdot\boldsymbol{n} \quad \text{and} \quad \bar{\boldsymbol{t}} = -\bar{p}_f(0)\boldsymbol{n}\,, \tag{8}$$

where $\bar{\boldsymbol{u}}$ and $\bar{\boldsymbol{t}}$, which are, respectively, the displacement and traction vectors along the solid surface, are related by the surface impedance tensor, *i.e.*

$$\bar{\boldsymbol{t}} = i\omega\boldsymbol{G}\bar{\boldsymbol{u}}\,. \tag{9}$$

By combining (9) with expressions (5), (7) and (8), and solving for the normal component of the particle velocity at the interface, we obtain

$$\left[\mathcal{Z}_f + \mathcal{Z}_s\right]\,\bar{\boldsymbol{v}}_f(0)\cdot\boldsymbol{n} = 0\,, \tag{10}$$

where $\mathcal{Z}_s = \hat{\mathcal{Z}}_s(\omega,kx)$ is the normal impedance of the solid surface, given by

$$\hat{\mathcal{Z}}_s(\omega,k_x) = (\boldsymbol{n}\cdot\boldsymbol{G}^{-1}\boldsymbol{n})^{-1}\,. \tag{11}$$

---

* The dependence of the impedance tensor as well as of $\bar{\boldsymbol{u}}$ and $\bar{\boldsymbol{t}}$ on the pair (ω,k_x) is implicitly assumed hereafter.

Equation (10) admits non-trivial solutions only if

$$\hat{\mathcal{Z}}_s(\omega, k_x) + \hat{\mathcal{Z}}_f(\omega, k_x) = 0 \ . \tag{12}$$

Hence, expression (12) above is the dispersion equation for the propagation of free harmonic waves along the fluid/solid interface.

4. THE EVALUATION OF THE SURFACE IMPEDANCE TENSOR

4.1 Homogeneous anisotropic media

In an unbounded homogeneous anisotropic body, the displacement field at a plane of coordinate z can be written as

$$\bar{\boldsymbol{u}}(z) = \bar{\boldsymbol{u}}_1(z) + \bar{\boldsymbol{u}}_2(z) \ , \tag{13}$$

where $\bar{\boldsymbol{u}}_1(z)$ is the displacement vector associated with waves propagating (or decaying) along the z-positive direction, while $\bar{\boldsymbol{u}}_2(z)$ correspond to waves propagating (or decaying) along the z-negative direction. Likewise, the total traction field can be written as the sum of two partial ("upgoing" and "downgoing") components

$$\bar{\boldsymbol{t}}(z) = \bar{\boldsymbol{t}}_1(z) + \bar{\boldsymbol{t}}_2(z) \ . \tag{14}$$

It can be shown that [10]

$$\bar{\boldsymbol{u}}_\alpha(z) = \boldsymbol{M}_\alpha(z)\bar{\boldsymbol{u}}_\alpha(0) \quad \alpha = 1,2 \tag{15}$$

where

$$\boldsymbol{M}_1(z) = \boldsymbol{A}_1\{\mathrm{diag}\,(e^{-ik_{z_1}z}, e^{-ik_{z_2}z}, e^{-ik_{z_3}z})\}\boldsymbol{A}_1^{-1} \tag{16a}$$

and

$$\boldsymbol{M}_2(z) = \boldsymbol{A}_2\{\mathrm{diag}\,(e^{-ik_{z_4}z}, e^{-ik_{z_5}z}, e^{-ik_{z_6}z})\}\boldsymbol{A}_2^{-1} \tag{16b}$$

The wave-numbers k_{z_i} and the (3×3) matrices $\boldsymbol{A}_\alpha$, which depend on the material properties as well as on the pair (ω,k_x), are obtained by solving the sextic eigenvalue problem [5]

$$\boldsymbol{N}\,\Xi = \Xi\,\{\mathrm{diag}\,(k_{z_1}, k_{z_2}, \ldots, k_{z_6})\} \tag{17}$$

where

$$\Xi = \begin{bmatrix} \boldsymbol{A}_1 & \boldsymbol{A}_2 \\ \boldsymbol{L}_1 & \boldsymbol{L}_2 \end{bmatrix} , \tag{18}$$

and $\boldsymbol{N}$ is the *fundamental elasticity tensor* for the homogeneous anisotropic medium. The eigenvalues of $\boldsymbol{N}$, which are the wave-numbers in the z-direction, are ordered such that the first three are associated with the "upgoing" waves (propagating or decaying in the z-positive direction) while the last three are associated with the "downgoing" waves (propagating or decaying in the z-negative direction). This ordering assures that

$$\left|\exp(-ik_{z_j}|z|)\right| \leq 1 \ , \quad \text{while} \quad \left|\exp(-ik_{z_{j+3}}|z|)\right| \geq 1 \quad (j = 1,2,3) \ . \tag{19}$$

The partial tractions $\boldsymbol{t}_1(z)$ and $\boldsymbol{t}_2(z)$ are given by

$$\bar{\boldsymbol{t}}_1(z) = i\omega\,\boldsymbol{Z}\,\bar{\boldsymbol{u}}_1(z) \quad \text{and} \quad \bar{\boldsymbol{t}}_2(z) = -i\omega\,\boldsymbol{Z}^T\,\bar{\boldsymbol{u}}_2(z) \ , \tag{20}$$

where

$$\boldsymbol{Z} = -(1/\omega)\,\boldsymbol{L}_1\,\boldsymbol{A}_1^{-1} \tag{21}$$

is the *local* impedance tensor for the anisotropic medium [11]. The values of the partial displacements at $z = 0$, $\bar{\boldsymbol{u}}_1(0)$ and $\bar{\boldsymbol{u}}_2(0)$, are unknown constants which must be obtained by imposing proper continuity or boundary conditions at planes normal to the z-axis.

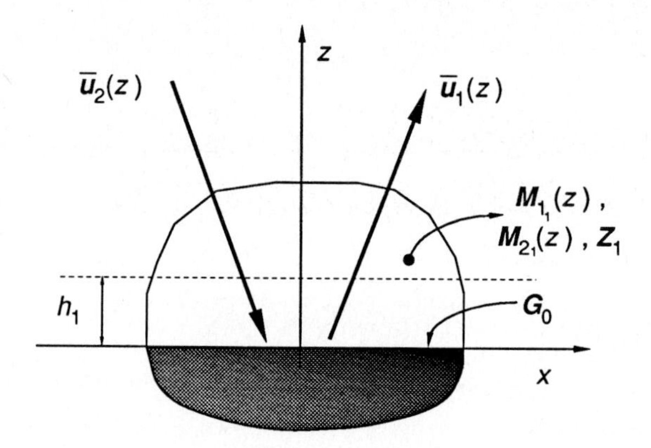

Figure 1. Homogeneous Anisotropic half-space bonded to another semi-infinite medium with surface impedance tensor $\boldsymbol{G}_0$.

We now employ the formalism presented above to solve the following problem :

Problem I. An elastic homogeneous anisotropic half-space, occupying the region $z > 0$, is bonded at $z = 0$ to the plane surface of another semi-infinite medium (Fig. 1), which is characterized by the surface impedance tensor $\boldsymbol{G}_0$ (assumed known). The frequency ω and wave-number k_x are fixed, and the fundamental elasticity tensor for the homogeneous half-space ($z > 0$) is denoted by $\boldsymbol{N}_1$.

find

1) The *reflection tensor* $\boldsymbol{R}_0$ such that $\bar{\boldsymbol{u}}_1(0) = \boldsymbol{R}_0\, \bar{\boldsymbol{u}}_2(0)$.
2) The *impedance tensor* at $z = h_1$,*i.e.* the tensor $\boldsymbol{G}_1$ such that $\bar{\boldsymbol{t}}(h_1) = i\omega \boldsymbol{G}_1\, \bar{\boldsymbol{u}}(h_1)$.

the solution [10]

$$\boldsymbol{R}_0 = (\boldsymbol{Z}_1 - \boldsymbol{G}_0)^{-1}(\boldsymbol{Z}_1^T + \boldsymbol{G}_0) \tag{22a}$$

$$\boldsymbol{G}_1 = (\boldsymbol{Z}_1\boldsymbol{H}_1 - \boldsymbol{Z}_1^T)(\boldsymbol{I} + \boldsymbol{H}_1)^{-1} \ , \tag{22b}$$

where

$$\boldsymbol{H}_1 = \boldsymbol{M}_{1_1}(h_1)\boldsymbol{R}_0\boldsymbol{M}_{2_1}^{-1}(h_1) \ . \tag{22b}$$

In equations (22), $\boldsymbol{Z}_1$, $\boldsymbol{M}_{1_1}(h_1)$ and $\boldsymbol{M}_{2_1}(h_1)$ are obtained from expressions (16) and (21), and by solving the sextic eigenvalue problem (17) for the fundamental elasticity tensor $\boldsymbol{N}_1$. The solution of Problem I can be used as a building block in a scheme for the evaluation of the surface impedance tensor for multilayered media.

4.2. Multilayered media

We now consider an infinite laminated plate, made of n homogeneous anisotropic layers, and bonded to an inhomogeneous elastic half-space with surface impedance $\boldsymbol{G}_0$ (Fig. 2). If the bottom of the plate is traction-free, we let $\boldsymbol{G}_0 = \boldsymbol{0}$, if, instead, the plate is in contact with an acoustic fluid of impedance $\hat{z}_{f_0}(\omega, k_x)$, we take $\boldsymbol{G}_0 = -\hat{z}_{f_0}(\omega, k_x)[\boldsymbol{n} \otimes \boldsymbol{n}]$. The layers have thickness h_1, h_2, ... h_n, and the total thickness of the plate is d. We want to evaluate the surface impedance tensor at the top of the plate, denoted by $\boldsymbol{G}_n$.

We start by using equations (22) to obtain the surface impedance tensor $\boldsymbol{G}_1$ at the top of

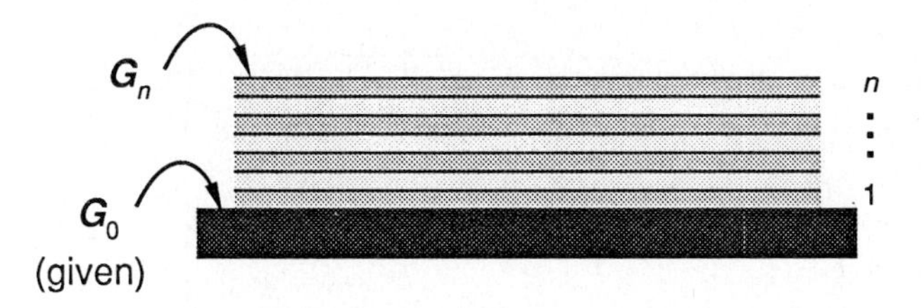

Figure 2. Laminated plate bonded to a plane solid surface with acoustic impedance tensor $\boldsymbol{G}_0$.

the first layer, then proceed by using this result to calculate the surface impedance tensor at the top of the second layer and so on, up to the nth layer. This procedure is summarized in the following algorithm :

Given $\boldsymbol{G}_0$

For $j = 1$ to n **Repeat**

$$\begin{aligned} \boldsymbol{R}_{j-1} &= (\boldsymbol{Z}_j - \boldsymbol{G}_{j-1})^{-1}(\boldsymbol{Z}_j^T + \boldsymbol{G}_{j-1}) \\ \boldsymbol{H}_j &= \boldsymbol{M}_{1_j}(h_j)\boldsymbol{R}_{j-1}\boldsymbol{M}_{2_j}^{-1}(h_j) \\ \boldsymbol{G}_j &= (\boldsymbol{Z}_j\boldsymbol{H}_j - \boldsymbol{Z}_j^T)(\boldsymbol{I} + \boldsymbol{H}_j)^{-1} \end{aligned} \tag{23}$$

It can be shown that, for any frequency, as a consequence of the way the wave-numbers k_{z_i} in the homogeneous layers are ordered, the magnitudes of the eigenvalues of both $\boldsymbol{M}_{1_j}(h_j)$ and $\boldsymbol{M}_{2_j}^{-1}(h_j)$ are less or equal to 1. For that reason, the algorithm (23) is stable at all frequencies. A more complete discussion of this statement will be presented in a forthcoming paper by the authors.

5. FREE WAVES AT THE FLUID/COMPOSITE INTERFACE

The recursive algorithm presented in Section 5 has been implemented in a computer program, which has been applied to the study of the propagation of free waves along the interface between an acoustic fluid and a laminated composite plate. As discussed in Section 3, (12) is the dispersion equation for such waves. The task here is to find the pairs (ω, k_x) for which equation (12) is satisfied. In this paper we will concentrate on the Scholte-Gogoladze-like wave, which, in contrast with the leaky wave, has received little attention in the recent literature.

Since the Scholte-Gogoladze-like wave propagates unattenuated along the fluid/solid interface, it may become another useful tool in ultrasonic NDT. A method for the generation of such waves in a submerged plate has been recently reported by Luppé and Doucet [12]. Furthermore, in problems of structural acoustics, this subsonic wave carries the most important contribution to the transfer impedance of fluid-loaded plates [13]. The results presented in this paper are for a cross-ply [0/90/0] laminate in contact with water. The plate is traction-free at the bottom. Fig 3 shows the dispersion curves for different orientations of the plane of propagation with respect to the direction of the reinforcement of the layer in contact with the fluid. We observe that, as the angle θ approaches 90°, and the layer in contact with the fluid becomes softer, the speed of propagation of the

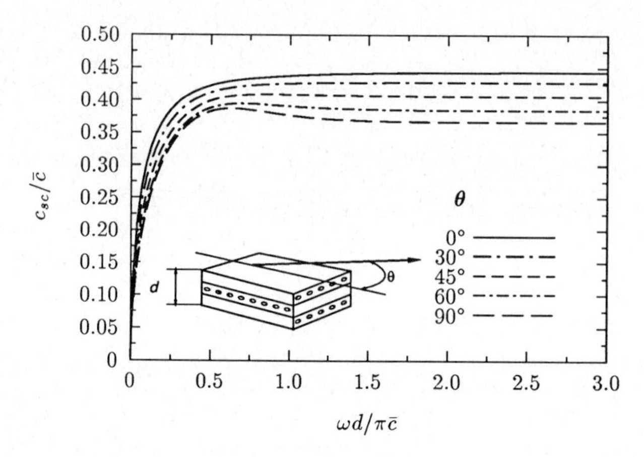

Figure 3. Phase velocity × frequency diagram ($\bar{c}$ = 2550 m/sec is a reference wave speed).

Scholte-Gogoladze-like wave decreases, showing the anisotropy of the laminate. As the frequency increases, the speed of propagation approach that of the Scholte-Gogoladze wave in a homogeneous half-space of the same material as the layer in contact with the fluid.

ACKNOWLEDGEMENTS

During the course of this work A. M. B. Braga was partially supported by the Brazilian Government through a grant CAPES Proc. 6579-84/5 , and by the Catholic University of Rio de Janeiro. Support through ONR contract N00014-85-K-0471 to Stanford University is also gratefully acknowledged.

REFERENCES

[1] Überall, H., "Surface Waves in Acoustics," in: Mason, W. P. and Thurston, R. N., (eds.), *Physical Acoustics Vol. X*, Academic Press, (1973), pp. 1–60.

[2] Scholte, J. G., *Proc. K. ned Akad. Wet.*, **51**, (1948), pp. 65–72.

[3] Gogoladze, V. Z., *Trudy Seismologicheskogo Instituta*, **127**, (1948), pp. 26–32.

[4] Nayfeh, A. H. and Chimenti, D. E., *J. Acoust. Soc. Am.*, **83**, (1988), pp. 1736–1746.

[5] Braga, A. M. B. and Herrmann, G., "Plane Waves in Anisotropic Layered Composites," in: Mal, A. K. and Ting, T. C., (eds.), *Wave Propagation in Structural Composites*, ASME-AMD Vol. 90, (1988), pp. 81–98.

[6] Kundu, T. and Mal, A. K., *Int. J. Engn. Sci.*, **24**, (1986), pp. 1819-1829.

[7] Dunkin, J. W., *Bull. Seism. Soc. Am.*, **55**, (1965), pp. 335–358.

[8] Mal, A. K., *Wave Motion*, **10**, (1988), pp. 257–266.

[9] Hager, W. W. and Rostamian, R., *Wave Motion*, **10**, (1988), pp. 333-348.

[10] Braga, A. M. B., *Wave Propagation in Fluid-Loaded Composites*, Doctoral Dissertation, Division of Applied Mechanics, Stanford University, in preparation.

[11] Ingebrigtsen, K. A. and Tonning, A., *Phys. Rev.*, **184**(3), (1969), pp. 1276–1279.

[12] Luppé, F. and Doucet, J., *J. Acoust. Soc. Am.*, **83**, (1988), pp. 1276–1279.

[13] Crighton, D. G., *J. Sound and Vibration*, **63**(2), (1979), pp. 225–235.

Elastic Waves and Ultrasonic Nondestructive Evaluation
S.K. Datta, J.D. Achenbach and Y.S. Rajapakse (Editors)

FOCUSING OF AN ULTRASONIC BEAM BY A CONCAVE INTERFACE

Hyung-Chul CHOI and John G. HARRIS
104 South Wright Street
Urbana, IL 61801, USA

We describe the compressional wavefield near the caustics and their cusp (focal region) formed when a well collimated, ultrasonic beam penetrates a concave fluid-solid interface.

1. INTRODUCTION

The need to transmit an ultrasonic beam across a curved fluid-solid interface arises frequently in nondestructive testing. One example is the use of ultrasound to detect cracks beneath the surface of a bore hole in an aircraft engine disc (Elsley, et al. [1]). The purpose of the present paper is to explore the effects that curvature has upon a transmitted compressional beam.

The incident beam is modeled as a time-harmonic, two-dimensional wavefield whose initial profile is rectangular. The curvature of the interface is concave to the source and its radius, measured in wavelengths, is very large. The interface is assumed to lie well within a Fresnel distance from the source so that it is struck by a well collimated beam and not by a plane or cylindrical wave. The beam can be thought of as composed of a plane wave emitted from the face of the aperture and diffracted waves emitted by its edges. The plane wave is reflected and refracted by the concave interface forming caustics and cusps exactly as ray tracing would indicate. However, to understand the wavefield quantitatively, the diffracted waves must also be included. In the present paper our aim is to describe some of the simpler aspects of this problem. A more detailed discussion can be found in [2] and [3].

2. DESCRIPTION OF THE PROBLEM

2.1 The Geometry

The geometry for the concave fluid-solid interface is shown in Fig. 1. The interface has a radius R_0; the fluid has a density ρ_0 and wave speed c_0; and the solid has a density ρ, a compressional wave speed c_L and a shear wave speed c_T. A two-dimensional ultrasonic beam emitted from an aperture whose half-width is a and whose center is located at (r_0, θ_0) strikes the interface, located a distance l_0 away, at an angle α. The observation point is located at (r, θ). To describe each beam a local coordinate system is used, namely, (l_I, d_I), where l_I measures the distance from the interface along the central axis of the scattered beam and d_I measures the transverse distance. The subscript I is replaced by R for the reflected beam, by L for the transmitted compressional beam and by T for the transmitted shear beam. The direction of each central axis is determined by Snell's law. The coordinate system (l_R, d_R) is shown in Fig. 2. The parameter t is used to locate a point on the aperture.

We are concerned with only a single interaction and thus do not consider the possibility that other waves, excited by subsequent reflections or refractions, strike the interface. Further α is small because we are concerned here only with situations in which the incident beam is transmitted across the interface.

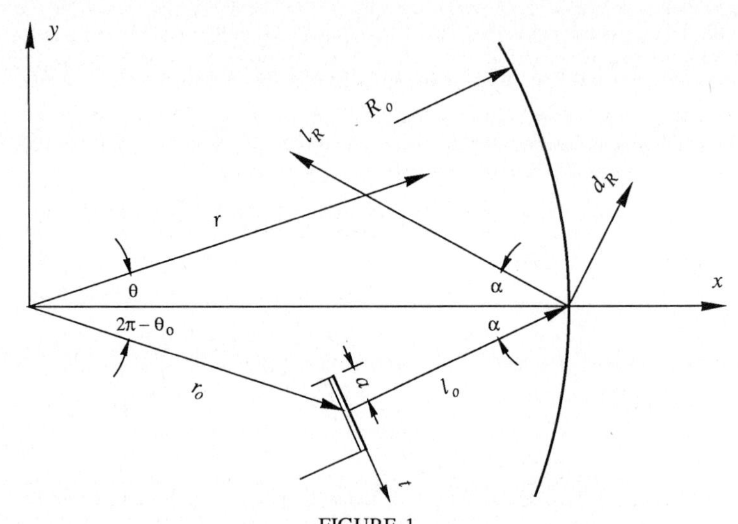

FIGURE 1
The geometry of the problem

2.2 Integral Representations

To calculate the beams scattered from the concave interface we (1) decompose the incident beam into an integration over cylindrical waves and then decompose each cylindrical wave into an integration over plane waves, (2) calculate the geometrical part of the wavefield scattered from the concave interface when struck by a plane wave, and (3) compose the scattered beam fields by integrating over the plane wave responses and then over the cylindrical wave responses. These integrals are then evaluated by asymptotically approximating each cylindrical wave response and numerically approximating the integration over the cylindrical wave responses. In this way a uniform approximation to the wavefields in the neighborhood of a caustic or cusp is obtained. Because the details of these calculations are complicated they are not given here and we refer the reader to [2] and [3]. The scattered wavefields are described by

$$k_I\Phi_I/A \;=\; B\int_{-1}^{1}\int_S\int_{S'} G_I(\eta)g_I(\eta)\exp[ik_0\Omega_I F_I(\eta,\nu,t)]\,d\eta\,d\nu\,dt. \tag{1}$$

The symbol Φ_I represents a displacement potential, $G_I(\eta)$ represents a plane-wave reflection or transmission coefficient and $g_I(\eta)$ represents a geometrical factor. The function $F_I(\eta,\nu,t)$ is the phase, that, at its stationary point, equals the (scaled) distance traveled by a ray that leaves a point t on the aperture and ends at a field point (r,θ). Explicit expressions for G_I, g_I and F_I are somewhat lengthy and are given in [2] and [3]. The parameters A, B and k_I represent the magnitude of the particle displacement at the aperture, a complex constant and the wave number in the fluid (note that $k_0 = k_R$) or the solid, respectively. The contours S and S' are Sommerfeld contours. The particle displacements are obtained from the potentials; in particular, the compressional-wave displacement $\mathbf{u}^L$ is given by the gradient of Φ_L. The selection of the scaling parameter Ω_I remains to be discussed.

2.3 Evaluation of the Wavefields

The transmitted compressional beam is commonly used for detecting defects, so that we shall limit our discussion to this case. Figure 2 illustrates the geometrical ray solution to the problem. The equations for the geometrical rays come from calculating the stationary point of F_L for the inner two contour integrals in Eq. 1. Moreover, by noting where the stationary phase approximation breaks down the equations describing the caustics can be found. To draw Fig. 2 we assumed that $a = 0.8cm$, $R_0 = 10cm$ and the frequency equals 50 MHz. Further, we assumed that $c_0 = 1.5 \times 10^3 m/s, c_T/c_0 = 2, c_L/c_0 = 3.742$ and $\rho_0/\rho = 0.125$. These values correspond to a water-metal interface. Two angles of incidence α were chosen, 0° and 5°, making $\gamma = 16.89°$. The distance l_0 equals 7 cm at 0° and 7.09 cm at 5°.

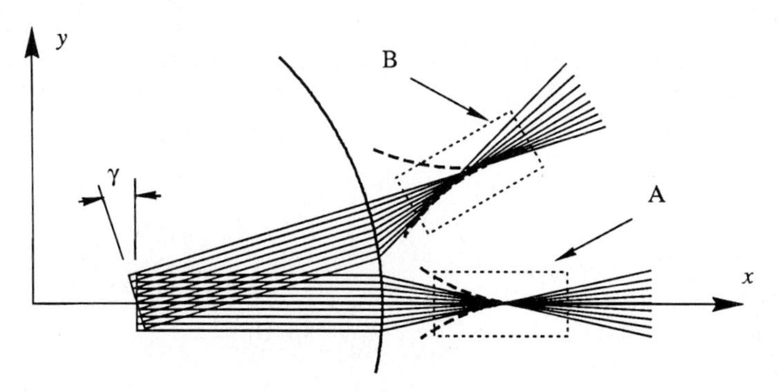

FIGURE 2
The refraction of the geometrical rays that form the transmitted compressional beam. The dashed lines indicate the caustics.

Note that the central ray of the refracted beam is always tangent to a caustic. We define a geometrical focal length as the distance travelled by the central ray from the interface to the tangent point on the caustic. This distance is given by

$$l_{fL} = -R_0 \cos\alpha_L / [1 - c_L \cos\alpha / c_0 \cos\alpha_L] \tag{2}$$

where

$$\sin\alpha_L = (c_L/c_0)\sin\alpha . \tag{3}$$

Using $\alpha = 5°$ and the values of the parameters used in Fig. 2, $l_{fL} = 3.21cm$. Referring to Fig. 2 and Eq. 2 we now choose Ω_L to be the apparent path length taken by the central ray from the aperture to the interface and then to the caustic. Thus

$$\Omega_L = l_0 + (c_0/c_L) l_{fL} . \tag{4}$$

The Fresnel length is much larger than Ω_L and Ω_L is larger than a. Having determined the scaling parameter we evaluate Φ_L as indicated earlier.

3. NUMERICAL RESULTS

We investigate numerically the focused compressional wavefield. Unless otherwise specified, the values of the parameters used are identical to those used in Fig. 2.

In Fig. 3(a) we have plotted the magnitude of the compressional beam, at $\alpha = 0°$, along the central axis. The normalized aperture size a/l_{fL}^0 is varied; it equals 0.11 (long-short dashed line), 0.22 (solid line) and 0.44 (dashed line). In Fig. 3(b) we have plotted cross-sections of the magnitude near and at the focal plane. The normalized aperture size is again varied. At $(l_L - l_{fL}^0)/l_{fL}^0 = -0.3$ it equals 0.11 (long-short dashed line) and 0.44 (solid line), and at $(l_L - l_{fL}^0)/l_{fL}^0 = 0$ it again equals 0.11 (dashed line) and 0.44 (dotted line). The distance l_{fL}^0 is the focal length given by Eq. 2 and equals 3.65 cm; the superscript 0 indicates that $\alpha = 0°$.

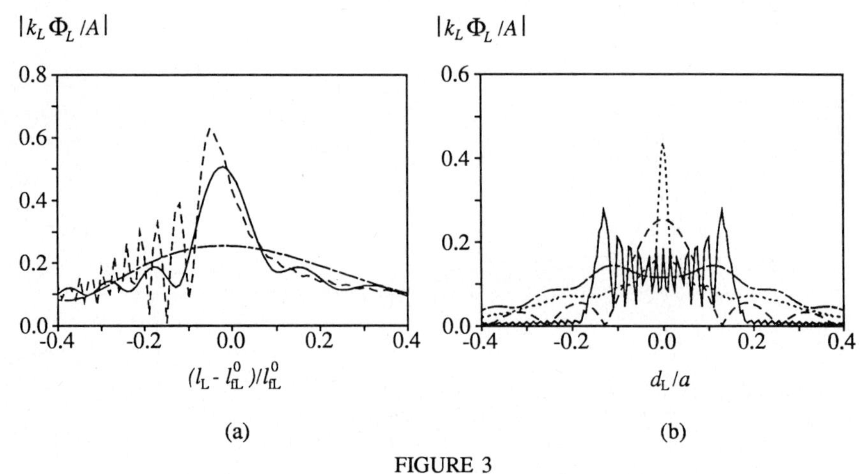

FIGURE 3
Magnitude of the compressional beam at $\alpha = 0°$. Along the central axis (a) and in transverse planes (b).

In Figs. 4 and 5 we have examined regions A and B, respectively, of Fig. 2. In Figs. 4(a) and 5(a) contour plots showing the relative magnitudes of the compressional beam in the neighborhood of the cusp are given. The larger the number the smaller the magnitude. Also plotted are the caustics (dashed lines) and the shadow boundaries (dotted lines). Note that near the cusp the effects of the caustics dominate, while elsewhere the effects of the shadow boundaries dominate. In Figs. 4(b) and 5(b) cross-sections of the magnitude are plotted. In Fig. 4(b) $(l_L - l_{fL}^0)/l_{fL}^0 = -0.6$ (dashed line), 0 (solid line) and 0.2 (dotted line). In Fig. 5(b) $(l_L - l_{fL}^5)/l_{fL}^5$ takes the same values with the same line conventions. The effect of the caustic in Fig. 5(b) is quite striking.

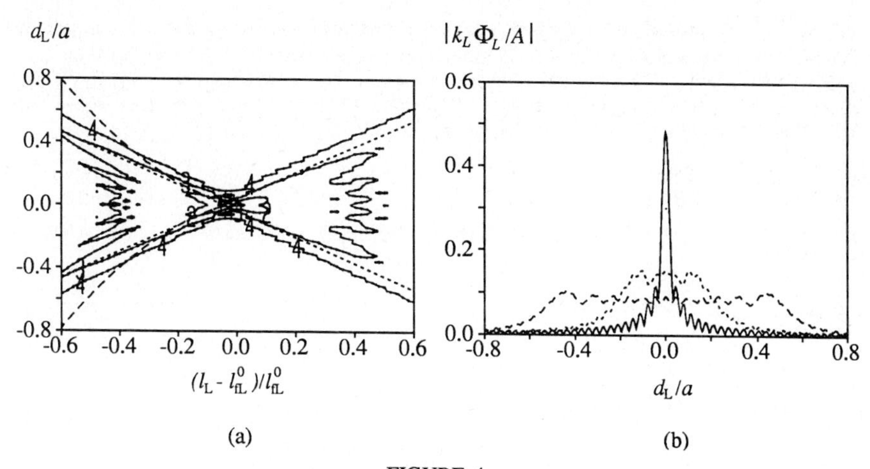

FIGURE 4
Contours of constant magnitude (a) and cross-sections of the magnitude (b) for the compressional beam at $\alpha = 0°$.

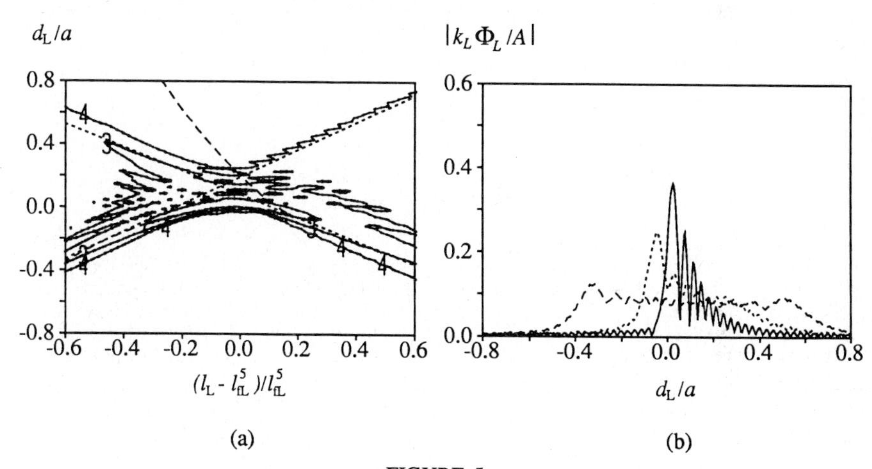

FIGURE 5
Contours of constant magnitude (a) and cross-sections of the magnitude (b) for the compressional beam at $\alpha = 5°$.

6. CLOSING REMARKS

Figures 2 through 5 indicate that the effect of the curvature of the interface can be quite significant at higher frequencies. Note that the depth at which the transmitted beam is distorted by the formation of caustics is not great.

Figure 3 illustrates how effectively the caustics direct the beam field into the cusp. Not surprisingly, the larger a/l_{fL}^0 the stronger the effect of the caustics and cusp. Examining Fig. 3(b) we can see that for $a/l_{fL}^0 = 0.44$ the formation of the caustics causes the beam to fall-off abruptly at each side. Beyond the cusp the beam spreads rapidly.

Figures 4 and 5 show the competing influences of diffraction from the aperture and diffraction from a caustic [the reader will find it instructive to imagine (b) and (a) as a single three dimensional figure]. Figure 5 is of particular interest. The wavefield is negligible on the silent side of the caustic. This suggests that a misalignment of a transducer could result in a silent region that lay, in the solid, at a point almost directly below the transducer.

ACKNOWLEDGMENT

This material is based, in part, on work supported by the National Science Foundation (MSM-8513928).

REFERENCES

[1] Elsley, R. K., Addison, R. C. and Graham, L. T., Detection of Flaws Below Curved Surfaces, in: Thompson, D. O. and Chimenti, D. E. (eds.), Review of Progress in Quantitative Nondestructive Evaluation, Vol. 2A (Plenum, New York, 1983) pp. 113-128.

[2] Choi, H. C. and Harris, J. G., Wave Motion 11 (1989) 383.

[3] Choi, H. C. and Harris, J. G., Focusing of an Ultrasonic Beam, submitted for review.

Elastic Waves and Ultrasonic Nondestructive Evaluation
S.K. Datta, J.D. Achenbach and Y.S. Rajapakse (Editors)

NONDESTRUCTIVE EVALUATION OF DAMAGE DEVELOPMENT IN COMPOSITE MATERIALS

Isaac M. Daniel, Shi-Chang Wooh, and Jae-Won Lee

Robert R. McCormick School
of Engineering and Applied Science
Northwestern University
Evanston, IL 60208

Nondestructive methods were developed and applied to the characterization of progressive damage in crossply composite laminates subjected to monotonic and cyclic tensile loading. The objective of the study was to study damage evolution, characterize the damage and correlate the damage/NDE output with stiffness.

1. INTRODUCTION

Damage in composite laminates consists of the development and accumulation of numerous defects. The basic failure mechanisms, i.e., intralaminar and interlaminar matrix failures, have been observed and identified [1–3]. In the case of crossply laminates fatigue damage development consists of transverse matrix cracking, followed by longitudinal matrix cracking, local delaminations at crack intersections, and ultimately longitudinal fiber breakage in the load carrying plies [4]. The state of damage is related to the three most important properties of the material, stiffness, strength and life. Some analytical procedures have been developed for relating stiffness reduction to damage state for some forms of damage [5–10]. Given such relationships it may be possible and desirable to monitor directly and nondestructively variations in stiffness, current strength and remaining life.

The objectives of the present investigation were to develop methods for characterization of damage in composite laminates, study damage evolution and correlate damage/NDE output with stiffness.

2. DAMAGE CHARACTERIZATION AND DEVELOPMENT

The material investigated was AS-4/3501-6 (Hercules, Inc.) graphite/epoxy of $[0/90_2]_S$ layup. Under axial tensile loading, whether monotonic or cyclic, the first stage of damage consists of transverse matrix cracks. These cracks are nearly equidistant and increase in density up to a characteristic limiting value. It is important to determine this crack density non-destructively. Penetrant-enhanced X-radiography is an effective means of characterizing matrix cracking, but it would be desirable to do so by ultrasonic techniques which do not require a penetrant.

In the case of a $[0/90_2]_S$ graphite/epoxy specimen loaded up to the first stage of damage, oblique incidence backscattering was used as shown in Fig. 1. Different backscattering waveforms were obtained at various locations having different crack densities (Fig. 2). The total backscattered energy obtained at each point was plotted versus location and compared with the crack density variation obtained from an X-radiograph (Fig. 3). The backscattered energy was also correlated with the total length of cracks within the elliptical window viewed by the transducer at oblique incidence. The correlation among all these quantities, i.e.,

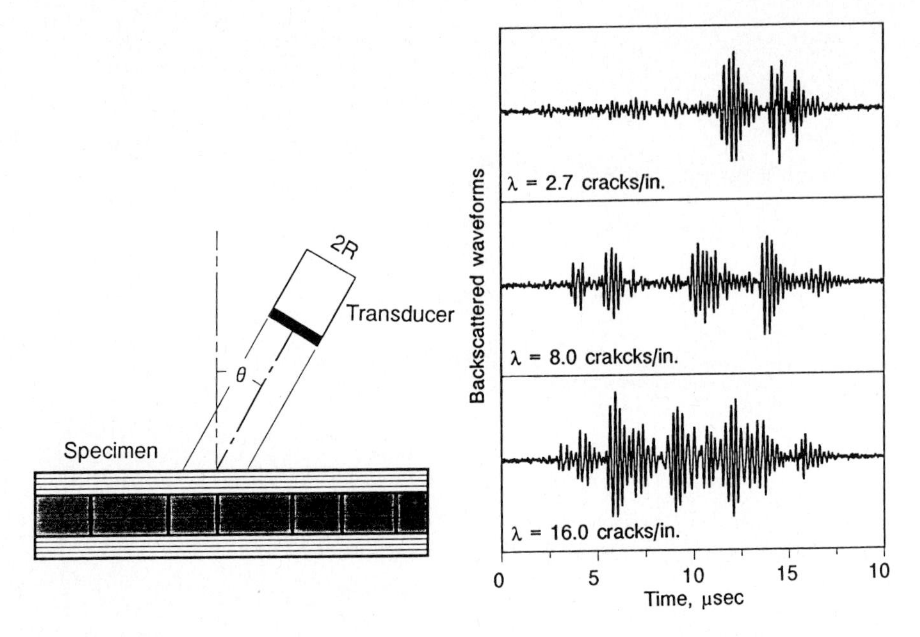

Fig. 1 Ultrasonic inspection of cracked laminate by oblique incidence backscattering technique.

Fig. 2 Ultrasonic backscattering waveforms for various crack densities in composite laminate.

crack density, backscattered energy and total crack length, is satisfactory. Thus, it is possible to obtain a one-to-one correlation between an ultrasonic indication and crack density.

The stress-strain curve of the $[0/90_2]_S$ crossply laminate shown in Fig. 4 displays characteristic features which reflect the various events and stages of damage initiation and development. The initial linear elastic region up to a strain of approximately 0.0045 corresponds to the stage before any damage initiation. The second region of the stress-strain curve, between strains of approximately 0.0045 and 0.0100, is one of decreasing stiffness and corresponds to damage development in the form of initiation, multiplication and saturation of transverse matrix cracking. The third region is a nearly linear one of stabilized stiffness following the saturation of transverse cracking and reflects the stress-strain behavior of the undamaged 0° plies. Superimposed on the same figure is the variation of crack density with applied stress, showing the direct correlation of damage stages and features of the stress-strain curve.

Damage development under fatigue loading is not only a function of number of loading cycles but also of the cyclic stress amplitude. The various damage mechanisms can be monitored by means of penetrant-enhanced X-radiography as well as ultrasonic backscattering. Figure 5 shows an X-radiograph and an ultrasonic backscatter image enhanced by one-dimensional segmentation for a $[0/90_2]_S$ laminate subjected cyclic loading [11]. The first stage of damage development consists of transverse matrix cracking which generally increases in density with number of cycles as shown in Fig. 6. It is seen that the crack density reaches or approaches a limiting value, the so-called *characteristic damage state* (CDS). At some point longitudinal matrix cracks develop with local delaminations at the intersections with transverse cracks, followed by fiber fractures. It was observed, at least for the higher cyclic stress amplitudes, that longitudinal matrix cracking starts at approximately 60 to 70 % of the normalized logarithmic lifetime ($\log n / \log N$), before transverse cracking reaches the CDS level at approximately 80 % of this lifetime (Fig. 7).

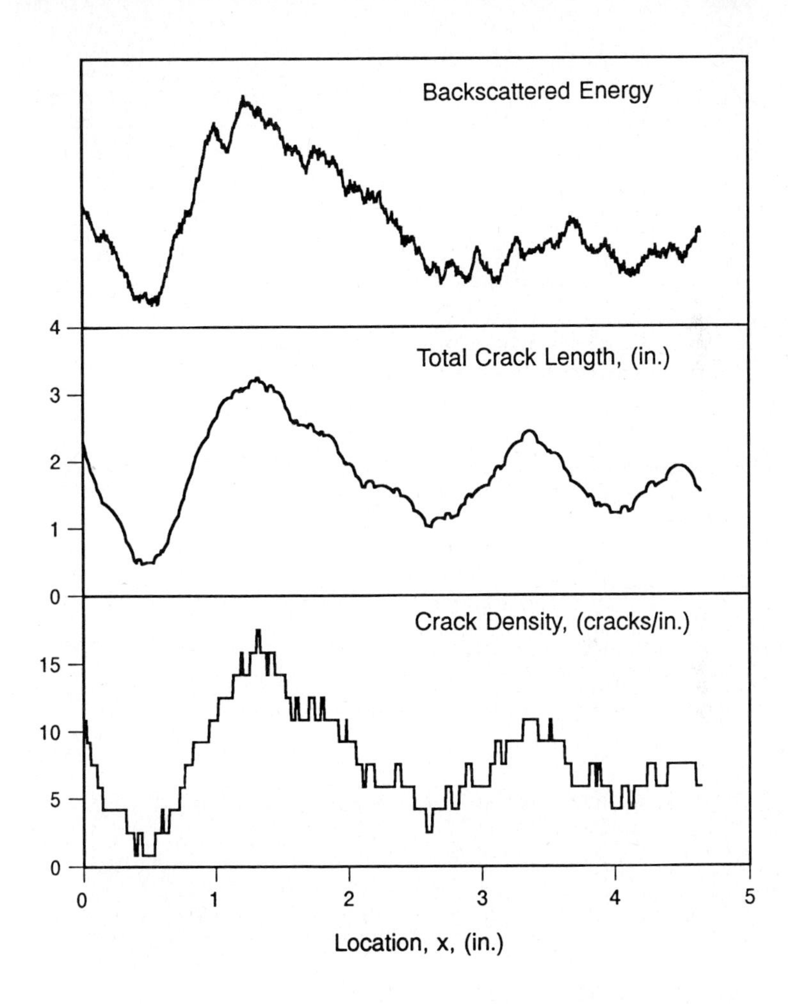

Fig. 3 Measured backscattered energy, total crack length viewed through moving window and crack density compared with X-radiograph for statically loaded crossply graphite/epoxy specimen.

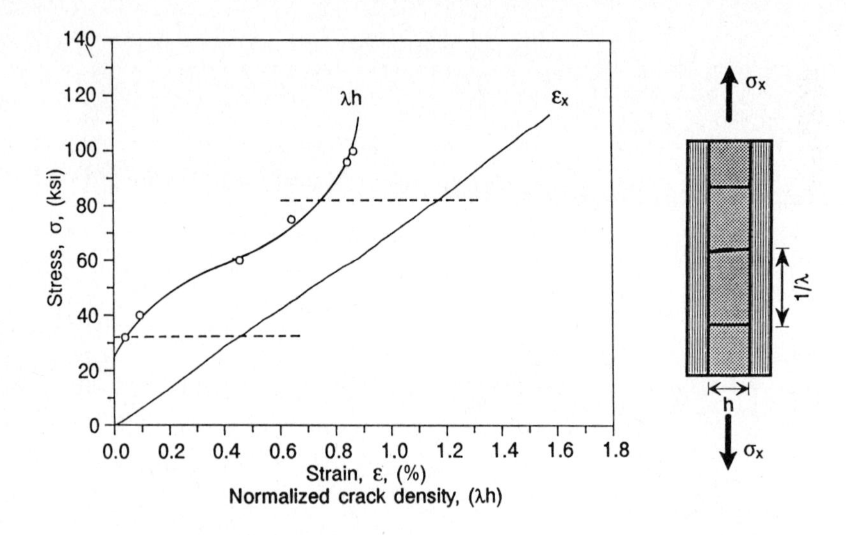

Fig. 4 Stress-strain and stress-crack density curves for $[0/90_2]_s$ graphite/epoxy specimen under uniaxial tensile loading.

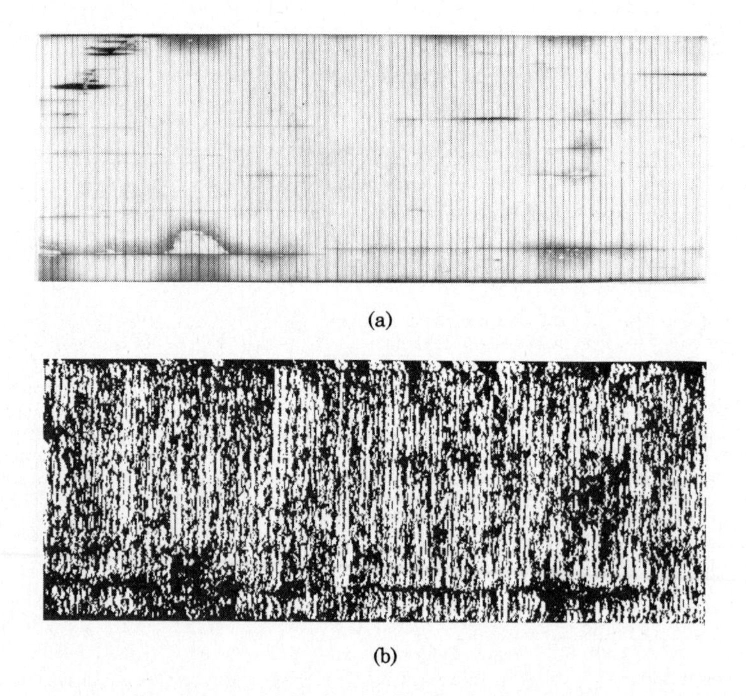

(a)

(b)

Fig. 5 Fatigue damage in $[0/90_2]_s$ graphite/epoxy laminate. (a) X-radiograph, (b) Ultrasonic backscatter image enhanced by one-dimensional segmentation.

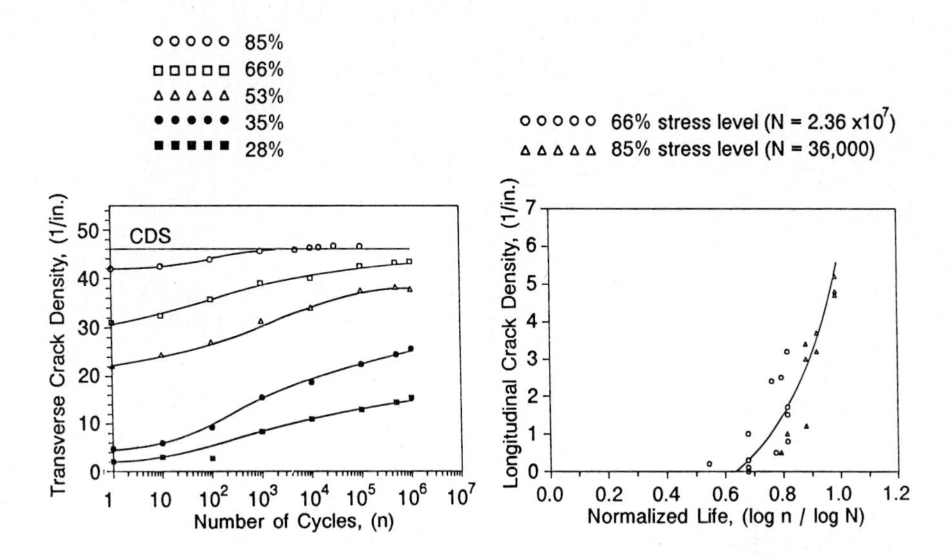

Fig. 6 Variation of transverse crack density with number of cycles at various cyclic stress levels for $[0/90_2]_S$ graphite/epoxy laminate.

Fig. 7 Longitudinal crack density as a function of normalized logarithmic lifetime for specimens cycled at 66 % and 85 % of the static strength.

3. STIFFNESS DEGRADATION

Stiffness reduction is expected to follow the observed damage development. However, different stiffness parameters show different sensitivities to the various damage mechanisms. For example, the axial modulus is sensitive to transverse matrix cracking and fiber fractures but not very sensitive to longitudinal cracking and delamination. The in-plane shear modulus is much more sensitive to longitudinal cracks than the axial modulus.

The degradation of axial modulus was studied for the $[0/90_2]_S$ as a function of fatigue cycles and the cyclic stress amplitude [12]. Results were presented in the form of curves of normalized residual modulus as a function of normalized logarithmic lifetime ($\log n / \log N$) (Fig. 8). These curves reflect the three stages of damage development. The characteristic features of the stiffness reduction curve are: (1) initial drop corresponding to transverse matrix cracking during the first loading cycle, (2) gradual reduction or plateau corresponding to crack multiplication or saturation at the CDS level, and (3) accelerated reduction corresponding to longitudinal matrix cracking, delaminations and fiber fractures in the last 20 % of normalized logarithmic lifetime.

The in-plane shear modulus shows a similar behavior as the axial modulus [13]. The main difference is that its accelerated reduction starts earlier, when longitudinal cracking starts, and is faster than that of the axial modulus as shown in Fig. 9.

4. SUMMARY AND CONCLUSIONS

Damage mechanisms and damage development in crossply graphite/epoxy laminates were characterized and studied by nondestructive methods. The ultrasonic backscattering method was used to characterize matrix crack density under monotonic and cyclic loading and monitor the damage evolution.

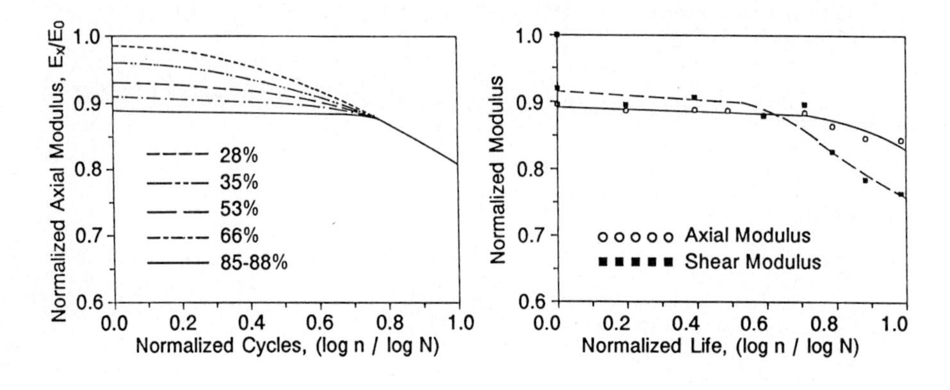

Fig. 8 Normalized axial modulus curves for $[0/90_2]_S$ graphite/epoxy laminate.

Fig. 9 Comparison between normalized axial and simple shear modulus degradation as a function of normalized logarithmic life for cyclic stress level of 85 % of static strength.

Under monotonic loading a correlation was illustrated between the various stages of damage development and characteristics of the stress-strain curve. Under cyclic tensile loading damage development depends on the applied cyclic stress level and consists of three stages: (1) damage occurring during the first fatigue cycle consisting of transverse matrix cracking, (2) damage developing during the first 80 % of the logarithmic lifetime of the material and consisting of multiplication of transverse matrix cracks up to the CDS level and initiation of longitudinal matrix cracking at approximately 65 % of the logarithmic lifetime, and (3) damage occurring in the last 20 % of the logarithmic lifetime and consisting primarily of longitudinal matrix cracking, local delaminations and fiber fractures.

The associated reduction in axial modulus was found to be closely related to the damage development. The curves of normalized residual axial modulus were found to depend on the cyclic stress level. They all show an initial drop, a gradual reduction and accelerated reduction at the end. The in-plane shear modulus shows similar characteristics with an increased sensitivity to longitudinal cracking. Thus, by correlating stiffness degradation with damage mechanisms and evolution, it would be possible to determine residual stiffness, and thereby remaining life, nondestructively.

ACKNOWLEDGEMENTS

The work described here was sponsored by the Office of Naval Research. We are grateful to Dr. Y. Rajapakse of ONR for his encouragement and cooperation.

REFERENCES

[1] Reifsneider, K. L., Henneke, E. G. II and Stinchcomb, W. W., "Defect Property Relationships in Composite Materials," AFML-TR-76-81 Part IV, 1979.

[2] Wang, A. S. D. and Crossman, F. W., "Initiation and Growth of Transverse Cracks and Edge Delamination in Composite Laminates — Part I: An Energy Method," *Journal of Composite Materials*, Special Issue, June, 1980.

[3] Crossman, F. W., Warren, W. J., Wang, A. S. D. and Law, G. L., "Initiation and Growth of Transverse Cracks and Edge Delamination in Composite Laminates — Part II: Experimental Correlation," *Journal of Composite Materials*, Special Issue, June 1980.

[4] Charewicz, A. and Daniel, I. M., "Damage Mechanisms and Accumulation in Graphite/Epoxy Laminates," in *Composite Materials: Fatigue and Fracture,* ASTM STP 907, H. T. Hahn ed., American Society for Testing and Materials, Philadelphia, PA, 1986, pp. 274-297.

[5] Ryder, J. T. and Crossman, F. W., "A Study of Stiffness, Residual Strength and Fatigue Life Relationships for Composite Laminates," NASA CR-172211, Oct. 1983.

[6] Talreja, R., "Transverse Cracking and Stiffness Reduction in Composite Laminates," *Journal of Composite Materials*, 1985, Vol. 19, pp. 355-375.

[7] Dvorak, G. J., Laws, N. and Hejazi, M., "Analysis of Progressive Matrix Cracking in Composite Laminates — I. Thermoelastic Properties of a Ply with Cracks," *Journal of Composite Materials*, May 1985, Vol. 19, pp. 216-234.

[8] Hashin, Z., "Analysis of Cracked Laminates: A Variational Approach," *Mechanics of Materials*, 1985, Vol. 4, pp. 121-136.

[9] Ogin, S. L., Smith, P. A. and Beaumont, P. W. R., "Matrix Cracking and Stiffness Reduction During the Fatigue of a [0/90]s GFRP Laminate," *Composites Science and Technology*, Vol. 22, 1985, pp. 23-31.

[10] Lee, J. W., and Daniel, I. M., "Progressive Damage Analysis of Crossply Composite Laminates," submitted for publication, 1989.

[11] Wooh, S. C. and Daniel, I. M., "Enhancement Techniques for Ultrasonic Nondestructive Evaluation of Composite Materials," *Journal of Engineering Materials and Technology*, Oct. 1989.

[12] Daniel, I. M., Yaniv, G. and Lee, J. W., "Damage Mechanisms and Stiffness Degradation in Graphite/Epoxy Composites,"" ICCM-VI and ECCM-II (*Sixth International Conf. on Composite Materials and Second European Conf. on Composite Materials*), Elsevier, London, 1987, Vol. 4, pp.4.129-4.138.

[13] Yaniv, G., Lee, J. W., and Daniel, I. M., "Damage Development and Shear Modulus Degradation in Graphite/Epoxy Laminates," presented at *ASTM Symposium on Composite Materials: Testing and Design*, Sparks, NV, Apr. 1988.

Elastic Waves and Ultrasonic Nondestructive Evaluation
S.K. Datta, J.D. Achenbach and Y.S. Rajapakse (Editors)
Elsevier Science Publishers B.V. (North-Holland), 1990

NONDESTRUCTIVE EVALUATION OF DEFECTS IN BEAMS BY GUIDED WAVES

Jurg DUAL*, Mahir SAYIR, Alexander WINKER, Markus STAUDENMANN

Institute of Mechanics
ETH Zurich
Zurich, Switzerland

Reflection characteristics of *guided waves* can be used to assess defects in structural elements. Here, *bending waves*, incident at a lateral notch in a beam, are considered. The incident wave generates reflected and transmitted bending, longitudinal and torsional waves and local bending vibrations at the discontinuity. From the arrival time of the reflected pulse and the relative amplitude of the different waves produced, information regarding location, size and nature of the notch may be gathered. *Experimental results* were obtained from interferometric displacement measurements of narrow band pulses travelling along aluminum beams of circular cross - section, which were excited piezoelectrically. These results are compared to a *theoretical model*, where the discontinuity is represented by a quasistatic flexibility matrix. Edge notches with a depth of a tenth of the beam diameter could be detected and quantitatively analyzed without special precautions regarding precision of excitation and measurement.

1. INTRODUCTION

In accordance with classical scattering theory, bulk waves with wavelengths that are of the order of the defect to be investigated are customarily used for NDE. However, for structural elements like beams, plates or shells, guided waves might also be employed. The wavelengths might be several times the thickness of the structure and therefore at least an order of magnitude larger than the size of the defect. Nevertheless, defects can be detected, because waves scattered at the discontinuity will be reflected at the boundaries of the structure and superimpose to form reflected and transmitted pulses.

Very little attention has been devoted to this type of NDE: Wong et al. [1] have performed experiments on the reflection of bending waves at a partial cut in beams and human tibia with the aim of monitoring the healing process of partially cracked bones. Phillips et al. [2] have conducted similar experiments and have also developed a theoretical model of a partially cracked beam. In both papers, a striker is used to excite a *broad - band pulse*. However, due to the strongly dispersive nature of the bending waves, quantitative evaluation is difficult and the experimental results indicate, that only serious cracks (i.e. with a depth greater than a third of the bar thickness) can be detected. Using the measured signals to *quantitatively* determine the size of the cut would be extremely difficult.
Wendtland [3] and Gudmundson [4] have investigated the *change of eigenfrequencies* of slender structures due to cracks. Gudmundson was able to detect cracks with a size of 20% of the thickness of the beam. The crack region is modeled as a spring element. However, boundary conditions, which are difficult to assess, strongly influence eigenfrequencies. This limits the applicability of the method to quantitative NDE. Boundary problems are eliminated, if transient waves are considered.

* Dept. of Theoretical and Applied Mechanics, Cornell University, Ithaca NY, USA

In the present paper, *narrow band pulses* are used, to detect a partial lateral notch in an aluminum beam with circular cross-section. Thereby, the effects of dispersion are minimized. By evaluating arrival times and amplitude ratios, much more information is obtained as compared to [1] and [2], that can be used to characterize the discontinuity.

2. A QUASISTATIC MODEL FOR THE BEHAVIOR OF WAVES AT THE NOTCH

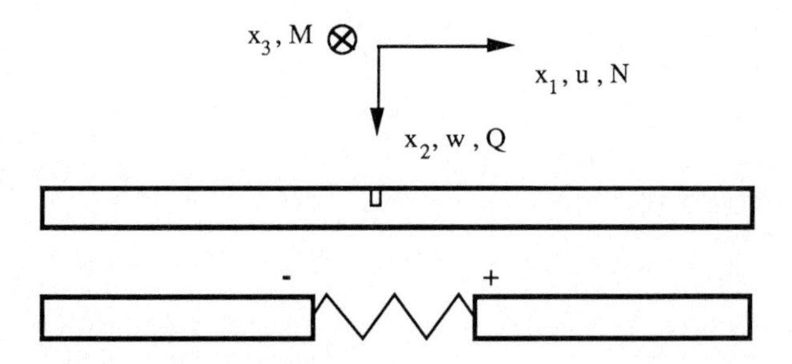

Fig. 1: Beam with a Partial, Lateral Cut and Definition of the Coordinate System

In order to evaluate the behavior of the waves at the crack, Gudmundson's model [4] together with one-dimensional theories for longitudinal and bending waves is adopted. Due to the opening of the crack and the shift between its faces, displacement and rotation jumps

$$\delta = u_+ - u_- \, , \quad \varphi = \psi_+ - \psi_- \quad , \varepsilon = w_+ - w_- \tag{1}$$

are expected there. u and w are the displacements of the beam center-line in the x_1 and in the x_2 directions, respectively, ψ is the rotation of the cross - section due to bending. The + and the - sign denote quantities at the left and at the right-hand side of the crack, respectively. The jumps may be related to the normal force N, the bending moment M and the shear force Q with the help of a matrix **A** according to

$$\begin{pmatrix} \delta \\ \varphi \\ \varepsilon \end{pmatrix} = A \begin{pmatrix} N \\ M \\ Q \end{pmatrix} \tag{2}$$

This linear relation corresponds to a " generalized spring " behavior and may be expected to hold for relatively long wavelengths as compared with the thickness of the beam. The Matrix **A** is obtained using Irwin's law [5]

$$\frac{\delta\theta}{\delta C} = - \frac{1}{E} \left(K_I^2 + K_{II}^2 \right). \tag{3}$$

where C is the area of the crack, and θ is the potential energy. The stress intensity factors K_I and K_{II} are expressed in terms of the loads at the notch

$$K_I = \frac{N}{A}\sqrt{a}\ f_t + \frac{M}{I}\ r\sqrt{a}\ f_b$$

$$K_{II} = \frac{Q}{A}\sqrt{a}\ f_s \tag{4}$$

where A and I are cross-sectional area and moment of inertia of the intact beam, a is the crack depth and f are dimensionless stress - functions, which for a circular bar with radius r are approximated by

$$f_t = 0.719 + 6.83\,\xi - 6.01\,\xi^2 - 5.14\,\xi^3 + 7.34\xi^4$$
$$f_b = 1.333 + 0.491\,\xi - 1.184\,\xi^2 - 1.45\,\xi^3 \tag{5}$$
$$\xi = a / r$$

if the bending motion is orthogonal to the base of the crack. [6 , 7] Because the stress function f_s could not be found in the literature, it was set to zero in this first order model. Thereby, the jump in the lateral displacement ε is set to zero.
Eq. 3 is integrated to the desired crack depth taking into account the circular cross - section of the beam. Using the quadratic expression for θ in terms of N and M from Eq.2 yields the matrix **A**.

A bending wave (w_i) incident at the crack will produce reflected and transmitted longitudinal (u_r, u_t) and bending waves (w_r, w_t) as well as local bending vibrations (w_{rs}, w_{ts}). Because the reflection and transmission characteristics are frequency dependent, harmonic waves of the kind

$$\{ w_i , w_r , w_t \} = \{ W_i , W_r , W_t \} e^{i(\pm k_b x_1 - \omega t)}$$
$$\{ w_{rs} , w_{ts} \} = \{ W_{rs} , W_{ts} \} e^{\pm k_s x_1} e^{-i\omega t} \tag{6}$$
$$\{ u_r , u_t \} = \{ U_r , U_t \} e^{i(\pm k_L x_1 - \omega t)}$$

are assumed. Here, ω is the circular frequency, and k_b , k_L and k_s are the corresponding wavenumbers for the propagating bending and longitudinal waves and the localized bending vibration, respectively. Timoshenko's beam theory [8] or an asymptotic theory [9] might be used to obtain wavenumbers and relations between displacements w and bending moment M, shear force Q, and angle of rotation ψ due to bending. The boundary conditions at the crack are Eqs. 1 and

$$Q_+ = Q_- , \qquad M_+ = M_- , \qquad N_+ = N_- \tag{7}$$

They are solved for the six unknown complex magnitudes $\{ W_r , W_t\ W_{rs} , W_{ts} , U_r , U_t \}$.
The results are presented in Fig. 4 together with experimental values.

3. EXPERIMENTS

The experiments were performed on aluminum rods (Ac-112 Anticorodal, density ρ = 2680 kg/m^3) of circular cross-section with a length of 3 m and diameters of 0.008 and 0.01 m. The bar velocity was measured as 5120 m/s. Assuming a literature value of 0.33 for Poisson's ratio, the shear velocity can then be calculated to be 3140 m/s.
Bending waves were excited by means of piezoelectric shear elements (Philipps, PXE 71), which were sandwiched between one end of the beam and an inertial disk of the same diameter as the beam

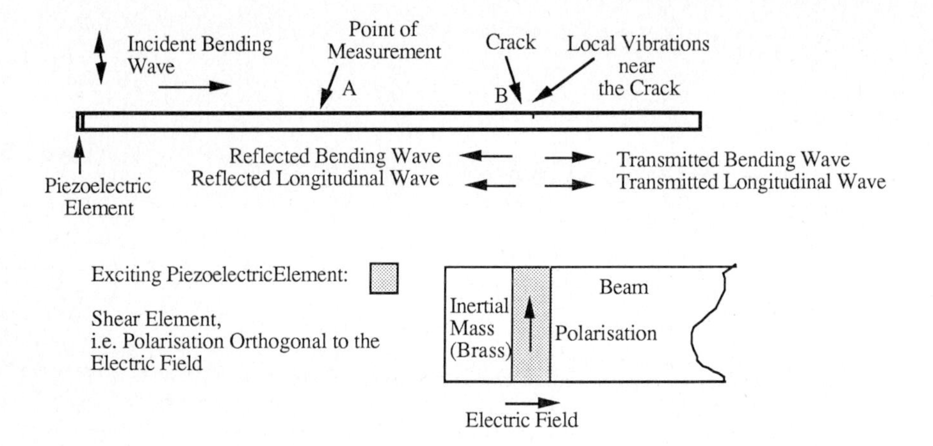

Fig. 2: Experimental Set-Up and Piezoelectric Element

and length of 0.003 m (Fig. 2). The beams were supported horizontally on two pieces of rubberfoam, which did not produce any reflections, because of their low mechanical impedance. The experiments were completely controlled by an Olivetti M24 PC, which was equipped with a digital arbitrary function generator. Its signal output was amplified to about 100 V by a Krohn Hite KH 7500 Amplifier and repetitively applied to the piezoelectric transducer, leading to lateral displacements of the order of 10^{-9} m. The repetition rate was chosen such that the pulses of the previous experiment were completely damped out, when the next signal was applied. The displacement was measured using a heterodyne laser interferometer. [10] A small piece of retro-reflecting tape (3M) was attached to the beam at the point of measurement to provide sufficient amplitude of the reflected light. The signal was phase-demodulated and fed through a Krohn-Hite analog filter (KH 3550) to a Le Croy 9400 digital oscilloscope. To increase the signal to noise ratio, the measured signals were averaged several 100 times. With a typical repetition rate of 10 experiments per second, a very clean signal is obtained within about a minute. (Fig.3) The result was then transferred back to the PC and analyzed by the Fast Fourier Transform.
The notches of a width of approximately 10^{-3}m were produced by sawing. Only notches orthogonal to the axis of the beam were considered. The depth was varied between 1/8 and 1/2 of the diameter of the beam.

Fig.3 shows two typical waveforms measured at the point A (Fig.2) with a notch of depth r at B. Because of narrow band excitation, individual pulses can be clearly discerned: The first pulse is the incoming wave, the second pulse is the longitudinal wave generated at the notch, followed very closely by a bending pulse reflected at the notch. Further pulses are waves reflected from the distant end of the beam. Longitudinal and bending waves can be distinguished either by their arrival times or using the 45° measurement, where the longitudinal motion is amplified as compared to the lateral motion. The amplitude of the longitudinal wave was best determined from the 90 degree measurement: From lateral contraction, the longitudinal amplitude can be determined if Poisson's number is known.
If the direction of sawing is parallel to the motion of the incoming wave (as shown in Fig. 3), only bending and longitudinal waves are generated. However, for other directions *torsional waves* are also generated and observed. They are difficult to measure quantitatively, because their velocity is very close to the group velocity of the bending waves for the frequency ranges considered here.

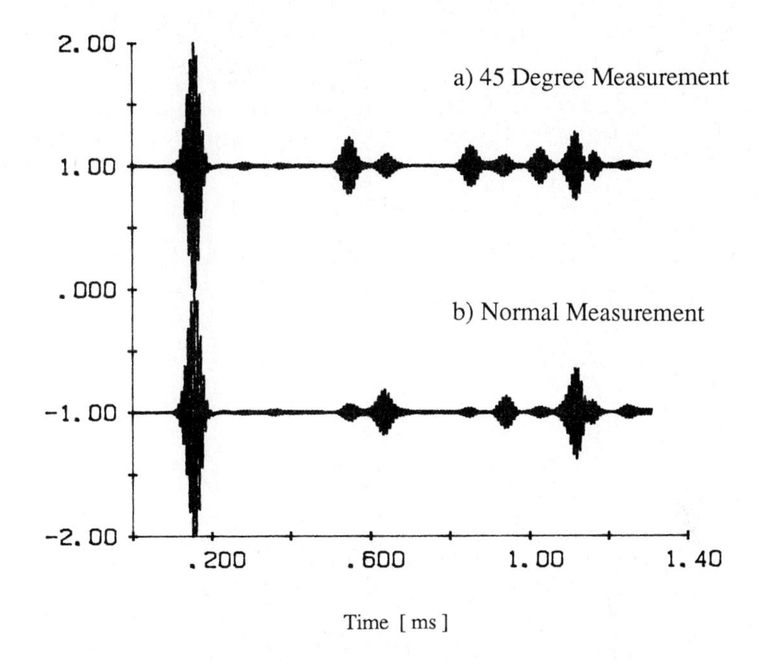

Fig.3: Waveforms Measured at Point A as a Result of a Notch of Depth r at B
a) Measurement at 45° with Respect to the Surface and the Axis of the Beam
b) Measurement Normal to the Surface
Excitation Signal: 10 Periods of a 120 kHz Sinewave multiplied by a Hanning Window

4. COMPARISON OF THEORY AND EXPERIMENTS

For comparison of theory and experiments, amplitude ratios were determined. Individual pulses were cut out of the total waveform and Fourier analyzed separately. The magnitudes of the Fourier-transforms of the reflected bending wave and the reflected longitudinal wave were divided by the spectrum magnitude of the incoming bending wave and are plotted in Fig. 4 together with the theoretical results. The agreement between theory and experiment is quite good for low frequencies. Considerable deviations occur for higher frequencies at large values of ξ. It was not possible to obtain the amplitude of the longitudinal wave for the lowest value of ξ.

5. CONCLUSIONS

Edge notches with a depth of a tenth of the diameter of the beam, which corresponds to about 5% of the total area, could be detected without special precautions regarding precision of excitation and measurements. The authors believe, that the resolution can be improved considerably by refining the measurement system. Current limitations of the setup consist in amplitude inaccuracies of the interferometer, the excitation of undesired modes of waves by unsymmetries in the attachments of transducers and leads and difficulties in separating individual pulses of different modes. Theory and experiment are in good agreement for low frequencies. For higher frequencies, the crack probably does not behave quasistatically anymore and a more refined model has to be developed.

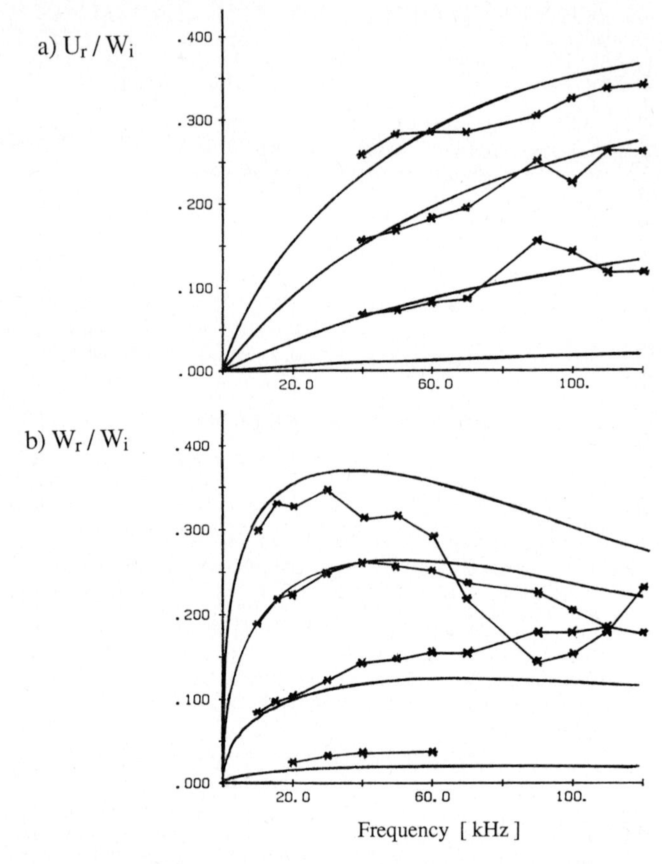

Fig.4: Amplitude Ratios as a Function of Frequency for an Incoming Bending Wave in a Circular Aluminum Bar (Diameter 0.01 m) Hitting a Notch of Dimensionless Depth ξ
$\xi = a/r$: 0.25, 0.5, 0.75, 1.0, (——) Theory, (∗——∗) Experiment
a) Reflected Longitudinal Wave U_r / W_i, b) Reflected Bending Wave W_r / W_i

REFERENCES

[1] Wong, A.T.C., Goldsmith,W., Sackman, J.L., J. Biomechanics, 9, (1976), p. 813

[2] Phillips, J.W., Mak, A.F., Ashbaugh, N.E., Int. J. Solids Structures, 14, (1978), p. 142

[3] Wendtland, D., Aenderung der Biegeeigenfrequenzen einer idealisierten Schaufel durch Risse, Ph.D. Thesis, University of Karlsruhe, 1972

[4] Gudmundson, P., J. Mech. Phys. Solids, 30, (1982), p. 339

[5] Irwin, G.R., Fracture Mechanics, Structural Mechanics, (Pergamon 1960), p. 557

[6] Bush, A.J., J. Testing and Evaluation, 9, (1981), p. 216

[7] Bush, A.J., Experimental Mechanics, 16, (1976), p.249

[8] Timoshenko, S.P., Phil. Mag., Ser. 7, 12, (1943), p. 125

[9] Sayir, M., Ingenieurarchiv, 49, (1980), p.309

[10] Dual, J., Diss ETH Nr. 8659, (1988) ETH Zurich, Switzerland

Elastic Waves and Ultrasonic Nondestructive Evaluation
S.K. Datta, J.D. Achenbach and Y.S. Rajapakse (Editors)

RAY THEORY FOR WAVE PHENOMENA IN THE PRESENCE OF FLUID LOADED THIN CURVED ELASTIC SHELLS WITH TRUNCATIONS

L.B. Felsen, I.T. Lu and J.M. Ho

Department of Electrical Engineering/Computer Science
Weber Research Institute, Polytechnic University
Farmingdale, New York 11735

Many structures of current interest contain as elements thin elastic shells which may be curved, truncated and(or) joined to other structural elements. When such composite structures are in vacuum, the wave response to excitation from an oscillating forcing function applied on the surface is confined to the structure *per se,* whereas fluid loading permits coupling of the elastic motion on the structure to the acoustic field in the fluid. While the wave phenomena on an elastic shell conglomerate are exceedingly complex, in general, the localization of these phenomena at high enough frequencies, due to constructive and destructive wave interference, permits the analytical modeling to be organized around tractable canonical problems with simplifying *global* symmetries that can be matched to the *local* features on the structure. The localization of the wave phenomena is systematized around ray trajectories that transport the constructively interfering wave groups from the source to the observer via various encounters (reflections, transmissions, diffractions, etc.) with the structural environment. Implementation of this scenario is based on the rigorous analysis and high frequency asymptotic treatment of the canonical problems.

A prototype for the present study is a fluid-loaded thin, homogeneous, infinitely long circular cylindrical elastic shell excited by an acoustic pressure source located in either the exterior or interior fluid media. After completion of the rigorous and asymptotic analysis, the localized wavefields derived for this canonical configuration are generalized to accommodate truncations and weak deviations from cylindrical curvature. The outcome is a methodology, akin to the geometrical theory of diffraction, for treating the acoustic response of elastic structures with shaped, thin elastic plate elements.

1. INTRODUCTION AND SUMMARY

The acoustic response from submerged elastic structures is of interest in various applications. If the structural configuration is a conglomerate, as is frequently the case, prediction of the response poses a problem of substantial complexity. For acoustic wavelengths that are larger than, or even comparable with, the structural dimensions, direct numerical modeling is feasible. However, at short wavelengths compared to these dimensions, a purely computational approach becomes prohibitive, and it also obscures wave phenomena, which are now operative, and which can furnish much insight into the interpretation of the data. Since waves are characterized by propagation, it is suggestive to build a wave-based scattering model of a composite structure around more elementary substructures, which interact with one another via appropriate propagators. Each substructure, in turn, is modeled by canonical prototypes that are analytically and numerically tractable, and incorporate the essential wave phenomena thereon.

With this motivation as background, the present study is concerned with substructures in the form of thin, truncated, curved elastic shells. The frequency range is assumed to be such that the shell thickness is very much smaller than the acoustic wavelength λ in the surrounding fluid, but that the lateral extent of the shell and its local principal radii of curvature are large compared to λ. Subject to these restrictions, one may appeal to the localization of shell-related wave phenomena caused by constructive and destructive interference. Constructive interference, which causes strong fields, takes place around ray trajectories. Local wavefields can then be tracked along these trajectories to encounters with edges or other obstructions, thereby giving rise to reflected, transmitted, and diffracted constituents. The resulting formulation is usually referred to as the geometrical theory of diffraction (GTD), which plays an important role in electromagnetic and underwater sound wave scattering [1,2], but it has also been applied to scattering by obstacles in elastic unbounded media, and in layered media with special symmetries [1]. The purpose of the present investigation is to generalize these treatments so as to accommodate more complicated shapes, taking advantage of simplifications permitted by the thin shell assumption.

Cylindrical and spherical shells are curved canonical prototypes, for which rigorous analytical solutions can be developed. Those that are available [2-4] have been specifically for the closed shell geometry, with its implied angular periodicity. For the problems to be addressed here, which involve sections excised from the closed prototype, the wave problem has to be parametrized in terms of *traveling* waves that are most readily described in infinitely extended angular domains [5,6] devoid of the periodicity constraint. Although the traveling wave format can be derived from the usual angularly periodic standing wave forms employed for the closed structure [4], the infinite-domain approach systematizes the ray parametrization [7] and subsequent treatment of deformations and truncations [8].

The canonical Green's function problems of determining the acoustic response of fluid-immersed thin cylindrical and spherical shells excited by an acoustic point source have been re-examined from these new perspectives. The analysis for the infinite cylindrical shell (Fig. 1) has been completed [9,10], while that for the spherical shell is in progress. The starting point in both has been synthesis of the rigorous multidimensional solution in terms of one-dimensional characteristic Green's functions defined in complex spectral wavenumber domains; this decomposition is made possible by the separability of the problems in the canonical coordinate systems, extended to infinity in the angular domains. For the cylinder, this extension is straightforward (Fig. 2). The presence of the shell in the radial domain is accounted

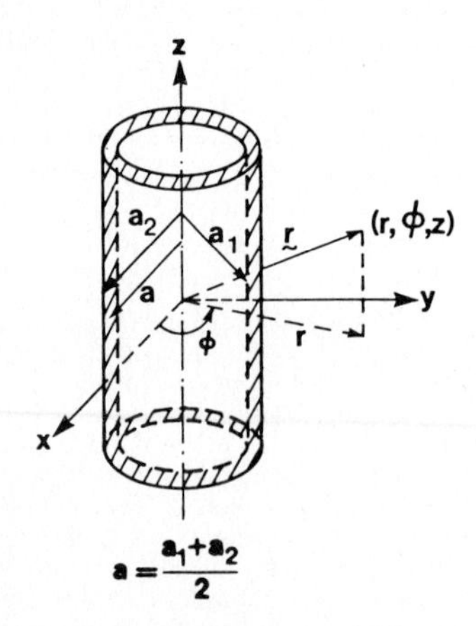

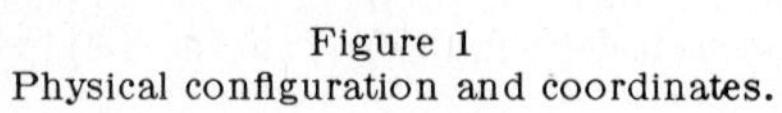
Figure 1
Physical configuration and coordinates.

Figure 2
Section through unbounded ϕ-space. The closed cylinder has been developed into an infinitely sheeted cylindrical plate.

for by continuity conditions derived from the equations governing thin shell dynamics [2]. The resulting explicit solution for the acoustic pressure field in the fluids (either external or internal) is in the form of a double integral over the two complex spectral wavenumbers that represent the separation parameters. By contour deformation, residue and branch cut calculus pertaining to spectral pole or branch point singularities, steepest descent (saddle point) techniques and asymptotics, etc., one may derive from this most general format of the three-dimensional Green's function a host of alternative representations with desirable convergence properties tailored to various parameter regimes. Details of the formal analysis and the derivation of ray acoustic approximations may be found in [9] and [10], respectively. The ray acoustic results for source and observation points outside the shell are summarized below.

The canonical problem for a truncated section of the cylinder is the semi-infinite structure terminated in the infinite angular domain along a helical cut (Fig. 3) which maps in a rectilinearly "rolled out" angular coordinate space into a straight edge. The resulting wave reflection, diffraction and coupling phenomena introduced by the truncation can be parametrized formally in an edge-adapted spectral representation through the inclusion of spectral reflection, diffraction and coupling coefficients that are phase-matched (along the edge) to the incident wave spectra. Asymptotic reduction leads to corresponding ray acoustic field contributions which, by recourse to the principle of locality, can then be adapted to other edge contours. Moreover, weak deviations from circularity in the cylindrical shell prototype can be accommodated approximately by spectral scaling that modifies, by localizing asymptotics, the globally valid spectra on the perfect cylinder. These considerations have been detailed elsewhere [8], and they are not summarized here because of space limitations. Corresponding results for the spherical shell are being developed.

2. FORMULATION AND SOLUTION

The problem geometry for the canonical circular cylindrical thin shell is shown in Fig. 1. The inner and outer shell radii are a_1 and a_2, respectively; the mean radius "a" will appear in most of what follows. The interior and exterior regions are assumed to be filled with dissimilar homogeneous fluids described by densities $\rho_{1,2}$ and propagation speeds $v_{1,2}$, respectively. This configuration is excited by a pressure source located at $\underset{\sim}{r}'=(r',\phi',z')$ in the cylindrical (r,ϕ,z) coordinate system. Suppressing a time dependence $\exp(-i\omega t)$, the resulting acoustic pressure, the three-dimensional Green's function $G(\underset{\sim}{r},\underset{\sim}{r}')$, satisfies in each fluid the wave equation

$$(\nabla^2+k_i^2)\, G(\underset{\sim}{r},\underset{\sim}{r}') = -\rho_i \frac{1}{r'} \delta(r-r')\delta(\phi-\phi')\delta(z-z') \quad , \quad k_i = \omega/v_i \tag{1}$$

where $i=1$ and 2 refers to $r<a_1$ and $r>a_2$, respectively. The pressure must be

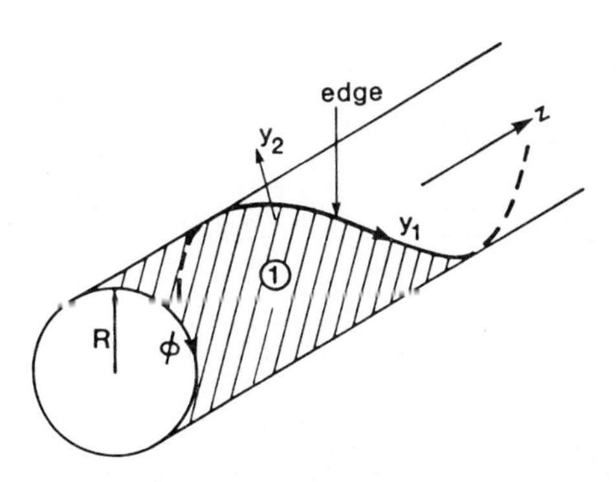

Figure 3
Helical truncation (along y_1) of infinitely sheeted cylindrical plate. (y_1,y_2) are edge-based orthogonal coordinates in the (ϕ,z) "plane".

bounded at r= 0, satisfy a radiation condition at $|\underset{\sim}{r}| \to \infty$, and satisfy across the shell boundaries a jump condition on the pressure

$$G(a_2,\phi,z;\underset{\sim}{r}') - G(a_1,\phi,z;\underset{\sim}{r}') = \underset{\approx}{Z}_a \frac{1}{\rho_1} \frac{\partial G}{\partial r}\bigg|_{a_1} \tag{2a}$$

and a continuity condition on the radial displacement

$$\frac{1}{\rho_1} \frac{\partial G}{\partial r}\bigg|_{a_1} = \frac{1}{\rho_2} \frac{\partial G}{\partial r}\bigg|_{a_2} \tag{2b}$$

In (2a), $\underset{\approx}{Z}_a$ is a nonlocal (integral) operator impedance, which becomes scalarized into the explicitly impedance $Z_a(\lambda_\phi,\lambda_z)$ in the one-dimensional radial problem that remains after elimination of the (ϕ,z) dependence by spectral decomposition in the separation-of-variables procedure; λ_ϕ and λ_z are the corresponding spectral parameters.

Details of the generalized spectral technique employed in the solution may be found in [5] and [9]. Extending the ϕ- domain bilaterally to $\pm\infty$, with the corresponding nonperiodic Green's function denoted by G^∞, the formal result after spectral decomposition in bases pertaining to the (ϕ,z) coordinates is as follows:

$$G^\infty(\underset{\sim}{r},\underset{\sim}{r}_q') = \frac{1}{(-2\pi i)^2} \oint_{C_z} d\lambda_z \oint_{C_\phi} d\lambda_\phi g_r(r,r';\lambda_\phi,\lambda_z) g_\phi^\infty(\phi,\phi_q';\lambda_\phi) g_z(z,z';\lambda_z) \tag{3}$$

where $\underset{\sim}{r}_q'=(r',\phi_q',z')$ specifies source location at any angular point ϕ_q' in $-\infty<\phi<\infty$. The angularly periodic Green's function is recovered by summing over an infinite array of image solutions

$$G(\underset{\sim}{r},\underset{\sim}{r}') = \sum_{q=-\infty}^{\infty} G^\infty(\underset{\sim}{r},\underset{\sim}{r}_q') \ , \tag{3a}$$

and it has a spectral representation as in (3), with $g_\phi^\infty(\phi,\phi_q';\lambda_\phi)$ replaced by the periodic one-dimensional angular Green's function $g_\phi(\phi,\phi';\lambda_\phi)$. In (3), g_r, g_ϕ^∞ and g_z are the one-dimensional characteristic Green's functions in the r,ϕ,z coordinates respectively, and C_z and C_ϕ are integration contours in the complex λ_z and λ_ϕ planes that surround the spectral singularities (poles and branch points) of g_z and g_ϕ^∞, respectively (see Fig. 4). Explicit results for $g_{r,\phi,z}$ may be found in [9]. Alternative representations of the solution may be derived by contour deformation, etc., as stated in Sec. 1.

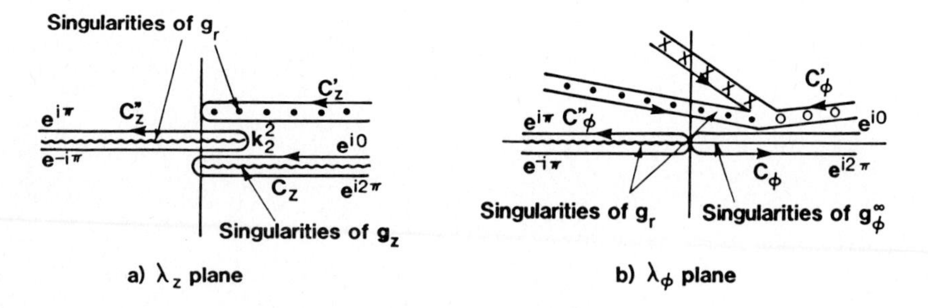

Figure 4
One-dimensional (spectral) Green's functions $g_r(r,r';\lambda_\phi\lambda_z)$, $g_\phi^\infty(\phi,\phi_q';\lambda_\phi)$, $g_z(z,z';\lambda_z)$, singularities and associated contours. · · leaky wave poles, xx creeping wave poles, oo trapped surface wave poles; ~~~ branch cut. For interpretation of the pole categories, see Figs. 6 and 7.

3. RAY ASYMPTOTICS

By invoking high frequency approximations in alternative representations derived from the exact spectral integral in (3), one may develop explicit *localized* wavefield solutions which have a cogent ray interpretation. Space limitations permit only a rough sketch of the procedure; details may be found in [10]. Considering the case where source and observer are both located in the exterior fluid medium, it may be anticipated that the high frequency response at the observer comprises a direct ray contribution which has not interacted with the shell, a geometrically reflected ray contribution, and a diffracted ray contribution. The reflected ray contribution interacts with the shell locally only near the specular reflection point, and both the direct and reflected contributions exist only in the geometrically illuminated region which is bounded by the ray family that grazes the shell. The diffracted contribution interacts with the shell more globally along surface ray trajectories which, by continuous curvature-induced leakage, can shed fields toward an observer either in the illuminated or shadow zones. Detailed asymptotics confirms this phenomenology and, furthermore, supplies the precise rules that allow construction of the ray field amplitude and phase. Being localized near the ray trajectories, the resulting wavefield parametrization is robust with respect to gradual deviations (over the scale of the local wavelengths) of the shell from its prototype circular cylindrical shape.

The direct and geometrically reflected ray field contributions arise from stationary (saddle) points (SP) in the phase of the asymptotically approximated spectral integrand in (3). For evaluation of the integral, the integration paths must be deformed through the SP into steepest descent paths (SDP) in the complex λ_ϕ, λ_z planes; this can be done sequentially. Any spectral singularities encountered during the path deformations must be accounted for, and they furnish the diffracted ray field; only the pole singularities are found to be relevant in this process. Because direct and reflected field contributions occur only in the region illuminated directly by the actual source at $\phi_q' = \phi_o' \equiv \phi'$, it may be anticipated (and is confirmed) that no stationary points are generated in the physical observation interval by image sources with $q \neq 0$ in (3a). However, image sources do contribute to the diffracted field because the surface-guided ray fields launched at any ϕ_q' can reach the observer in $0 \leq \phi \leq 2\pi$ via multiple nonperiodic excursions around the infinitely sheeted circular cylindrical plate. Being guided along the shell, surface ray fields are associated with *each* of the wave species that can propagate inside the shell. Within the confines of thin shell dynamics,

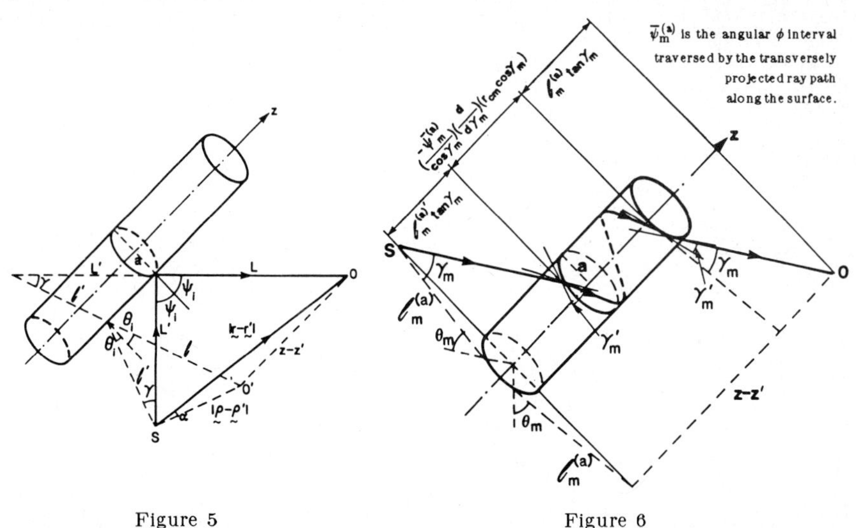

Figure 5
Reflected ray trajectory.

Figure 6
Leaky ray trajectory.

the relevant wave species are quasi-compressional, quasi-flexural, and quasi-shear [2,7].

Selected results from the asymptotic analysis in [10] are summarized below and schematized ray acoustically in Figs. 5 to 7.

3.1 Direct field

$$G_o^{(i)}(\underset{\sim}{r},\underset{\sim}{r}') \sim \frac{\rho_2 e^{ik_2|\underset{\sim}{r}-\underset{\sim}{r}'|}}{4\pi\,|\underset{\sim}{r}-\underset{\sim}{r}'|}\quad,\quad k_2|\underset{\sim}{r}-\underset{\sim}{r}'|\cos\alpha >> 1 \tag{4}$$

$|\underset{\sim}{r}-\underset{\sim}{r}'|$ is the distance from source to observer.

3.2 Reflected field (Fig. 5)

$$G_o^{(r)} \sim \frac{\Gamma\rho_2 e^{ik_2(L+L')}}{4\pi\,(L+L')}\sqrt{\frac{(\ell+\ell')\,a\cos\theta_i}{2\ell\ell'+(\ell+\ell')\,a\cos\theta_i}}\quad,\quad k_2 a\cos\gamma >> 1 \tag{5}$$

Γ is the local reflection coefficient of the shell derived from the full radial impedance of the shell. Geometrical parameters are identified in Fig. 5, which also provides the ray acoustic interpretation. L' and L are the incident and specularly reflected ray segments, respectively, ℓ' and ℓ are their projections on the transverse cross section plane $z=z'$, θ_i is the projected incidence and reflection angle, and γ is the elevation angle between L' and ℓ' or between L and ℓ.

3.3 Diffracted field (Fig. 6)

Depending on the pole location (see Fig. 4b), which is established by the radial resonance condition for the shell, one may distinguish three categories with distinct wave behavior dictated by whether the real part of the phase propagation speed in the surface-guided pole wave is greater than, approximately equal to, or smaller than the wave propagation speed in the exterior fluid. In turn, these conditions describe (a) leaky waves, (b) creeping waves, and (c) trapped surface waves, as established by phase matching between the surface-projected incident ray field and the wavefield guided along the shell. For the mth leaky wave (case a), the contribution at the observer from the qth image source (which models in *physical* space the contribution after q encirclements of the cylinder), is given by

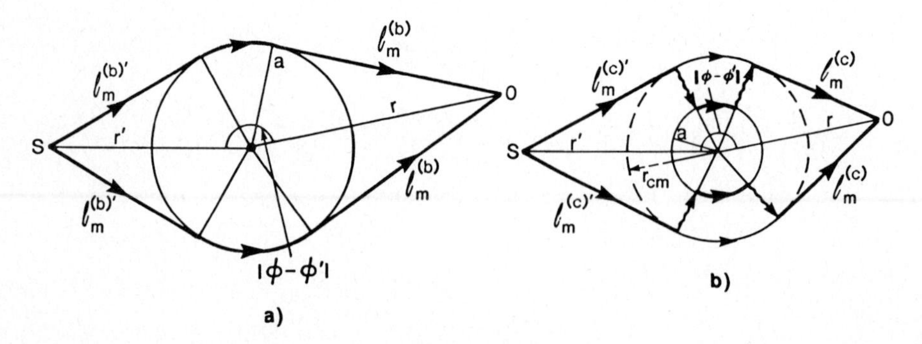

Figure 7
Projected ray paths in transverse plane; (a) creeping waves, (b) trapped surface waves; the wiggly trajectories schematize evanescent tunneling.

$$G_m^{\infty}(\underset{\sim}{r},\underset{\sim}{r}'_q) \sim \sum_m - \frac{\rho_2 A_m e^{\pm i\pi/4} \exp[-(\mathrm{Im}\,\nu_m)\overline{\psi}_m^{(a)}] e^{i\hat{\phi}_m^{(a)}}}{2\left[2\pi\left|\dfrac{d^2\hat{\phi}_m^{(a)}}{d\gamma_m^2}\right|\right]^{1/2}(r^2-r_{cm}^2)^{1/4}(r'^2-r_{cm}^2)^{1/4}}\,, \quad \frac{d^2\hat{\phi}_m^{(a)}}{d\gamma_m^2} \gtrless 0 \qquad (6)$$

$$\hat{\phi}_m^{(a)} = k_2\left[(z-z')\sin\gamma_m + \left(\ell_m^{(a)}+\ell_m^{(a)\prime} + r_{cm}\overline{\psi}_m^{(a)}\right)\cos\gamma_m\right] \qquad (6a)$$

Here, $\nu_m=\lambda_{\phi_m}^{1/2}$ denotes the complex pole location and $r_{cm}=\nu_m(k_2^2-\lambda_z)^{1/2}$ is the radius of a complex caustic whose real part generates approximately the system of launching and detaching rays. The ray interpretation is shown in Fig. 6. For the creeping waves and trapped surface waves, transverse projections of the ray paths in the source plane $z=z'$ are shown in Fig. 7. The analytic forms are analogous to those in (6) and (6a). The wave motion associated with the surface rays in Figs. 6 and 7 is *anisotropic,* due to the dependence of ν_m on the pitch angle of the helical rays [7,8,10].

ACKNOWLEDGEMENT

This work was supported by the Office of Naval Research under Contract No. N00014-88-K-0511, and by the David Taylor Research Center.

REFERENCES

[1] D.O. Thompson and D.E. Chimenti, "Review of Progress in Quantitative Nondestructive Evaluation," Plenum Press, NY. This series of volumes contains representative examples. See also the special issues of The IEEE Journal of Ocean Engineering, *OE-12,* (May 1987), and the Proceedings of the Institute of Electrical and Electronic Engineers (IEEE), *77* (May 1989).

[2] M.C. Junger and D. Feit, *Sound, Structures, and Their Interaction,* 2nd ed., Chapts. 6, 7 and 12, Cambridge, MA: MIT Press, 1986.

[3] G.C. Gaunaurd, "Elastic and Acoustic Resonance Wave Scattering," in *Applied Mechanics Reviews,* Vol. 42, pp. 143-192, New York: The American Society of Mechanical Engineers, 1989.

[4] H. Überall, "Surface Waves in Acoustics," in *Physical Acoustics,* W.P. Mason and R.N. Thurston eds., Vol. 10, pp. 1-60, New York: Academic Press, 1973.

[5] L.B. Felsen and N. Marcuvitz, *Radiation and Scattering of Waves,* Sec. 3.4b, Prentice Hall, Englewood Cliffs, NJ, 1973.

[6] F.G. Friedlander, "Diffraction of Pulses by a Circular Cylinder," Comm. Pure Appl. Math, *7,* (1954), pp. 705-732.

[7] A.D. Pierce, "Wave Propagation on Thin-Walled Elastic Cylindrical Shells," *Elastic Wave Propagation,* (M.F. McCarthy and M.A. Hayes, eds.), Elsevier, New York, pp. 205-210, 1989.

[8] L.B. Felsen and I.T. Lu, "Ray Treatment of Wave Propagation on Thin-Walled Curved Elastic Plates with Truncations," J. Acoust. Soc. Am. *86,* 1989, pp. 360-374.

[9] L.B. Felsen, J.M. Ho and I.T. Lu, "Three-Dimensional Green's Function for Fluid-Loaded Thin Elastic Cylindrical Shell: Formulation and Solution," to be published in J. Acoust. Soc. Am.

[10] L.B. Felsen, J.M. Ho and I.T. Lu, "Three-Dimensional Green's Function for Fluid-Loaded Thin Elastic Cylindrical Shell: Alternative Representations and Ray Acoustic Forms," to be published in J. Acoust. Soc. Am.

Elastic Waves and Ultrasonic Nondestructive Evaluation
S.K. Datta, J.D. Achenbach and Y.S. Rajapakse (Editors)

A RAY ANALYSIS OF ACOUSTIC PULSES REFLECTED FROM A THICK COMPOSITE PLATE IMMERSED IN A FLUID

A.K. MAL and C.-C. YIN
University of California, Los Angeles, California, U.S.A.

Y. Bar-Cohen
Douglas Aircraft Company, Long Beach, California, U.S.A.

The physical nature of the waves produced within a unidirectional fiber-reinforced composite plate immersed in a fluid and insonified by an obliquely incident plane acoustic wave is discussed by means of the ray analysis. For incidence beyond the smallest critical angle, only a small number of mode converted rays are shown to have significant amplitude after reflection and refraction at the two interfaces. The successive mode-converted pulses are shown to be easily identifiable in the overall reflected signal.

1. INTRODUCTION

We consider a composite laminate immersed in a fluid and insonified by a plane acoustic wave. It has been demonstrated earlier that, through the selection of proper transmission frequency and angle of incidence, guided Lamb type waves can be induced in the laminate and that the dispersion curves for guided waves in general, multilayered isotropic and anisotropic laminates can be determined accurately in a broad range of frequencies and velocities by means of this so-called leaky Lamb wave (LLW) technique [1-3]. The dispersion curves have been shown to be useful in the nondestructive material characterization and damage evaluation of structural components [2, 3] of relatively small thickness.

For laminates of larger thickness, if the incident wave is a relatively short pulse, the reflected signal consists of a series of well separated pulses [4], which cannot be interpreted as guided waves. Moreover, the reflected pulses may be unevenly spaced, and the amplitude of the successive reflections may have irregular variations. Although the theoretical results based on the solution of the associated elastodynamic boundary-value problem are in excellent agreement with these observations [1, 4], the source of these unusual phenomena could not be identified from such calculations. We present an alternative analysis based on ray theory, which not only identifies the source of the anomaly, but also gives a clear and simple picture of the wave phenomena that are associated with the experiment for a unidirectional fiber-reinforced plate.

The ray diagram for an anisotropic medium is, in general, far more complex than that for the corresponding isotropic case. It is well known that for a bulk wave propagating in an arbitrary direction in a unidirectional composite plate, three new bulk waves are generated on reflection at the top or bottom interface resulting in 3^n rays after a total of n reflections. There are two special cases, where only 2^n rays are created after n reflections, namely, when the projection of the incident ray on the plate surface is either parallel or perpendicular to the fiber direction. The overall reflected field in the fluid is the superposition of an infinite number of rays refracted into the surrounding fluid from those within the solid. The contribution of each ray to the reflected field is calculated by means of the theory outlined below; the details can be found in [5]. The ray contributions are compared to the exact calculations based on the theory developed in [1].

2. FORMULATION

Consider a unidirectional fiber-reinforced composite plate of thickness H immersed in a fluid and insonified by a plane, time harmonic acoustic wave of unit amplitude and frequency ω. The time dependence $e^{-i\omega t}$ in all field variables is suppressed throughout. The frequency domain calculations are used to obtain time histories of the reflected field for incident pulses through inversion (by FFT).

Within the frequency range normally used in ultrasonic nondestructive evaluation, the material of the plate can be assumed to be transversely isotropic with its symmetry axis parallel to the fibers. Referred to a Cartesian coordinate system with x_1- axis parallel to the symmetry axis, as shown in Fig. 1, the constitutive equation for the material can be written in terms of five complex-valued stiffness constants C_{11}, C_{12}, C_{22}, C_{23} and C_{55} which are functions of the five real stiffness constants, c_{ij} and two material dissipation parameters p_o and a_o [2, 5].

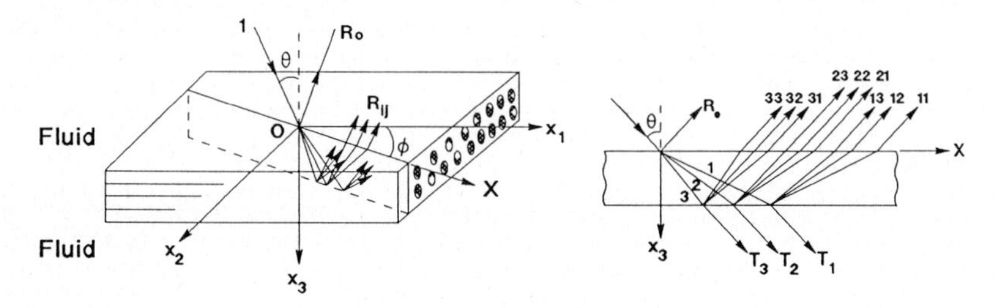

Fig.1. A ray diagram of the leaky Lamb wave (LLW) experiment.

For convenience, we introduce five constants a_1, a_2, a_3, a_4, a_5 related to the stiffness constants C_{ij} through $a_1 = C_{22}/\rho$, $a_2 = C_{11}/\rho$, $a_3 = (C_{12} + C_{55})/\rho$, $a_4 = (C_{22} - C_{23})/2\rho$, $a_5 = C_{55}/\rho$, where ρ is material density. Let the displacement vector $\{u_1, u_2, u_3\}$ be expressed in terms of three scalar potentials Φ_j (j = 1, 2, 3) through

$$\begin{Bmatrix} u_1 \\ u_2 \\ u_3 \end{Bmatrix} = \begin{bmatrix} \partial/\partial x_1 & 0 & 0 \\ 0 & \partial/\partial x_2 & \partial/\partial x_3 \\ 0 & \partial/\partial x_3 & -\partial/\partial x_2 \end{bmatrix} \begin{Bmatrix} \Phi_1 \\ \Phi_2 \\ \Phi_3 \end{Bmatrix} \tag{1}$$

The potential functions satisfy a system of differential equations derived from Cauchy's equation of motion [5]. The plane acoustic waves in the fluid can be similarly represented in terms of two scaler potentials, Φ_o and Φ_b and the boundary conditions at the top and bottom surfaces of the plate can be expressed in the form

$$[\ u_1\ u_2\ u_3\ \sigma_{13}\ \sigma_{23}\ \sigma_{33}\ ]$$

$$= [\ U_o\ V_o\ i\eta_o\Phi_{o,3}\ \ 0\ \ 0\ \ -\rho_o\omega^2\Phi_o\ ], \quad x_3 = 0 \tag{2a}$$

$$= [\ U_1\ V_1\ i\eta_o\Phi_{b,3}\ \ 0\ \ 0\ \ -\rho_o\omega^2\Phi_b\ ], \quad x_3 = H \tag{2b}$$

where U_o, V_o and U_1, V_1 are the slips in the tangential plane at $x_3 = 0$ and H respectively, ρ_o is the mass density and η_o the wavenumber in the x_3- direction, of the surrounding fluid. The potentials satisfy the Helmholtz equation with wavenumber η_o; their expressions will be given later.

(1) The first reflection at the upper interface.

Let the potential in the upper fluid corresponding to the incident wave and the wave reflected from the upper fluid-solid interface be expressed in the form

$$\Phi_o = (e^{i\eta_o x_3} + R_o e^{-i\eta_o x_3})e^{i\xi X} \tag{3}$$

where R_o is the as yet unknown reflection coefficient at the top interface, ξ is the "horizontal" component of the wavenumber and $X = x_1 \cos\phi + x_2 \sin\phi$. Similarly, the potentials within the plate associated with the elastic waves transmitted through the upper interface are given by

$$\begin{Bmatrix} \Phi_1 \\ \Phi_2 \\ \Phi_3 \end{Bmatrix} = \begin{bmatrix} q_{11} & q_{12} & 0 \\ q_{21} & q_{22} & 0 \\ 0 & 0 & 1 \end{bmatrix} \mathrm{Diag}\left[e^{i\zeta_1 x_3}\ e^{i\zeta_2 x_3}\ e^{i\zeta_3 x_3}\right] \begin{Bmatrix} A_{11}^+ \\ A_{12}^+ \\ A_{13}^+ \end{Bmatrix} e^{i\xi X} \tag{4}$$

The unknown constants A_{1j}^+ are associated with the three bulk waves refracted through the top interface and the superscript "+" indicates a "downgoing" wave. ζ_j (j = 1, 2, 3) are the wavenumbers and q_{ij} (i, j = 1, 2) are certain proportionality factors defined in [1, 5]. It should be noted that in (4) all rays have been assumed to be confined to the vertical plane containing the x_3 axis and the incident ray. Moreover, in contrast to the isotropic case, the direct use of Snell's law to determine the angles of refraction of the bulk waves is not possible in the anisotropic case, since the wave speed itself is a function of the directions of propagation of the ray. The direction of propagation of each ray is determined from the wavenumbers ξ, ζ_1, ζ_2 and ζ_3. By taking into account the boundary conditions at the top surface, the following system of linear equations is obtained for the determination of the three unknown coefficients A_{1j}^+ and the surface reflection coefficient R_o:

$$\left[\, P_{21} \;\middle|\; B_1 \,\right] \begin{Bmatrix} A^+ \\ \hline R_o \end{Bmatrix} = \begin{Bmatrix} B_2 \end{Bmatrix} \tag{5}$$

where $\{A^+\} = \{A_{11}^+, A_{12}^+, A_{13}^+\}$. The matrix $[P_{21}]$ and the vectors $\{B_1\}$, $\{B_2\}$ are known; their detailed expressions are given in [5]. The system of equations (5) can be solved to determine the amplitude and phase of the reflected and refracted rays at the upper fluid-solid interface.

(2) The second reflection at the bottom interface.

Each bulk wave that is refracted at the top surface generates three new bulk waves at the bottom surface reflection and refracts one acoustic wave into the surrounding fluid. The potentials associated with these rays are given by

$$\begin{Bmatrix} \Phi_1 \\ \Phi_2 \\ \Phi_3 \end{Bmatrix} = \begin{bmatrix} q_{11} & q_{12} & 0 \\ q_{21} & q_{22} & 0 \\ 0 & 0 & 1 \end{bmatrix} \Bigg[\mathrm{Diag}\left[\delta_{j1} e^{i\zeta_1 x_3}\ \delta_{j2} e^{i\zeta_2 x_3}\ \delta_{j3} e^{i\zeta_3 x_3}\right] \begin{Bmatrix} A_{11}^+ \\ A_{12}^+ \\ A_{13}^+ \end{Bmatrix}$$

$$+ \mathrm{Diag}\left[e^{i\zeta_1 (H-x_3)}\ e^{i\zeta_2 (H-x_3)}\ e^{i\zeta_3 (H-x_3)} \right] \begin{Bmatrix} A_{j1}^- \\ A_{j2}^- \\ A_{j3}^- \end{Bmatrix} \Bigg] e^{i\xi X} \tag{6}$$

$$\Phi_b = T_j e^{i\eta_o (x_3 - H) + i\xi X} \tag{7}$$

where the index j (= 1, 2, 3) corresponds to the order of a specific bulk wave, and δ_{ji} is the Kronecker delta. The superscript "−" denotes an "upgoing" wave and the unknown vector $\{A_j^-\} = \{A_{j1}^-, A_{j2}^-, A_{j3}^-\}$ is associated with the waves reflected from the bottom surface. As in the previous case, the boundary conditions at the bottom surface yield the system of linear equations:

$$\left[P_{22} \;\middle|\; B_3 \right]\left\{\begin{array}{c} A_j^- \\ \hline T_j \end{array}\right\} = -\left[P_{21} \right]\mathrm{Diag}\left[e^{i\zeta_1 H}\; e^{i\zeta_2 H}\; e^{i\zeta_3 H}\right]\left\{ A^+ \right\} \tag{8}$$

where the matrix $[P_{22}]$ and vector $\{B_3\}$ are known [5]. The system of equations (8) can be solved to determine the amplitude and phase of the second set of rays.

(3) Higher order reflections and exact reflection coefficient.

The rays generated by subsequent reflections and refractions can be handled in a similar manner. As an example, each "upgoing" bulk wave refracts a reflected acoustic wave into the overlying fluid above the plate. The above procedure can be repeated to obtain the reflection coefficients R_{ji} (i,j = 1, 2, 3) for the associated rays. In this case, the right-hand side of (5) is replaced by a vector $-[P_{22}][E]\{A_j^-\}$ and the reflection coefficient R_o on the left-hand side by the components R_{ji}. The coefficient R_{ji} for each ray in the recursive derivation (5-8) corresponds to a reflected acoustic wave refracted by a bulk wave labelled "ji" at the top surface. In general, the total contribution from an infinite number of rays needs to be calculated in order to obtain the overall reflection coefficient. An alternative and more convenient procedure can be used to evaluate the exact reflection coefficient through the solution of the associated boundary-value problem. The details of this procedure can be found in a number of earlier papers, e.g. [1, 6] and will be omitted.

3. NUMERICAL RESULTS AND DISCUSSION

The numerical results are presented for incident pulses of given amplitude and frequency content. The time harmonic solution obtained by the above method is multiplied by the Fourier transform of the incident pulse and the result is inverted by means of FFT. The elastic and dissipative properties of the material of the plate are given in [2, 4 and 5].

The calculated time histories of the reflected pulses form a 25 mm thick unidirectional laminate oriented at 0°, 45° and 90° relative to the projection of the incident ray on the laminate surface are shown in Figs. 2-4. The angle of incidence is 15° and the incident pulse is similar to that produced by a flat transducer of 5 MHz center frequency for all the cases. The pulse generated by an individual ray is shown, only if its amplitude is significant as compared to the incident pulse. The results of the exact calculation are also shown for comparison. It can be seen from Fig. 2 that for waves propagating in the vertical plane parallel to the fibers, the reflected rays emerge sequentially at equal intervals. Moreover, the first reflected ray from the top surface of the plate and the converted shear wave (mode 22) dominate the reflected signals; the influence of all the other rays is negligible. This result is similar to that for an isotropic plate.

The corresponding results for propagation at 45° and 90° to the fibers are shown in Figs. 3 and 4. In both cases, the time-delay between the first two arrivals is longer than that between the later arrivals. As an example, for an incident wave at 45° orientation, the reflected signal of mode 22 arrives *after* the top surface reflected pulse and is followed by the converted modes 23 (or 32), 33, etc. The reflected signals associated with the other modes are of negligible amplitude. Moreover, as can be seen from Fig. 3, the signals carried by modes

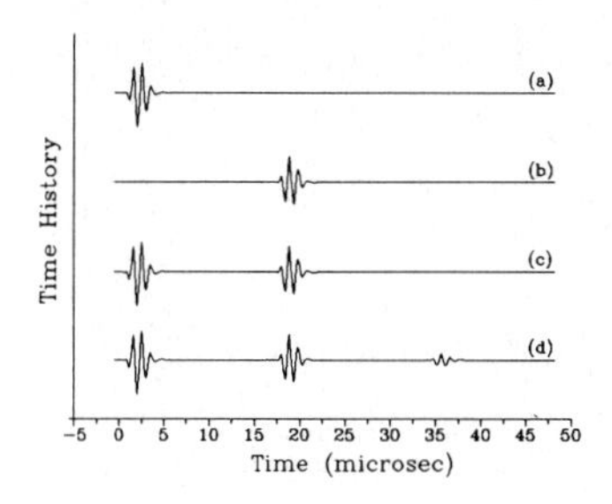

Fig. 2. Reflected time histories for a 25 mm thick unidirectional graphite/epoxy plate for angle of incidence of 15° and 0° fiber orientation. (a) Reflection from upper surface, (b) mode 22, (c) combination of the first 5 rays, and (d) the exact reflected signal.

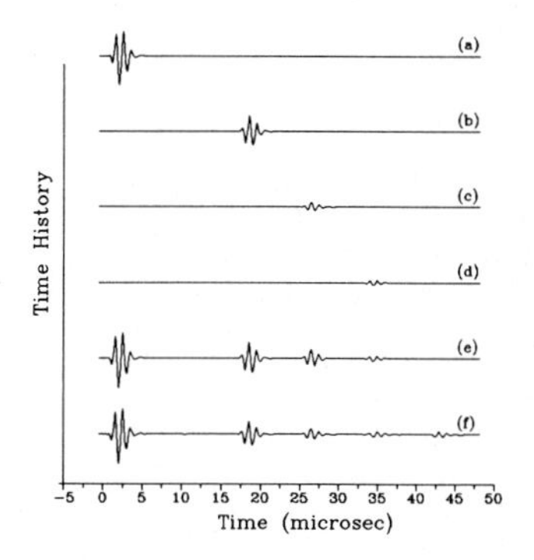

Fig.3. Same as in Fig. 2, but for 45° orientation. (a) Reflection from upper surface, (b) mode 22, (c) converted mode 23 or 32, (d) mode 33, (e) Combination of the first 10 rays, and (f) The exact reflected signal.

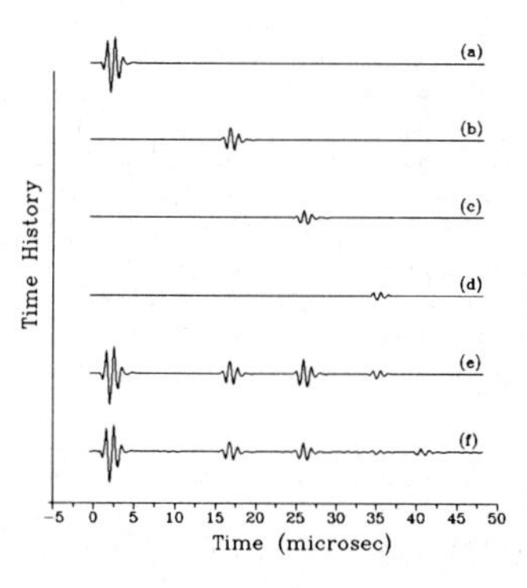

Fig.4. Same as in Fig. 2, but for 90° orientation. (a) Reflection from upper surface, (b) mode 11, (c) converted mode 12 or 21, (d) mode 22, (e) Combination of the first 5 rays, and (f) The exact reflected signal.

23 and 32 arrive simultaneously and in phase. In addition, for propagation at 90° to the fibers (in the quasi-isotropic plane), both longitudinal and transverse waves of modes 11, 12, 21, and 22, contribute to the overall reflected signal. Thus, unlike the isotropic case, all the bulk waves induced by reflection play a significant role in the reflected field. The signal from the fastest longitudinal longitudinal wave (mode 11) is followed by that from longitudinal-transverse (mode 12) or transverse-longitudinal (mode 21) wave. The latest arrival is from mode 22 (transverse-transverse wave).

In Fig. 5, we present the calculated reflected signal from a unidirectional graphite/epoxy plate of different thicknesses at 45° orientation. The incident acoustic pulse has a center frequency of 1 MHz. The time delay between any two successive pulses can be seen to be innversely proportional to the plate thickness. With decrease in plate thickness, the reflected signals move closer until they become indistinguishible, resulting in the formation of guided waves in the plate.

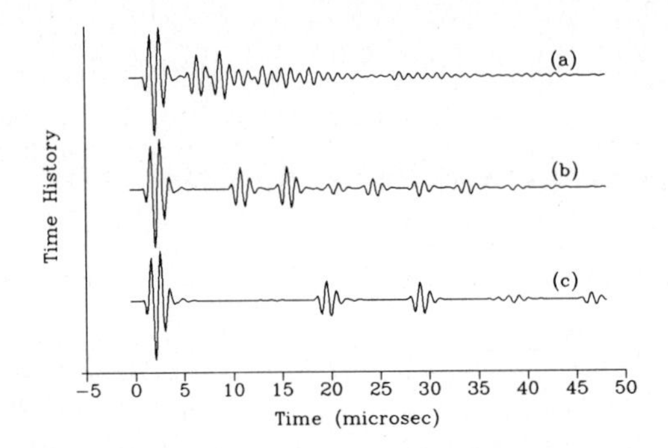

Fig.5. Reflected time histories for unidirectional graphite/epoxy plates of different thickness, H at 45° orientation. (a) H = 7.5 mm, (b) H = 15 mm, (c) 30 mm. The angle of incidence is 20°.

4. CONCLUDING REMARKS

The reflected signals from a unidirectional fiber-reinforced composite plate can be completely explained through mode conversion. The analysis presented here shows that a small number of rays contribute to the overall reflected signal and that the irregular behavior of the reflected field is caused by mode-converted waves, many of which are extinguished during reflection. The reflected acoustic field from the composite plate in the quasi-isotropic plane, perpendicular to fibers, exhibits quite different behavior from that in an isotropic plate. In contrast, the behavior of the reflected field for propagation parallel to the fibers is similar to that of an isotropic plate.

ACKNOWLEDGMENT

This research was supported by the Office of Naval Research under contract N00014-87-K-0351. We are thankful to Dr. Y. Rajapakse of ONR for his encouragement and support during the course of this work.

REFERENCES

[1] A.K. Mal and Y. Bar-Cohen, Ultrasonic characterization of composite laminates, Proceedings of the joint ASME and SES symposium on wave propagation in structural composites, A.K. Mal and T.C.T Ting, (eds.) ASME-AMD-Vol. 90 (1988) pp. 1-16.

[2] M.R. Karim, A.K. Mal and Y. Bar-Cohen, Inversion of leaky Lamb wave data by simplex algorithm, Submitted to J. Acoust. Soc. Am.

[3] B. Tang and E.G. Henneke, II, Lamb-wave monitoring of axial stiffness reduction of laminated composite plates, Materials Evaluation, 47 (1989) pp. 928-934.

[4] A.K. Mal and Y. Bar-Cohen, Ultrasonic NDE of thick composite laminates, Review of Progr. in QNDE, D. Tompson and D. Chimenti (eds.) 8B (1989) pp. 1551-1558 .

[5] A.K. Mal, C.-C. Yin and Y. Bar-Cohen, A Theoretical and experimental study of waves reflected from composite laminates immersed in a fluid, Submitted to J. Appl. Mech.

[6] A.H. Nayfeh and D.E. Chimenti, Ultrasonic wave reflection from liquid-coupled orthotropic plates with application to fibrous composites, J. Appl. Mech. 55 (1988) pp. 863-870 .

Elastic Waves and Ultrasonic Nondestructive Evaluation
S.K. Datta, J.D. Achenbach and Y.S. Rajapakse (Editors)

RAYLEIGH, LAMB, AND WHISPERING GALLERY WAVE CONTRIBUTIONS TO BACKSCATTERING FROM SMOOTH ELASTIC OBJECTS IN WATER DESCRIBED BY A GENERALIZATION OF GTD

Philip L. MARSTON,* Steven G. KARGL,* and Kevin L. WILLIAMS†

*Department of Physics
Washington State University
Pullman, Washington 99164

†Applied Physics Laboratory
University of Washington
Seattle, Washington 98195

Surface guided elastic waves on objects in water can significantly contribute to the scattering of sound. The present research concerns the development and testing of a quantitative ray representation of such contributions to backscattering from smooth elastic objects. Resulting resonance structure is compared with that from the exact partial wave series for solid and hollow spheres. Experimental tests used short tone bursts so that different contributions were separated in time. The analysis concerns leaky guided waves having phase velocities exceeding that of water.

1. INTRODUCTION

The usual geometrical theory of diffraction (GTD) gives a ray representation of scattering amplitudes for rigid objects [1] and is generalizable to certain bulk-transmitted-wave contributions for solid elastic objects [2]. One of the difficulties in extending GTD methods to elastic objects such as shells in water is that the contributions due to *surface guided elastic waves* (SEW) can be important. Examples of SEW include leaky Lamb waves on shells and Rayleigh and whispering gallery waves on spheres. The extension of GTD to represent backscattering from such objects should facilitate the partitioning of complicated high-frequency scattering problems into geometry and the local mechanics of the interaction of the sound field in water with the SEW. Such an extension should give a simple and *quantitative* understanding of the scattering process which could be useful both for inverse problems and for predicting how changes in an object will affect the scattering. The present paper summarizes the development of such an extension, directs the reader to detailed literature, and gives novel relevant derivations for amplitudes and phases.

Figure 1 illustrates a generic problem of interest. A plane wave is incident on a smooth empty shell having a circular profile. One contribution to backscattering is due to specular reflection from the region near point C'; however, that is not the contribution of primary concern here. The incident sound wave excites an elastic surface wave at point B which repeatedly circumnavigates the object radiating sound back towards the source from point B'. A heuristic model of this scattering process was put forth by Borovikov and Veksler [3]. In their analysis, however, the coefficient which describes the coupling of SEW with the acoustic field in water was determined by a fitting procedure. A rigorous analysis of backscattering from solid elastic spheres was carried out by Williams and Marston [4,5] based on the Watson transformation of the exact partial-wave series. That analysis gives a virtually exact expression for the required complex coupling coefficient G_l for the specific case of a solid elastic sphere. (The dependence of G_l on physically relevant parameters was, however, obscure.) That analysis predicts that the contribution to the form function for the farfield backscattering amplitude due to the lth class of SEW is simply

$$f_l = \frac{-G_l \exp\left[-2(\pi-\theta_l)\beta_l + i\eta_l\right]}{\left[1 + j\exp(-2\pi\beta_l + i2\pi ka/c_l)\right]}, \qquad \frac{p_{sca}}{p_{inc}} = \frac{fa\exp[i(kr - \omega t)]}{2r} \tag{1,2}$$

where c and k are the sound velocity and wavenumber in the surrounding water, a is the sphere's radius, η_l is a propagation phase delay and $j = 1$ for a sphere. Parameters which describe the surface wave are the phase velocity c_l and radiation damping parameter β_l (in np/radian). As reviewed in Sec. 3, f_l is part of the total form function f which relates the scattered and incident pressures by Eq. (2). A trace-velocity matching condition gives for $c_l > c$

$$\theta_l = \arcsin(c/c_l)\,, \quad \eta_l = 2ka\left[(c/c_l)\left(\pi - \theta_l\right) - \cos\theta_l\right] - (j+1)\pi/4 \tag{3a,b}$$

Equation (1) predicts a sequence of resonance peaks and is like the form used to describe Fabry-Perot resonators [5]. Superposition of such contributions for Rayleigh and whispering gallery waves with a specular reflection term synthesize accurately the backscattering computed directly from the partial-wave series for a solid sphere [5,6]. [It may be shown [6] that Eqs. (1) and (3) apply also to right circular cylinders by taking $j = -1$ though the exact expression for the coefficient G_l was not obtained.] Measurements of backscattering of short tone bursts due to Rayleigh waves on a solid sphere, reviewed below in Sec. 2, were also described by the theory [4,7].

To facilitate the description of how f_l depends on the interaction of the sound wave in water with the surface wave, it was desirable to express G_l in terms of the SEW properties c_l and β_l. An analysis [6] yields the following approximations for the sphere and circular cylinder cases,

$$|G_l^{sp}| \approx 8\pi\beta_l\, c/c_l, \qquad |G_l^{cy}| \approx 8\pi\beta_l/(\pi ka)^{1/2}, \tag{4,5}$$

where a is the radius of the cylinder. As discussed in Sec. 4, where the phase of G_l is also considered, these approximations were derived by comparing the form of Eq. (1) near a weakly damped resonance peak with standard (approximate) results [8] of resonance scattering theory (RST). It should be emphasized that Eq. (4) is much simpler than the exact result known for solid spheres [4] since the dependence on β_l is evident. The connection between (4) and (5) follows from ray arguments summarized in Ref. 6 and below in Sec. 4. The original confirmation of Eqs. (4) and (5) was based on numerical comparison with the exact $|G_l|$ for solid spheres [4] and Veksler's fitted result for symmetric Lamb waves on a hollow cylinder [3]. The simple form of Eqs. (4) and (5) facilitate quantitative ray tracing for surface wave contributions without use of a fitting procedure and should simplify the procedure for revising Eq. (1) for other objects. The important SEW parameters for a given frequency and radius of curvature a were originally obtained by locating the complex root ν of $D_\nu(ka) = 0$ associated with the lth SEW where $D_n(ka)$ is the denominator of the nth partial wave in the exact series [6]. For a leaky Lamb wave on a shell, c_l can sometimes be approximated from flat plate results by introducing curvature corrections [9].

2. MEASURED BACKSCATTERING OF SHORT TONE BURSTS FROM A SHELL AND A SOLID SPHERE

This section reviews theoretical results and supporting experiments from Refs. 4, 7, and 10 concerning the backscattering of short tone bursts. The wave incident on the spheres was approximately a 3 or 4 cycle sine wave burst from a distant source with a frequency $\omega/2\pi$ as low as 300 kHz. For the experiments with a solid tungsten carbide sphere [4,7], Fig. 1 is applicable with $b = 0$ and the major SEW contribution was due to a leaky Rayleigh wave designated by l = R. Figure 2(a) shows the observed sequence of backscattered echoes for such a sphere with a = 1.27 cm and $ka = \omega a/c = 43.2$. The initial response labeled S is due to the specular reflection from the region near C' in Fig. 1. The first echo labeled R is due to a Rayleigh wave which partially circumnavigates the sphere (from B to B' in Fig. 1) and will be assigned an index m = 0. The subsequent echoes are due to energy radiated by the Rayleigh wave burst as it circumnavigates the sphere m times, m = 1, 2, ... Echoes with large m are reduced in amplitude by radiation damping. The pressure amplitude of a distinct echo due to the lth class of surface wave has the predicted form [7,10]

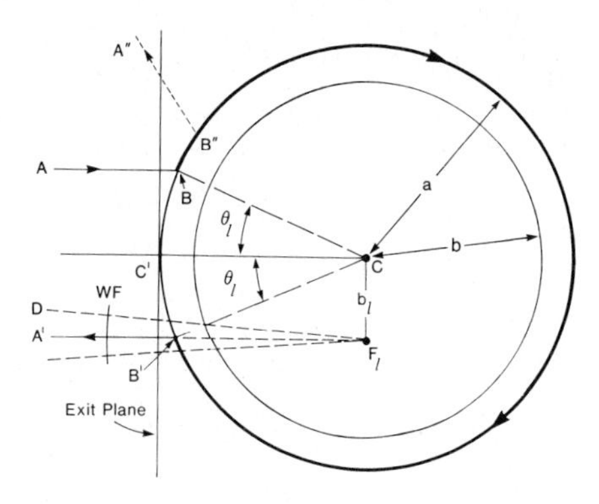

FIGURE 1
Ray diagram for contributions to backscattering due to a surface wave (of type l) excited near B on an elastic sphere or cylinder. WF designates part of the outgoing wavefront.

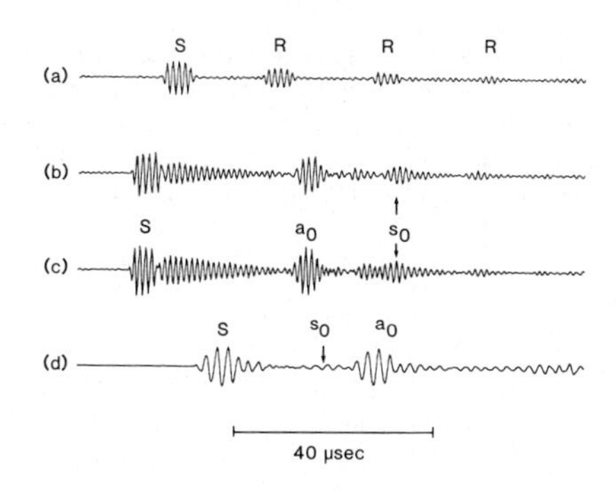

FIGURE 2
Measured echoes for backscattering from solid (a) and hollow (b-d) spheres. S denotes the specular echo. It is followed in (b) and (c) by a tail; here ka is near the thickness resonance and the onset of the $l = s_1$ wave.

$$|p_{ml}| = |p_i| A_{ml}\, a/2r, \qquad A_{ml} = |G_l| \exp\left[-2(\pi-\theta_l)\beta_l - 2\pi m\beta_l\right] |J_0(u) - i\beta_l\gamma J_1(u)|, \qquad (6,7)$$

where J_0 and J_1 are Bessel functions evaluated at $u = ka\,\gamma c/c_l$; $|p_i|$ is the amplitude of the incident wave, γ is the backscattering angle, and r is the distance from the center of the sphere. (The normalization in Eq. (6) is such that specular reflection from a large fixed rigid sphere produces a wave whose amplitude is $|p_i|\, a/2r$.) For scattering in the exactly backward direction, $\gamma = 0$ so that $|J_0(u) - i\beta_l\,\gamma J_1(u)| = 1$. Measured amplitudes with $\gamma = 0$ agreed with predictions for m = 0 and 1 over a range of ka [4]. As γ is shifted away from $\gamma = 0$, $|p_{ml}|$ decreases which is a manifestation of glory scattering [7]. The reduced amplitude was nearly proportional to $|J_0(u)|$ since $|\beta_R\, \gamma J_1(u)| << 1$.

Backscattering was measured from an empty stainless steel spherical shell [10]. The shell dimensions were: radius a = 19.05 mm; thickness h = a - b = 3.1 mm; and b/a = 0.838. Figure 2(b), (c), and (d) are representative records for the hydrophone placed on the backward axis for ka of 64.7, 68.8, and 36.4, respectively. For these ka, the SEW contributions are primarily the lowest antisymmetric (or flexural) and symmetric leaky Lamb waves for which case l becomes a_o and s_o, respectively. The identification of the m = 0 contributions in Fig. 2 is in accordance with their arrival time relative to the specular echo S. The times were in good agreement with delays based on group velocities c_{gl} and modeled path lengths. From echo amplitudes, A_{0l} of Eq. (6) was measured and compared with the predictions of Eqs. (4) and (7). This was done with $\gamma = 0$ for several ka from 24 to 75 except where a_o and s_o echoes overlapped or were weak. The agreement with theory was good for a laboratory scale experiment of this type. Detailed off-axis measurements of A_{0l} with $l = a_o$ were obtained for ka = 24.3 where the on-axis echo was strong ($A_{0l} = 1.1$). These measurements were in excellent agreement with Eqs. (4) and (7) and clearly display a minimum near the γ of the first zero of $J_0(ka\,\gamma c/c_l)$. Note that Eq. (7) neglects the reduction in echo amplitude due to dispersion. For the conditions of the experiments the reduction was estimated to be negligible [10].

3. SIMPLE APPROXIMATE SYNTHESIS OF BACKSCATTERING FORM FUNCTIONS

The synthesis of the form function for spheres previously confirmed for tungsten carbide [5] was greatly simplified and applied to solid and hollow spheres. In Eq. (1), the approximation $G_l \approx 8\pi\beta_l\, c/c_l$ was used from Eq. (3) and an approximation (see Sec. 4) yielding $\arg(G_l) \approx 0$. The specular contributions to backscattering may be approximated as [3,4]

$$f_S^{(solid)} = r\exp(-i2ka), \qquad r = (\rho_E c_L - \rho c)/(\rho_E c_L + \rho c), \tag{8,9}$$

$$f_S^{(shell)} = \mathcal{R}\exp(-i2ka) + f_{Scc}, \qquad \mathcal{R} = r - \frac{(1 - r^2)\exp(i2k_L h)}{1 - r\exp(i2k_L h)} \tag{10,11}$$

where ρ_E and c_L are the density and bulk longitudinal velocity of the solid, $\mathcal{R}$ is the complex reflection coefficient of a vacuum-backed flat plate of thickness h, $k_L = \omega/c_L$, and f_{Scc} is a curvature correction. The expression for $\mathcal{R}$ may be derived by summation of the ray geometric series for reverberations in a plate [3]. As was done in Ref. 3 for cylinders, the curvature correction will be omitted in the synthesis which follows so that for the shell $|f_S| \approx |\mathcal{R}| = 1$. The ray approximations to the total form functions for the solid and shell spheres are

$$f = f_S + f_{l=R} + f_{l=WG1}, \qquad f = f_S + \sum_{n=0}^{N_a-1} f_{l=a_n} + \sum_{n=0}^{N_s-1} f_{l=s_n}, \tag{12,13}$$

where WG1 designates the slowest whispering gallery wave and the number N_a and N_s of antisymmetric and symmetric Lamb wave contributions depends on ka through the existence of roots of $D_\nu(ka) = 0$. Figure 3 and 4 compare resulting $|f|$ with exact values from the partial wave series. The synthesis is verified in each case. The material parameters are as listed in Ref. 5 and 10, respectively, and for the shell $b/a = 0.838$. Notice that in addition to parameters in f_S, the material parameters only enter through the c_l and β_l and these are given from roots ν_l of $D_\nu(ka) = 0$. The finest structure in Fig. 3 is due to WG wave resonances and fine structure not synthesized is due to higher WG modes. In Fig. 4, it is due to the s_o wave resonances since β_l is less than that for $l = a_o$ in this region and $N_s = N_a = 1$. Whether a resonance causes a dip or a peak in $|f|$ can be understood from the relative phases of the contributions [5]. The spacing Δka of resonances for a given class of weakly dispersive SEW is roughly c_{gl}/c.

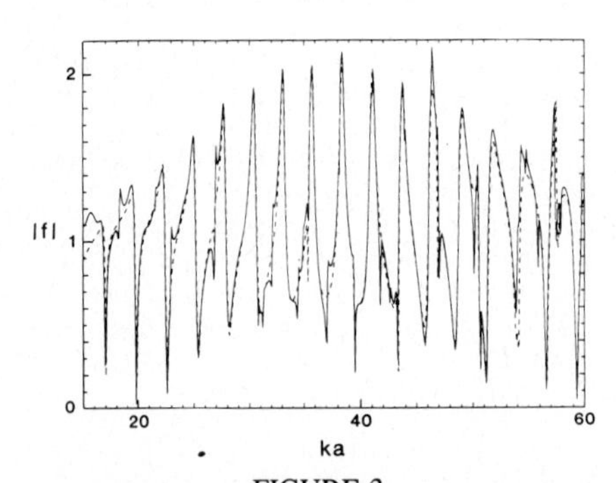

FIGURE 3
Form function for steady-state backscattering from a tungsten carbide sphere: exact (solid curve) and synthesis from Eq. (12), (dashed).

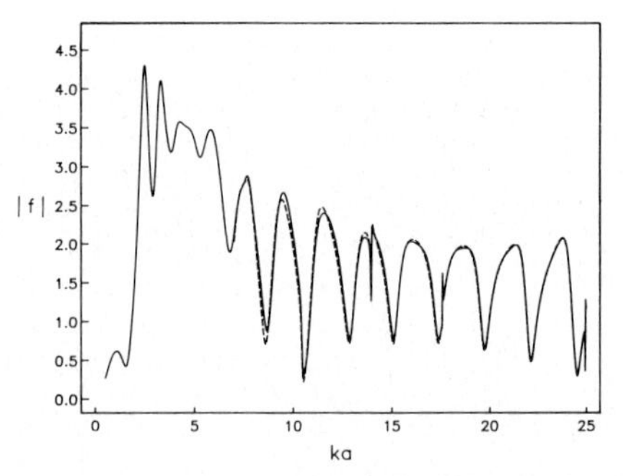

FIGURE 4
As in Fig. 3 but for a stainless-steel spherical shell and the synthesis (dashed) from Eq. (13), includes $l = a_o$ and s_o Lamb contributions.

The synthesis is terminated in Fig. 4 at $ka = 7$ since $c_l < c$ at lower ka for $l = a_o$. Felsen *et al.* [11] have developed a ray representation for cylinders which should facilitate the description of scattering contributions of subsonic or "trapped" waves such as flexural waves having $c_l < c$. Unlike Fig. 1, the coupling with the sound field involves tunneling through an evanescent region. Some *other* limitations of our approximations for the shell are found by extending the comparison in Fig. 4 to higher ka. The omitted f_{Scc} is significant near the lowest thickness resonance which occurs at $ka = \pi c_L a/ch \approx 77$. While the synthesis up through $ka = 100$ correctly describes additional fine structure due to the a_1 wave (which occurs when N_a becomes 2 for $ka > 40$), the ka region near 70 where N_s becomes 2 is not well approximated by Eq. (13). At the cutoff of the $l = s_1$ mode, $c_l/c \to \infty$ such that from Eq. (3a), $\theta_l \to 0$; the SEW radiates nearly backwards directly (along ray B"A" with B and B" shifted closer to C' than in Fig. 1) without having navigated the far side of the sphere. Such radiation is *not included* in the ray representation [5,6] leading to Eq. (1). The difficulties in the ka region between 70 and 80 were not present in a ray representation of the *forward* scattering. Using the optical theorem, the total scattering cross section is well approximated from the synthesized forward amplitude [12].

4. RAY REPRESENTATION FOR SPHERES FROM ANALYSIS FOR CYLINDERS

The connection between Eq. (1) and rays due to the repeated circumnavigation of SEW may be seen as follows [5]. For either the sphere or cylinder case, the ray representation of the f_l is

$$f_l = \sum_{m=0}^{\infty} f_{ml}, \qquad f_{ml} = -G_l \exp(i\eta_l) \exp\left[-2(\pi - \theta_l)\beta_l\right] \mathcal{Z}^m, \tag{14,15}$$

$$\mathcal{Z} = \exp(-2\pi\beta_l) \exp\left[-i\pi(j+1)/2\right] \exp(i2\pi ka\, c/c_l), \tag{16}$$

where as in Sec. 2, m is the circumnavigation index and the phase factor depending on $j = \pm 1$ accounts for the caustics at the poles of a sphere. Summation of Eq. (15) as a geometrical series yields Eq. (1). While the form in Eq. (1) is useful for describing resonances, Eq. (14) gives a simpler description of the scattering of short tone bursts. Note that $|f_{ml}| = A_{ml}(\gamma = 0)$ of Eq. (7).

While the coefficient G_l accounts for the local coupling of the sound field in the fluid with the SEW in the cylinder case, it also includes the effects of axial focusing [2,13] in the sphere case. For the cylinder case, Eq. (2) is replaced by Eq. (17) below:

$$\frac{p_{sca}}{p_{inc}} = \left(\frac{a}{2r}\right)^{1/2} f \exp[i(kr - \omega t)], \qquad |f_{ml}| = 2\sqrt{2}\, |p_{mle}|/|p_{inc}| \tag{17,18}$$

and Eq. (18), which follows from (17) and consideration of ray tubes, expresses $|f_{ml}|$ in terms of the local pressure amplitude $|p_{mle}|$ of the corresponding outgoing wave in the *exit plane* (Fig. 1)of the *cylinder* from either of two virtual line sources. For cylinders, Eqs. (15) and (18) may be combined to describe the amplitude of the wave diverging from the virtual line source at F_l. (The factor 2 in Eq. (17) accounts for the virtual source, not shown, displaced *above C by b_l*.) The corresponding relationship between $|f_{ml}|$ of a *sphere* and the local amplitude $|p_{mle}|$ in the exit plane is complicated by axial focusing. Approximation of the relevant diffraction integral [13] gives for the sphere case

$$|f_{ml}| \approx \frac{2k|p_{mle}|}{a|p_{inc}|} \left| \int_0^{\infty} s \exp\left[ik\left(s - b_l\right)^2/2a\right] ds \right| \approx (8\pi ka)^{1/2} \frac{|p_{mle}|}{|p_{inc}|} \left(\frac{b_l}{a} + O(ka)^{-1/2}\right), \tag{19}$$

where $b_l = ac/c_l$ is the focal circle radius and it is assumed that $ka >> 1$ and b_l/a is not $<< 1$. Given Eq. (5) for $|G_l|$ of a cylinder, the $|f_{ml}|$ for the sphere may be approximated as follows. Consider a cylinder having the same β_l, c_l, radius a, and p_{inc} as the sphere of interest. Consideration of the energy flux along ray tubes on the surface of each scatterer gives $|p_{mle}$ (sphere)$|$ $\approx |p_{mle}$ (cylinder)$|$. Hence $|p_{mle}/p_{inc}|$ may be eliminated from the right side of (19) by using Eqs. (18), (15), and (5). The result for the sphere is that $|f_{ml}|$ is as in Eq. (15) with $|G_l|$ given by Eq. (4). Hence G_l of a sphere in Eq. (4) accounts for the effects of axial focusing on the far field amplitude as approximated by Eq. (19) when b_l/a is not $<< 1$.

Extension of the ray arguments above gives $\arg(G_l^{sp}) \approx \arg(G_l^{cy}) - \pi/4$ when phase information is retained at all stages. Approximations for $\arg(G_l)$ for either case may be obtained by retaining phase information in the comparison of Eq. (1) with a Breit-Wigner form function from RST. (In the latter, a rigid background is assumed to be appropriate [8].) Because of limitations on RST, the comparison may only be valid if $\beta_l << 1$. The results are:

$$\arg(G_l^{sp}) \approx 0, \qquad \arg(G_l^{cy}) \approx \pi/4 \tag{20,21}$$

in agreement with the relationship noted above. It is evident from Figs. 3 and 4 that Eqs. (4) and (20) are applicable in the problems considered.

Consider now SEW contributions to the backscattering from a sphere which is only slightly deformed into the shape of a spheroid. The outgoing wavefront is no longer toroidal, but is perturbed. Consequently the axial caustic (which led to the γ dependence of Eq. (7)) is unfolded to give an astroid caustic which has a more complicated directional dependence [14].

ACKNOWLEDGMENTS

This research was supported by the U.S. Office of Naval Research.

REFERENCES

[1] Keller, J. B., IEEE Trans. Ant. Propag. AP-23 (1985) 123.
[2] Marston, P. L., Williams, K. L., and Hanson, T. J. B., J. Acoust. Soc. Am. 74 (1983) 605.
[3] Borovikov, V. A. and Veksler, N. D., Wave Motion 7 (1985) 143.
[4] Williams, K. L. and Marston, P. L., J. Acoust. Soc. Am. 78 (1985) 1093.
[5] Williams, K. L. and Marston, P. L., J. Acoust. Soc. Am. 79 (1986) 1702.
[6] Marston, P. L., J. Acoust. Soc. Am. 83 (1988) 25.
[7] Williams, K. L. and Marston, P. L., J. Acoust. Soc. 78 (1985) 722.
[8] Flax, L., Dragonette, L. R., and Uberall H., J. Acoust. Soc. Am. 63 (1978) 723.
[9] Marston, P. L., J. Acoust. Soc. Am. 85 (1989) 2663.
[10] Kargl, S. G. and Marston, P. L., J. Acoust. Soc. Am. 85 (1989) 1014.
[11] Felsen, L. B., Ho, J. M., and Lu, I. T., J. Acoust. Soc. Am. (to be published).
[12] Marston, P. L. and Kargl, S. G., in: Oceans '89 Conference Record (IEEE, New York, 1989) pp. 1194-1198.
[13] Marston, P. L. and Langley, D. S., J. Acoust. Soc. Am. 73 (1983) 1464.
[14] Arnott, W. P. and Marston, P. L., J. Acoust. Soc. Am. 85 (1989) 1427.

Elastic Waves and Ultrasonic Nondestructive Evaluation
S.K. Datta, J.D. Achenbach and Y.S. Rajapakse (Editors)

THIN INTERFACE LAYERS: ADHESIVES, APPROXIMATIONS AND ANALYSIS

P.A. MARTIN

Department of Mathematics, University of Manchester
Manchester M13 9PL, England

Linear models of imperfect interfaces between elastic bodies are studied; both inclusions and laminated structures are considered. After a brief review of the literature, the problem of scattering by an inclusion with an imperfect interface is analysed. Uniqueness theorems are obtained, and boundary integral equations over the interface are derived.

1. INTRODUCTION

There are several branches of mechanics in which two solids interact across a common interface. Examples are adhesive joints [1], frictional contacts [2], laminated structures and composite materials [3, 4]. The simplest situations are those where the two solids are welded together ('perfect interface'), so that the displacement and traction vectors are continuous across the interface. The opposite extreme is when there is no interaction ('complete debonding'). Intermediate situations arise when the two solids can slip or separate, or when there is a thin layer of a different material (such as glue or lubricant) between the solids. In this paper, we are especially interested in those intermediate situations that can be modelled by simple linear modifications to the perfect-interface continuity conditions.

2. IMPERFECT PLANE INTERFACES: A REVIEW

Consider a plane interface $x_3 = z = 0$ between two elastic solids. Let $u_i^\pm$ and $\tau_{ij}^\pm$ be the components of the displacement vector and stress tensor, respectively, in $\pm z > 0$. The traction vectors on $z = 0$ are given by $t_i^\pm = \tau_{i3}^\pm$. If $z = 0$ is a perfect interface, we have

$$[\mathbf{t}] = \mathbf{0} \qquad \text{and} \qquad [\mathbf{u}] = \mathbf{0}, \tag{2.1}$$

where square brackets denote discontinuities across the interface:

$$[\mathbf{u}] = \mathbf{u}^+ - \mathbf{u}^-, \qquad \text{evaluated on } z = 0.$$

The perfect-interface conditions (2.1) were first modified by Newmark in 1943 [5]. He explicitly allows slipping to occur, so that (2.1) are replaced by

$$[\mathbf{t}] = \mathbf{0}, \qquad [u_3] = 0 \qquad \text{and} \qquad [u_\alpha] = M t_\alpha \qquad (?\,?)$$

where $\alpha = 1$ or 2, and M is a positive constant. Note that $M = 0$ corresponds to a perfect interface, whereas $M = \infty$ corresponds to a 'lubricated interface', i.e. one where there is a thin layer of inviscid fluid between the two solids.

Newmark's theory is for two beams in two dimensions, but it has been generalized to

$$[\mathbf{t}] = \mathbf{0}, \qquad [u_3] = 0 \qquad \text{and} \qquad [u_\alpha] = M_{\alpha\beta} t_\beta \tag{2.3}$$

where summation of β over $\beta = 1$ and $\beta = 2$ is implied, and $M_{\alpha\beta}$ can be a function of t_γ and $[u_\gamma]$. For a review, see [6], where $M_{\alpha\beta}$ is assumed to be a positive diagonal matrix.

Similar boundary conditions have been used by Murty [7] to model the propagation of waves through a 'loosely-bonded interface'. In [8], (2.2) are derived by assuming that there is a thin interface layer of viscous fluid. Now, the parameter M is given by

$$M = ih/(\omega\eta), \tag{2.4}$$

where h is the thickness of the layer, η is the shear viscosity coefficient of the fluid, and a harmonic time-dependence of $e^{-i\omega t}$ is implied.

It is perhaps worth noting that the problem of wave propagation through a layer of viscous fluid (or of viscoelastic solid) between two elastic solids can be analysed exactly. However, the extraction of *approximate* conditions, connecting the displacements and tractions across a *thin* layer, is not usually the aim of such analyses [9].

Jones and Whittier [10] have modelled wave propagation through a 'flexibly-bonded interface' by allowing both slip and separation. They replace (2.1) with

$$[\mathbf{t}] = \mathbf{0}, \qquad [u_3] = M_n t_3 \qquad \text{and} \qquad [u_\alpha] = M_s t_\alpha, \tag{2.5}$$

where M_n and M_s are constants. Several authors have used (2.5) [11–16]. In [15, 16], the formulae

$$M_n = h/(\lambda + \mu) \qquad \text{and} \qquad M_s = h/\mu, \tag{2.6}$$

are given, where λ and μ are the Lamé moduli of a thin elastic layer modelling the bond.

3. SCATTERING BY AN INCLUSION

Let B_i denote a bounded domain, with a smooth closed boundary S and simply-connected exterior, B_e. We seek displacements $\mathbf{u}_e(P)$ and $\mathbf{u}_i(P)$ so that

$$L_e\mathbf{u}_e(P) = \mathbf{0}, \quad P \in B_e \qquad \text{and} \qquad L_i\mathbf{u}_i(P) = \mathbf{0}, \quad P \in B_i, \tag{3.1a, b}$$

where $\mathbf{u}(P) = \mathbf{u}_e(P) + \mathbf{u}_{inc}(P)$ for $P \in B_e$, $\mathbf{u}_{inc}$ is the given incident wave and $\mathbf{u}_e$ satisfies a radiation condition at infinity. In addition, we shall impose certain continuity conditions across S; these are specified below. The operator L_a is defined by

$$L_a\mathbf{u} = k_a^{-2} \text{ grad div } \mathbf{u} - K_a^{-2} \text{ curl curl } \mathbf{u} + \mathbf{u}$$

where $\rho_a\omega^2 = (\lambda_a + 2\mu_a)k_a^2 = \mu_a K_a^2$ and $a = e$ or i. ρ_a is the density of the solid in B_a, whereas λ_a and μ_a are the Lamé moduli. The traction operator T_a is defined on S by

$$(T_a\mathbf{u})_m(p) = \lambda_a n_m \text{ div } \mathbf{u} + \mu_a n_\ell(\partial u_m/\partial x_\ell + \partial u_\ell/\partial x_m)$$

where $\mathbf{n}(p)$ is the unit normal at $p \in S$, pointing into B_e.

4. INCLUSIONS WITH IMPERFECT INTERFACES: A REVIEW

Suppose that the interface S is imperfect. For example, the inclusion might be surrounded by a thin layer of a different elastic material. For simple geometries, such problems can be treated exactly [17, 18]. The first approximate treatment, using continuity conditions across S similar to those described in §2, was given by Mal and Bose [19]. They considered spherical inclusions with the following interface conditions:

$$[\mathbf{t}] = \mathbf{0}, \qquad [u_n] = 0, \qquad \text{and} \qquad [u_\alpha] = M t_\alpha. \tag{4.1}$$

Here, $\mathbf{t} = T_e \mathbf{u}$, $\mathbf{t}_i = T_i \mathbf{u}_i$, $[\mathbf{t}] = \mathbf{t} - \mathbf{t}_i$ is the discontinuity in the tractions across S and we decompose vectors as $\mathbf{u}(p) = u_1 \mathbf{s}_1 + u_2 \mathbf{s}_2 + u_n \mathbf{n}$ for $p \in S$, where $\mathbf{s}_1$ and $\mathbf{s}_2$ are unit vectors in the tangent plane at p, satisfying $\mathbf{s}_1 . \mathbf{s}_2 = 0$ and $\mathbf{n} = \mathbf{s}_1 \times \mathbf{s}_2$. The parameter M can be a complex function of ω: $M = 0$ for a perfect interface; M is given by (2.4) for a thin layer of viscous fluid; and $M = \infty$ for a lubricated interface, whence

$$[\mathbf{t}] = \mathbf{0}, \qquad t_\alpha = 0 \qquad \text{and} \qquad [u_n] = 0. \tag{4.2}$$

Similar interface conditions have been used in models of composite materials, where identical inclusions are arranged periodically prior to analysis using homogenization techniques. Let ϵ be a length scale associated with the periodicity. For elastostatics, Lene and Leguillon [20] take $M = k\epsilon$, where $k > 0$. For time-harmonic waves, Santosa and Symes [21] use $M = i\epsilon/(\omega c)$, where c is a 'viscous constant'; cf. (2.4).

Aboudi [22] has used flexibly-bonded interfaces in a different model of composites, with

$$[\mathbf{t}] = \mathbf{0}, \qquad [u_n] = M_n t_n, \qquad \text{and} \qquad [u_\alpha] = M_t t_\alpha. \tag{4.3}$$

He identifies M_n and M_t as h/E and h/μ, respectively, where the thin elastic interface layer has thickness h, Young's modulus E and shear modulus μ; cf. (2.6).

Kitahara, Nakagawa and Achenbach [23] consider an inclusion in 'spring contact' with the exterior solid. This is intended to model a thin compliant interface layer and leads to

$$[\mathbf{t}] = \mathbf{0} \qquad \text{and} \qquad [\mathbf{u}] = F.\mathbf{t}, \tag{4.4}$$

where the matrix F (called the 'flexibility matrix' [15]) is a positive diagonal matrix. Later, we shall allow F to be a full matrix, with elements that vary with position p on S.

More complicated interface conditions are obtained in [24, 25], where all terms to $O(h)$ are included for an elastic interface layer of thickness h. Non-local terms, involving various tangential derivatives, are present; see also [26]. The simplest (local) conditions discussed in [24, 25] are

$$[\mathbf{t}] = -\rho h \omega^2 \mathbf{u} \qquad \text{and} \qquad [\mathbf{u}] = \mathbf{0}, \tag{4.5}$$

where the layer has density ρ. A generalization of (4.5) is

$$[\mathbf{t}] = G.\mathbf{u} \qquad \text{and} \qquad [\mathbf{u}] = \mathbf{0}, \tag{4.6}$$

where the elements of the matrix G could vary with position p on S.

Finally, we consider a model that includes both (4.4) and (4.6), namely (cf. [27])

$$[\mathbf{t}] = G.\langle \mathbf{u} \rangle \qquad \text{and} \qquad [\mathbf{u}] = F.\langle \mathbf{t} \rangle, \tag{4.7}$$

where $\langle \mathbf{u} \rangle = \frac{1}{2}(\mathbf{u} + \mathbf{u}_i)$ is the average of $\mathbf{u}$ and $\mathbf{u}_i$ on S.

5. UNIQUENESS THEOREMS

Consider the problem of scattering by an inclusion with an imperfect interface. We can prove uniqueness theorems for interfaces characterized by (4.4), (4.6) or (4.7), by adapting standard arguments from [28]. Thus, surround S with a large sphere S_R of radius R. For $P \in B_e$, write

$$\mathbf{u}(P) = \mathbf{u}^{(p)} + \mathbf{u}^{(s)}, \qquad \text{where} \qquad \mathbf{u}^{(p)} = -k_e^{-2} \text{ grad div } \mathbf{u}, \qquad \mathbf{u}^{(s)} = \mathbf{u} - \mathbf{u}^{(p)},$$

and $\mathbf{u}_{inc} \equiv \mathbf{0}$. Then, an application of Betti's reciprocal theorem to $\mathbf{u}$ and its complex conjugate, $\overline{\mathbf{u}}$, in the region between S and S_R gives

$$k_e(\lambda_e + 2\mu_e) \lim_{R\to\infty} \int_{S_R} |\mathbf{u}^{(p)}|^2 \, ds + K_e \mu_e \lim_{R\to\infty} \int_{S_R} |\mathbf{u}^{(s)}|^2 \, ds + I = 0, \tag{5.1}$$

where

$$I = \frac{1}{2i} \int_S (\mathbf{u}.\overline{\mathbf{t}} - \overline{\mathbf{u}}.\mathbf{t}) \, ds = \mathcal{I}m \int_S \mathbf{u}.\overline{\mathbf{t}} \, ds, \tag{5.2}$$

$\mathcal{I}m$ denotes imaginary part and the radiation condition has been used (see [28], Chpt. 3, §2). If we can show that $I \geq 0$, we can deduce from (5.1) that $\mathbf{u}^{(p)} \equiv \mathbf{0}$ and $\mathbf{u}^{(s)} \equiv \mathbf{0}$, whence $\mathbf{u} \equiv \mathbf{0}$ in B_e. Then, (4.4) imply that $\mathbf{u}_i = \mathbf{0}$ and $\mathbf{t}_i = \mathbf{0}$ on S, whence $\mathbf{u}_i \equiv \mathbf{0}$ in B_i.

Now, applying Betti's theorem in B_i to $\mathbf{u}_i$ and $\overline{\mathbf{u}}_i$ gives

$$0 = \frac{1}{2i} \int_S (\mathbf{u}_i.\overline{\mathbf{t}}_i - \overline{\mathbf{u}}_i.\mathbf{t}_i) \, ds = \mathcal{I}m \int_S \mathbf{u}_i.\overline{\mathbf{t}}_i \, ds.$$

Hence, subtracting from (5.2) gives

$$I = \mathcal{I}m \int_S \overline{\mathbf{t}}.[\mathbf{u}] \, ds = \mathcal{I}m \int_S \overline{\mathbf{t}}.F\mathbf{t} \, ds,$$

after using (4.4). Thus, $I \geq 0$, provided that

$$F_{k\ell} = \overline{F}_{\ell k} \quad \text{for} \quad k \neq \ell \qquad \text{and} \qquad \mathcal{I}m(F_{kk}) \geq 0 \quad (\text{no sum}). \tag{5.3}$$

So, if the elements of F are finite and satisfy (5.3) (for all $p \in S$ if F varies with p), we have proved that the corresponding inclusion problem has at most one solution.

The above proof fails for lubricated interfaces, given by (4.2). We obtain $I = 0$, whence $\mathbf{u}(P) \equiv \mathbf{0}$ for $P \in B_e$. It follows that $\mathbf{t}_i = \mathbf{0}$ and $\mathbf{n}.\mathbf{u}_i = 0$ on S. But, it does not follow that $\mathbf{u}_i \equiv \mathbf{0}$ in B_i, as for some geometries and frequencies the interior solid can support free oscillations which do not couple to the exterior solid. See [29] for more details.

We can obtain similar results for interfaces characterized by (4.6). For finite G, we find uniqueness, subject to

$$G_{k\ell} = \overline{G}_{\ell k} \quad \text{for} \quad k \neq \ell \qquad \text{and} \qquad \mathcal{I}m(G_{kk}) \leq 0 \quad (\text{no sum}).$$

We can also prove uniqueness when (4.7) are used, provided that F and G are both real diagonal non-zero matrices.

6. BOUNDARY INTEGRAL EQUATIONS

We conclude by deriving (direct) boundary integral equations over S for inclusions with imperfect interfaces characterized by (4.4), in the plane case. First, we introduce two fundamental Green's tensors, $\mathbf{G}_a(P;Q)$ $(a = e, i)$:

$$(\mathbf{G}_a(P;Q))_{ij} = \frac{1}{\mu_a} \left\{ \Psi_a \delta_{ij} + \frac{1}{K_a^2} \frac{\partial^2}{\partial x_i \partial x_j} (\Psi_a - \Phi_a) \right\}$$

where $\Phi_a = -(i/2)H_0^{(1)}(k_a R)$, $\Psi_a = -(i/2)H_0^{(1)}(K_a R)$ and $R = |P - Q|$. Next, we define elastic single-layer and double-layer potentials by

$$(S_a \mathbf{u})(P) = \int_S \mathbf{u}(q).\mathbf{G}_a(q;P)\, ds_q \qquad \text{and} \qquad (D_a \mathbf{u})(P) = \int_S \mathbf{u}(q).T_a^q \mathbf{G}_a(q;P)\, ds_q,$$

respectively, where T_a^q means T_a applied at $q \in S$. Then, three applications of Betti's theorem (one in B_e to $\mathbf{u}_e$ and $\mathbf{G}_e$, one in B_i to $\mathbf{u}_{inc}$ and $\mathbf{G}_e$, and one in B_i to $\mathbf{u}_i$ and $\mathbf{G}_i$) yield the familiar representations

$$2\mathbf{u}_e(P) = (S_e \mathbf{t})(P) - (D_e \mathbf{u})(P), \qquad P \in B_e, \tag{6.1}$$

and

$$-2\mathbf{u}_i(P) = (S_i \mathbf{t}_i)(P) - (D_i \mathbf{u}_i)(P), \qquad P \in B_i. \tag{6.2}$$

Letting $P \to p \in S$, (6.1) and (6.2) give

$$(I + \overline{K_e^*})\mathbf{u} - S_e \mathbf{t} = 2\mathbf{u}_{inc} \tag{6.3}$$

and

$$(I - \overline{K_i^*})\mathbf{u}_i + S_i \mathbf{t}_i = \mathbf{0}, \tag{6.4}$$

respectively, where

$$\overline{K_a^*}\mathbf{u} = \int_S \mathbf{u}(q).T_a^q \mathbf{G}_a(q;p)\, ds_q.$$

($\overline{K_a^*}$ is a singular integral operator.) Using (4.4) in (6.4) gives

$$(I - \overline{K_i^*})\mathbf{u} + \{S_i - (I - \overline{K_i^*})F\}\mathbf{t} = \mathbf{0}. \tag{6.5}$$

The pair (6.3) and (6.5) is a system of four coupled singular integral equations for the four components of the two vectors $\mathbf{u}(p)$ and $\mathbf{t}(p)$, $p \in S$. It can be shown that this system is a *quasi-Fredholm* system [30], provided that F is a non-singular matrix (for all $p \in S$ if F varies with p). This means that all the usual Fredholm theorems hold. In particular, we can analyse the solvability of (6.3) and (6.5) by showing that the corresponding pair of homogeneous equations has only the trivial solution. These aspects will be considered elsewhere.

REFERENCES

[1] Mittal, K.L. (ed.), *Adhesive Joints*, Plenum, New York, 1984.
[2] Selvadurai, A.P.S. & Voyiadjis, G.Z. (ed.), *Mechanics of Material Interfaces*, Elsevier, Amsterdam, 1986.
[3] Metcalfe, A.G. (ed.), *Interfaces in Metal Matrix Composites*, Academic, New York, 1974.
[4] Christensen, R.M., *Mechanics of Composite Materials*, Wiley, New York, 1979
[5] Newmark, N.M., Siess, C.P. & Viest, I.M., 'Tests and analysis of composite beams with incomplete interaction', *Proc. Soc. for Experimental Stress Analysis* **9**, No.1 (1951) 75–92.
[6] Toledano, A. & Murakami, H., 'Shear-deformable two-layer plate theory with interlayer slip', *Proc. ASCE, J. Eng. Mech.* **114** (1988) 604–623.
[7] Murty, G.S., 'A theoretical model for the attenuation and dispersion of Stoneley waves at the loosely bonded interface of elastic half spaces', *Phys. of the Earth & Planetary Interiors* **11** (1975) 65–79.

[8] Banghar, A.R., Murty, G.S. & Raghavacharyulu, I.V.V., 'On the parametric model of loose bonding of elastic half spaces', *J. Acoust. Soc. Amer.* **60** (1976) 1071–1078.
[9] Rokhlin, S., Hefets, M. & Rosen, M., 'An elastic interface wave guided by a thin film between two solids', *J. Appl. Phys.* **51** (1980) 3579–3582.
[10] Jones, J.P. & Whittier, J.S., 'Waves at a flexibly bonded interface', *J. Appl. Mech.* **34** (1967) 905–909.
[11] Schoenberg, M., 'Elastic wave behaviour across linear slip interfaces', *J. Acoust. Soc. Amer.* **68** (1980) 1516–1521.
[12] Chonan, S., 'Vibration and stability of a two-layered beam with imperfect bonding', *J. Acoust. Soc. Amer.* **72** (1982) 208–213.
[13] Angel, Y.C. & Achenbach, J.D., 'Reflection and transmission of elastic waves by a periodic array of cracks', *J. Appl. Mech.* **52** (1985) 33–41.
[14] Mal, A.K., 'Guided waves in layered solids with interface zones', *Int. J. Eng. Sci.* **26** (1988) 873–881.
[15] Mal, A.K. & Xu, P.C., 'Elastic waves in layered media with interface features', in: *Elastic Wave Propagation* (ed. M.F. McCarthy & M.A. Hayes), North-Holland, Amsterdam, 1989, 67–73.
[16] Pilarski, A. & Rose, J.L., 'A transverse-wave ultrasonic oblique-incidence technique for interfacial weakness detection in adhesive bonds', *J. Appl. Phys.* **63** (1988) 300–307.
[17] Datta, S.K. & Ledbetter, H.M., 'Effect of interface properties on wave propagation in a medium with inclusions', in: [2], 131–141.
[18] Paskaramoorthy, R., Datta, S.K. & Shah, A.H., 'Effect of interface layers on scattering of elastic waves', *J. Appl. Mech.* **55** (1988) 871–878.
[19] Mal, A.K. & Bose, S.K., 'Dynamic elastic moduli of a suspension of imperfectly bonded spheres', *Proc. Camb. Phil. Soc.* **76** (1974) 587–600.
[20] Lene, F. & Leguillon, D., 'Homogenized constitutive law for a partially cohesive composite material', *Int. J. Solids Struct.* **18** (1982) 443–458.
[21] Santosa, F. & Symes, W.W., 'A model for a composite with anisotropic dissipation by homogenization', *Int. J. Solids Struct.* **25** (1989) 381–392.
[22] Aboudi, J., 'Wave propagation in damaged composite materials', *Int. J. Solids Struct.* **24** (1988) 117–138.
[23] Kitahara, M., Nakagawa, K. & Achenbach, J.D., 'On a method to analyze scattering problems of an inclusion with spring contacts', in: *Boundary Element Methods in Applied Mechanics* (ed. M. Tanaka & T.A. Cruse), Pergamon, Oxford, 1988, 239–244.
[24] Datta, S.K., Olsson, P. & Boström, A., 'Elastodynamic scattering from inclusions with thin interface layers', in: *Wave Propagation in Structural Composites* (ed. A.K. Mal & T.C.T. Ting), ASME, New York, 1988, 109–116.
[25] Olsson, P., Datta, S.K. & Boström, A., 'Elastodynamic scattering from inclusions surrounded by thin interface layers', *J. Appl. Mech.* to appear.
[26] Nayfeh, A.H. & Nassar, E.A.M., 'Simulation of the influence of bonding materials on the dynamic behavior of laminated composites', *J. Appl. Mech.* **45** (1978) 822–828.
[27] Baik, J. & Thompson, R.B., 'Long wavelength elastic scattering from a planar distribution of inclusions', *J. Appl. Mech.* **52** (1985) 974–976.
[28] Kupradze, V.D., Gegelia, T.G., Basheleishvili, M.O., & Burchuladze, T.V., Three-dimensional Problems of the Mathematical Theory of Elasticity and Thermoelasticity, North-Holland, Amsterdam, 1979.
[29] Jones, D.S., 'Low-frequency scattering by a body in lubricated contact', *Quart. J. Mech. Appl. Math.* **36** (1983) 111–138.
[30] Muskhelishvili, N.I., *Singular Integral Equations*, Noordhoff, Groningen, 1953.

Elastic Waves and Ultrasonic Nondestructive Evaluation
S.K. Datta, J.D. Achenbach and Y.S. Rajapakse (Editors)

Ultrasonic Nondestructive Evaluation of Graphite Epoxy Composite Laminates*

James G. Miller

Department of Physics
Washington University
St. Louis, Missouri

The purpose of this manuscript is to summarize our approach to quantitative non-destructive evaluation. This report focuses on impact damage, on the combined effects of fatigue and impact, and on the detection of porosity in composite laminates. In addition, we review the quantitative ultrasonic techniques developed to measure ultrasonic attenuation and backscatter in composites.

1. INTRODUCTION

This manuscript summarizes our approach to quantitative ultrasonic measurements and images for the nondestructive evaluation of composite laminates. We have had many collaborators in this work; their original contributions can be identified from the citations to published articles that appear at the end of this manuscript.

We summarize the quantitative ultrasonic techniques developed to measure frequency dependent attenuation and backscatter and then illustrate the use of these techniques for NDE. Specifically, we review the use of broadband substitution techniques to determine signal loss, attenuation coefficient, and slope of attenuation with respect to frequency from transmission measurements. Methods to reduce phase cancellation errors associated with the use of piezoelectric receiving transducers are also illustrated. We also review methods for obtaining the backscatter coefficient from the measured backscatter transfer function by compensating for the frequency dependent volume under interrogation and for attenuation. For measurements in which the frequency dependence is not required, the frequency average of the backscatter transfer function over the useful bandwidth, defined as the integrated backscatter, provides a useful index which serves to reduce fluctuations arising from the inhomogeneous nature of the interrogated medium and phase cancellation effects.

To illustrate the use of quantitative ultrasonic imaging as an approach to materials characterization in inherently inhomogeneous media, impact and fatigue damage was investigated in quasi-isotropic graphite epoxy composite laminates. Images were obtained in transmission-mode based on frequency dependent attenuation and in reflection-mode based on integrated backscatter with data acquired in the polar backscatter configuration. A similar approach was used to investigate the potential of these quantitative ultrasonic techniques for the study of porosity in composite laminates.

2. Results of Ultrasonic NDE of Composite Laminates

2.1. Impact and Fatigue Damage

To illustrate the potential of quantitative ultrasonic imaging as an approach to materials characterization in inherently inhomogeneous media, low velocity impact and fatigue damage were investigated alone and in combination in quasi-isotropic graphite epoxy composite laminates. Quantitative images based on the slope of the attenuation coefficient measured as a function of frequency over a broad bandwidth were obtained using a phase insensitive acoustoelectric receiving transducer. Results indicated that low velocity impact yielded increased values for the slope of the

---

* This research supported in part by NASA grant NSG 1601

attenuation. Fatigued samples which were subsequently subjected to impact suffered more significant damage that composites subjected to impact alone.[1]

The same set of fatigue and impact damaged graphite epoxy samples were investigated using the polar backscatter technique introduced by Bar-Cohen and Crane.[2] For each composite investigated, four images were generated based on the ultrasonic backscatter obtained at a fixed, non-perpendicular polar angle and azimuthal angles perpendicular to the 45, 0, -45, and -90 degree fiber orientation in the quasi-isotropic composites. Results of these investigations suggest that polar backscatter offers the potential of a ply-by-ply examination of the location and extent of damage.[3]

The apparent success of our initial polar backscatter imaging suggested the need to validate the results of our ply-by-ply ultrasonic images with a destructive method of evaluation. In collaboration with S. M. Freeman of Lockheed-Georgia Company we investigated impact damage in another set of graphite epoxy laminates. The nondestructive technique of polar backscatter was employed at Washington University to detect and assess area, configuration, and approximate interlaminar location of impact induced delamination. In this technique the insonifying beam is incident on the sample at a non-zero polar angle so that the specular echo from the water-composite interface does not dominate the backscattered signal. The destructive technique of deply[4] was employed at the Lockheed-Georgia Company on the same impact sites to determine areas and configurations of the delaminations at each interlaminar location. In this technique, the internal matrix damage is marked with a gold solution, the laminate is subjected to a partial pyrolysis and unstacked, and the areas of damage quantified with an image analyzer. Results of these investigations indicated an excellent correlation between the area and orientation of impact damage as measured by the polar backscatter technique and the lamina by lamina examination of the actual damage obtained with the destructive deply technique.[5]

2.2. Porosity in Composite Laminates

The detrimental effects of porosity on material strength are well-known. We have examined methods for characterizing porosity in composites based on both frequency dependent ultrasonic attenuation and backscatter. In a defect-free composite, polar backscatter is greatest when the insonifying beam is perpendicular to a major fiber direction, and falls off dramatically at other azimuthal angles. We investigated the hypothesis that scattering from small spherical voids will be relatively isotropic with respect to the azimuthal angle of incidence, so that the magnitude of variation of the azimuthal backscatter may be indicative of the porosity.[2] As a preliminary approach, glass fiber, epoxy matrix composites were fabricated with a range of "porosities" by the introduction of glass beads of a known size distribution. Results of this preliminary investigation revealed a good correlation between the average isotropic component of scatter and the volume fraction of "porosity", with a correlation coefficient of $r = 0.99$.[6]

Based on these encouraging results obtained on glass fiber composites, we investigated 5 uniaxial graphite epoxy composites with 1% to 8% volume fraction of solid glass inclusions to model "porosity". For polar backscatter measurements, the samples were insonified at a polar angle of 30 degrees and an azimuthal angle centered at zero degrees with respect to the fiber orientation. For each specimen, data were acquired at 121 sites by translating the sample over an 11 by 11 grid in 2 mm steps. At each site the azimuthal angle was varied in 5 degree steps from -10 to +10 degrees and the resulting spectra were averaged in order to remove background variations not attributable to porosity. Integrated polar backscatter was obtained by averaging over the useful bandwidth. The results obtained from polar backscatter correlated with the volume fraction of "porosity" with a correlation coefficient of $r = 0.98$.[7]

The frequency dependent attenuation methods were also evaluated for characterization of porosity in graphite epoxy composites. Previous theoretical and experimental work by Rose and his collaborators suggested that the volume fraction of porosity and the average pore radius could be obtained from the frequency dependence of the attenuation coefficient.[8, 9] We proposed an approach based on the slope of the attenuation coefficient and evaluated this approach on a set of 5 glass fiber, epoxy matrix test specimens with simulated porosity (glass beads) ranging from 0% to 12%. Good correlation was obtained between the measured slopes of attenuation and "porosity", with a correlation coefficient of $r = 0.9$.[10]

Encouraged by these results, we applied the method based on slope of attenuation to the set of five uniaxial composites with 1% to 8% simulated porosity described above. For the frequency dependent attenuation method, data were acquired at 441 sites on a 21 by 21 grid in 1 mm steps. Signal loss relative to a water-only path was obtained as a function of frequency using the method of log spectral subtraction. The normalized data were analyzed by performing a two-parameter polynomial fit about the center frequency of the useful bandwidth. The rate of increase with frequency of excess attenuation exhibited a good correlation with the volume fraction of "porosity", with a correlation coefficient r = 0.98.[7]

Investigations by Bar-Cohen[11] and our Laboratory[12] indicate that the presence of the bleeder cloth impressions substantially influences the degree of anisotropy measured. For relatively thin samples in which selective time gating is not feasible, not only the state of the insonified surface but also the state of the opposite surface influences the received signal. Results of measurements performed with the bleeder cloth impressions intact were compared with the corresponding results obtained after their complete removal. Without bleeder cloth impressions the integrated backscatter difference was 16 dB for a "pore-free" region and 3 dB for a region containing ≈3.4% volume fraction of "porosity". In contrast, measurements made on the same sample prior to the removal of the bleeder cloth impression, exhibited differences of 5.5 dB for a "pore-free" region and 5.6 dB for the region containing ≈3.4% "porosity".

To extend the ultrasonic polar backscatter technique to increasingly complex fiber orientations we investigated the relationship between contrast (difference between the spatially averaged integrated backscatter value of the "porous" and pore-free regions for a specific azimuthal angle ϕ) and azimuthal angle of insonification.[13] As the azimuthal angle of insonification was varied from 0° to 90° the contrast decreased monotonically from a value of 9.1 dB to a value of 0.8 dB.

The results summarized above suggest that quantitative ultrasonic measurements may provide a practical and reliable approach to the nondestructive evaluation and characterization of composites. In the next section we summarize the methodological developments that underpin our approach to quantitative ultrasonic measurements.

3. Methods for Quantitative Ultrasonic Measurements

3.1. Ultrasonic Attenuation

Transmission Measurements of Attenuation:

Review of broadband substitution technique; slope of attenuation determined from least square line fit to attenuation coefficient versus frequency data [14]

Comparison of apparent attenuation coefficient versus frequency obtained with pointlike and finite-aperture piezoelectric receivers, focused-focused pairs of piezoelectric receivers, and acoustoelectric receivers [14, 15, 16]

Reconstructive Tomography Based on Attenuation:

Quantitative images obtained using an acoustoelectric receiver to eliminate phase cancellation errors and reconstructing frequency derivative (slope) of the attenuation coefficient to minimize reflection losses and reduce refraction errors[17, 18]

Compensation for the decrease of transmitted beamwidth with frequency prior to computing the slope [19]

Large-aperture phase-insensitive receiver used to avoid partial loss of beam due to refraction [19, 20]

Consequences of anisotropy for reconstructive tomography [21][21]

Estimating Attenuation from Backscattered Ultrasound:

Attenuation was estimated without requiring access to both sides of a composite by using the fact that propagation of a broadband pulse through a medium exhibiting a frequency dependent attenuation results in systematic changes in the backscattered spectrum, including changes in the lowest order moments (total energy, spectral centroid, and spectral variance).[22]

3.2. Ultrasonic Backscatter

Measurement of Backscatter:

Broadband substitution technique to obtain backscatter transfer function; backscatter coefficient obtained by compensating backscatter transfer function for frequency-dependent volume under interrogation and for attenuation; spatial averaging to reduce variance due to effects of interference and phase cancellation effects at piezoelectric receiver arising from the inhomogeneous nature of the medium.[23, 24]

Alternate Approaches to the Measurement of Backscatter:

Frequency average of the backscatter transfer function (integrated backscatter) estimated directly from the time domain signal using analog processing; integrated backscatter employed to reduce fluctuations arising from the inhomogeneous nature of the medium.[25]

Backscatter transfer function obtained as a function of frequency using an analog spectrum analyzer; backscatter coefficient and integrated backscatter obtained from backscatter transfer function [26, 27]

Integrated backscatter estimated in real time using an acoustoelectric energy detector [28]

Compensation for the Attenuation of Intervening Media:

Backscatter transfer function and integrated backscatter obtained by compensating for the frequency-dependent attenuation of intervening medium [26]

Contrast Optimization:

An approach for determining the limits of detectability of porosity in composites of increasingly complex fiber orientations, the contrast available at a specific azimuthal angle for a given ply lay-up[13]

3.3. Phase-sensitive and Phase-insensitive Detection of Ultrasound

Phase Cancellation Effects at Piezoelectric Receiving Transducers:

Apparent attenuation coefficients overestimated with large variances due to phase-sensitive detection of distorted ultrasonic wavefronts arising from transmission through inhomogeneous media; root mean square deviation of attenuation coefficient versus frequency curve from least square fit line used to identify measurements exhibiting substantial error; phase cancellation errors reduced by reducing area of receiver aperture [14, 15]

Phase cancellation errors reduced in transmission measurements of attenuation by requiring an entirely homogeneous (i.e. water) propagation path except for an inhomogeneous sample sufficiently thin to lie entirely within the depth of field of the focal zones of a matched pair of focused piezoelectric receivers [16]

Large variances observed in ultrasonic backscatter measurements and speckle observed in two-dimensional ultrasonic imaging arise not only because of interference effects that occur in the ultrasonic field independent of the type of receiver but also because of phase cancellation effects at piezoelectric receivers; reduced variance can be achieved by spatial averaging or by frequency averaging (e.g., integrated backscatter) [23, 24, 29, 30]

Phase-insensitive Detection of Ultrasound:

Acoustoelectric effect in cadmium sulfide provided an approach to the construction of a finite-aperture phase-insensitive receiver; attenuation coefficients obtained in transmission experiments with an acoustoelectric receiver exhibited lower average values and substantially reduced variances than measurements carried out with piezoelectric receivers [15, 29, 31]

Response characteristics of an acoustoelectric receiver; criteria for optimizing ultrasonic bandwidth, response time, sensitivity, and electrical and mechanical loading [32]

Large-aperture phase-insensitive detector based on spatial moments of ultrasonic intensity distribution obtained from a two-dimensional array [33, 34]

Comparison of Results of Phase-sensitive and Phase-insensitive Detection:

Apparent attenuation coefficient versus frequency of deterministic and random test objects measured with pointlike and finite-aperture piezoelectric receivers, focused-focused pairs of piezoelectric receivers, and acoustoelectric receivers [14, 15, 17, 18, 29, 30, 33]

Laboratory and computer simulation comparisons of signals scattered from deterministic and random specimens using single-element and focusing arrays of phase-sensitive (i.e., piezoelectric) and phase-insensitive (e.g., acoustoelectric) receivers [29, 30, 34]

Phase-insensitive and phase-sensitive analysis of phase-distorted scattered ultrasound with two-dimensional focused arrays.[35]

REFERENCES

1. Thomas A. Shoup, J.G. Miller, Joseph S. Heyman, and Walter Illg, "Ultrasonic Characterization of Fatigue and Impact Damage in Graphite Epoxy Composite Laminates", *Proc. IEEE Ultrasonics Symposium*, Vol. 82 CH 1823-4, pp. 960-964, (1982).
2. Y. Bar-Cohen, and R.L. Crane, "Acoustic-Backscattering Imaging of Subcritical Flaws in Composites", *Materials Evaluation*, Vol. 40, pp. 970-975, (1982).
3. Lewis J. Thomas III, Eric I. Madaras, and J.G. Miller, "Two-Dimensional Imaging of Selected Ply Orientations in Quasi-Isotropic Composite Laminates Using Polar Backscattering", *Proc. IEEE Ultrasonics Symposium*, Vol. 82 CH 1823-4, pp. 965-970, (1982).
4. S.M. Freeman, "Correlation of X-Ray Radiograph Images with Actual Damage in Graphite-Epoxy Composites by the Deply Technique", *Composites in Manufacturing 3 Conference*, Vol. EM84-101, pp. 1-13, (1984).
5. Earl D. Blodgett, S.M. Freeman, and J.G. Miller, "Correlation of Ultrasonic Polar Backscatter With the Deply Technique for Assessment of Impact Damage in Composite Laminates", *Review of Progress in Quantitative Nondestructive Evaluation*, Vol. 5B, pp. 1227-1238, (1986).
6. Earl D. Blodgett, Lewis J. Thomas III, and J.G. Miller, "Effects of Porosity on Polar Backscatter From Fiber Reinforced Composites", *Review of Progress in Quantitative Nondestructive Evaluation*, Vol. 5B, pp. 1267-1274, (1986).
7. S.M. Handley, M.S. Hughes, J.G. Miller, and E.I. Madaras, "Characterization of Porosity in Graphite Epoxy Composite Laminates With Polar Backscatter and Frequency Dependent Attenuation", *IEEE Ultrasonics Symposium*, Vol. 87CH2492-7, pp. 827-830, (1987).
8. J.H. Rose, D.K. Hsu, and L. Adler, "Ultrasonic Characterization of Porosity Using the Kramers-Kronig Relations", *Journal De Physique*, Vol. Colloque C10, supplement au n 12, Tome 46, pp. 787-790, (1985).
9. Laszlo Adler, James H. Rose, and Carroll Mobley, "Ultrasonic Method to Determine Gas Porosity in Aluminum Alloy Castings: Theory and Experiment", *J. Appl. Phys.*, Vol. 59, pp. 336-347, (1986).
10. M.S. Hughes, S.M. Handley, J.G. Miller, and E.I. Madaras, "A Relationship Between Frequency Dependent Ultrasonic Attenuation and Porosity in Composite Laminates", *Review of Progress in Quantitative Nondestructive Evaluation*, Vol. 7B, pp. 1037-1044, (1988).
11. Yoseph Bar-Cohen, *Nondestructive Characterization of Defects in Multilayered Media Using Ultrasonic Backscattering*, McDonnell-Douglas Corp., (1987). Douglas Paper 7781. Unpublished.
12. S.M. Handley, J.G. Miller, and E.I. Madaras, "Effects of Bleeder Cloth Impressions on the Use of Polar Backscatter to Detect Porosity", *Review of Progress in Quantitative Nondestructive Evaluation*, Vol. 8B, pp. 1581-1587, (1989).
13. S.M. Handley, J.G. Miller, and E.I. Madaras, "An Investigation of the Relationship Between Contrast and Azimuthal Angle for Imaging Porosity in Graphite/Epoxy Composites with Ultrasonic Polar Backscatter", *IEEE Ultrasonics Symposium*, Vol. 88 CH 2578-3, pp. 1031-1034, (1988).
14. J.G. Miller, *et al.*, "Ultrasonic Tissue Characterization: Correlation Between Biochemical and Ultrasonic Indices of Myocardial Injury", *Proc. IEEE Ultrasonics Symposium*, Vol. 76 CH 1120-5SU, pp. 33-43, (1976).
15. L.J. Busse, *et al.*, "Phase Cancellation Effects: A Source of Attenuation Artifact Eliminated by a CdS Acoustoelectric Receiver", *Ultrasound in Medicine*, Vol. 3, pp. 1519-1535, (1977).

16. M. O'Donnell, J.W. Mimbs, B.E. Sobel, and J.G. Miller, "Ultrasonic Attenuation in Normal and Ischemic Myocardium", *Proc. Second International Symposium on Ultrasonic Tissue Characterization*, Vol. NBS Special Publication 525 (1979), pp. 63-71, (1977).
17. J.R. Klepper, G.H. Brandenburger, L.J. Busse, and J.G. Miller, "Phase Cancellation, Reflection, and Refraction Effects in Quantitative Ultrasonic Attenuation Tomography", *Proc. IEEE Ultrasonics Symposium*, Vol. 77 CH 1264-1SU, pp. 182-188, (1977).
18. J.G. Miller, *et al. Reconstructive Tomography Based on Ultrasonic Attenuation* , pp. 151-164, North Holland Pub. Co., New York, (1979).
19. J.R. Klepper, G.H. Brandenburger, J.W. Mimbs, B.E. Sobel, and J.G. Miller, "Application of Phase Insensitive Detection and Frequency Dependent Measurements to Computed Ultrasonic Attenuation Tomography", *IEEE Trans. on Biomedical Engineering*, Vol. BME-28, pp. 186-201, (1981).
20. G.H. Brandenburger, J.R. Cox, J.R. Klepper, and J.G. Miller, "Computer Simulation to Evaluate Strategies for Enhancing Accuracy in Attenuation Tomography", *Proc. IEEE Ultrasonics Symposium*, Vol. 82 CH 1823-4, pp. 685-690, (1982).
21. G.H. Brandenburger, J.R. Klepper, J.G. Miller, and D.L. Snyder, "Effects of Anisotropy in the Ultrasonic Attenuation of Tissue on Computed Tomography", *Ultrasonic Imaging*, Vol. 3, pp. 113-143, (1981).
22. Earl D. Blodgett, Patrick H. Johnston, and J.G. Miller, "Estimating Attenuation in Composite Laminates Using Backscattered Ultrasound", *Proc. IEEE Ultrasonics Symposium*, Vol. 84 CH 2112-1, pp. 748-753, (1984).
23. M. O'Donnell, J.W. Mimbs, and J.G. Miller, "The Relationship Between Collagen and Ultrasonic Backscatter in Myocardial Tissue", *J. Acoust. Soc. Am.*, Vol. 69, pp. 580-588, (1981).
24. M. O'Donnell, and J.G. Miller, "Quantitative Broadband Ultrasonic Backscatter: An Approach to Non-Destructive Evaluation in Acoustically Inhomogeneous Materials", *J. Appl. Phys.*, Vol. 52, pp. 1056-1065, (1981).
25. M. O'Donnell, D. Bauwens, J.W. Mimbs, and J.G. Miller, "Broadband Integrated Backscatter: An Approach to Spatially Localized Tissue Characterization In Vivo", *Proc. IEEE Ultrasonics Symposium*, Vol. 79 CH 1482-9, pp. 175-178, (1979).
26. R.D. Cohen, Jack G. Mottley, J.G. Miller, Peter B. Kurnik, and B.E. Sobel, "Detection of Ischemic Myocardium In Vivo Through the Chest Wall by Quantitative Ultrasonic Tissue Characterization", *Am. J. Cardiology*, Vol. 50, pp. 838-843, (1982).
27. Eric I. Madaras, B. Barzilai, J.E. Perez, B.E. Sobel, and J.G. Miller, "Changes in Myocardial Backscatter Throughout the Cardiac Cycle", *Ultrasonic Imaging*, Vol. 5, pp. 229-239, (1983).
28. Lewis J. Thomas III, S.A. Wickline, J.E. Perez, B.E. Sobel, and J.G. Miller, "A Real-Time Integrated Backscatter Measurement System for Quantitative Cardiac Tissue Characterization", *IEEE Trans. Ultrasonics, Ferroelectrics, and Frequency Control*, Vol. UFFC-33, pp. 27-32, (1986).
29. L.J. Busse, and J.G. Miller, "Detection of Spatially Nonuniform Ultrasonic Radiation with Phase Sensitive (Piezoelectric) and Phase Insensitive (Acoustoelectric) Receivers", *J. Acoust. Soc. Am.*, Vol. 70, pp. 1377-1386, (1981).
30. L.J. Busse, and J.G. Miller, "A Comparison of Finite Aperture Phase Sensitive and Phase Insensitive Detection in the Near Field of Inhomogeneous Material", *Proc. IEEE Ultrasonics Symposium*, Vol. 81 CH 1689-9, pp. 617-626, (1981).
31. J.G. Miller, Joseph S. Heyman, D.E. Yuhas, and Alan N. Weiss, "A Power Sensitive Transducer for Echocardiography and Other Medical Ultrasonic Applications", *Ultrasound in Medicine*, Vol. 1, pp. 447-453, (1975).
32. L.J. Busse, and J.G. Miller, "Response Characteristics of a Finite Aperture, Phase Insensitive Ultrasonic Receiver Based Upon the Acoustoelectric Effect", *J. Acoust. Soc. Am.*, Vol. 70, pp. 1370-1376, (1981).
33. Thomas A. Shoup, G.H. Brandenburger, and J.G. Miller, "Spatial Moments of the Ultrasonic Intensity Distribution for the Purpose of Quantitative Imaging in Inhomogeneous Media", *Proc. IEEE Ultrasonics Symposium*, Vol. 80 CH 1602-2, pp. 973-978, (1980).
34. Patrick H. Johnston, and J.G. Miller, "Phase-Insensitive Detection for Measurement of Backscattered Ultrasound", *IEEE Trans. Ultrasonics, Ferroelectrics, and Frequency Control*, Vol. UFFC-33, pp. 713-721, (1986).
35. Mark R. Holland, and J.G. Miller, "Phase-Insensitive and Phase-Sensitive Quantitative Imaging of Scattered Ultrasound Using a Two-Dimensional Pseudo-Array", *IEEE Ultrasonics Symposium*, Vol. 88 CH 2578-3, pp. 815-819, (1988).

Elastic Waves and Ultrasonic Nondestructive Evaluation
S.K. Datta, J.D. Achenbach and Y.S. Rajapakse (Editors)

NEW ULTRASONIC TECHNIQUES TO EVALUATE INTERFACES

Peter B. NAGY and Laszlo ADLER

Department of Welding Engineering
The Ohio State University
Columbus, Ohio USA

This paper discusses some of the most recent developments in ultrasonic evaluation of interface properties. The finite boundary stiffness model is used to compare different types of imperfections such as kissing bond, partial bond, and slip bond. It is shown that the ratio of the extensional and transverse boundary stiffnesses can be used to identify the nature of the detected imperfection, which is the crucial first step in interface characterization. An analytical technique is presented to calculate the frequency dependent guided wave velocity along such interfaces, and a novel experimental technique is introduced to generate and detect such guided waves via readily accessible bulk modes at grazing angle.

1. INTRODUCTION

Ultrasonic assessment of interface properties has been and continues to be one of the most important problems in quantitative nondestructive evaluation. Although we have been witnessing rapid development in both theoretical and experimental fields for many years, it becomes more and more clear that none of the known techniques can offer a single universal solution for all those different problems encountered in various bonds. What we need to succeed is a great variety of analytical and experimental tools so that we can address each particular problem in the most effective way. In pursuit of new, more efficient techniques to characterize interface imperfections, we studied the feasibility of using interface waves to inspect the bonded region. We considered "thin" imperfections only, when the overall thickness of the affected interface region is negligible with respect to the wavelength of the ultrasonic wave used to inspect the boundary. Such essentially two-dimensional interfaces can be easily modeled by the well-known finite boundary stiffness technique. Figure 1 shows the geometrical configuration of the interface problem. The boundary conditions require that both tangential (T_{xy}) and normal (T_{yy}) components of the stress must be continuous at the interface, but there is a local discontinuity of the displacement components. The resulting displacement jump through the interface is proportional to the corresponding stress components. At z = 0,

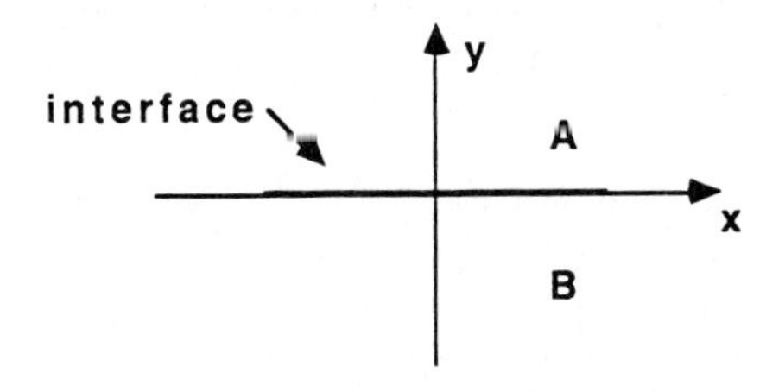

FIGURE 1
Coordinate system

$$T_{yy}^{A} = T_{yy}^{B} = S_L(u_y^B - u_y^A) \tag{1}$$

$$T_{xy}^{A} = T_{xy}^{B} = S_T(u_x^B - u_x^A) \tag{2}$$

where S_L and S_T are the extensional and transverse interfacial stiffness constants of the imperfect interface [1-8].

The reflection and transmission coefficients of such an interface can be calculated from Eqs. 1 and 2 for any combination of longitudinal and transverse waves at arbitrary incidence [3,7,9]. For the simplest case of normal incidence at an interface between similar material, the following well-known results can be obtained:

$$R_{L,T} = \frac{-\dfrac{i\omega}{\Omega_{L,T}}}{1 + \dfrac{i\omega}{\Omega_{L,T}}} \tag{3}$$

$$T_{L,T} = \frac{1}{1 + \dfrac{i\omega}{\Omega_{L,T}}} \tag{4}$$

where subscripts L and T correspond to longitudinal and transverse waves, respectively.

Figure 2 shows these reflection and transmission coefficients as a function of frequency. At very low frequencies, the interface appears to be perfect. As the frequency increases, the reflection increases to 100% while the transmission drops to zero. The characteristic frequeny (Ω) where the transition from the apparently perfect interface to the apparently delaminated one occurs is somewhat different for longitudinal and transverse incidence. Actually, the ratio of these frequencies (Ω_T/Ω_L) can be shown to be closely related to the physical nature of the interface imperfection, and therefore, plays a very important role in the evaluation of the measured ultrasonic signals.

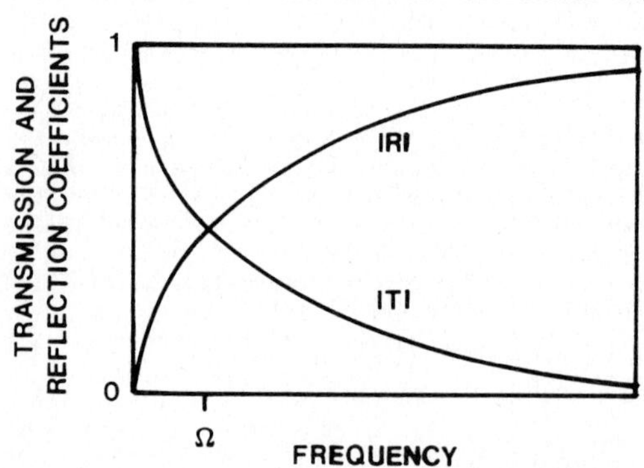

FIGURE 2
Reflection and transmission coefficients of an imperfect interface as a function of frequency at normal incidence.

2. DIFFERENT TYPES OF IMPERFECTIONS

The finite boundary stiffness approximation of Eqs. 1 and 2 can be used to model imperfect interfaces of very different physical natures, e.g. partial bond, kissing bond, slip bond, etc. A partial bond consists of a number of unbonded areas in the interface plane which are not resolved individually by the interrogating ultrasonic beam. The overall effect of such an array of cracks on ultrasonic transmission and reflection can be used to measure certain statistical properties of the partial bond [4-7,9-12], and the measured characteristic frequencies can be easily related to the interface stiffnesses.

$$\Omega_{L,T} = \frac{2S_{L,T}}{c_{L,T}\rho} \tag{5}$$

where c_L and c_T are the longitudinal and transverse velocities, and ρ denotes the density of the material. For certain simple partial bond patterns, the stiffness constants can be obtained using deformation results tabulated by Tada, et al. [13]. For a periodic array of cracks running parallel or normal to the polarization of the transverse wave [6,9]

$$S_{TP} = (1 - \nu)S_L \tag{6}$$

$$S_{TN} = S_L \tag{7}$$

where ν is the Poisson ratio. For the more realistic model of periodic distribution of circular cracks, the exact solution is not available. Assuming a constant proportionality connecting S_L with $S_{TP} = S_{TN}$, Margetan, et al. [9] obtained the following approximate formula from Eqs. 6 and 7

$$S_T \approx \frac{2 - \nu}{2} S_L . \tag{8}$$

The sought ratio between the transverse and longitudinal characteristic frequencies can be calculated from Eqs. 5 and 8:

$$\frac{\Omega_T}{\Omega_L} = \frac{S_T}{S_L}\frac{c_L}{c_T} \approx \frac{2 - \nu}{2}\sqrt{\frac{2(1 - \nu)}{1 - 2\nu}} \tag{9}$$

For most structural materials, ν varies between 0.3 and 0.33, and Ω_T/Ω_L changes from 2.24 to 1.65. In other words, at a certain frequency, a partial bond produces stronger reflection by longitudinal inspection simply because Ω_T is considerably higher than Ω_L.

An even more dangerous type of interface imperfection, a so-called kissing bond, can occur as a result of plastic contact between smooth, or slightly rough surfaces. Besides some very weak "sticking" effects, a kissing bond practically has no strength, but because of the intimate mechanical contact between the contacting parts, it produces very low ultrasonic contrast, i.e. low reflection and high transmission. Haines found that such a kissing bond can also be modeled by the finite boundary stiffness assumption, and the ratio of the characteristic frequencies can be approximated by

$$\frac{\Omega_T}{\Omega_L} \quad \frac{G}{E}\frac{c_L}{c_T} = \frac{1}{2(1 + \nu)}\sqrt{\frac{2(1 - \nu)}{1 - 2\nu}} \tag{10}$$

where E and G are the Young and shear moduli, respectively [14]. In this case, when ν varies between 0.3 and 0.33, Ω_T/Ω_L changes from 0.72 to 0.75. In other words, at a certain frequency, a kissing bond produces stronger reflection by shear inspection simply because Ω_T is lower than Ω_L.

Another often applied interface model is the so-called viscous slip condition which is particularly useful in connection with partially polymerized adhesive bonds [5,8,16]. Schoenberg showed that the finite boundary stiffness approach leads to reflection and transmission coefficients of the form of Eqs. 1 and 2 [3]. Assuming that the acoustic impedance of the layer is much less than that of the bonded half-spaces and the layer thickness is negligible with respect to the wavelength, the boundary stiffness constants can be calculated as follows

$$S_L = \frac{B' + \frac{4}{3}\mu'}{h} \tag{11}$$

and

$$S_T = \frac{\mu'}{h} \tag{12}$$

where h is the layer thickness, and μ' and B' are the rigidity and bulk modulus of the viscous layer. Substantial transverse slip occurs when the thickness of the hydrodynamic boundary layer $(\eta/\omega)^{1/2}$ is much smaller than the thickness of the interface layer h, and the ultrasonic frequency is low with respect to the shear relaxation time of the fluid. Then

$$\mu' \simeq i\omega\eta\rho' \tag{13}$$

where η is the kinematic viscosity of the layer and ρ' is the fluid density.

$$\frac{S_T}{S_L} \simeq \frac{i\omega\eta\rho'}{B'} = \frac{i\omega\eta}{c'^2} \tag{14}$$

where c' is the compressional wave velocity in the fluid. For ordinary water at 15 C°, the kinematic viscosity $\eta \simeq 10^{-6}$ m²/s, c' ≃ 1500 m/s, and S_T/S_L is as low as 3 10^{-5} at 10 MHz, i.e. almost pure transverse slip occurs. Other couplant materials, such as oil or glycerin, have much higher viscosity around $\eta \simeq 2\ 10^{-3}$ m²/s, and S_T/S_L is approximately 5%, while commercial ultrasonic couplants feature high viscosity of 4 10^{-2} m²/s or more, and S_T/S_L is about one. We can conclude that a typical viscous slip corresponds to $\Omega_T/\Omega_L \lesssim 0.1$, i.e. shear wave inspection yields at least ten times higher interface reflection at normal incidence than longitudinal one.

Figure 3 summarizes the above results by showing the shear wave reflection coefficients (solid lines) of different boundary imperfections resulting in identical longitudinal wave reflection (dashed line). A partial bond usually has reduced, but still substantial strength, while a kissing or slip bond has no strength at all. In order to evaluate the measured reflection coefficient in terms of interface properties, we must distinguish between these different imperfections. While a single, either longitudinal or shear measurement cannot facilitate such identification, the ratio of the characteristic frequencies can do it. Compared to a partial bond, a kissing bond is offset towards the slip boundary case, but a "strong" kissing bond still represents almost rigid boundary conditions, therefore it is very difficult to detect.

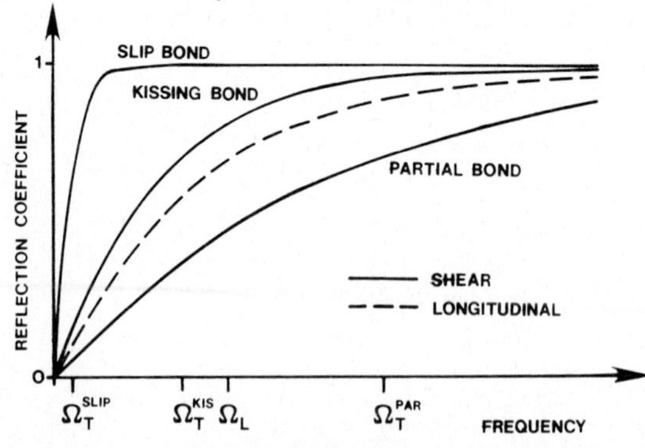

FIGURE 3
Comparison of different boundary imperfections with identical longitudinal wave reflection.

3. INTERFACE WAVE TECHNIQUE

The reflection and transmission coefficients of a plane interface with finite boundary stiffness have been calculated by many authors [3,7,9]. Based on the four boundary conditions given by Eqs. 1 and 2, the following matrix equations can be obtained for longitudinal and transverse wave incidence at arbitrary angle

$$\mathbf{A}\begin{bmatrix} R_{LL} \\ R_{LT} \\ T_{LL} \\ T_{LT} \end{bmatrix} = \mathbf{A}_L \tag{15}$$

and

$$\mathbf{A}\begin{bmatrix} R_{TL} \\ R_{TT} \\ T_{TL} \\ T_{TT} \end{bmatrix} = \mathbf{A}_T \tag{16}$$

where $\mathbf{A}$ is a four-by-four matrix, and $\mathbf{A}_L$ and $\mathbf{A}_T$ are both four-component vectors describing the normal and tangential stress and displacement components at the interface due to incident longitudinal and transverse waves, respectively. Elements of $\mathbf{A}$, $\mathbf{A}_L$, and $\mathbf{A}_T$ can be found in the literature along with calculated values of the reflection and transmission coefficients [3,7,9].

The same formalism can be readily used to study guided mode propagation along the interface. In order to obtain the free vibrations of the interface with finite boundary stiffness, both incidence waves are set to zero, and as a condition on the existence of nontrivial solution for Eqs. 15 and 16, the determinant of $\mathbf{A}$ must vanish. Considerable simplification can be achieved by separation of the field into symmetric and antisymmetric components [7]. The corresponding secular determinants can be written as follows

$$\Delta_s = D - \frac{2S_L}{\mu}\frac{\sqrt{k^2 - k_L^2}}{k_T^2} \tag{17}$$

and

$$\Delta_a = D - \frac{2S_T}{\mu}\frac{\sqrt{k^2 - k_T^2}}{k_T^2} \tag{18}$$

where μ is the rigidity of the substrate and

$$D = \frac{(2k^2 - k_T^2)^2 - 4k^2\sqrt{k^2 - k_T^2}\,\sqrt{k^2 - k_L^2}}{k_T^4} \tag{19}$$

Here k, k_T, and k_L are the wave-numbers of the guided interface mode, and the shear and longitudinal bulk modes in the substrate, respectively. It is interesting that the symmetric mode depends on the extensional spring constant (S_L) only, since it does not generate any shear stress at the interface. On the other hand, the antisymmetric mode depends on the transverse spring constant only, since it does not generate any normal stress at the interface. At zero boundary stiffness, both modes degenerate into a simple Rayleigh mode on the free surface of the substrate, since the numerator of D is the well-known Rayleigh wave characteristic equation. The same thing happens at very high frequency when the second terms in Eqs. 17 and 18 diminish as ω^{-1}, and the dispersive interface wave velocity approaches the Rayleigh velocity.

At low frequencies the situation is more complicated. A simple analysis shows that D increases from 0 to 1 as c covers the feasible range between the Rayleigh velocity and the shear one. The second term in Δ_a can reach down to zero, therefore there is always an antisymmetric solution. On the other hand, the minimum of the second term in Δ_s can be higher than the maximum of D:

$$\frac{2S_L}{\mu}\frac{\sqrt{k_T^2 - k_L^2}}{k_T^2} \geq 1 \tag{20}$$

therefore there is a cut-off frequency

$$\Omega_c = \frac{2S_L c_T}{\mu}\sqrt{1 - \frac{c_T^2}{c_L^2}} \tag{21}$$

below which the symmetric mode becomes nonpropagatory. A quick comparison between Ω_c and the formerly introduced characteristic frequency for normal incidence longitudinal wave inspection, Ω_L, from Eq. 5 reveals that

$$\Omega_c = \Omega_L\sqrt{\frac{c_L^2}{c_T^2} - 1} \; . \tag{22}$$

From Eq. 22 we can conclude that the symmetric interface mode is of little practical importance since it does not propagate in the crucial frequency range of $\omega \leq \Omega_L$. On the other hand the antisymmetric mode remains propagatory at all frequencies and it approaches the shear mode at low frequencies or for an increasingly rigid boundary. Similar results were presented by Rokhlin, et al. for the special case of a thin viscous film between two solids in connection with ultrasonic evaluation of adhesive joints [15,16]. Finally, it should be mentioned that a horizontally polarized Love-type interface wave can also propagate along a plane boundary of finite stiffness.

Figure 4 shows the dispersion curves of the predicted interface wave between two aluminum half-spaces under different compressive pressures. The boundary stiffness constants were calculated by Haines' technique [14] assuming 1 µm surface roughness on both parts and 50 ksi flow pressure in the aluminum. The transverse boundary stiffness constants were S_T = 0.15, 0.43, 1.1, 2.6, and 4 x 10^{14} N/m^3 for 0.1, 0.3, 1, 3, and 10 ksi pressures, respectively. In good agreement with our expectations, the interface wave velocity drops from the shear velocity of the substrate to the Rayleigh velocity as the frequency increases and the transition frequency increases with the compressional pressure.

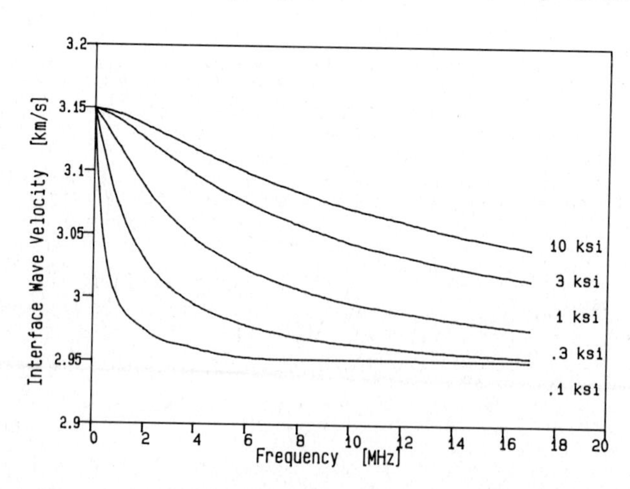

FIGURE 4
Dispersion curves of the predicted interface wave between aluminum half-spaces under different compressive pressures.

The transition frequency Ω_I can be defined as the point where the interface velocity reaches the $\frac{1}{2}(c_T + c_R)$ value. From Eq. 17 we can obtain for $\nu = 0.3$

$$\Omega_I = 2.03 \frac{S_T}{c_T \rho} \tag{23}$$

almost exactly the same characteristic frequency as Ω_T for shear wave inspection at normal incidence.

4. EXPERIMENTAL RESULTS

Interface waves are especially sensitive to boundary imperfections since they are effectively confined to the immediate vicinity of the crucial bond region. Unfortunately, this very guided nature which confines them to the interface makes them particularly unaccessible, i.e. difficult to excite and detect. Figure 5 shows the schematic diagram of the most often used Rayleigh wave coupling technique. Surface waves propagating along the free surface of one of the specimens are generated by conventional angle-beam contact transducers. Such waves are relatively strongly coupled to interface modes along the boundary between the two counterparts thereby providing a relatively simple, indirect access to the interface under study.

Lee and Corbly were the first to experimentally observe the gradual transformation of a Rayleigh wave into a Stoneley wave due to strong pressing together of two flat, polished solids [17]. Although this technique was shown to have a potential application in ultrasonic NDE of solid-state bonds [18], the required awkward geometry renders it useless in most practical applications. Figure 6 shows an alternative geometrical configuration for direct generation and detection of interface waves in a bonded structure. A contact ultrasonic transducer is placed directly over the boundary region so that it can generate both bulk and interface modes. A longitudinal transducer could be used to generate symmetric modes, while a shear transducer produces vertically or horizontally polarized antisymmetric modes depending on the transducer orientation. In this particular experiment, we used vertical polarization, and the interface stiffness was externally controlled by changing the compressive pressure between the aluminum counterparts.

Figure 7 shows the reflected ultrasonic signal from the back wall of the specimen. At zero compressive pressure, two separate signals can be observed, the first arrival is a shear-type bulk wave while the second one is a Rayleigh-type interface wave. At 4000 psi compressive pressure, the two signals are not separated sufficiently to directly measure their respective time delays, but appropriate frequency analysis can still readily reveal the sought separation. Figure 8 shows the corresponding frequency spectra of the detected signals shown in Fig. 7. Distinct minima occur as a result of destructive interference between

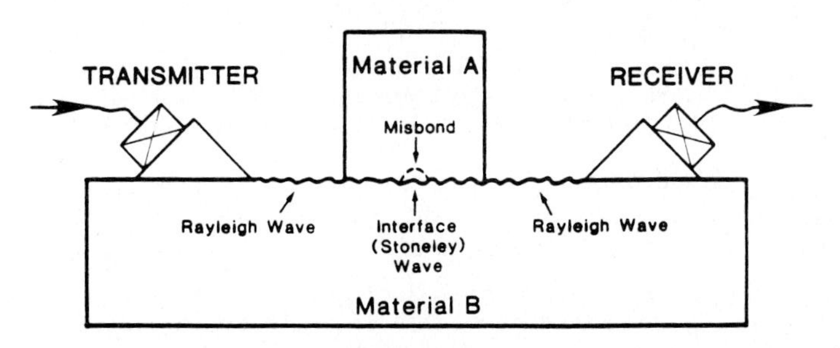

FIGURE 5
Interface wave generation and detection by Rayleigh wave coupling.

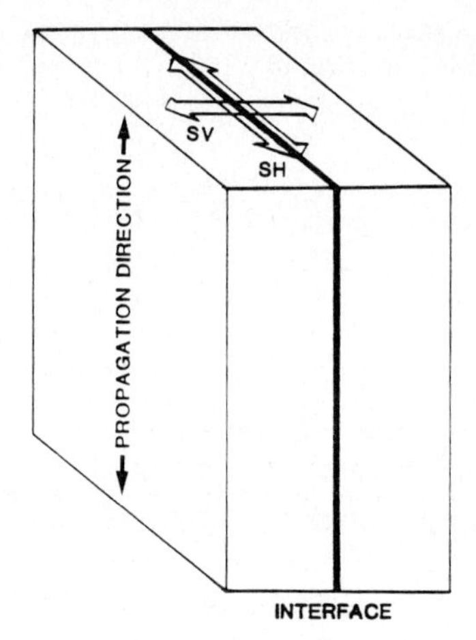

FIGURE 6
Schematic diagram of direct interface wave generation.

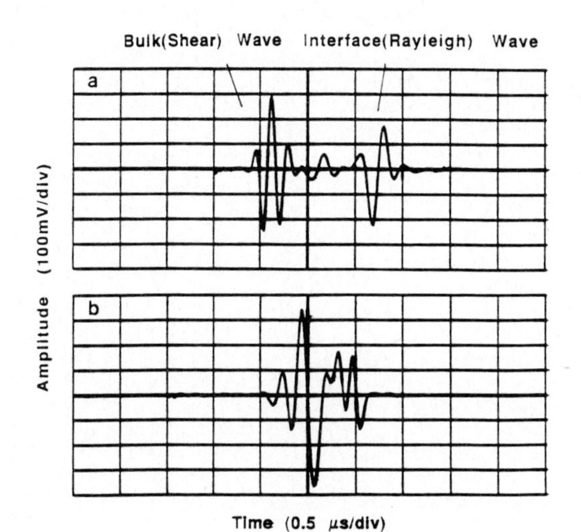

FIGURE 7
Received ultrasonic signals along a dry interface between aluminum surfaces under (a) zero and (b) 4000 psi compressive pressure.

the bulk and interface waves, and the periodicity of the observed frequency modulation can be used to determine the interface wave velocity with respect to the shear velocity.

In connection with the frequency spectra shown in Fig. 8, one detail of interest needs further explanation. At low compressive pressure, the observed minima at the center of the ultrasonic bandwidth are much sharper than the others at lower and higher frequencies. The explanation of this phenomenon lies in the complementary nature of the individual frequency spectra of the bulk and

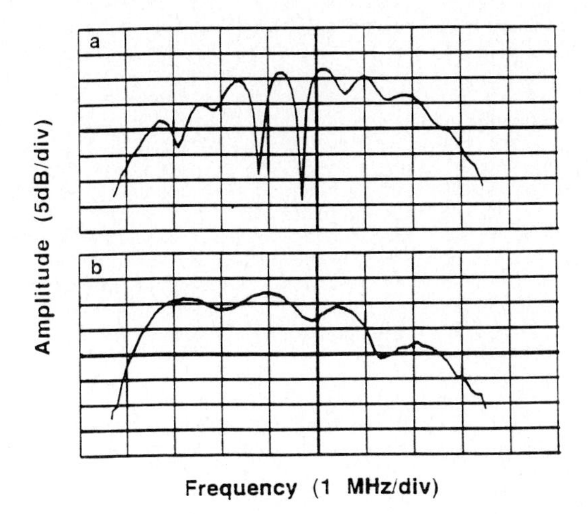

FIGURE 8
Frequency spectra of the detected ultrasonic signals of Figure 7.

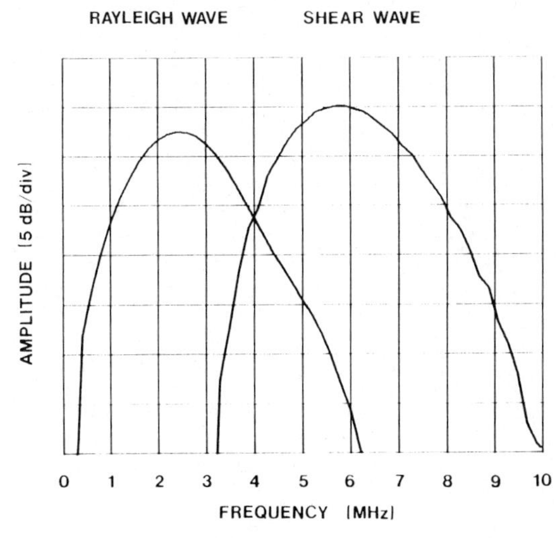

FIGURE 9
Frequency spectra of the Rayleigh and shear pulses at zero compressive pressure.

interface modes shown in Fig. 9. The center frequency of the ¼"-diameter shear transducer was approximately 5 MHz. That part of the transducer which is within one wavelength of the bondline generates mostly the interface mode while the remaining part which is farther than one wavelength from the interface generates mostly the bulk wave. Consequently, each component can get stronger only at the expense of the other one. At low frequency, practically all energy goes into the interface mode, while at high frequency all energy goes into the bulk one. Strong interference resulting in the observed very sharp minima can occur when the two components are of more-or-less the same strength, i.e. around the crossing frequency of 4 MHz.

Besides the above explained differences in the depth of the modulation of the interference spectrum, its periodicity apparently does not change over the useful frequency range of 2 - 8 MHz. This result is in good agreement with Fig. 4 showing negligible dispersion of the Rayleigh-type interface wave along the boundary between solid half-spaces in light contact. At 4000 psi, only three minima can be observed at approximately 3, 4.9, and 6.3 MHz, i.e. the separation is changing with frequency. In this case, the evaluation of the frequency modulation is much more complicated, but roughly we can assume that the measured separations belong to the center frequencies between the minima. In this way, 1.9 MHz separation gives 3075 m/s at 3.95 MHz, and 1.4 MHz separation gives 3049 m/s at 5.6 MHz, i.e. the interface velocity decreases with increasing frequency.

In order to study the interface velocity as a function of compressive pressure, we adopted a similar approach for the whole frequency range of 2 - 8 MHz by relating the average periodicity to the 5 MHz center frequency. Figure 10 shows the measured interface wave velocity as a function of compressive pressure as calculated from the overall periodicity. There is a good quantitative agreement between the calculated and measured interface wave velocities except at high compressive pressure-to-flow pressure ratios where Haines' model is not expected to be very accurate.

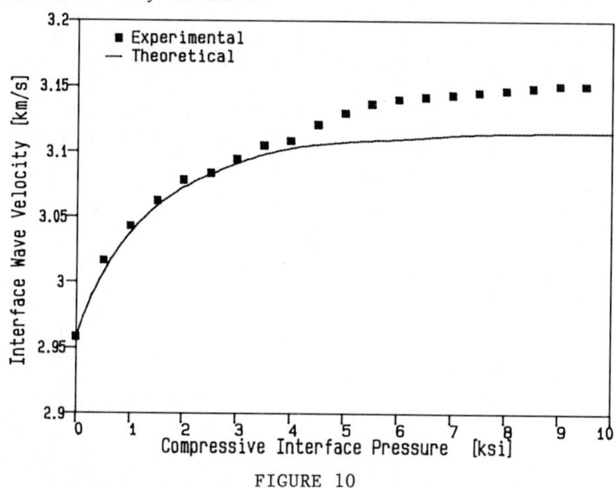

FIGURE 10
Measured and calculated interface wave velocity as a function of compressive pressure between aluminum counterparts at 5 MHz.

5. CONCLUSIONS

It was shown that the ratio of the normal and transverse boundary stiffness constants could be used to identify the nature of the detected interface imperfections. A new experimental technique was introduced to generate and detect interface waves along otherwise unaccessible plane boundaries. Theoretical and experimental results for the dispersive interface wave along kissing bonded imperfect boundaries were found to be in good agreement.

ACKNOWLEDGEMENTS

This work was supported by the Office of Naval Research under contract N0014-C-2129.

REFERENCES

[1] Jones, J.P. and Whittier, J.S., J. Appl. Mech. 34 (1967) 905.
[2] Tattersall, H.G., J. Phys. 6 (1973) 819.
[3] Schoenberg, M., J. Acoust. Soc. Am. 68 (1980) 1516.
[4] Baik, J.M. and Thompson, R.B., J. Nondestr. Eval. 4 (1984) 177.
[5] Thompson, R.B. and Fiedler, C.J., in: Thompson, D.O. and Chimenti, D.E., (eds), Review of Progress in Quantitative Nondestructive Evaluation (Plenum, New York, 1984) Vol. 3A, pp. 207-215.
[6] Angel, Y.C. and Achenbach, J.D., J. Appl. Mech. 107 (1985) 33.
[7] Angel, Y.C. and Achenbach, J.D., in: Thompson, D.O. and Chimenti, D.E., (eds), Review of Progress in Quantitative Nondestructive Evaluation (Plenum, New York, 1985) Vol. 4A, pp. 83-89.
[8] Pilarski, A. and Rose, J.L., J. Appl. Phys. 63 (1988) 300.
[9] Margetan, F.J., Thompson, R.B., and Gray, T.A., J. Nondestr. Eval. 7 (1988) 131.
[10] Angel, Y.C. and Achenbach, J.D., Wave Motion 7 (1985) 375.
[11] Sotiropoulos, D.A. and Achenbach, J.D., J. Acoust. Soc. Am. 84 (1988) 752.
[12] Sotiropoulos, D.A. and Achenbach, J.D., J. Nondestr. Eval. 7 (1988) 123.
[13] Tada, H., Paris, P., and Irwin, G., The Stress Analysis of Cracks Handbook (Del Research, St. Louis, 1973).
[14] Haines, N.F., The Theory of Sound Transmission and Reflection at Contacting Surfaces (Berkeley Nuclear Laboratories, RD-B-N4744, 1980).
[15] Rokhlin, S.I., Hefets, M., and Rosen, M., J. Appl. Phys. 51 (1980) 3579.
[16] Rokhlin, S.I., Hefets, M., and Rosen, M., J. Appl. Phys. 52 (1981) 2847.
[17] Lee, D.A. and Corbley, D.M., IEEE Trans. Son. Ultrason. SU-24 (1977) 206.
[18] Nagy, P.B. and Adler, L., J. Nondestr. Eval. 7 (1988) 199.

Elastic Waves and Ultrasonic Nondestructive Evaluation
S.K. Datta, J.D. Achenbach and Y.S. Rajapakse (Editors)

ELASTIC WAVE PROPAGATION IN MULTILAYERED ANISOTROPIC MEDIA

Adnan H. NAYFEH

Department of Aerospace Engineering
and Engineering Mechanics
University of Cincinnati
Cincinnati, OH 45221

Dale E. CHIMENTI*

Air Force Materials Laboratory
Wright-Patterson Air Force Base, OH 45433

Theoretical investigations supported by extensive experimental comparisons are carried out on the interaction of ultrasonic waves with multilayered media. It is assumed that each constituent of the plate can posses up to as low as monoclinic symmetry. The plate is assumed to be immersed in a fluid and subjected to incident acoustic waves at arbitrary angles from the normal as well as at arbitrary azimuthal angles. Reflection and transmission coefficients are derived from which all characteristic behavior of the system is identified. Solutions are obtained for the individual layers which relate the field variables at the upper and lower layer surfaces. The response of the total plate proceeds by satisfying appropriate interfacial conditions across the layers. Our results are rather general and contain a wide variety of special cases.

1. INTRODUCTION

In a recent paper [1] we used the transfer matrix method and derived exact expressions for the reflection and transmission coefficients from liquid-loaded multilayered anisotropic plates. The waves are allowed to propagate at arbitrary direction from the normal to the layers interfaces as well as at arbitrary azimuthal angles. The procedure used in [1] constitutes an extension of our earlier analysis conducted on a single layer plate [2].

General studies of the propagation of elastic waves in layered media have long interested researchers in geophysics, ocean acoustics, saw devices, and more recently, nondestructive evaluation (see, for example, Brekhovskikh [3] and Ewing et al. [4]).

Typically, a layered medium consists of two or more material components joined at their interface in some fashion. For wave propagation in such media, solutions are generally obtained by expressing the displacements and stresses in each component in terms of its wave amplitudes. By satisfying appropriate interfacial conditions, characteristic equations are constructed which involve the amplitudes of all layers. This procedure constitutes the direct approach and has been taken by Kulkarni and Pagano [5], and Kundu and Mal [6]. The degree of complication in the algebraic manipulation of the analysis will thus depend upon the number of components and layers. For relatively few layers, the direct approach is appropriate. However, as the number of components increases the direct approach becomes cumbersome, and

* Present address: Materials Science and Engineering Department
Johns Hopkins University, Baltimore, MD 21218

one may resort to the alternative matrix transfer technique introduced originally by Thomson [7] and somewhat later on by Haskell [8]. For further relevant literature we refer the reader to [1]. According to this technique one constructs the propagation matrix for a stack of an arbitrary number of layers by extending the solution from one layer to the next while satisfying the appropriate interfacial continuity conditions.

In this paper we present comparisons of our theoretical model with the results of numerous ultrasonic reflection measurements on plates of laminated composites in which the angle of incidence of the acoustic wave and its frequency have been varied. These measurements have been carried out on biaxially laminated graphite-epoxy specimens utilizing water as a fluid coupling medium. The stacking sequence of the samples investigated is 0°/90° lamination. In general, good agreement is found between prediction and experiment.

2. SUMMARY OF THEORY

Consider a plate consisting of an arbitrary number n of anisotropic layers, each exhibiting a monoclinic symmetry, rigidly bonded at their interfaces and stacked normal to the x_3 axis of a global orthogonal Cartesian system $x_i = (x_1, x_2, x_3)$. Hence the plane of each layer is parallel to the $x_1 - x_2$ plane which is also chosen to coincide with the bottom surface of the layered plate. To maintain generality we shall assume each layer to be arbitrarily oriented in the $x_1 - x_2$ plane. To describe the relative orientation of the layers, we assign an index k for each layer, k = 1,2,...,n, and a local cartesian coordinate system $(x_i')_k$ such that its origin is located in the midplane of the layer with $(x_3')_k$ normal to it. Thus, layer k extends from $-d^{(k)}/2 \leq (x_3')_k \leq d^{(k)}/2$ where $d^{(k)}$ is its thickness and d equals the total thickness of the plate. The material orientation in the kth layer is given by a rotation angle ϕ_k between $(x_1')_k$ and x_1.

We assume that a plane wave is incident in the x_1-x_3 plane on the plate from the upper fluid at an arbitrary angle. The problem here is to study the reflected and transmitted fields. We therefore conduct our analysis in a coordinate system formed by incident and reflected planes rather than by material symmetry axes. Accordingly, the primed system $(x_i')_k$ rotates with one set of material symmetry axes while the global unprimed system x_i remains invariant. This approach leads to significant simplification in our algebraic analysis and computations [1,2].

In this section we construct a transfer matrix for each layer k. This matrix relates the displacements $(u_i)_k$ and stresses $(\sigma_{ij})_k$ of one face to those of the other of layer k. These relations are given in [1] as

$$F_p^+ = a_{pq} F_q^- \quad , \; p,q = 1,2,\ldots,6 \tag{1}$$

Here, $F_p^{\pm}$ stands for the variables column $[u_1, u_2, u_3, \sigma_{33}, \sigma_{13}, \sigma_{23}]_{\pm}^T$ specialized to the upper and lower surfaces of the kth layer, and

$$[a_{pq}]_k = [T_{ps} B_{sq}]_k, \tag{2}$$

where

$$T_{ps} = \begin{vmatrix} 1 & i & 1 & i & 1 & i \\ V_1 & iV_1 & V_3 & iV_3 & V_5 & iV_5 \\ iW_1T_1 & W_1T_1' & iW_3T_3 & W_3T_3' & iW_5T_5 & W_5T_5' \\ D_{11} & iD_{11} & D_{13} & iD_{13} & D_{15} & iD_{15} \\ iD_{21}T_1 & D_{21}T_1' & iD_{23}T_3 & D_{23}T_3' & iD_{25}T_5 & D_{25}T_5' \\ iD_{31}T_1 & D_{31}T_1' & iD_{33}T_3 & D_{33}T_3' & iD_{35}T_5 & D_{35}T_5' \end{vmatrix} \tag{3}$$

In Eq. (3), B_{sq} is obtained from the above matrix T_{ps} by replacing i with -i as it explicitly appears and inverting the resulting matrix, where

$$T_r' = T_r^{-1} = \cos(\xi\alpha_r d/2^{(k)}) \quad , \quad r = 1,3,5. \tag{4}$$

The matrix $[a_{pq}]_k$ constitutes the most general transfer matrix for the kth monoclinic layer. It allows the wave to be incident on layer k at an arbitrary angle from the normal x_3, or equivalently $(x_3')_k$ and at any azimuthal angle ϕ.

The matrix transfer technique then yields, through the layer boundary conditions, the response vector at $x_3 = d$ in terms of the response vector at $x_3 = 0$

$$F_p(d) = A_{pq}\ F_q(0) \tag{5}$$

where

$$[A_{pq}] = [a_{pj}]_n\ [a_{jk}]_{n-1} \cdots [a_{rq}]_1\ . \tag{6}$$

The plate we consider is assumed to be immersed in a fluid. The input wave is assumed to be periodic and originating in the upper portion of the fluid and incident on the plate at an arbitrary angle from the normal. Following the proceedure of [1] we caculate the reflection and transmission coefficients as

$$R = \frac{(M_{21} + QM_{22}) - Q(M_{11} + QM_{12})}{(M_{21} + QM_{22}) + Q(M_{11} + QM_{12})} \tag{7a}$$

$$T = \frac{2Q(A_{51}A_{62} - A_{61}A_{52})}{(M_{21} + QM_{22}) + Q(M_{11} + Q_bM_{12})} \tag{7b}$$

where

$$M_{11} = \det \begin{vmatrix} A_{31} & A_{32} & A_{33} \\ A_{51} & A_{52} & A_{53} \\ A_{61} & A_{62} & A_{63} \end{vmatrix}, \quad M_{12} = \det \begin{vmatrix} A_{31} & A_{32} & A_{34} \\ A_{51} & A_{52} & A_{54} \\ A_{61} & A_{62} & A_{64} \end{vmatrix} \tag{8a}$$

$$M_{21} = \det \begin{vmatrix} A_{41} & A_{42} & A_{43} \\ A_{51} & A_{52} & A_{53} \\ A_{61} & A_{62} & A_{63} \end{vmatrix}, \quad M_{22} = \det \begin{vmatrix} A_{41} & A_{42} & A_{44} \\ A_{51} & A_{52} & A_{54} \\ A_{61} & A_{62} & A_{64} \end{vmatrix} \tag{8b}$$

and

$$Q = \frac{\rho_f c^2}{\alpha_f}, \quad \alpha_f = (c^2/c_f^2 - 1)^{1/2} \tag{8c}$$

3. EXPERIMENTAL TECHNIQUE

The analysis of wave propagation in terms of reflection properties of the fluid-coupled structure under study is a well established approach Schoch [9]. It has been used by us and others in work on both Rayleigh and Lamb waves. To obtain the reflection characteristics of fluid-coupled plates we direct a sound beam from an ultrasonic piston transducer (9.5 mm diameter) onto the surface of the plate at a chosen angle θ. The plate normal vector and the incident sound wavevector define the incident plane of the ultrasound. A second transducer of nominally identical frequency bandwidth and sensitivity characteristics, whose axis is also in the incident plane, receives the reflected signal from the plate. The receiver is oriented at the negative incident angle, and the point of intersection of its axis with the surface of the plate may either coincide with that of the incident transducer or may be displaced from it in the direction of plate wave propagation. Detailed descriptions of the experimental geometry and the excitation waveforms for these measurements have been presented elsewhere [1,2].

In these experiments we have investigated one sample of biaxially laminated graphite-epoxy composite. This sample has been fabricated from Hercules AS/4 fibers and 3501 epoxy. Fiber volume fraction is 0.64, from acid digestion tests. We have taken literature values [10] for the elastic constants, appropriately adjusted for any difference in fiber volume fraction. The sample has the lay-up geometry [0, 90] with 8 plies. Sample thickness is 1.15 mm, where we have estimated the sonic thickness from the physical thickness by subtracting 0.02 mm for the impressions left by the bleeder cloths. The 8-ply sample was carefully prepared to maximize elastic and dimensional uniformity, and only small variations in elastic behavior and thickness have been observed in ultrasonic and mechanical survey scans across these samples.

4. RESULTS AND DISCUSSION

We present our results as direct comparisons between normalized experimental frequency spectra and the plane-wave reflection coefficient. An example of experimental and theoretical curves for the biaxial composite is shown in Fig. 1. The solid curve is the measurement, while the calculation is represented as a dashed curve. Only the relative amplitudes of the two curves have been scaled, since absolute reflectance has not been measured. An incident angle of 16° is selected, and the fiber direction in the upper layer makes an angle of 29° with respect to the incident plane.

Comparison of the prediction with the experimental data is very good; nearly all details of the data are reproduced in the calculation. These data may be compared to the results of our earlier measurements [11] made with the incident plane containing a principal material axis in each layer. In that case the reflection spectrum structure was much simpler than seen here, and the additional complication may be ascribed to the coupling, off principal axes, between in-plane and out-of-plane particle motion. We have noted similar behavior in uniaxial plates [2].

Figure 2 shows the result of rotating the incident plane so that it makes an angle of 50° with the fiber direction in the uppermost layer, while maintaining a 16° incident angle. The experimental reflection spectrum has changed markedly compared to Fig. 1. In fact, even small azimuthal variations can significantly change the appearance of the reflection spectrum. While we do not find exact agreement between experiment and prediction in Fig. 1, nearly every feature in the measurement has a corresponding expression in the theory curve. The lack of detailed agreement may be due to several sources. First, we are comparing plane-wave reflection predictions with finite aperture measurements, although this approximation is relatively accurate. More importantly, as we have seen in previous studies, our samples are rather imperfect realizations of the ideal morphology assumed in the model calculation. Similar small disparities, likely attributable to the same sources, will also be noted in the balance of the results reported here.

This case, and many others we have examined in other specimens, exemplify the complex nature of ultrasonic reflection in layered anisotropic materials. Clearly, a simpler model in which the elastic properties were averaged through the plate thickness would be unable to account for the differences between the data in Figs. 1 and 2. Further comparisons of plate reflection spectra at an incident angle of 12° are presented in Figs. 3 and 4. In Fig. 3 the azimuthal angle is 0° and the comparisons is shown from 0 to 12 MHz. These data indicate the relative lack of complication observed in plate wave propagation along a principal material axis. Figure 4 shows the same data for an azimuthal angle of 60°. More rapid, and less regular, variations in the magnitude of the reflection coefficient are seen here, noting the change in the frequency scale. Similar effects have been observed in our previous studies on uniaxial plates [2,11], and we assume that the behavior observed here also arises from the coupling, off principal axis, of x_1 and x_2 particle displacements. In the case of propagation in a general azimuthal angle, the vertically polarized and horizontally polarized (SH) wave motions will not be independent.

5. CONCLUSIONS

We have presented the results of numerous ultrasonic reflection measurements on plates of laminated composites in which the angle of incidence of the acoustic wave and its frequency have been varied. The measurements have been carried out on 0°/90° laminated graphite-epoxy specimens, utilizing water as a fluid coupling medium. A plane-wave reflection coefficient for the laminate has been calculated by combining an exact analytical treatment of wave propagation in a fluid-coupled monoclinic elastic layer with a transfer matrix approach to accommodate the plate lamination. Frequency spectra of the reflection coefficient magnitude have been compared to experimental measurements at several angles of incidence and with the fibers in the uppermost ply at arbitrary angles to the incident plane. We have seen that despite unavoidable departures in the composite morphology from the ideal behavior modelled in the calculation, rather good detailed agreement is observed in most of the data presented.

ACKNOWLEDGEMENT

This work has been supported by AFOSR

REFERENCES

1. Nayfeh, A. H., Taylor, T. W. and Chimenti, D. E., Theoretical Wave Propagation in Multilayered Orthotropic Media, in Wave Propagation in Structural Composites, ed. by A.K. Mal and T.C.T. Ting, AMD-Vol. 90, ASME, pp. 17-27, (1988).
2. Nayfeh, A. H. and Chimenti, D. E., Ultrasonic Wave Reflection From Liquid-Loaded Orthotropic Plates With Applications to Fibrous Composites, Journal of Applied Mechanics, Vol. 15, p. 863, (1988).
3. Brekhovskikh, L. M., Waves in Layered Media, Academic Press, New York, pp. 30-45, (1960).
4. Ewing, W. M., Jardetsky, W. S. and Press, F., Elastic Waves in Layered Media , McGraw-Hill, New York (1957).
5. Kulkarni, S. V. and Pagano, N.J., Dynamic Characteristics of Composite Laminates, Journal of Sound and Vibration, Vol. 123, pp. 127-143 (1972).
6. Kundu, T. and Mal, K., Elastic Waves in a Multilayered Solid Due to a Dislocation Source, Wave Motion, Vol. 7, pp.459-471 (1985).
7. Thomson, W. T., Transmission of Elastic Waves Through a Stratified Solid Medium, Journal of Applied Physics, 21, p. 89, (1950).
8. Haskell, N. A., The Dispersion of Surface Waves in Multilayered Media, Bulletin of the Seismological Society of America, 43, p. 17, (1953).
9. Schoch, A., Acoustica 2, 1 (1952)
10. Kriz, R. D. and Stinchcomb, W. W., Exp. Mech. 19, 41 (1978)
11. Chimenti, D. E. and Nayfeh, A. H., Ultrasonic Reflection and Guided Wave Propagation in Biaxially Laminated Composite Plates, to appear in the Journal of the Acoustical Society of America.

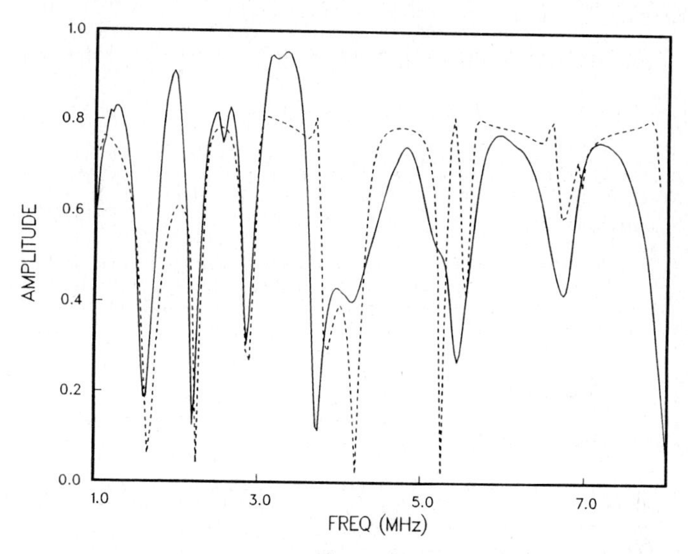

Figure 1
Experimental (solid) and theoretical (dashed) reflection spectrum for the $[0_2,90_2]s$ laminate with an incident angle $\theta = 16^{0}$ and with the fiber direction of the uppermost ply at $\phi = 29^{0}$.

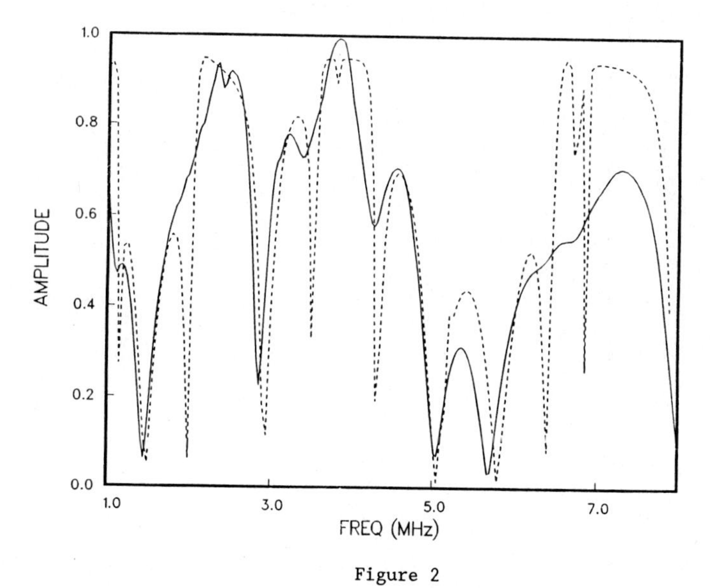

Figure 2

Same as Figure 1 with $\phi = 50^{0}$.

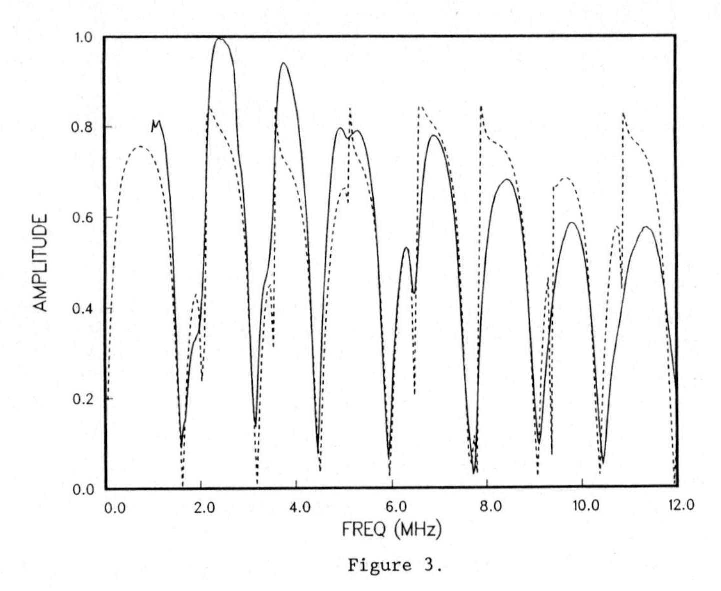

Figure 3.

Same as Figure 1 with $\theta = 12^{0}$ and $\phi = 0^{0}$.

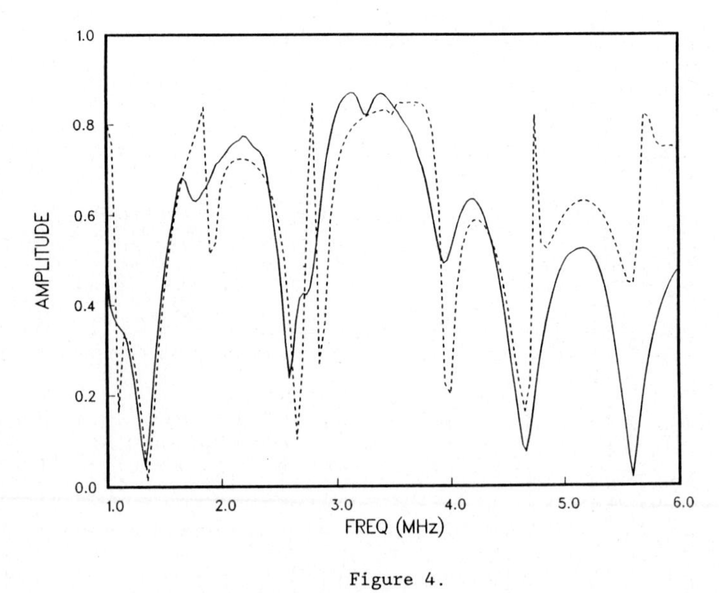

Figure 4.

Same as Figure 3 with $\phi = 60^{0}$.

Elastic Waves and Ultrasonic Nondestructive Evaluation
S.K. Datta, J.D. Achenbach and Y.S. Rajapakse (Editors)

NONDESTRUCTIVE MEASUREMENTS OF ELASTIC CONSTANTS OF COMPOSITE MATERIALS BY ULTRASONIC METHODS

S. I. Rokhlin and W. Wang

The Ohio State University
Department of Welding Engineering
Columbus, OH 43210

In this work three ultrasonic methods are considered: 1) critical angle measurement for in-plane property evaluation, 2) double transmission time-delay measurement for out-of-plane property evaluation, and 3) transmission resonance measurement for in-plane and out-of-plane property evaluations. The first two methods are utilized for thick section composite materials; the last method is devoted to thinner ones. Reconstruction of elastic constants from the results of the measurements is done by a nonlinear least square optimization algorithm which minimizes deviations between theoretical predictions and experimental data by adjusting elastic constants in multidimensional space.

1. INTRODUCTION

Evaluation of elastic properties of composite materials using ultrasound is important for the generation of input data for the design of composites. The full set of elastic constants was destructively measured by Kriz and Stinchcomb [1] on samples cut out in different directions from a composite plate. It is extremely important to develop a nondestructive method for elastic property evaluation of the composites. The problem was addressed in the seventies [2-4] when Markham [2] suggested using the time-delay through transmission technique with obliquely incident ultrasonic waves from water onto a composite plate.

In this paper three nondestructive ultrasonic techniques for reconstruction of the elastic properties of composite materials are considered: 1) critical angle method, 2) double transmission method based on time delay measurements, and 3) a new double transmission method based on the measurement of the incident angles at which the interference between symmetric and antisymmetric guided waves results in a sharp depression of the transmitted signal. While the former two methods are utilized to evaluate thick composite materials, the last method is especially useful for thinner ones.

The elastic constants are reconstructed by a least-square optimization technique from the measured slowness surfaces in critical angle and transmission time-delay measurements. Previous investigators [5,6] used reconstruction in planes of symmetry while our reconstruction method is valid for an arbitrary plane in a composite material. Reconstruction of elastic constants from transmission minimum data also uses a least square optimization method but the algorithm relating elastic constants and experimental data is much more complicated in this case. Elastic constants of a composite layer may also be reconstructed by using experimental data on the zeros of the reflection coefficient [7]. Application of this last technique may be useful in combination with the transmission minima method.

2. EXPERIMENTAL APPARATUS

All measurements have been done in a novel type of goniometer as shown in Fig. 1. It is distinguished by its ability to make simultaneous measurements of both doubly-transmitted (through the sample) and doubly-reflected (from the sample) ultrasonic signals with the use of only one ultrasonic transducer. A plane sample is mounted in the center of an aluminum cylinder. The sample is rotated around a cylinder axis which lies in the frontal plane of the sample.

Incident and reflected ultrasonic beams propagate along cylinder radii. After being reflected from the cylinder wall the ultrasonic signal is reflected a second time from the sample surface and returns to the receiving transducer. The transmitted ultrasonic signal is reflected and returned to the transducer after the second through-transmission. The received signals are digitized and averaged by a LeCroy 9400 125MHz digital oscilloscope and then collected by a computer as a function of the angle of rotation.

For sample rotation around the cylinder axis a computer controlled DC motor is used. Angle resolution and repeatibility of 0.01° are achieved in our setup. In addition, the sample may be rotated in its own plane by an angle α, so different orientations of the incident plane relative to fiber orientation may be selected. The water tank is temperature stabilized to $29.8 \pm 0.1^\circ C$.

3. CRITICAL ANGLE MEASUREMENTS (IN-PLANE PROPERTIES)

Our recent theoretical work [8] has shown that a strong, sharp maximum in the reflection coefficient appears at the critical angles for quasi-longitudinal and fast quasi-transverse waves. The third critical angle, which corresponds to the slow quasi-transverse wave, also manifests itself strongly in the theoretical dependence of the reflection coefficient on the incident angle. When a composite sample is rotated around its surface normal, the value of the critical angles changes in correspondence with the in-plane slowness surfaces.

The last statement requires special consideration. The critical angle may be defined as the angle of incidence for which the direction of energy flow for the refracted wave is parallel to the sample surface. Since in anisotropic material the energy flow (group velocity) direction and the wave vector direction do not coincide, critical angle measurement in general may not be used for phase velocity measurement in anisotropic materials. However, for two dimensional composite materials with fibers lying in the plane of the plate, the situation is exceptional. Although for a general direction of propagation in such composite materials the deviation angle between phase and group velocities has both in-plane and out-of-plane components [8], when the wave is propagated in the in-plane direction the out-of-plane component vanishes. This is clear from symmetry considerations since the in-plane elastic properties of a composite plate are invariant under change of direction of the normal to that plane. This means that at the critical angle both the wave vector (phase velocity vector) and the group velocity vector will be parallel to the plate surface. Therefore, Snell's law can be used at the critical angle for the calculation of the in-plane phase velocity.

Typical traces of the amplitudes of a doubly-reflected ultrasonic signal for a unidirectional composite plate is shown in Fig. 2 as a function of incident angle. The data are shown for only one angle $\alpha = 55^\circ$ of deviation of the fiber direction from the plane of incidence. By rotating the sample along the normal to the sample surface one can determine the critical angle and phase velocity as functions of the angle α. The corresponding results are shown in Fig. 3, where the experimental data of the quasi-longitudinal and two quasi-transverse waves are shown as discrete points.

To extract elastic properties from the phase velocity data for different propagation angles a nonlinear least square optimization is conducted for best data fitting. Such an optimization is done by using Cardan's solution of the Christoffel equation, which relates phase velocity with elastic constants.

$$\rho V_k^2 = Z_k - a/3; \tag{1}$$

$$Z_k = 2\sqrt{-p/3}\cos((\psi+2k\pi)/3), \quad k=1,2,3 \tag{2}$$

where

$$\psi = \cos^{-1}(-q/[2(p/3)^{3/2}]); \quad p=a^2/3-b; \quad q=c-ab/3+2(a/3)^3; \quad a=-G_{ii}$$

$$b = -(G_{12}^2+G_{13}^2+G_{23}^2-G_{11}G_{22}-G_{11}G_{33}-G_{22}G_{33});$$

$$c = -(G_{11}G_{22}G_{33}+2G_{12}G_{13}G_{23}-G_{11}G_{23}^2-G_{22}G_{13}^2-G_{33}G_{12}^2)$$

$G_{im} = C_{ijlm}n_jn_l$; C_{ijlm} are the elastic constants in tensor form.

The program minimizes the sum of all squares of the deviations between the experimental and calculated velocities considering the elastic constants as variables in a multi-dimensional space. The result of the fitting is shown as solid lines in Fig. 3. It is seen that comparison of the calculated and the experimental data is satisfactory.

4. OUT-OF-PLANE PROPERTIES: THROUGH-TRANSMISSION MEASUREMENTS

The through-transmission method is illustrated schematically in Fig. 4. An ultrasonic wave is obliquely incident on the composite sample. The angle of refraction and therefore the time of flight inside the material depends on the velocity in the direction of the refraction angle. In our system we use a plane reflector to reflect all transmitted waves to the transmitter-receiver. This technique has advantages in precision over Markham's method since it removes the necessity of shifting the receiving transducer to the central axis of a refracted beam with change of incident angle.

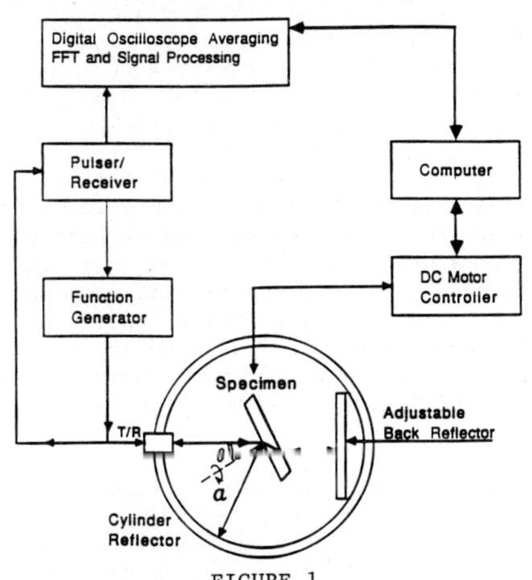

FIGURE 1
Schematic diagram of the experimental system.

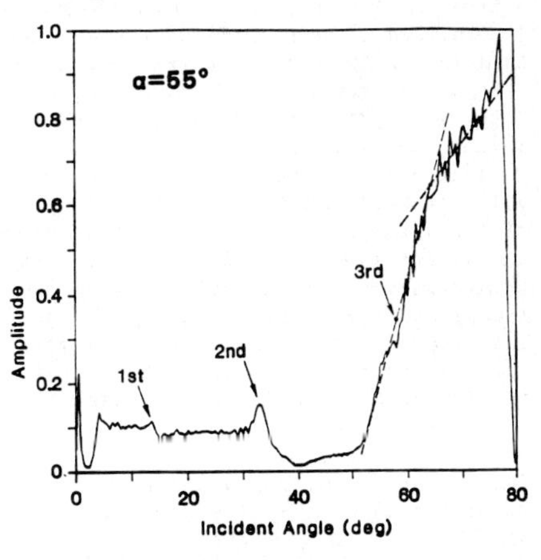

FIGURE 2
Typical trace of the amplitude of a double-reflected ultrasonic signal. Three critical angles are indicated by arrows. α is the angle of fiber deviation from the incident angle.

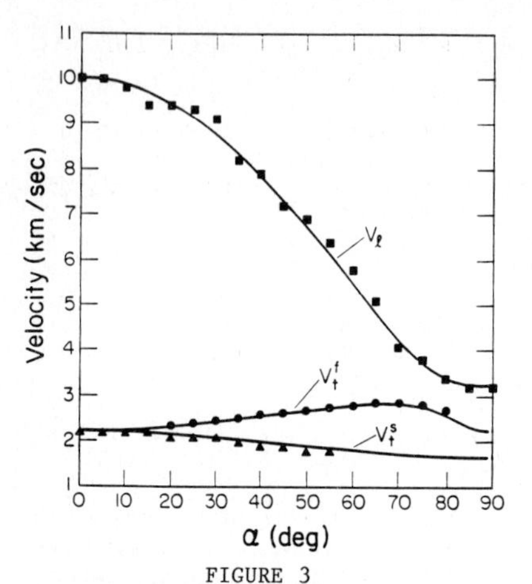

FIGURE 3
Experimental velocity data points are obtained from critical angle measurements. Solid lines are best fit to experimental data by nonlinear square optimization technique.

Specimen
T/R
slow shear
fast shear
longitudinal
A
Reflector

FIGURE 4
Schematic illustration of three refracted waves for anisotropic sample. Wave vector and group velocity directions are shown by dashed and solid lines, respectively.

It should be noted, however, that because of the deviation in angle between phase and group velocity, the direction of the wave vector (dashed line in Fig. 4) and the direction of the group velocity (solid line in Fig. 4) do not coincide. Since the time of flight in a composite material measured by this technique corresponds to the group velocity while the equations for calculations use the phase velocity [5], one may suspect that the results of measurement are somewhat incorrect. This question has been clarified by a direct analytical calculation [9] which shows that the (imaginary) time of arrival at the receiving transducer for the phase velocity is exactly the same as the actual time of flight of the signal propagating at the group velocity. Therefore, the method utilized here gives absolutely correct results for arbitrary orientation of the composite material relative to the incident beam.

The measurements were performed by the method discussed here for different angles of orientation of the fibers relative to the incident plane. As an example the experimental results for $\alpha=45°$ are shown in Fig. 5 as discrete points. The solid lines in Fig. 5 are velocities calculated by taking the set of elastic constants reconstructed from the critical angle measurement. It is seen that the agreement between the calculated and the experimental data is satisfactory, indicating that this particular material may be considered to be transversely isotropic for planes perpendicular to the fiber direction.

5. MEASUREMENT OF ELASTIC CONSTANTS OF THIN COMPOSITE PLATES

While the preceding two methods are used for thick section composite material evaluations, the following method is devoted especially to thin section composite materials.

The equations for calculations of the transmission coefficient for a plane wave obliquely incident on an arbitrary oriented liquid coupled orthotropic composite plate have been obtained in paper [10]:

$$T = iY(S+A)/[(S+iY)(A-iY)] \tag{3}$$

where S and A, respectively, are characteristic functions for symmetric and antisymmetric Lamb waves, Y is a term due to fluid loading.

Apparently, zero transmission coefficient occurs when S+A=0. This indicates the fact that the interference between symmetric and antisymmetric waves at some conditions can result in minimum normal displacement on the back side of the plate and, hence, minimum transmission coefficient. One can also expect that at higher frequency and thickness which allows more wave modes to exist a larger number of transmission minima can be observed. More importantly, since the characteristic functions S and A are both functions of elastic constants of the material, the change of the elastic properties in the incident plane due to rotation of sample in its own plane should give rise to changes of the transmission coefficient and the locations of the transmission minima. Inversely, from measuring the transmission coefficient and the locations of transmission minima for different fiber orientations and frequencies, one can evaluate the elastic constants.

A typical trace of the amplitude of a doubly-transmitted ultrasonic signal for a thin unidirectional composite plate is shown in Fig. 6 as a function of incident angle. The data are shown for only one angle $\alpha=20°$ and one frequency f=1.5MHz. By rotating the sample along the normal to the sample surface one can obtain the minimum-transmission angle as a function of the fiber orientation α and by repeating the measurements at different frequencies the dependence on ultrasonic frequency can be measured. Examples of the measured dependencies of minimum-transmission angle on fiber orientation and wave frequency are shown in Figures 7 and 8, respectively.

Finally, by using the same algorithm for nonlinear least square optimization as that described in previous sections together with equation (3) as the analytical function, the elastic constants can be reconstructed from the measured minimum-transmission angles. The best fit results are indicated in Figures 7 and 8 by solid lines. In addition, the transmission coefficient calculated by using the reconstructed elastic constants is shown as a solid line in Fig. 5. As can be seen in the figures, the agreement between the calculated results and the measured data is satisfactory.

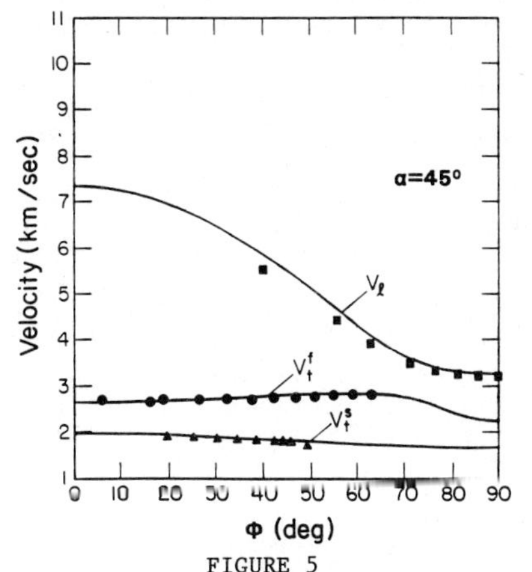

FIGURE 5
Phase velocities measured from through-transmission measurement (points) and calculated from elastic constants reconstructed from critical angle measurement. $\alpha=45°$. Angle $\Phi=90-\theta_r$ where θ_r is angle of refraction.

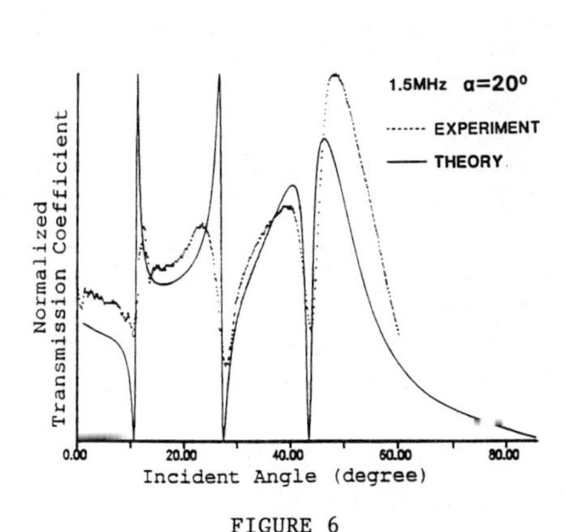

FIGURE 6
Experimental and theoretical transmission coefficient vs. incident angle at 1.5 MHz and $\alpha=20°$.

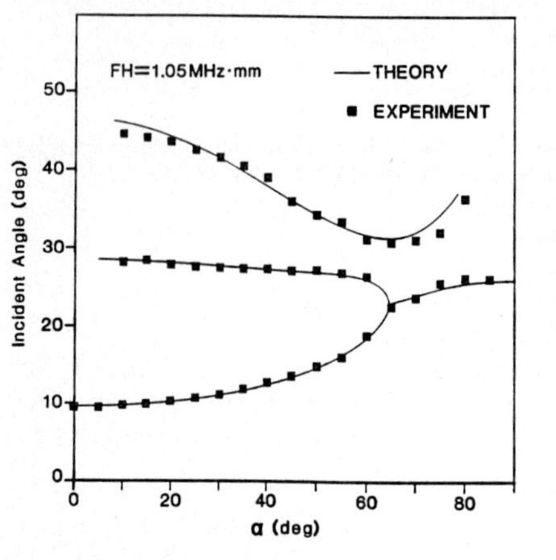

FIGURE 7
Experimental and theoretical dependence of transmission minima as a function of fiber orientation. f=1.5 MHz.

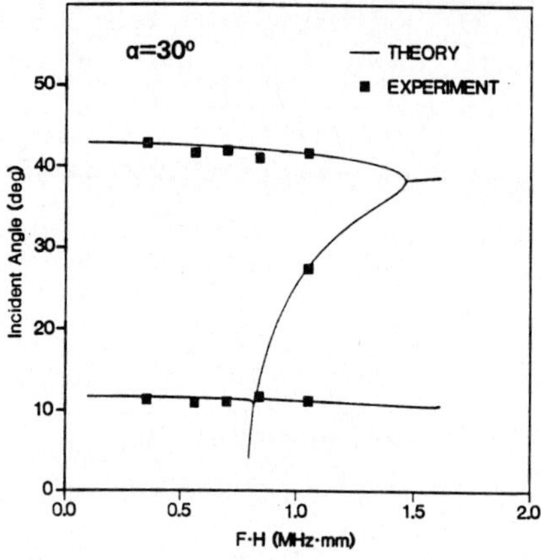

FIGURE 8
Experimental and theoretical dependence of transmission minima as a function of frequency. α=30°.

ACKNOWLEDGEMENTS

This work was in part sponsored by the Center for Advanced NDE, operated by the Ames Laboratory, USDOE, for the Air Force Wright Aeronautical Laboratories/ Materials Laboratory under contract #SC-88-1488. The assistance of Mr. L. Wang with calculations and Mr. C. Y. Wu with experiment is also appreciated.

REFERENCES

1. R. D. Kriz and W. W. Stinchcomb, Experimental Mechanics, 49, 41-49 (1979).
2. M. F. Markham, Composites 1, 145 (1970).
3. R. E. Smith, J. Appl. Phys., 43, 2555-2562 (1972).
4. J. M. Gieske and R. E. Allred, Experimental Mechanics, 14, 158-165 (1974).
5. L. H. Pearson and W. J. Murri, in Review of Progress in Quantitative Nondestructive Evaluation, D. O. Thompson and D. E. Chimenti eds., Vol. 6B, 1093-1101, Plenum Press, N.Y. (1987).
6. B. Hosten, M. Deschamps and B.R. Tittman, J. Acoust. Soc. Am., 82, 1763-1770 (1987).
7. S. I. Rokhlin and D. E. Chimenti, in Review of Progress in Quantitative Nondestructive Evaluation, D. O. Thompson and D. E. Chimenti, eds., Plenum Press, N.Y. (1990).
8. S. I. Rokhlin, T. K. Bolland and L. Adler, in Review of Progress in Quantitative Nondestructive Evaluation, D. O. Thompson and D. E. Chimenti eds., Vol. 6B, 1093-1101, Plenum Press, N.Y. (1987).
9. S. I. Rokhlin and W. Wang, in Review of Progress in Quantitative Nondestructive Evaluation, D. O. Thompson and D. E. Chimenti eds., Vol. 8B, 1489-1496, Plenum Press, N.Y. (1988).
10. A. H. Nayfeh and D. E. Chimenti, J. Appl. Mech., 55, 863-870 (1988).

Elastic Waves and Ultrasonic Nondestructive Evaluation
S.K. Datta, J.D. Achenbach and Y.S. Rajapakse (Editors)

DYNAMIC EFFECTIVE PROPERTIES OF TRANSVERSELY ISOTROPIC ELASTIC SPHERES IN AN ISOTROPIC ELASTIC MATRIX

Federico J. SABINA

Instituto de Investigaciones en Matemáticas Aplicadas y en Sistemas
Universidad Nacional Autónoma de México
Apartado Postal 20-726
01000 México, D.F., México

Wave propagation in a particle-reinforced composite medium is studied here. An isotropic medium containing a random distribution of spherical inclusions of transversely isotropic material is considered. The self-consistent equations for two transversely isotropic phases are fully displayed. An approximate solution is obtained when the matrix phase is isotropic. Numerical results of dispersion and attenuation of quasi-longitudinal waves in epon 828Z epoxy and cobalt spheres are shown for .15 concentration.

1. INTRODUCTION

There has been lately a lot of attention to the elastic wave propagation problem in particulate materials from both the theoretical and experimental points of view. For references, see [1–4]. Here a two-phase composite is studied where the particulate phase is spherical but has the elastic properties of a transversely isotropic medium. The matrix phase, however, is isotropic as in many of the previous studies. The self-consistent equations for the effective moduli and density derived in [4] are given in an expanded form in section 2, which is convenient when the two phases considered are transversely isotropic. When the matrix phase is isotropic an iterative scheme to solve the implicit equations is proposed, which uses the updated properties of a comparison material insofar as if it were isotropic. It is possible to obtain, by a few iterations, dynamic effective properties of this two-phase composite. From these properties, the speed and attenuation of waves may be easily calculated.

2. FORMULATION

A two-phase particulate composite is studied here. It consists of a matrix with tensor of elastic moduli L_2 and mass density ρ_2 with embedded spherical inclusions, all of the same radius a, having tensor of elastic moduli L_1 and mass density ρ_1. The inclusions are homogeneously and isotropically distributed at random in the matrix. Thus a typical inclusion, which occupies the domain $x' + \Omega_1$, where Ω_1 is a sphere of radius a, is such that its 'centre' is distributed randomly. The probability density for finding the inclusion centered at x' is equal to n_1, i.e., the number density of inclusions, which is independent of x'. The probability that a point lies within the inclusion is then the volume fraction $c_1 = n_1|\Omega_1|$, where $|\Omega_1|$ is the volume occupied by an inclusion. The volume fraction of the matrix material is $c_2 = 1 - c_1$.

Sabina and Willis developed in [4] a simple self-consistent scheme applicable to a composite comprising a matrix containing inclusions when spatial correlations are unknown, which by-passes the use of the full machinery of stochastic wave propagation. The self-consistent equations, derived by them, for the dynamic effective tensor of elastic moduli L_0 and mass density ρ_0 are given by the expressions:

$$L_0 = L_2 + c_1 h_1(k) h_1(-k)(L_1 - L_2)[I + \bar{S}_x(L_1 - L_0)]^{-1}, \tag{2.1}$$

$$\rho_0 = \rho_2 + c_1 h_1(k) h_1(-k)(\rho_1 - \rho_2)[I + \bar{M}_t(\rho_1 - \rho_0)]^{-1}. \tag{2.2}$$

where k is the wave number of the mean harmonic plane wave in the effective media. To solve equations (2.1–2), a starting value of L_0 and ρ_0 is taken on the right-hand side of the equations equal to, say, the matrix properties; the kernel functions $\bar{S}_x$ and $\bar{M}_t$, and the function $h_1(k)$ are computed for those properties. Then first values of $L_0^{(1)}$ and $\rho_0^{(1)}$ are obtained. Next $\bar{S}_x, \bar{M}_t$ and $h_1(k)$ should be updated for the new matrix properties $L_0^{(1)}$ and $\rho^{(1)}$ in order to get $L_0^{(2)}$, $\rho_0^{(2)}$ and so on, until convergence is achieved.

For the spherical inclusion.

$$h_1(k) = 3(\sin ka - ka \cos ka)/(ka)^3. \tag{2.3}$$

The kernel functions $\bar{S}_x$ and $\bar{M}_t$, when the matrix material is taken isotropic are given by Sabina and Willis [4] in Appendix B. The term $\bar{S}_x$ is an isotropic fourth-order tensor with the same symmetries as L_2. Using the symbolic notation [5]

$$L_2 = (3\kappa_2, 2\mu_2), \tag{2.4}$$

where κ_2 is the bulk modulus and μ_2 is the shear modulus of the matrix, the corresponding terms for $\bar{S}_x$ are given by

$$\bar{S}_x = \left(\frac{\epsilon_\alpha}{3\kappa_2 + 4\mu_2}, \frac{1}{5}\left[\frac{2\epsilon_\alpha}{3\kappa_2 + 4\mu_2} + \frac{\epsilon_\beta}{\mu_2} \right] \right), \tag{2.5}$$

where

$$\epsilon_\gamma = 3(1 - ik_\gamma a)[\sin(k_\gamma a) - k_\gamma a \cos(k_\gamma a)] e^{ik_\gamma a}/(k_\gamma a)^3. \tag{2.6}$$

Here

$$k_\gamma = \omega/\gamma, \tag{2.7}$$

where ω is the frequency of the wave and $\gamma = \alpha$ or β as required, given by

$$\alpha = \left(\frac{\kappa_2 + \frac{4}{3}\mu_2}{\rho_2} \right)^{1/2}, \beta = \left(\frac{\mu_2}{\rho_2} \right)^{1/2} \tag{2.8}$$

since the properties of the comparison material are taken to be equal to those of the matrix material.

The term $\bar{M}_t$ is an isotropic tensor of the second order given by

$$(\bar{M}_t)_{ij} = \delta_{ij}(3 - \epsilon_\alpha - 2\epsilon_\beta)/3\rho_2, \tag{2.9}$$

where $\delta_{ij} = 1$ if $i = j$ and 0 otherwise.

With regard to the properties of the inclusion, the fourth-order tensor L_1 shall be taken as transversely isotropic and the symbolic notation for such tensors introduced by Hill [6] facilitates the algebra. A transversely isotropic tensor L is expressed in the form

$$L = (2k', l, q, n, 2m, 2p), \tag{2.10}$$

if the stress-strain relation $\sigma = Le$, where σ and e are the stress and strain tensors, implies

$$\frac{1}{2}(\sigma_{11} + \sigma_{22}) = k'(e_{11} + e_{22}) + le_{33}, \tag{2.11}$$

$$\sigma_{33} = q(e_{11} + e_{22}) + ne_{33}, \tag{2.12}$$

$$\sigma_{11} - \sigma_{22} = 2m(e_{11} - e_{22}), \tag{2.13}$$

$$\sigma_{12} = 2me_{12}, \tag{2.14}$$

$$\sigma_{23} = 2pe_{23}, \tag{2.15}$$

$$\sigma_{31} = 2pe_{31}. \tag{2.16}$$

Note that, when $l = q$, the tensor L is diagonally symmetric. With this notation

$$L^{-1} = \left(\frac{n}{2\Delta}, -\frac{l}{2\Delta}, -\frac{q}{2\Delta}, \frac{k'}{\Delta}, \frac{1}{2m}, \frac{1}{2p}\right), \tag{2.17}$$

where

$$\Delta = k'n - lq. \tag{2.18}$$

The product rule of two such tensors L and $\bar{L} = (2\bar{k}', \bar{l}, \bar{q}, \bar{n}, 2\bar{m}, 2\bar{p})$ is given by

$$L\bar{L} = (4k'\bar{k}' + 2q\bar{l}, 2l\bar{k}' + n\bar{l}, 2k'\bar{q} + q\bar{n}, 2l\bar{q} + n\bar{n}, 4m\bar{m}, 4p\bar{p}). \tag{2.19}$$

It is worthwhile to mention that the result of the product of two diagonally symmetric tensors does not have the same symmetry.

Also, if L is isotropic, with bulk κ and shear μ moduli, it has the form

$$L = \left(2(\kappa + \frac{1}{3}\mu), \kappa - \frac{2}{3}, \kappa - \frac{2}{3}, \kappa + \frac{4}{3}\mu, 2\mu, 2\mu)\right) \tag{2.20}$$

Thus, in terms of "bulk" κ_x and "shear moduli" μ_x, the isotropic tensor $\bar{S}_x$ becomes

$$\begin{aligned}\bar{S}_x &= (2k'_x, l_x, q_x, n_x, 2m_x, 2p_x) \\ &= (2(\kappa_x + \frac{1}{3}\mu_x), \kappa_x - \frac{2}{3}\mu_x, \kappa_x - \frac{2}{3}\mu_x, \kappa_x + \frac{4}{3}\mu_x, 2\mu_x, 2\mu_x).\end{aligned} \tag{2.21}$$

The equations (2.1) can now be written explicity in terms of the components of the effective transversely isotropic tensor L_0,

$$L_0 = (2k'_0, l_0, q_0, n_0, 2m_0, 2p_0), \tag{2.22}$$

Using the above rules it is easily shown that the resulting equations are:

$$\begin{aligned}k'_0 = k'_2 + c_1 h_1(k) h_1(-k) \{&(k'_1 - k'_2)[1 + 2l_x(q_1 - q_0) + n_x(n_1 - n_0)] \\ &- (q_1 - q_2)[2l_x(k'_1 - k'_0) + n_x(l_1 - l_0)]\}/D,\end{aligned} \tag{2.23}$$

$$\begin{aligned}l_0 = l_2 + c_1 h_1(k) h_1(-k) \{&-2(n_1 - n_2)[2l_x(k'_1 - k'_0) + n_x(l_1 - l_0)] \\ &+ (l_1 - l_2)[1 + 4n_x(n_1 - n_0) + 2l_x(q_1 - q_0)]\}/D,\end{aligned} \tag{2.24}$$

$$q_0 = q_2 + c_1 h_1(k) h_1(-k) \{(q_1 - q_2)[1 + 2q_x(l_1 - l_0) + 4k'_x(k'_1 - k'_0)]$$

$$- 2\,(k_1' - k_2')[2k_x'(q_1 - q_0) + q_x(n_1 - n_0)]\}\,/D, \tag{2.25}$$

$$n_0 = n_2 + c_1 h_1(k) h_1(-k)\,\{-2(l_1 - l_2)[2k_x'(q_1 - q_0) + q_x(n_1 - n_0] + (n_1 - n_2)[1 + 4k_x'(k_1' - k_0') + 2q_x(l_1 - l_0)]\}\,/D, \tag{2.26}$$

$$m_0 = m_2 + c_1 h_1(k) h_1(-k)(m_1 - m_2)/[(1 + 4m_x(m_1 - m_0)], \tag{2.27}$$

$$p_0 = p_2 + c_1 h_1(k) h_1(-k)(p_1 - p_2)/[1 + 4p_x(p_1 - p_0)], \tag{2.28}$$

where

$$D = 1 + 4k_x'(k_1' - k_0') + 2l_x(q_1 - q_0) + 2q_x(l_1 - l_0) + n_x(n_1 - n_0) + 4(k_x n_x - l_x q_x)[(k_1' - k_0')(n_1 - n_0) - (l_1 - l_0)(q_1 - q_0)]. \tag{2.29}$$

It is clear that $l_0 \neq q_0$ so the effective elastic tensor is not diagonally symmetric. This property is lost according to the product rule (2.19).

The equation for the effective mass density (2.2) becomes

$$\rho_0 = \rho_2 + c_1 h_1(k) h_1(-k)(\rho_1 - \rho_2)/[1 + (3 - \epsilon_\alpha - 2\epsilon_\beta)/3\rho_2(\rho_1 - \rho_0)], \tag{2.30}$$

using (2.9).

For self-consistency, each effective parameter in the implicit equations (2.23–30), except the first iterated ones, would require the evaluation of the kernel and h_1 functions for a transversely isotropic matrix, which are not available yet. However, an approximate solution is simply obtained by exploiting the structure of an isotropic tensor using (2.21). The kernel and h_1 functions are updated as if they were calculated for a new isotropic matrix with bulk and shear moduli taken from the updated values of $n_0(= \kappa + \frac{4}{3}\mu)$ and $p_0(= \mu)$ only.

It is clear that the updated material although transversely isotropic is close in some sense to the updated isotropic one chosen. For low concentration values it could be expected to be a fair approximation. Comparison with experiments is desirable to test the theory, which, as far as it is known, are not available for this kind of composites.

3. RESULTS

Equations (2.23–30) have been solved by the iteration scheme described above for cobalt spheres embedded in epon 828Z epoxy. Their properties are: $\rho_1 = 8.9\,\mathrm{g/cm^3}$; $k_1' = 23.6$, $l_1 = q_1 = 10.27$, $n_1 = 35.81$, $m_1 = 7.1$, $p_1 = 7.55$ (all 10^{11} dyn/cm^2) [7] and $\rho_2 = 1.202\,\mathrm{g/cm^3}$; $\alpha_2 = 2.64$, $\beta_2 = 1.20$ (all mm/μs) (See [4]). The concentration of the spheres is $c_1 = 0.15$. Here the $z-$axis is the axis of transverse isotropy.

Plots of normalized phase velocity and attenuation against normalized frequency for quasi-longitudinal waves are shown in figures 1 and 2. The phase velocity is taken as $\omega/\,\mathrm{Re}\,\{k_\gamma\}$, where $\gamma = \alpha$, is the speed of the quasi-longitudinal wave in the transversely isotropic effective medium. The phase velocity is normalized to the phase velocity of the longitudinal wave in the matrix material. Normalized frequency is $k_2 a$, where $k_2 = \mathrm{Re}\,\{k_\alpha\}$ evaluated for $L_0 = L_2$ and $\rho_0 = \rho_2$. The measure of attenuation that is used is Im $\{k_\alpha a\}$.

The results of the first and last iteration are shown in figures 1 and 2 for phase velocity and attenuation, respectively. The continuous line refers to the first iteration for a quasi-longitudinal wave propagating in the $x-$direction and the dashed line corresponds to the last iterated value.

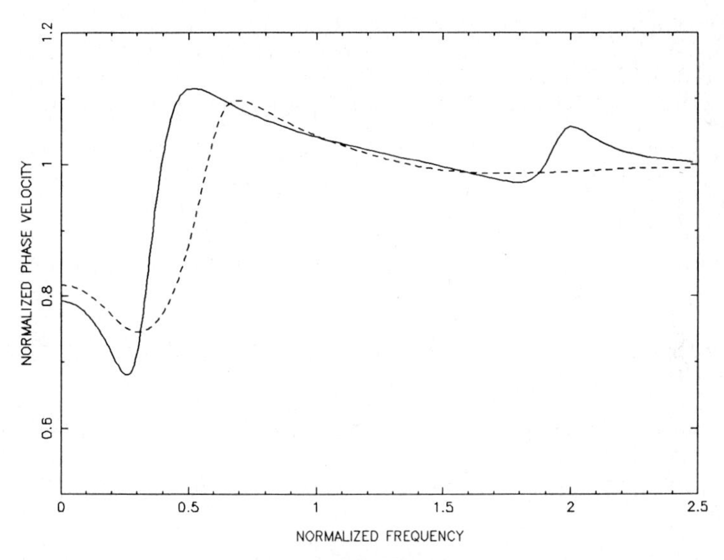

FIGURE 1

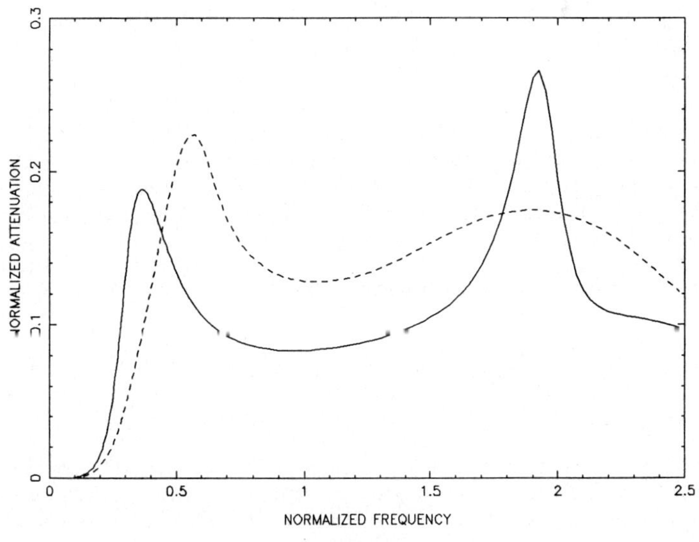

FIGURE 2

ACKNOWLEDGEMENTS

I should like to thank Prof. J.R. Willis for providing useful comments which make possible the improvement of this work.

REFERENCES

[1] Beltzer, A.I., Brauner, N., Acustica 65 (1988) 156.
[2] Datta, S.K., Ledbetter, H.M., Shindo, Y., and Shah, S.H., Wave Motion 10 (1988) 171.
[3] Lewandowski, J., Acta Mechanica, 68 (1987), 1 and 21.
[4] Sabina, F.J. and Willis, J.R., Wave Motion 10 (1988) 127.
[5] Hill, R., J. Mech. Phys. Solids 13 (1965) 89.
[6] Walpole, L.J., Elastic Behavior of Composite Materials: Theoretical Foundations, in: Yih, C.S., (ed.), Advances in Applied Mechanics (Academic Press, New York, 1981) pp. 169–242.
[7] Buchwald, V.T., Proc. R. Soc. London A 253 (1959) 563.

Elastic Waves and Ultrasonic Nondestructive Evaluation
S.K. Datta, J.D. Achenbach and Y.S. Rajapakse (Editors)

HARMONIC WAVES IN A PERIODICALLY BILAMINATED ANISOTROPIC ELASTIC COMPOSITE

T. C. T. Ting
Department of Civil Engineering, Mechanics and Metallurgy
University of Illinois at Chicago
P.O. Box 4348, Chicago, IL, 60680, USA

P. Chadwick
School of Mathematics
University of East Anglia
Norwich, NR4 7TJ, England

In previous work the dispersion relation for harmonic waves propagating at normal incidence to the interfaces of an infinite periodically layered elastic composite of general anisotropy was derived in the form of a 6x6 determinant equated to zero. It is shown here that when the unit cell of the composite consists of only two layers the dispersion relation reduces to the vanishing of a 3x3 determinant. Two cases of acoustical matching between the layers are briefly considered.

1. INTRODUCTION

It has recently been shown by us [1] and by Braga and Herrmann [2] that the sextic formalism of steady plane elasto–dynamics, allied to the use of propagator matrices, provides an economical method of obtaining dispersion relations for harmonic waves propagating in a periodically layered composite of general anisotropy. In the present paper we continue the discussion initiated in the final section of [1], of waves travelling normal to the layering and study in detail the situation in which the structure of the composite is of maximal simplicity, with just two layers in the unit cell. Matrix manipulations are described in Section 2 which reduce the dispersion relation to a 3x3 determinantal equation, and attention is drawn in Section 3 to further simplifications that occur when different degrees of acoustical matching apply at the interfaces. The argument employed in Section 2 has parallels with matrix calculations presented in [3].

We are concerned, then, with the special case $n = 2$ of the analysis developed in Section 6 of [1]. The notation is eased in this instance by omitting the superscript (1), replacing (2) by a prime and putting $h_1 = 2h$, $h_2 = 2h'$. Quantities relating to the layers of thickness $2h$ and $2h'$ constituting the unit cell are accordingly denoted by unprimed and primed symbols, respectively. In view of equations (4.5) and $(3.9)_3$ of [1], we can express the dispersion relation (4.7) for harmonic waves at normal incidence to the interfaces of the bilaminate as

$$\det\left\{\mathbf{M}'(h')\mathbf{M}'(h')\mathbf{M}(h)\mathbf{M}(h) - e^{2i\kappa d}\mathbf{I}\right\} = 0 . \tag{1.1}$$

Here κ is the Floquet wave number,

$$d = h + h' , \tag{1.2}$$

and $\underset{\sim}{I}$ is the 6x6 identity matrix. The propagator matrix $\underset{\sim}{M}(h)$ has the representation

$$\underset{\sim}{M}(h) = \underset{\sim}{U}\underset{\sim}{E}(h)\underset{\sim}{U}^{-1} , \tag{1.3}$$

where

$$\underset{\sim}{U} = (2v)^{-1/2}\begin{bmatrix} \underset{\sim}{G}\underset{\sim}{Z}^{-1/2} & i\underset{\sim}{G}\underset{\sim}{Z}^{-1/2} \\ v\underset{\sim}{G}\underset{\sim}{Z}^{1/2} & -iv\underset{\sim}{G}\underset{\sim}{Z}^{1/2} \end{bmatrix},$$

$$\underset{\sim}{U}^{-1} = (2v)^{-1/2}\begin{bmatrix} v\underset{\sim}{Z}^{1/2}\underset{\sim}{G}^T & \underset{\sim}{Z}^{-1/2}\underset{\sim}{G}^T \\ -iv\underset{\sim}{Z}^{1/2}\underset{\sim}{G}^T & i\underset{\sim}{Z}^{-1/2}\underset{\sim}{G}^T \end{bmatrix}, \tag{1.4}$$

and

$$\underset{\sim}{E}(h) = e^{-ikh}\begin{bmatrix} \underset{\sim}{C}+i\underset{\sim}{S} & \underset{\sim}{0} \\ \underset{\sim}{0} & \underset{\sim}{C}-i\underset{\sim}{S} \end{bmatrix}, \tag{1.5}$$

with

$$\underset{\sim}{C} = \text{diag}\left[\cos(\omega h/c_1),\ \cos(\omega h/c_2),\ \cos(\omega h/c_3)\right],$$
$$\underset{\sim}{S} = \text{diag}\left[\sin(\omega h/c_1),\ \sin(\omega h/c_2),\ \sin(\omega h/c_3)\right]. \tag{1.6}$$

In (1.4), v is the speed of propagation of the wave and $\underset{\sim}{G}$ and $\underset{\sim}{Z}$ are the 3x3 polarization and impedance matrices defined on p. 77 of [1]. In (1.5) and (1.6), k is the wave number and $\omega = kv$ the angular frequency of the wave and c_1, c_2, c_3 are the speeds of the co–directional plane body waves. Equations (1.3) to (1.6), which relate to a layer of thickness 2h, have primed analogues for a layer of thickness 2h′. The detailed background to these results is given in Sections 3 and 6 of [1].

2. REDUCED DISPERSION RELATION FOR HARMONIC WAVES PROPAGATING NORMAL TO THE LAYERING OF A PERIODIC BILAMINATE

Cyclic reordering of the propagator matrices in (1.1) is permissible and leads to

$$\det\left\{\underset{\sim}{M}(h)\underset{\sim}{M}'(h')\underset{\sim}{M}'(h')\underset{\sim}{M}(h) - e^{2i\kappa d}\underset{\sim}{I}\right\} = 0 . \tag{2.1}$$

Let $\underset{\sim}{A}$ and $\underset{\sim}{B}$ be arbitrary invertible 6x6 matrices. Then equation (2.1) is unaltered by the insertion of $\underset{\sim}{A}$ at the beginning and $\underset{\sim}{A}^{-1}$ at the end of the matrix product and the interpolation of $\underset{\sim}{B}\underset{\sim}{B}^{-1}$ between any pair of consecutive factors. Exploiting this invariance, we replace the product in (2.1) by

$$\left\{\underset{\sim}{Y}\underset{\sim}{U}^{-1}\underset{\sim}{M}(h)\underset{\sim}{M}'(h')\underset{\sim}{U}'\underset{\sim}{Y}'^{-1}\right\}\left\{\underset{\sim}{Y}'\underset{\sim}{U}'^{-1}\underset{\sim}{M}'(h')\underset{\sim}{M}(h)\underset{\sim}{U}\underset{\sim}{Y}^{-1}\right\}, \tag{2.2}$$

where

$$\underset{\sim}{Y} = \left[\frac{1}{2}v\right]^{1/2}\begin{bmatrix} \underset{\sim}{Z}^{1/2} & i\underset{\sim}{Z}^{1/2} \\ \underset{\sim}{Z}^{1/2} & -i\underset{\sim}{Z}^{1/2} \end{bmatrix},$$

$$\underset{\sim}{Y}^{-1} = (2v)^{-1/2}\begin{bmatrix} \underset{\sim}{Z}^{-1/2} & \underset{\sim}{Z}^{-1/2} \\ -i\underset{\sim}{Z}^{-1/2} & i\underset{\sim}{Z}^{-1/2} \end{bmatrix}, \tag{2.3}$$

and $\underset{\sim}{Y}'$ and its inverse are defined similarly. With the use of (1.3) and the corresponding representation of $\underset{\sim}{M}'(h')$, (2.2) can be rewritten as

$$\left\{\underset{\sim}{Y}\underset{\sim}{E}(h)\underset{\sim}{U}^{-1}\underset{\sim}{U}'\underset{\sim}{E}'(h')\underset{\sim}{Y}'^{-1}\right\}\left\{\underset{\sim}{Y}'\underset{\sim}{E}'(h')\underset{\sim}{U}'^{-1}\underset{\sim}{U}\underset{\sim}{E}(h)\underset{\sim}{Y}^{-1}\right\}.$$

The dispersion relation (2.1) thus becomes

$$\det\left[\underset{\sim}{\Omega}\underset{\sim}{\Omega}^* - e^{2iKd}\underset{\sim}{I}\right] = 0, \tag{2.4}$$

where $K = k + \kappa$,

$$\underset{\sim}{\Omega} = e^{ikd}\underset{\sim}{Y}\underset{\sim}{E}(h)\underset{\sim}{U}^{-1}\underset{\sim}{U}'\underset{\sim}{E}'(h')\underset{\sim}{Y}'^{-1}, \tag{2.5}$$

and $*$ signifies the interchange of unprimed and primed symbols.

With a view to simplifying $\underset{\sim}{\Omega}$ we first observe, from (1.4) and (2.3), that

$$\underset{\sim}{U}^{-1}\underset{\sim}{U}' = \underset{\sim}{Y}^{-1}\underset{\sim}{\Lambda}\underset{\sim}{Y}', \tag{2.6}$$

where

$$\underset{\sim}{\Lambda} = \begin{bmatrix} \underset{\sim}{\Delta} & \underset{\sim}{0} \\ \underset{\sim}{0} & \underset{\sim}{\Gamma} \end{bmatrix}, \quad \underset{\sim}{\Lambda}^{-1} = \begin{bmatrix} \underset{\sim}{\Delta}^{-1} & \underset{\sim}{0} \\ \underset{\sim}{0} & \underset{\sim}{\Gamma}^{-1} \end{bmatrix}, \tag{2.7}$$

and

$$\underset{\sim}{\Gamma} = \underset{\sim}{G}^T\underset{\sim}{G}', \quad \underset{\sim}{\Gamma}^{-1} = \underset{\sim}{G}'^T\underset{\sim}{G} = \underset{\sim}{\Gamma}^T = \underset{\sim}{\Gamma}^*,$$

$$\underset{\sim}{\Delta} = \underset{\sim}{Z}\underset{\sim}{\Gamma}\underset{\sim}{Z}'^{-1}, \quad \underset{\sim}{\Delta}^{-1} = \underset{\sim}{Z}'\underset{\sim}{\Gamma}^{-1}\underset{\sim}{Z}^{-1} = \underset{\sim}{\Delta}^*. \tag{2.8}$$

Secondly, from (2.3) and (1.5),

$$\underset{\sim}{Y}\underset{\sim}{E}(h) = e^{-ikh}\underset{\sim}{W}\underset{\sim}{Y}, \quad \underset{\sim}{E}(h)\underset{\sim}{Y}^{-1} = e^{-ikh}\underset{\sim}{Y}^{-1}\underset{\sim}{W}, \tag{2.9}$$

where

$$\underset{\sim}{W} = \begin{bmatrix} \underset{\sim}{C} & i\underset{\sim}{S} \\ i\underset{\sim}{S} & \underset{\sim}{C} \end{bmatrix}, \quad \underset{\sim}{W}^{-1} = \begin{bmatrix} \underset{\sim}{C} & -i\underset{\sim}{S} \\ -i\underset{\sim}{S} & \underset{\sim}{C} \end{bmatrix}. \tag{2.10}$$

Substitution into (2.5) of $(2.9)_1$, (2.6) and the primed counterpart of $(2.9)_2$ yields

$$\underset{\sim}{\Omega} = \underset{\sim}{W}\underset{\sim}{\Lambda}\underset{\sim}{W}', \tag{2.11}$$

use being made of (1.2). It follows from (2.7) and (2.8) that $\underset{\sim}{\Lambda}^* = \underset{\sim}{\Lambda}^{-1}$. The result of applying the * operation to (2.11) is therefore

$$\underset{\sim}{\Omega}^* = \underset{\sim}{W}'\underset{\sim}{\Lambda}^{-1}\underset{\sim}{W}. \tag{2.12}$$

Equation (2.4) implies the existence of a non–zero column vector $\underset{\sim}{\gamma}$ satisfying

$$\underset{\sim}{\Omega}\underset{\sim}{\Omega}^*\underset{\sim}{\gamma} = e^{2iKd}\underset{\sim}{\gamma} \tag{2.13}$$

and

$$(\underset{\sim}{\Omega}\underset{\sim}{\Omega}^*)^{-1}\underset{\sim}{\gamma} = e^{-2iKd}\underset{\sim}{\gamma}. \tag{2.14}$$

We infer from the sum of (2.13) and (2.14) that

$$\det\left[\underset{\sim}{\Omega}\underset{\sim}{\Omega}^* + \underset{\sim}{\Omega}^{*-1}\underset{\sim}{\Omega}^{-1} - 2\cos 2Kd\,\underset{\sim}{I}\right] = 0,$$

or, invoking the identity

$$\underset{\sim}{\Omega}\underset{\sim}{\Omega}^* + \underset{\sim}{\Omega}^{*-1}\underset{\sim}{\Omega}^{-1} = \left[\underset{\sim}{\Omega} + \underset{\sim}{\Omega}^{*-1}\right]\left[\underset{\sim}{\Omega}^* + \underset{\sim}{\Omega}^{-1}\right] - 2\underset{\sim}{I},$$

that

$$\det\left\{\frac{1}{4}\left[\underset{\sim}{\Omega} + \underset{\sim}{\Omega}^{*-1}\right]\left[\underset{\sim}{\Omega}^* + \underset{\sim}{\Omega}^{-1}\right] - \cos^2 Kd\,\underset{\sim}{I}\right\} = 0. \tag{2.15}$$

In combination with (2.7) and (2.10), equations (2.11) and (2.12) give

$$\begin{gathered} \underset{\sim}{\Omega} = \begin{bmatrix} \underset{\sim}{\Phi} & i\underset{\sim}{\chi} \\ i\underset{\sim}{\Theta} & \underset{\sim}{\Psi} \end{bmatrix}, \quad \underset{\sim}{\Omega}^{*-1} = \begin{bmatrix} \underset{\sim}{\Phi} & -i\underset{\sim}{\chi} \\ -i\underset{\sim}{\Theta} & \underset{\sim}{\Psi} \end{bmatrix}, \\ \underset{\sim}{\Omega}^* = \begin{bmatrix} \underset{\sim}{\Phi}^* & i\underset{\sim}{\chi}^* \\ i\underset{\sim}{\Theta}^* & \underset{\sim}{\Psi}^* \end{bmatrix}, \quad \underset{\sim}{\Omega}^{-1} = \begin{bmatrix} \underset{\sim}{\Phi}^* & -i\underset{\sim}{\chi}^* \\ -i\underset{\sim}{\Theta}^* & \underset{\sim}{\Psi}^* \end{bmatrix}, \end{gathered} \tag{2.16}$$

where

$$
\begin{aligned}
&\underset{\sim}{\Phi} = \underset{\sim}{C}\underset{\sim}{\Delta}\underset{\sim}{C}' - \underset{\sim}{S}\underset{\sim}{\Gamma}\underset{\sim}{S}' , && \underset{\sim}{\Phi}^* = \underset{\sim}{C}'\underset{\sim}{\Delta}^{-1}\underset{\sim}{C} - \underset{\sim}{S}'\underset{\sim}{\Gamma}^{-1}\underset{\sim}{S} , \\
&\underset{\sim}{\Psi} = \underset{\sim}{C}\underset{\sim}{\Gamma}\underset{\sim}{C}' - \underset{\sim}{S}\underset{\sim}{\Delta}\underset{\sim}{S}' , && \underset{\sim}{\Psi}^* = \underset{\sim}{C}'\underset{\sim}{\Gamma}^{-1}\underset{\sim}{C} - \underset{\sim}{S}'\underset{\sim}{\Delta}^{-1}\underset{\sim}{S} , \\
&\underset{\sim}{\Theta} = \underset{\sim}{C}\underset{\sim}{\Gamma}\underset{\sim}{S}' + \underset{\sim}{S}\underset{\sim}{\Delta}\underset{\sim}{C}' , && \underset{\sim}{\Theta}^* = \underset{\sim}{C}'\underset{\sim}{\Gamma}^{-1}\underset{\sim}{S} + \underset{\sim}{S}'\underset{\sim}{\Delta}^{-1}\underset{\sim}{C} , \\
&\underset{\sim}{\chi} = \underset{\sim}{S}\underset{\sim}{\Gamma}\underset{\sim}{C}' + \underset{\sim}{C}\underset{\sim}{\Delta}\underset{\sim}{S}' , && \underset{\sim}{\chi}^* = \underset{\sim}{S}'\underset{\sim}{\Gamma}^{-1}\underset{\sim}{C} + \underset{\sim}{C}'\underset{\sim}{\Delta}^{-1}\underset{\sim}{S} ,
\end{aligned}
\tag{2.17}
$$

The dispersion relation (2.15) thus takes the form

$$
\det \begin{bmatrix} \underset{\sim}{\Phi}\underset{\sim}{\Phi}^* - \cos^2 Kd\, \underset{\sim}{I} & \underset{\sim}{0} \\ \underset{\sim}{0} & \underset{\sim}{\Psi}\underset{\sim}{\Psi}^* - \cos^2 Kd\, \underset{\sim}{I} \end{bmatrix} = 0 ,
$$

$\underset{\sim}{I}$ now denoting the 3x3 identity matrix, and we arrive at the alternatives

$$
\det (\underset{\sim}{\Phi}\underset{\sim}{\Phi}^* - \cos^2 Kd\, \underset{\sim}{I}) = 0 , \quad \det (\underset{\sim}{\Psi}\underset{\sim}{\Psi}^* - \cos^2 Kd\, \underset{\sim}{I}) = 0 . \tag{2.18}
$$

To complete the reduction of (2.1) we prove that equations (2.18) are equivalent. Among the relations found by multiplying $(2.16)_1$ by $(2.16)_4$, and in the reverse order, are

$$
\underset{\sim}{\Theta}\underset{\sim}{\Phi}^* = \underset{\sim}{\Psi}\underset{\sim}{\Theta}^* , \qquad \underset{\sim}{\Theta}^*\underset{\sim}{\Phi} = \underset{\sim}{\Psi}^*\underset{\sim}{\Theta} .
$$

Hence

$$
\underset{\sim}{\Theta}\underset{\sim}{\Phi}^*\underset{\sim}{\Phi} = \underset{\sim}{\Psi}\underset{\sim}{\Psi}^*\underset{\sim}{\Theta} .
$$

The only property of the wave entering into $\underset{\sim}{\Theta}$ is the frequency ω and $\underset{\sim}{\Theta}$ is consequently non–singular except, possibly, at certain discrete values of ω. Elsewhere,

$$
\begin{aligned}
\det (\underset{\sim}{\Phi}\underset{\sim}{\Phi}^* - \cos^2 Kd\, \underset{\sim}{I}) &= \det\left\{\underset{\sim}{\Theta}(\underset{\sim}{\Phi}^*\underset{\sim}{\Phi} - \cos^2 Kd\, \underset{\sim}{I})\underset{\sim}{\Theta}^{-1}\right\} \\
&= \det (\underset{\sim}{\Psi}\underset{\sim}{\Psi}^* - \cos^2 Kd\, \underset{\sim}{I}) ,
\end{aligned}
$$

so that $(2.18)_1$ implies $(2.18)_2$ and vice versa.

3. DISPERSION RELATIONS FOR HARMONIC WAVES PROPAGATING NORMAL TO THE LAYERING OF ACOUSTICALLY MATCHED PERIODIC BILAMINATES

We take note in this section of the further simplification of the dispersion relation $(2.18)_1$ which takes place when the polarizations $\underset{\sim}{g}_i$ and $\underset{\sim}{g}_i'$ in adjacent layers coincide. This occurs, for example, when both materials are orthorhombic and the planes of material symmetry are everywhere parallel to the coordinate planes (see [1], Fig. 1). The orthogonal matrices $\underset{\sim}{G}$ and $\underset{\sim}{G}'$ are then equal and $(2.8)_{1,5}$ condense to

$$\underset{\sim}{\Gamma} = \underset{\sim}{I}\,, \qquad \underset{\sim}{\Delta} = \underset{\sim}{Z}\underset{\sim}{Z}'^{-1}\,, \tag{3.1}$$

implying, via $(2.17)_{1,2}$, that $\underset{\sim}{\Phi}$ and $\underset{\sim}{\Phi}^*$ are diagonal. Equation $(2.18)_1$ therefore factors into three uncoupled dispersion relations associated with displacements in the directions of the common polarizations. Known results for a bilaminated isotropic composite [4] emerge when the appropriate specializations are made.

The expression $(3.1)_2$ for $\underset{\sim}{\Delta}$ is a measure of the impedance mismatch between the two materials. In the absence of such a discontinuity, that is when $\underset{\sim}{\Delta} = \underset{\sim}{I}$, we find from $(2.17)_{1-4}$ and (1.6) that $\underset{\sim}{\Phi}$, $\underset{\sim}{\Psi}$, $\underset{\sim}{\Phi}^*$ and $\underset{\sim}{\Psi}^*$ are all equal to

$$\mathrm{diag}\left[\cos\left\{\omega\left[\frac{h}{c_1} + \frac{h'}{c_1'}\right]\right\},\ \cos\left\{\omega\left[\frac{h}{c_2} + \frac{h'}{c_2'}\right]\right\},\ \cos\left\{\omega\left[\frac{h}{c_3} + \frac{h'}{c_3'}\right]\right\}\right].$$

The factors of $(2.18)_1$ then yield

$$Kd = \omega\left[\frac{h}{c_i} + \frac{h'}{c_i'}\right], \qquad i = 1,2,3\,. \tag{3.2}$$

In each of equations (3.2) the resultant wave number K depends linearly on the frequency, indicating, as would be expected, that the wave is not dispersed when the polarizations in adjoining layers are aligned and the impedances are matched.

REFERENCES

[1] Ting, T. C. T. and Chadwick, P., "Harmonic Waves in Periodically Layered Anisotropic Elastic Composites," Wave Propagation in Structural Composites. eds., A. K. Mal and T. C. T. Ting, AMD–Vol. 90, G00426, American Society of Mechanical Engineers, New York, 1988, pp. 69–79.

[2] Braga, A. M. B. and Herrmann, G., "Plane Waves in Anisotropic Layered Composites," Wave Propagation in Structural Composites, eds. A. K. Mal and T. C. T. Ting, AMD–Vol. 90, G00426, American Society of Mechanical Engineers, New York, 1988, pp. 81–98.

[3] Tang, Zhijing and Ting, T. C. T., "Transient Waves in a Layered Anisotropic Elastic Medium," Proc. R. Soc. Lond. A 397, 67–85, 1985.

[4] Rytov, S. M., "Acoustic Properties of a Thinly Laminated Medium," Soviet Physics Acoustics, 2, 68–80, 1956.

Elastic Waves and Ultrasonic Nondestructive Evaluation
S.K. Datta, J.D. Achenbach and Y.S. Rajapakse (Editors)

GENERALIZED SCALING TRANSFORM FOR THE POINT SOURCE SOLUTION OF THE WAVE EQUATION IN ANISOTROPIC MEDIA

Andrey TVERDOKHLEBOV and Joseph L.ROSE

Department of Mechanical Engineering and Mechanics
Drexel University, Philadelphia, PA 19104, USA.

Conditions for the stiffness matrix of an orthotropic media are derived for an assumption that the quasilongitudinal mode slowness profile is an exact elliptical surface. The validity of the retarded potential approximation, introduced earlier for the point source solution of the wave equation in arbitrary anisotropic, homogeneous media can therefore be confirmed. The ellipticity of the velocity profile may be verified experimentally by conventional ultrasonic NDE techniques through the measurements of the quasi-longitudinal mode velocity as a function of direction. When established, it simplifies tremendously an inverse problem for elastic constant evaluation.

1. INTRODUCTION

The Green's function for elastic wave propagation in mildly anisotropic media was obtained earlier in the frequency domain [1]. Transferred into time dependence it yields the generalized retarded potentials approximation with the Green's function

$$G \sim A\delta(t-r/w)$$

where t is time, $\delta(x)$ is the delta-function, $r = |\mathbf{r}|$, $\mathbf{r}$ is the radius-vector from the source point to the observation point, and $w = |\mathbf{w}|$, $\mathbf{w}$ is the group velocity as a function of direction $\mathbf{r}/r$ in anisotropic media. For an isotropic medium $w = \text{const.}$, and the approximation is reduced to the exact, regular retarded potentials representation [2]. Mild anisotropy roughly means that the slowness profile of a given mode differs very little from an average sphere. In this presentation we introduce a technical means to validate this approach further, namely for materials with the longitudinal slowness profile being close to an ellipsoid. The main idea is to use a scaling change of the original Cartesian spatial coordinates in order to handle any aspect ratios of the ellipsoid.

Generally, two different scaling factors along two principal axes of the material may be applied, for example, to equalize the values of the longitudinal slowness along all three principal axes. Subsequently, in a scaled space, the slowness profile becomes very close to spherical, hence satisfying the conditions of mild anisotropy. It was shown elsewhere [3,4] that a standard, unidirectional graphite-epoxy composite satisfies this condition. Here we derive the rules that establish the validity of the generalized retarded potentials model to the arbitrary anisotropic, homogeneous materials of orthotropic symmetry.

2. FACTORIZING THE QUASILONGITUDINAL MODE IN THE CHARACTERISTIC EQUATION FOR THE ORTHOTROPIC MEDIUM

The conventional characteristic equation for an arbitrary anisotropic, homogeneous medium may be presented in terms of the slowness vector S_j as follows

$$C_{jklm} S_k S_m U_l - U_j = 0 \qquad (1)$$

where U_l is the amplitude of the plane wave solution of the wave equation, $Q_{jl} = \lambda_{jklm} S_k S_m$ and λ_{jklm} is the elastic tensor of the material.

Eq.(1) is simply a wave equation for a linear, homogeneous anisotropic medium when an unknown displacement vector assumes a harmonic, plane wave solution with the amplitude U_l. The frequency of this harmonic solution cancels as a common factor due to the use of the slowness vector instead of the wave vector. (The density of the material is taken to be unity, for the sake of simplicity).

Eq.(1) has non-trivial solutions when

$$\det\{Q(S) - I\} = 0 \quad . \tag{2}$$

We consider now a general orthotropic medium with stiffness matrix components C_{MN}, $M,N = 1,\ldots, 6$, as they appear in a Cartesian coordinate system with axes along the principal directions of symmetry. To make concise a tedious, though elementary algebra, we present now the elements of the matrix Q as it results into Q_{scaled}, after a scaling transform is applied with the factors of contraction k_M

$$k_M = (C_{MM})^{1/2} \quad , \quad M = 1, 2, 3 \, ,$$

along the corresponding principal axes. (No summation assumed with respect to the subscript M).

For the first row of the matrix Q_{scaled} one finds

$$X^1 + B_{62}X_2 + B_{53}X_3 \, , \qquad D_3(X_1X_2)^{1/2} \quad , \qquad D_2(X_1X_3)^{1/2}$$

where X_j are the squares of the slowness components in a new scaled space, and

$$\begin{aligned} B_{62} &= C_{66}/C_{22} \, , \quad & D_3 &= (C_{12}+C_{66})/(C_{11}C_{22})^{1/2} \, , \\ B_{53} &= C_{55}/C_{33} \, , \quad & D_2 &= (C_{13}+C_{55})/(C_{11}C_{33})^{1/2} \, . \end{aligned} \tag{3}$$

The next two rows could be obtained with the corresponding permutations of the subscripts 1,2,3. Matrix Q_{scaled} shall have equal all three values of the quasilongitudinal slowness along the three principal axes (equal to one in our choice, if that matters). Note, that matrix Q_{scaled} is symmetrical and, therefore, still has the three mutually orthogonal eigen vectors - the quasilongitudinal and two quasitransverse modes of propagation. If one requires the quasilongitudinal slowness profile to be an ellipsoid in the original, non-scaled coordinate system, then, in the scaled space, Eq.(2) has to be factorized as

$$\det\{Q_{scaled}(X_j) - I\} = (1 - X_1 - X_2 - X_3)\cdot P_4(X_j) = 0 \, , \tag{4}$$

where $P_4(X_j)$ is some polynomial of fourth degree with respect to the components X_j. Eq.(4) means that the quasilongitudinal slowness profile in the scaled space is an exact sphere.

If Eq.(4) is true, then the direct calculations lead to the following conditions that has to be imposed on the values of the stiffness matrix in an original, non-scaled medium.

$$C_{22}\,C_{33} - (C_{23})^2 - C_{44}(2C_{23} + C_{22} + C_{33}) = 0 \tag{5}$$

$$C_{33}\,C_{11} - (C_{31})^2 - C_{55}(2C_{31} + C_{33} + C_{11}) = 0 \tag{6}$$

$$C_{11}C_{22} - (C_{12})^2 - C_{66}(2C_{12} + C_{11} + C_{22}) = 0 \tag{7}$$

$$(C_{11}-C_{55})(C_{22}-C_{66})(C_{33}-C_{44}) + (C_{11}-C_{66})(C_{22}-C_{44})(C_{33}-C_{55}) - 2(C_{23}+C_{44})(C_{31}+C_{55})(C_{12}+C_{66}) = 0. \tag{8}$$

The conditions (5-8) are sufficient, though not necessary. There exist other, more exotic rules that result into the same, factorized form (4) of the characteristic equation. However, they seem not to be important in practical applications. In particular case of unidirectional (hexagonal) symmetry, all four Eqs.(5-8) are reduced to a single one that was obtained earlier in [2].

Actually, Eqs.(5-8) will never hold exactly, of course. The relative deviations could be obtained as the ratios of Eqs.(5-8) to the $C_{22}C_{33}$, $C_{33}C_{11}$, $C_{11}C_{22}$ and $C_{11}C_{22}C_{33}$, respectively. These ratios make an estimate for the order of magnitude of the relative error in the retarded potentials approximation, when applied.

3. CONCLUSIONS

The practical significance of conditions (5-8) is revealed best when one faces an inverse problem in ultrasonic NDE. As long as an experimenter established that the quasilongitudinal velocity profile is close to an elliptical surface, then we may say that Eqs.(5-8) hold. Therefore, the inverse problem of ultrasonic evaluation of the stiffness matrix will be simplified tremendously by reducing the number of unknowns by four.

ACKNOWLEDGEMENT

The work was partially funded by the EPRI, Inc., Palo Alto, CA USA. (Project RP# 2405-25).

REFERENCES

[1] A.Tverdokhlebov, J.Rose,"On Green's function for elastic waves in anisotropic media", J.Acoust.Soc.Am., vol.80, p.118 (1988).

[2] J.D.Achenbach, "Wave Propagation in elastic solids", (Elsevier, Amsterdam, 1984), p.93

[3] A.Tverdokhlebov, J.Rose,"On application domain of Green's function approximation for the mild anisotropic media", accepted for publication in J.Acoust.Soc.Am.

[4] J.L.Rose, A.Pilarski, K.Balasubramaniam, A.Tverdokhlebov, J.Ditri, "Ultrasonic Wave Considerations for the Development of an NDE Feature Matrix for Anisotropic Media", J.Eng.Mat.&Tech., ASME, Vol.111(3), p.255, July 1987.

C

WAVE PROPAGATION IN NONHOMOGENEOUS AND PRESTRESSED MEDIA AND ACOUSTIC EMISSION

Elastic Waves and Ultrasonic Nondestructive Evaluation
S.K. Datta, J.D. Achenbach and Y.S. Rajapakse (Editors)

CHARACTERIZATION OF MATERIAL DEFORMATION AND FAILURE RESPONSES FROM ULTRASONIC MEASUREMENTS

R. Bruce Thompson

Ames Laboratory, Center for Nondestructive Evaluation
and College of Engineering
Iowa State University
Ames, Iowa, U.S.A.

ABSTRACT

Recent advances in ultrasonic techniques to characterize the structure of materials are presented. Observable parameters include the velocity, attenuation and scattering of ultrasonic waves and their variation with propagation direction and frequency. Applications are described in which information obtained from such measurements is used to characterize the deformation and failure responses of metals, metal-metal bonds, heavily deformed metal-metal composites and thick, filament-wound composites. In each case, unsolved mechanics problems whose solutions are needed to improve the application results are identified.

1. INTRODUCTION

Ultrasound interacts with the structure of materials by scattering from inhomogeneities. In traditional applications, ultrasonic scattering has been a means towards the detection, characterization and sizing of discrete cracks, voids and inclusions [1-3]. However, the need to characterize distributed microstructural inhomogeneities has become increasingly evident [4-7]. Often these inhomogeneities have been distributed throughout the microstructure in particular ways in the material design process to achieve performance goals associated with parameters such as strength, fracture toughness, etc. Modern designs which operate close to the limits of those performance parameters dictate the need for nondestructive techniques to characterize microstructure. Ultrasound responds to the microstructural inhomogeneities through attenuation, scattering, and velocity shifts. The purpose of this paper is to review recent examples of the use of such measurements to study the mechanical properties of various materials. In each case, the physical basis of the relationship of the measurement to a deformation or failure related material property is emphasized. The potential for failure prediction based on ultrasonic measurements has already been discussed in this volume by Achenbach [8].

Sections 2, 3 and 4 present the details of the applications in metals, heavily deformed metal-metal composites, and thick, filament-wound composites. Included are examples of techniques to characterize planar arrays of cracks, volumetric distributions of porosity, grain structure and reinforcing fibers. At the conclusion of the discussion of each application, theoretical problems whose solutions would enable further developments are identified.

2. METALS

Metal parts deform and fail through a rich variety of processes. When complex part shapes are produced by forming operations, large scale plastic deformation occurs. When parts are subjected to large static loads, creep can lead to changes in component shape and ultimately to failure. Under cyclic loading, the growth of fatigue cracks is the most important failure mechanism. Under conditions of imperfect metal-metal bonding, local microstructural changes near the bond plane may produce dramatic changes in strength with respect to that of the base metal. Application of ultrasonic measurements to characterize microstructural changes associated with each of these deformation/failure modes is discussed below.

2.1 Texture as it Influences Formability

The texture of a metal sheet plays an important role in determining its ability be formed into parts of complex shapes such as automobile components or beverage cans. Work in the early 1970's established that measurement of elastic properties, such as Young's modulus, and their anisotropy correlated well with plastic strain ratios [9,10], parameters important in predicting metal drawability [11]. Analysis has shown that these correlations can be interpreted in terms of a set of constants, known as orientation distribution coefficients (ODC's), which define the texture (preferred grain orientation) of a polycrystal [12]. In particular, the ODC's are coefficients in the representation of the probability density function for grain orientation as a series of generalized spherical harmonics [13]. For polycrystalline sheets composed of cubic metal crystallites, the only independent ODC's which can be determined from measurements of the elasticity response are W_{400}, W_{420} and W_{440} [14,15]. Davies et al. noted that the coefficient W_{400} is closely related to the average plastic strain ratio, while W_{420} and W_{440} are indications of the tendency to form two-fold and four-fold ears, respectively, in deep drawing [12].

These relationships have led to a commercial instrument for prediction of formability parameters of rolled sheet. In that device, vibrational techniques are used to measure Young's modulus of coupons cut from the sheet and orientated at 0°, 45° and 90° with respect to the rolling direction [16]. The correlation of Stickels and Mould is then the basis for the prediction of formability parameters [9-10]. More recent research has concentrated on inference of similar information from ultrasonic measurements. In one approach, the elastic constants can be written as a function of the ODC's [14,15]. Theories of elastic wave propagation in anisotropic media can then be used to relate wave speeds to these texture parameters. Particular attention has been recently focussed on uses of Lamb mode propagation characteristics to gather this information [17,18]. The ultrasonic predictions of the ODC's are based on a comparison of theoretical predictions of the angular dependence of the velocities to experimental data, with the ODC's being adjusted for best fit. The predictions of W_{420} and W_{440} depend only on angular dependences, i.e., on the relative agreement of theory and experiment, while the predictions of W_{400} depend on an absolute comparison. As an example, Fig. 1 presents a comparison of values of W_{400} inferred from absolute velocities of the S_o mode at long wavelength to those inferred from neutron or x-ray diffraction [18].

In most samples, the agreement is quite good. However, in sample #2, an 1100 aluminum alloy, this is not the case. This lack of agreement in aluminum has been interpreted in terms of microstructural influences on the elastic moduli, which compete with the texture effect. This problem appears to be particularly severe in aluminum because of the large propagation of errors implied by its small elastic anisotropy. Fewer difficulties are obtained in the prediction of W_{420} and W_{440}, which depend only on the relative angular variations in velocity.

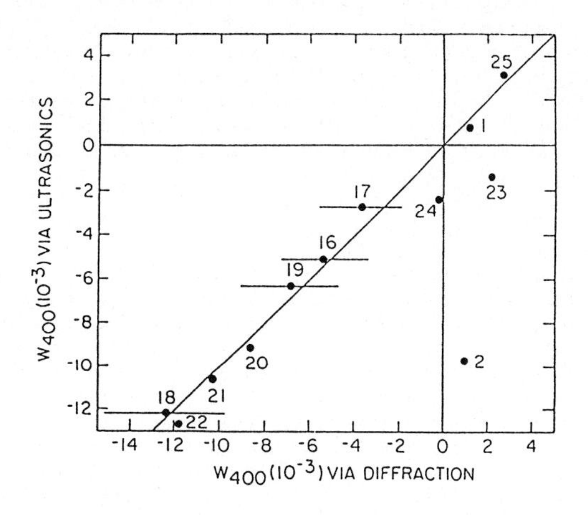

FIGURE 1

Comparison of ultrasonic and diffraction (x-ray or neutron) predictions of the ODC W_{400}.

There are several mechanics questions whose solutions would be expected to improve the performance of this technique, in particular the accuracy of predictions of W_{400}. The central issue is our ability to model the absolute values of elastic moduli of polycrystals. One issue is the effect of dislocations. In the theory of texture measurement [17], based on the elastic moduli of polycrystals [14,15], dislocation contributions are neglected. However, their presence is known to introduce additional changes in effective elastic moduli moduli which would influence both relative and absolute wave speed measurements [19]. More research is needed to merge our present microscopic theories for dislocation-ultrasound interactions [20] with proper continuum theories [21,22]. Another important consideration is the effect of alloying, second phases, or other metallurgical variables on the elastic moduli. These metallurgical conditions can be expected to produce small moduli shifts, and experience suggests that they are most important when using absolute velocities to predict W_{400}. It is believed that the error in the prediction of W_{400} in sample #2, Fig. 1, is a result of this problem.

2.2 Porosity as it Influences Ductile Fracture

It is well known that the presence of porosity causes a decrease in the ultrasonic velocity. Figure 2 presents, as an example, results obtained by Spitzig, et al. on a set of iron compacts [23,24], Here the data is compared to the theoretical predictions of Sayers and Smith [25] for porosities up to 11%. The fact that the measured velocities were lower than the theoretical predictions at the higher porosities was interpreted as a consequence of non-spherical pore shape [26,27].

Using the theoretical expectations, it is possible to estimate porosity from velocity measurements. These estimates can then be used as inputs into models predicting the effect of porosity on mechanical properties. Figure 3 presents the results of predictions of the yield stress of the same set of iron compacts based on ultrasonically inferred initial porosities coupled with a normalized deformation model [24]. Good agreement is again obtained, with the error at 60% yield being a consequence of the previously identified influence of non-spherical pore shape.

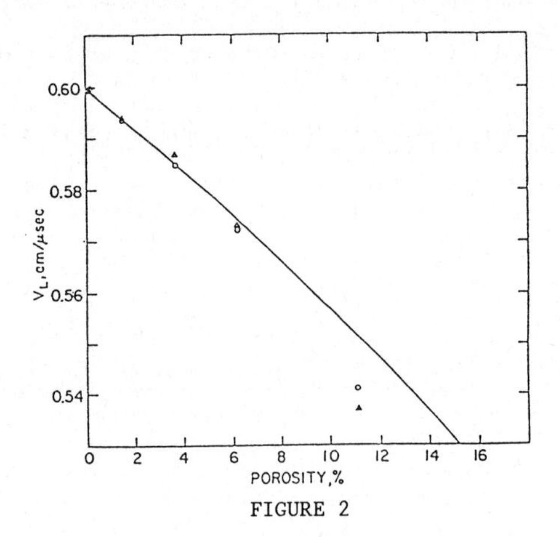

FIGURE 2

Comparison of the ultrasonic velocity measured in a set of iron compacts to the theoretical predictions of Sayers and Smith.

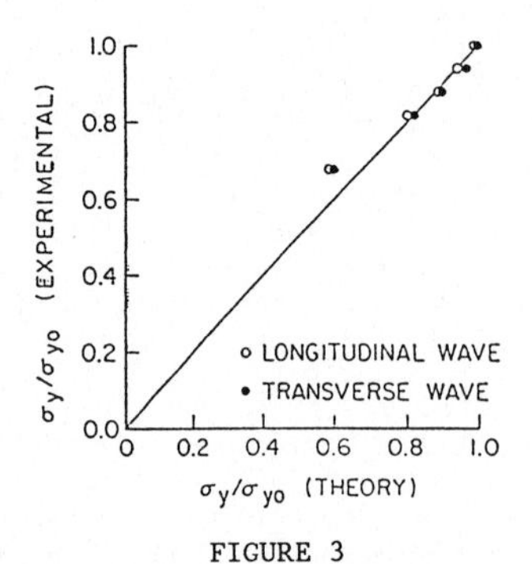

FIGURE 3

Comparison of measured yield stress on a set of iron compacts to theoretical predictions based on a deformation model coupled with ultrasonically based predictions of initial porosity.

It is believed that additional work is required on modeling the effect of non-spherical pore shapes on ultrasonic velocities, particularly at high volume fractions of porosity at which pore interconnection becomes significant. Such models will be important in evaluating other potential applications such as monitoring the progress of hot isostatic pressing or other densification processes.

2.3 Asperity Contact as it Influences Fatigue Crack Propagation

When a metal experiences cyclic rather than monotonically increasing loads, failure often occurs through the process of fatigue crack propagation. In this case the distributed homogeneity may take the form of a crack plane bridged by a set of contacting asperities. Buck et al. have discussed the ultrasonic evaluation of such a structure elsewhere in this book [28]. Therein they show how contact separation and densities can be inferred from ultrasonic scattering data.

2.4 Microcrack Distributions as they Influence Metal-Metal Bond Strength

A quite similar ultrasonic problem is encountered when one attempts to monitor solid-state bonding operations such as friction welding, inertial welding, diffusion bonding, etc. In these operations, application of temperature and pressure to two mating metal surfaces leads to plastic deformation of asperities until full contact is achieved, followed by diffusion and recrystallization processes which ideally render the bond plane indistinguishable from the base metal. An important defect structure that may remain in the bond plane if bonding is incomplete is an array of microcracks. Such an array interacts with ultrasound in a fashion that is quite analogous to the partially contacting fatigue crack discussed in section 2.3.

The theoretical interpretation of measurements on such samples has been aided by a distributed spring model of the interface [29], applicable when the ultrasonic wavelength is large with respect to the microcrack dimensions and separations. The two sides of the interface are viewed as being joined by a distributed spring, whose stiffness κ is a function of the microcrack topology. The values of this stiffness can be deduced from ultrasonic transmission and reflection measurements.

Figure 4 presents the results of an experimental confirmation of the model [30].

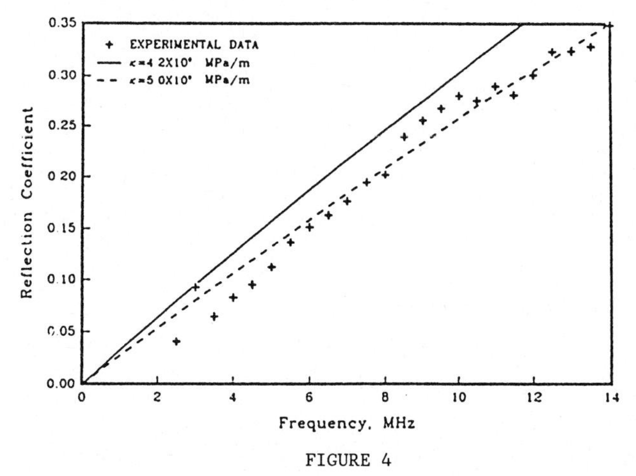

FIGURE 4

Comparison of ultrasonic reflectivities measured on a set of diffusion bonded copper samples as a function of frequency to predictions of the distributed spring model. The predictions described by the solid line are based on estimates of the spring stiffness κ from fracture surface micrographs. The dashed line presents the results of fitting to a 20% higher spring constant.

Here the ultrasonic reflectivity is plotted as a function of frequency for a diffusion bonded copper sample. The solid line indicates the linear increase with frequency predicted by the theory when the microcrack topology is inferred from examination of micrographs of the bond plane after tensile fracture. The experimental data points are in agreement with this prediction, although they are better fit by a spring stiffness that is 20% higher than that estimated from the micrographs. Figure 5 summarizes a set of experiments performed on nine diffusion bonded samples. Here the reflection coefficient at 10 MHz is plotted as a function of the spring stiffness κ. The solid line gives the theoretical expectations, which compare favorably with the data over a wide range of stiffnesses. Such measurements have been found to be quite useful in the study of fracture of such interfaces [30].

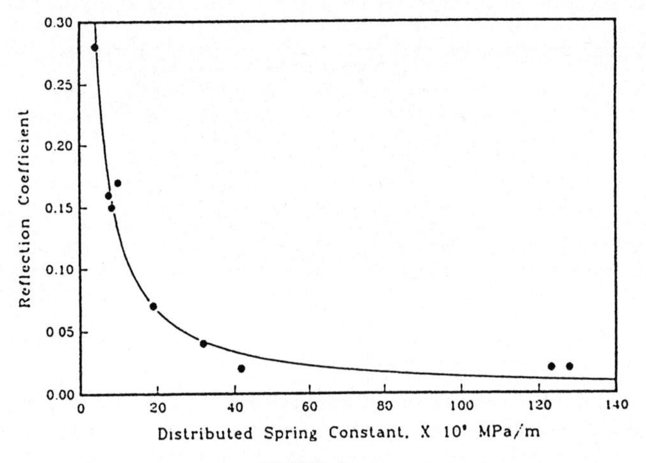

FIGURE 5

Comparison of ultrasonic reflections measured on a set of diffusion bonded copper samples as a function of spring constant to predictions of distributed spring model with κ estimated from fracture surface micrographs.

The models used in the interpretation of scattering from partially contacting fatigue crack faces and metal-metal bonding interfaces often neglect interactions of the asperities and/or microcracks. Theories taking these into account are needed to allow a more complete and accurate interpretation of the experimental results.

3. HEAVILY-DEFORMED METAL-METAL COMPOSITES

Heavily deformed metal-metal composites are advanced materials combining high mechanical strength with high electrical and/or thermal conductivities. In a typical example of a copper-niobium composite [31,32], processing begins with a casting procedure which produces an array of randomly oriented niobium dendrites in a copper matrix. Subsequent deformation processing such as wire drawing or plate rolling leads to a structure of aligned niobium filaments or platelets in a copper matrix. Figure 6 presents an example of the latter structure. An important deformation parameter is the draw ratio, η, defined as the logarithm of the ratio of cross-sectional areas before and after reduction. As shown in Fig. 7, increases in the draw ratio lead to dramatically increased strengths, a result which has been interpreted in terms of the decrease in the spacing of the reinforcing elements. Strengths equal to those of high strength steels have been demonstrated in materials which retain a large fraction of the electrical conductivity of copper. These composites show promise for use in a number of structures, including combustion chambers of rocket engines.

The mechanisms for this strengthening are not fully understood. One unknown, which is needed to differentiate competing theories is the degree of preferred grain orientation, particularly in the Cu. The small sample sizes and two phase structure have introduced difficulties with x-ray and neutron diffraction measurements. As an alternative, precise velocity measurements based on acoustic microscopy have been employed [35]. In this procedure, a highly focussed ultrasonic beam is created by a cylindrical lens in a water coupling medium and directed at the sample.

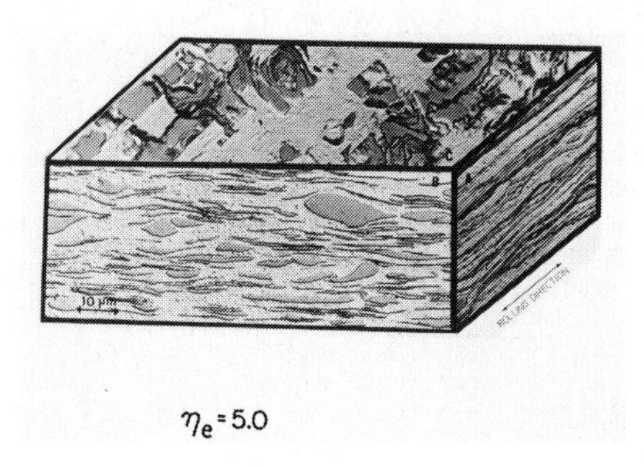

FIGURE 6

Micrograph of Cu-20% Nb plate showing highly aligned, Nb platelets.

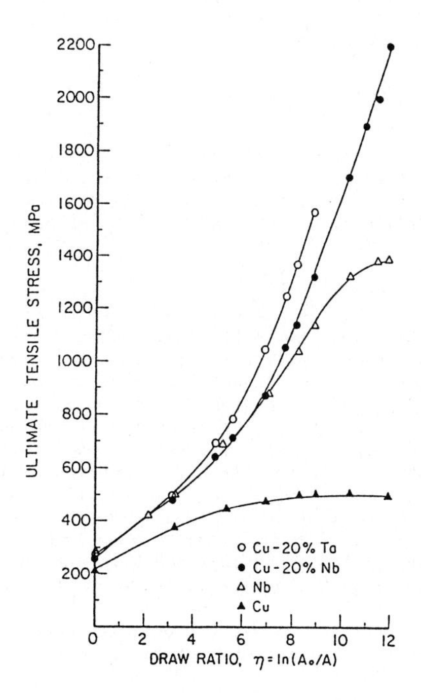

FIGURE 7

Strength of Cu-Nb composite as a function of draw ratio.

It is well known that the reflected beam can be viewed as the superposition of a ray specularly reflected from the sample surface and one which has entered the sample at the Rayleigh critical angle, propagated along the surface as a Rayleigh wave, and reradiated back to the lens. By varying the lens-solid distance and observing the constructive and destructive interferences of energy propagating along these two paths, it is possible to determine the velocity of a Rayleigh wave, propagating along the metal surface in a direction perpendicular to the generating elements of the lens. By rotating the lens, one can then obtain the angular variation of the Rayleigh velocity. This technique has the advantage that the data can be obtained on relatively small samples, although the sample surface must be very flat. A typical measurement area is a circle of a few millimeters diameter.

Figure 8 presents the angular dependence of the Rayleigh velocity measured on two Cu-20% Nb specimen, one having a draw ratio of 3.6 and the other having a draw ratio of 5.4. The obvious increase in anisotropy has been interpreted as a result of increasing texture in the copper [36]. From such data, preliminary estimates of the variation of Cu texture with draw ratio have been made [36].

Essential to the above argument is the assumption that the primary source of anisotropy is the texture of the sample. However, it is obvious from Fig. 6 that additional anisotropy may be induced by the platelike morphology of the reinforcing elements, although this may only have a weak effect on variations of the wave speed in the plane of the plate. Further work is required to develop theories fully incorporating both of these sources of anisotropy.

4. FILAMENT-WOUND, THICK COMPOSITES

Filament-wound composite structures exhibit a number of desirable structural features. There is considerable need for nondestructive techniques for both in-situ predictions of their anisotropic elastic properties as well as their internal defect structures. The anisotropic elastic properties can be obtained by extensions of techniques such as those which have been discussed previously in this paper. Here, techniques for characterization of internal porosity will be discussed.

It has recently been shown that the attenuation of ultrasonic waves passing through a porous composite varies linearly with frequency over a wide range of frequencies and that the slope of this plot is proportional to the porosity [37-39], with the proportionality constant being determined by pore morphology. Figure 9 presents an extension of this result to a 1.5" (3.81 cm) thick filament-wound composite. Again, the linear variation of attenuation is observed, with a slope of 1.76 $(\text{cm MHz})^{-1}$. From the previously determined proportionality constants, a porosity of 3.42% is estimated [40].

Considerable additional theoretical work is required to complete the understanding of this technique. In addition to questions involving the effects of pore shape and sample anisotropy, further work is required to fully explain the reasons that attenuation is proportional to frequency over such a wide range. Simple scattering models have developed a preliminary understanding [39]. However, they do not fully explain the apparent robustness of the technique.

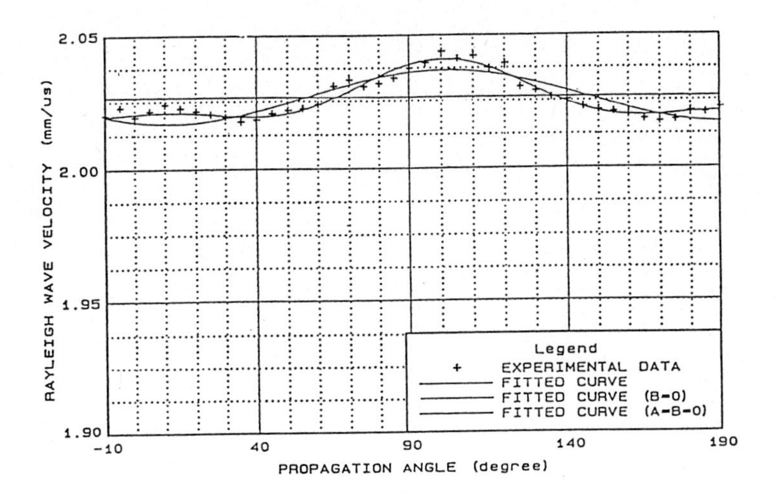

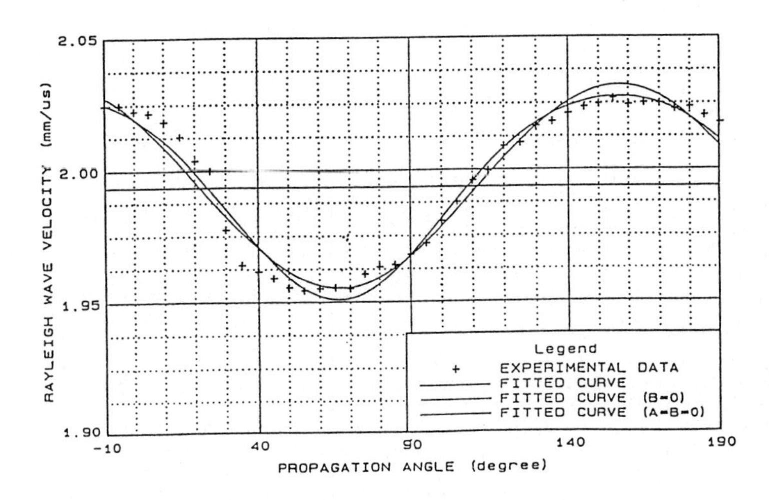

FIGURE 8

Angular variation of Rayleigh wave velocity in a heavily deformed Cu-20% Nb plate as measured with cylindrically focussed acoustic microscopic. Data has been fitted to the relationship $V_R = V_o + A \cos(2\theta - X_o) + B \cos(4\theta - X_o)$, with V_o, A, B and X_o as variables. Also included is a second theoretical fit with B = 0.

a) η - 3.6 b) η - 5.4

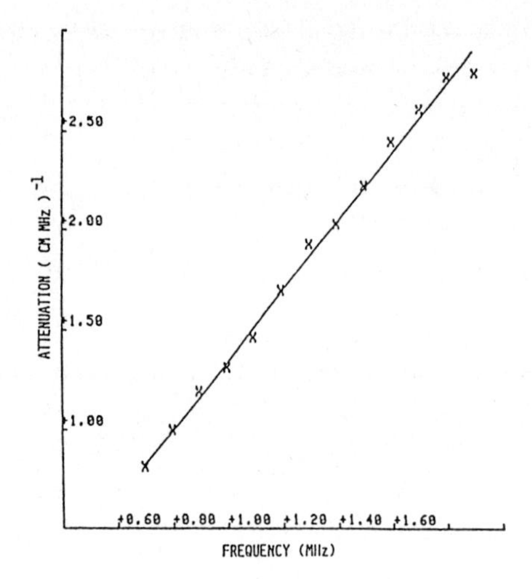

FIGURE 9

Attenuation of longitudinal waves in a 1.5" (3.81 cm) thick specimen of filament-wound composite as a function of frequency.

5. CONCLUSIONS

The above discussions are examples of the importance of ultrasonic measurements in general, and understanding of the elastic wave scattering process in particular, in studying the deformation and failure responses of structural materials. As more advanced materials continue to be developed, there will be an increasing need for such studies because of more stingent demands to detect complex damage processes in inhomogeneous microstructures.

ACKNOWLEDGEMENTS

The work on metals and metal-metal composites was supported by the USDOE, Office of Basic Energy Sciences at the Ames Laboratory, Iowa State University. The studies of metal-metal bonding and thick composites were performed in the Center for Nondestructive Evaluation at Iowa State University, with the latter supported by the Office of Naval Research. The Ames Laboratory is operated for the U.S. Department of Energy by Iowa State University under contract No. W-7405-ENG-82.

REFERENCES

1. R. B. Thompson, Trans. ASME/J. Appl. Mech. 50 (1983) 1191-1201.

2. R. B. Thompson and D. O. Thompson, Proc. IEEE 73 (1985) 1716-1755.

3. R. B. Thompson and H. N. G. Wadley, CRC Critical Reviews in Solid State and Material Sciences (in press).

4. Analytical Ultrasonics in Materials Research and Engineering, A. Vary, Ed., NASA Conference Publication 2383 (NASA, Scientific and Technical Information Branch, 1986).

5. C. O. Ruud and R. E. Green, Jr., Nondestructive Methods for Material Property Determination (Plenum Press, N.Y., 1984).

6. J. F. Bussiere, J.-P. Monchalin, C. O. Ruud and R. E. Green, Jr., Nondestructive Characterization of Materials II (Plenum Press, N.Y., 1987).

7. Nondestructive Characterization of Materials III (Springer-Verlag, Berlin, in press).

8. J. D. Achenbach, these proceedings.

9. C. A. Stickels and P. R. Mould, Metall. Trans. 1 (1970) 1303-1312.

10. P. R. Mould and T. R. Johnson, Jr., Sheet Met. Ind. 50 (1973) 328.

11. S. Kalpakjian, Manufacturing Processes for Engineering Materials, (Addison-Wesley, Reading, MA., 1985) 426-439.

12. G. J. Davies, D. J. Goodwill and J. S. Kallend, Metall. Trans. 3 (1972) 1627-1631.

13. R.-J. Roe, J. Appl. Phys. 36 (1965) 2024-2031.

14. P. R. Morris, J. Appl. Phys. 40 (1969) 447-448.

15. C. M. Sayers, J. Phys. D: Appl. Phys. 15 (1982) 2157.

16. Modul-$\overline{r}$, Tinius-Olsen, Willow Grove, PA.

17. R. B. Thompson, S. S. Lee and J. F. Smith, Ultrasonics 25 (1987) 133-137.

18. R. B. Thompson,. J. F. Smith, S. S. Lee and G. C. Johnson, Metall. Trans. (in press).

19. H. M. Ledbetter and M. W. Austin, Physica Status Solidi 104 (1987) 203-212.

20. A. Granato and K. Lücke, J. Appl. Phys. 7 (1956) 789.

21. G. C. Johnson, J. Acoust. Soc. Am. 70, 591-595 (1981). Also J. Appl. Mech. 50 (1983) 689-691.

22. Y.-H. Pao and V. Gamer, Materials Science Center Technical Report 5121 (Cornell University, Ithaca, NY, 1983).

23. R. B. Thompson, W. A. Spitzig and T. A. Gray, in Review of Progress in Quantitative Nondestructive Evaluation, Vol. 5B, D. O. Thompson and D. E. Chimenti, Eds. (Plenum Press, NY, 1986) 1643-1653.

24. W. A. Spitzig, R. B. Thompson and D. C. Jiles, Metall. Trans. 20A (1989) 571-578.

25. C. M. Sayers and R. L. Smith, Ultrasonics 19 (1982) 201-205.

26. A. N. Norris, in Review of Progress in Quantitative Nondestructive Evaluation, Vol. 5B, D. O. Thompson and D. E. Chimenti, Eds. (Plenum Press, NY, 1986) 1609-1616.

27. H. M. Ledbetter, R. J. Fields and S. K. Datta, Acta Metall. 35 (1987) 2393-2398.

28. O. Buck, R. B. Thompson, D. K. Rehbein and L. Van Wyk, these proceedings.

29. J.-M. Baik and R. B. Thompson, J. Nondestr. Eval. 4, (1984) 177-196.

30. D. D. Palmer, D. K. Rehbein, J. F. Smith and O. Buck, J. Nondestr. Eval. 7 (1988) 167-174.

31. J. D. Verhoeven, F. A. Schmidt, E. D. Gibson and W. A. Spitzig, J. of Metals 38 (1986) 20.

32. W. A. Spitzig, A. R. Pelton, F. C. Laabs, Acta Met. 35, (1987) 2427.

33. W. A. Spitzig, Scripta Met. 23 (1989) 1177.

34. W. A. Spitzig, P. D. Krotz, Acta Met. 36 (1988) 1709.

35. G. A. D. Briggs, An Introduction to Scanning Acoustic Microscopy, Royal Microscopical Society Microscopy Handbook 12 (Oxford University Press, Oxford, 1985).

36. R. B. Thompson, W. A. Spitzig, Y. Li, G. A. D. Briggs, A. Fagan and J. Kushibiki, in Review of Progress in Quantitative Nondestructive Evaluation Vol. 9, D. O. Thompson and D. E. Chimenti, Eds. (Plenum Press, NY, in press).

37. D. K. Hsu and S. M. Nair, in Review of Progress in Quantitative Nondestructive Evaluation Vol. 6B, D. O. Thompson and D. E. Chimenti, Eds. (Plenum Press, NY, 1987) 1185-1193.

38. D. K. Hsu, in Review of Progress in Quantitative Nondestructive Evaluation Vol 7B, D. O. Thompson and D. E. Chimenti, Eds. (Plenum Press, NY, 1988) 1063-1069.

39. S. M. Nair, D. K. Hsu and James H. Rose, J. Nondestr. Eval. 8 (in press).

40. D. K. Hsu and A. Minachi, in Review of Progress in Quantitative Nondestructive Evaluation, D. O. Thompson and D. E. Chimenti, Eds. (Plenum Press, NY, in press).

Elastic Waves and Ultrasonic Nondestructive Evaluation
S.K. Datta, J.D. Achenbach and Y.S. Rajapakse (Editors)

RECENT DEVELOPMENTS IN QUANTITATIVE AE

Wolfgang SACHSE

Department of Theoretical and Applied Mechanics
Cornell University
Ithaca, New York - 14853, U. S. A.

The aim of most *quantitative* acoustic emission measurements is to obtain a solution to the *inverse source* problem by using signal processing algorithms to recover from the signals detected on the surface of a specimen, the location of the emitting source as well as its temporal and spatial characteristics. This paper presents a brief summary of recent results obtained from such measurements. It is demonstrated in this paper, however, that in some cases, quantitative information about the mechanical properties of a material can be determined from conventional AE measurements provided that the data can be interpreted with a suitable micro-mechanical model. Also described in this paper is a measurement system incorporating a well-characterized AE source via the *point-source/point-receiver* technique to determine the wave propagation characteristics and elastic properties of the medium. An alternative development to these direct inverse approaches is described which utilizes a testing procedure and a processing scheme resembling an intelligent system to process the signals like a *neural network.*

1.0 INTRODUCTION

Acoustic emissions (AE) are the transient elastic waves emitted when there is a sudden, localized change of stress in a material. Several advances in the last decade have given impetus to the further development of such measurements. These include the development of well-characterized, high-fidelity point receivers, a theory for analyzing the propagation of transient elastic waves through a bounded medium and the development and implementation of appropriate signal processing algorithms. By incorporating into the measurement system an artificial point source, the *point-source/point-receiver*, *(PS/PR)*, technique results by which the propagation of elastic waves from the source to the receiver can be studied and the properties of the material determined.

All active as well as passive ultrasonic systems are comprised of three components - *source*, *structure* or *test specimen* and *sensor*. Truly *quantitative* ultrasonic measurements are possible provided that the transfer characteristics of two of the three elements comprising the system are known so that the characteristics of the third can be recovered by signal processing techniques. For quantitative, active ultrasonic measurements, the source and receiver characteristics must be known if the detailed spatial variation of the properties of the medium are to be recovered and the *inverse medium* problem solved. For quantitative, passive acoustic emission measurements, the transfer characteristics of the medium between the source and the receiver and the characteristics of the sensor must be known if a solution to the *inverse source* problem is to be found. It needs, however, to be emphasized that for many applications, the detailed solution to an inverse problem is not sought and for these, the signal processing is usually greatly simplified.

In the following sections, we briefly summarize the recent advances obtained for locating and characterizing sources of emission in materials. We describe procedures for locating sources in a ***single fiber composite*** specimen and using these to determine the interfacial strength between the fiber and its matrix. Also summarized are the recent procedures for locating sources of emission in anisotropic solids. We then outline the procedure for recovering the parameters by which an AE source can be described which permits investigation of the dynamic fracture behavior of materials. In the third section, we describe the characteristics of the *PS/PR* technique and illustrate its application to determine the frequency-dependent wavespeeds and attenuation values of a material and the elastic constants of an anisotropic composite from wave arrival measurements made in non-principal directions in it. In the fourth section we briefly review the principal elements of the *neural-like* signal processing procedure by which ultrasonic signals can be analyzed. The example presented deals with the location and features of a simulated AE source. Some concluding remarks are in the final section.

2.0 CHARACTERIZING SOURCES OF EMISSION

2.1 Source Location

Conventional AE source location techniques are based on the triangulation procedure utilizing measurements of the arrival time of a particular wave mode at a sensor [1]. The basis of the algorithm is that if a source of emission is located at $\mathbf{x}'$, (i.e. x', y', z') in a structure and if the AE is detected by Q sensors whose locations are specified by $\mathbf{x}^{(q)}$ ($q = 1, 2, 3, \ldots, Q$), then the distance between the q-th sensor and the source can be determined from measurements of the arrivals, $t^{(q)}$, of particular waves at the sensor according to $d^{(q)} = |\mathbf{x}^{(q)} - \mathbf{x}'| = (t^{(q)} - t_0)\, c$, where c is the speed of propagation of the wave and t_0 is the unknown time origin of the source. In the general problem, there are three unknown source coordinates, and hence, the arrival times determined from the waveforms detected at four sensor positions are required to uniquely locate a source. Recent work has utilized the signals detected at more than four sensors resulting in an overdetermined system of equations which can be optimally solved via a least-squares processing algorithm. Critical to the above procedure is knowledge of the speed of propagation of a particular wave mode. When the signals are detected at widely different distances from the source point, geometric wave dispersion effects may dominate the signals. The influence of these can be minimized by utilizing a sensor array such that the source of emission is always exterior to the array [2].

Determination of the location of a source of emission in an anisotropic material is complicated by the fact that the speed of propagation of each wave mode is dependent on the propagation direction in the material. An algorithm for solving this problem has recently been formulated [3]. By including the orientation dependence of the wavespeed in the triangulation equation, a system of non-linear transcendental equations is formed which are functions of the source coordinates, the geometric parameters related to the array and the structure as well as the measured wave arrival-times. The source location algorithm is based on minimizing the Euclidian functional of these equations and using the *Newton-Raphson* method to recover the coordinates of the source. The procedure has been successfully demonstrated with simulated data in transversely isotropic as well as orthorhombic materials and, as Fig. 1 demonstrates, with real data which was obtained with artificial sources of AE in a plate of transversely isotropic fiberglass/Polyester, *Extren* [3].

Another recent development is the use of source location information to determine the fiber/matrix interfacial shear strength, IFSS. The procedure involves monitoring the pro-

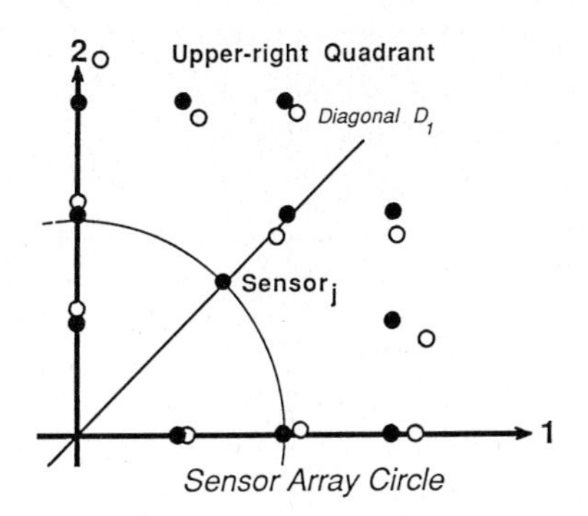

Figure 1: Source location in a transversely isotropic material; • - Actual; ∘ - Retrieved.

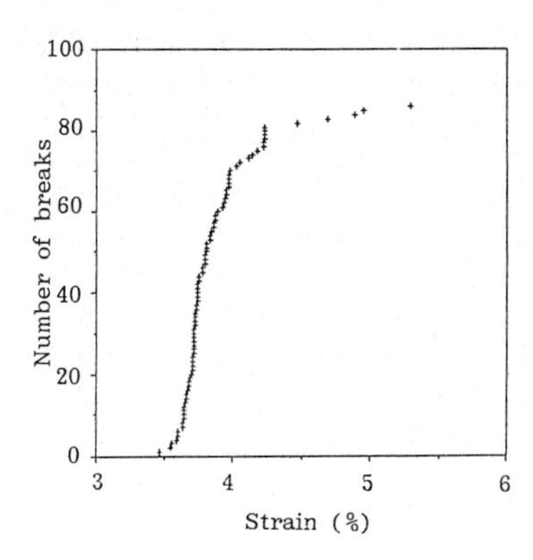

Figure 2: Fiber breaks in a single fiber composite specimen as a function of strain.

gression of fiber failures of a single fiber imbedded in a ductile matrix under tensile loading. Load transfer occurs between the fiber segments by shear until the segments become too short to break. This fiber length corresponds to the ***critical fragmentation length*** or the ***effective load transfer length***, l_c, of the fiber. Using a simple shear lag analysis, one can compute the failure stress in shear τ_c of the interface or matrix [4]. This is given by

$$\tau_c = \frac{\sigma_c d}{2l_c} \tag{1}$$

where d and σ_c are the fiber diameter and fiber failure stress, respectively. When the stress transfer between the fiber and the matrix is controlled by the fiber/matrix interfacial strength, then τ_c is a measure of adhesion of the fiber/matrix interface [5]-[7]. Recent work has considered the dependence of the fiber failure strength on the mean critical fragment length as well as the statistical variation of the failure strength which results in a modification of Eq. (1) [8,9]. The critical fragmentation length of the fiber can be determined by using a one-dimensional AE source location procedure to monitor the progression of fiber failure in the single fiber composite specimen [10]. The advantages of the AE method include its use for measurements in non-transparent matrix materials, its application to study fibers whose breaks are fibrillar and spread over a length of 20-50 fiber diameters (such as Kevlar) and the possibility of automating the measurements.

A study of E-glass fibers in epoxy is reported in Refs. [10,11]. Here are summarized the results obtained with Kevlar/epoxy specimens. The fibers were about 12 μm in diameter and they were left untreated or treated with NH_3 plasma for 1 or 10 minutes. The number of fiber breaks as a function of strain is shown in Fig. 2. Two distinct waveforms were detected in these tests with the maximum amplitude of the signal in one group four to five times greater than that of the other. These signals correspond, respectively, to complete and partial breaks of the Kevlar fibers [12]. The interfacial shear strength, IFSS, obtained in the untreated specimens ranged from 42.8 MPa when both the total and the partial breaks are included in the fragmentation length determination and 37.0 MPa if only the total breaks are. After one minute of NH_3 plasma treatment, the IFSS increases to 50.6 MPa and 42.7 MPa, respectively. However, if the fiber is treated for 10 minutes, the strength decreases to 47.1 MPa and 43.1 MPa, respectively. These results demonstrate a potential role of such a test in the development of new composites.

2.2 Quantitative Source Characterization

In the quantitative AE source characterization procedure, one parameterizes the source by forces and moments, which are directly related to the micromechanics of the source, c.f. [13]. The displacement signals u_i detected at a receiver location, $\mathbf{r}$, in a structure from an arbitrary source $f(\mathbf{r}',t)$ located at $\mathbf{r}'$ having source volume V_0 can be written as [14]

$$u_i(\mathbf{r},\ t) \ = \int_{V_0}\int_0^t f_j(\mathbf{r}',\tau)G_{ij}\ (\mathbf{r}|\mathbf{r}',t-\tau)\ d\tau\ dV' \tag{2}$$

The integrals appearing in this equation extends over space and time. The term G_{ij} represents the dynamic Green's function of the structure which can be computed [15] or determined experimentally for particular sources and source-receiver configurations [16]. In the inverse source problem, the force density is recovered from the measured AE displacement signals and inverting Eq. (2). This is, however, an ill-posed inverse problem for which only approximate methods of solution have been developed. Multi-component *vector* and *tensor* point sources, such as a small crack in a material, for which the time and spatial components of the source are usually separable, require the inversion of the signal detected at the q-th sensor as

$$v^{(q)}(t) \ = \ \sum_{j=1}^{3}\sum_{k=1}^{3} m_{jk}\cdot S(t) * G_{ij,k}^{(q)}(t) * T_i^{(q)}(t) \qquad \text{(No summation over } i) \tag{3}$$

where the m_{jk} are called the *moment tensor coefficients*, $S(t)$ is the time function of the source, $G_{ij,k}^{(q)}(t)$ are the spatial derivatives of the Green's functions corresponding to a particular component of m_{jk} and a source-receiver geometry and $T_i^{(q)}(t)$ is the u_i-displacement–voltage transfer function of the sensor. To recover the m_{jk} as well as the time-function $S(t)$ of the source, several processing algorithms have been developed, which, in general, require that the number of measurement channels equals or exceeds the total number of unknowns. The various algorithms by which Eq. (3) has been inverted are reviewed in Ref. [17].

Once the moment tensor components have been determined, several parameters describing the formation of the crack can be recovered. By diagonalizing m_{jk}, three eigenvalues and their corresponding eigenvectors are determined and, as shown by Hsieh [18], from these, the crack orientation and mode - tensile, shear or mixed - can be recovered. An example is shown in Fig. 3. By combining the results of AE measurements with a fracture mechanics model, such as that of *Dugdale-Barenblatt*, all the important parameters related to the formation of a crack in a brittle solid can be recovered [19]. Included are the critical rupture stress, crack tip opening displacement, the effective microstructural gauge length and the time-dependent velocity of the crack [19].

The problem of characterizing a sensor of finite aperture is equivalent to characterizing a source of finite extent. A solution of this problem has been given in Ref. [20] and its application to characterize a sensor of finite aperature is in Ref. [21].

3.0 *PS/PR* MEASUREMENTS FOR CHARACTERIZING MATERIALS

The development of a theory of transient ultrasonic signals in an ideal, bounded medium has permitted its application to interpret the signals propagating in a real material from a point source to a receiver of known transfer characteristics and small aperture. The result has been a powerful, new *quantitative* ultrasonic *PS/PR* measurement technique capable of recovering the characteristics of waves propagating through a material. Simultaneous

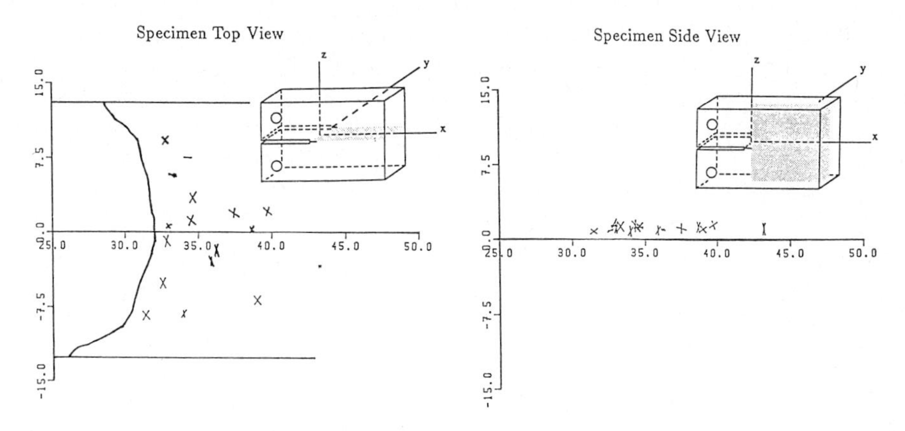

Figure 3: 7075-T6 Aluminum crack orientation and mode identification

longitudinal and shear wave measurements can be extracted from one waveform, and if an array of sensors is used, the orientation dependence of the attenuations and wavespeeds can be determined from just one test. Furthermore, the technique shows particular promise for measurements in thick and highly absorptive composite materials since broadband excitations rich in low frequencies can be generated. Also, by scanning the source and/or receiver, it can be used to image spatial variations of the attenuations and wavespeeds in inhomogeneous materials.

The wavespeeds are determined from measurement of the arrival-time of particular rays in the detected signals and knowledge of the propagation distance. An alternative to this, which is especially useful when the wavespeeds are functions of frequency, is a Fourier phase spectral analysis of particular wave arrivals. The signals from a point-source exhibit significant geometric dispersion effects. However, the analysis of the propagation of transient ultrasonic waves in a viscoelastic plate has permitted the formulation of an equation relating the Fourier phase function of the displacement signal of a particular wave arrival and the dispersion relation of the material in which these effects are included [22]. Similarly, if one compares the amplitude of a particular ray arrival with that of the ideal, un-attenuated waveform, one can recover the materials attenuation coefficient [22]. An alternative to the procedure outlined above which does not require processing of a velocity (or displacement) signal, but rather utilizes the simulated AE signals detected at two positions from the source has recently been described [23]. A comparison between the input phase velocity and attenuation and that recovered from synthetic waveforms computed on the basis of a Voigt model shows good agreement, demonstrating the correctness of the procedure.

An important application of PS/PR wavespeed measurements made along non-principal directions of a specimen of an anisotropic material is to the determination of the matrix of elastic constants which are used to describe the elastic properties of the material. An algorithm for recovering this matrix for a material of arbitrary symmetry has recently been described [24,25]. By constructing the characteristic equation associated with the *Green-Christoffel* tensor, a functional is found which is related to the unknown elastic constants. An algorithm has been developed based on the *Newton-Raphson* method which minimizes this functional, thus permitting recovery of the optimized elastic constants corresponding to the *PS/PR* measured wavespeeds. Use of this algorithm with synthetic data has been to the

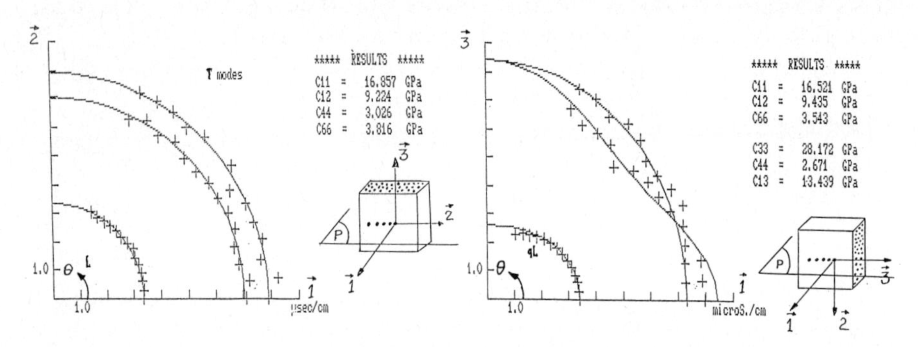

Figure 4: Measured, recovered slowness values in fiberglass/polyester; (a) Scan in the isotropic plane (1,2); (b) Scan in the anisotropic plane (1,3).

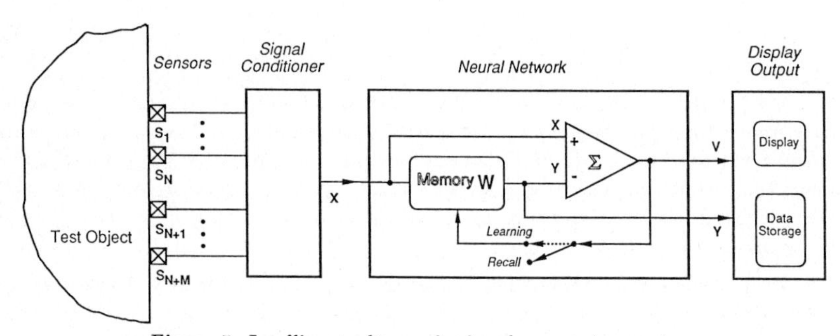

Figure 5: Intelligent ultrasonic signal processing system.

characterization of transversely isotropic and orthorhombic materials. The simulations were used to demonstrate the convergence of the procedure even when the measured data was noise-corrupted. The results of actual measurements made with laser-generated ultrasonic signals in a transversely isotropic material are shown in Figs. 4(a) and (b) [26].

4.0 INTELLIGENT, NEURAL-LIKE PROCESSING OF ULTRASONIC SIGNALS

The quantitative source characterization studies described above are experimentally and computationally a prohibitive task for most practical applications. It is for this reason that there has been a great deal of interest in the application of neural-like processing schemes as an alternative. In this approach, a system is taught and a *memory* is developed corresponding to known sources and known propagation conditions. Signals with missing information related either to the source or the material can then be processed to recover this missing information. It has been shown that such processing yields an optimal solution to an inverse problem from the given data [27].

Our application of this approach is shown schematically in Fig. 5. Its features, which have been detailed in several publications [27]-[29], uses some of the fundamental principles of neural networks [30]. Briefly, we assume that an AE event can be characterized by a finite set of data supplied from an array of sensors together with selected features of the source. These data constitute a N-dimensional *pattern* vector given by $\mathbf{X} = (X_1, X_2, \ldots, X_N)$. When there are Q sensors, the n-th component of this vector is specified by

$$X_n = \{v_n^{(1)}(t),\ v_n^{(2)}(t), \ldots, v_n^{(Q)}(t);\ g_n\} \tag{4}$$

which is applied as input to the processor shown in Fig. 5. For each input, the system responds with an output vector $\mathbf{Y}$ which is an estimator of the input $\mathbf{X}$ and is of the same dimension. It can be determined from the linear matrix equation

$$\mathbf{Y} = \mathbf{W} \cdot \mathbf{X} \tag{5}$$

where the matrix $\mathbf{W}$ represents the response function or *memory* of the system. In order to obtain an *associative* operation of the processing system, we assume that the system adapts to the input vectors such that the *discrepancy* or *novelty* between the input and output vectors given by

$$\mathbf{V} = \mathbf{X} - \mathbf{Y} = \mathbf{X} - \mathbf{W} \cdot \mathbf{X} \tag{6}$$

is reduced with a repetition of the inputs. This is possible with the feedback loop shown in the neural network in the figure. The adaptive law of the system which governs how the memory develops is similar to that used in other applications and is expressed by [31]

$$\Delta \mathbf{W} = C \mathbf{V} \otimes \mathbf{X}^T \tag{7}$$

Here C is an *adaptation* constant and $\mathbf{X}^T$ denotes the transposed pattern vector. It is seen that as input vectors are presented to the system, the result is the formation of the memory matrix $\mathbf{W}$ from an initially empty state, i. e. $\mathbf{W}_0 = 0$. This resembles a *learning* process during which, the system adapts to the input so that the output vectors resemble the input vectors. If a new input vector is presented to the system, the generated output vector is a linear mixture of all the previously presented pattern vectors which most closely correlate to the new one.

On the other hand, if the processing system is presented a new pattern vector which is similar to a previously presented one but with some of the components missing, then the output vector will contain those components from that previously presented pattern vector which most closely resembles the unknown. Such a recall is characteristic of an *associative* operation of intelligent systems [32]. Hence, one can use this processing procedure to obtain a solution to the forward elastodynamic problem from input of the source characteristics, or one can use it to obtain a solution to the inverse source problem from the ultrasonic signals presented to it.

To date, the operation of such a processing system has been demonstrated with several simple source location and source characterization problems [27]-[29]. They have included: Linear and two-dimensional source location problems; combined one-dimensional source location and scalar source characterization problems and the characterization of simple vector sources. The example, depicted in Fig. 6, was obtained from measurements made in a thick, anisotropic graphite-epoxy composite for which no Green's function is yet available. The system was trained with two source parameters: The location of the point of impact and the size of a steel ball causing the impact.

Two miniature, broadband piezoelectric transducers were mounted 6 in. apart on the surface of the slightly curved composite specimen which was 1.5 in. thick. The signals from the two sensors were amplified and recorded by a multi-channel waveform recording system.

The data record length of the waveforms was reduced so that it would not exceed 128-points. To form the pattern vector, encoded information about the location and size of the source was added. Learning was performed by 15 signals generated by three different sized balls impacting at five points 1-inch apart on the line between the two sensors. The training signals are shown in Fig. 6(a). The developed memory is in Fig. 6(b) and ten pattern vectors which were used test the system and the recovered encoded source information is in Fig. 6(c). A tabulation of the results is given in Fig. 6(d). It is seen that while in some cases, there are differences between the actual and recovered data, the results look promising. Furthermore, the last two cases are for impacts at positions which were not previously learned by the system. Even for these, the system was able to recover an approximate value of the source location and ball size. The application of neural-like intelligent processing of AE waveforms appears to be a promising means for processing ultrasonic waveforms.

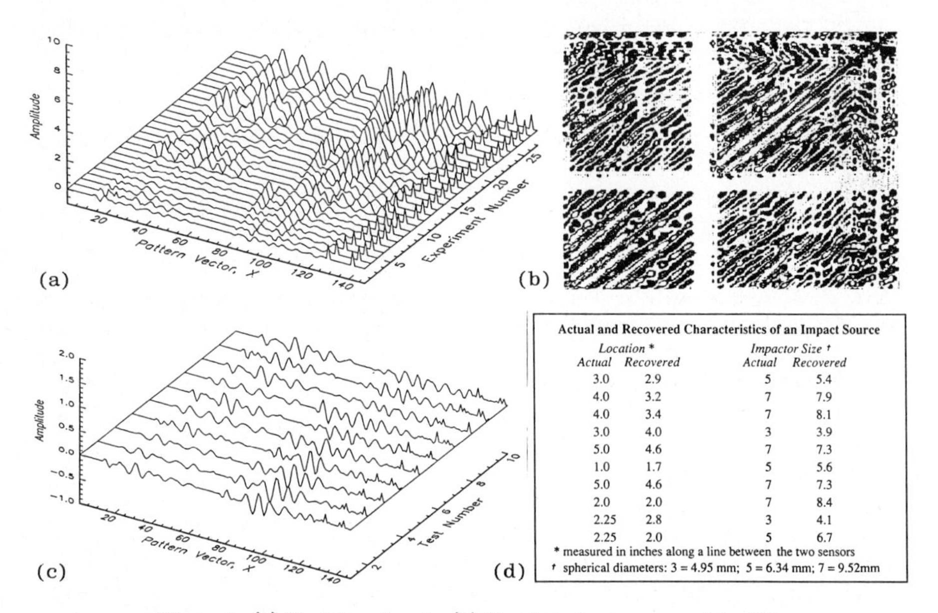

Actual and Recovered Characteristics of an Impact Source

Location *		Impactor Size †	
Actual	*Recovered*	*Actual*	*Recovered*
3.0	2.9	5	5.4
4.0	3.2	7	7.9
4.0	3.4	7	8.1
3.0	4.0	3	3.9
5.0	4.6	7	7.3
1.0	1.7	5	5.6
5.0	4.6	7	7.3
2.0	2.0	7	8.4
2.25	2.8	3	4.1
2.25	2.0	5	6.7

* measured in inches along a line between the two sensors
† spherical diameters: 3 = 4.95 mm; 5 = 6.34 mm; 7 = 9.52mm

Figure 6: (a) Training signals; (b) Developed memory matrix, **W**; (c) Input test waveforms, recovered source information; (d) Measurement results.

5.0 SUMMARY

This paper has presented a brief summary of recent results obtained from quantitative AE measurements. It was demonstrated however, that in some cases, quantitative information about the interfacial shear properties of a fiber/matrix interface could be determined from conventional AE measurements when these were coupled with a suitable micro-mechanical model. Also described were the source location results obtained in an anisotropic material. Some of the characteristics of the *PS/PR* technique were described by which the characteristics of the wave propagation and the elastic constants of an anisotropic solid can be determined. Further, an alternative development to these direct inverse approaches was described which utilizes a testing procedure and a processing scheme resembling an intelligent system to process the signals like a *neural network* which shows great promise in processing ultrasonic signals.

ACKNOWLEDGEMENTS

I am pleased to acknowledge my colleagues Drs. B. Castagnede, I. Grabec, P. N. Hsieh, K. Y. Kim, P. Petrina, A. Netravali and R. Weaver whose work I have cited. The support of the Office of Naval Research (Physical Acoustics and Solid Mechanics Programs) is also gratefully acknowledged. Portions of this work were supported by the National Science Foundation through grants to the College of Engineering and the Materials Science Center at Cornell University.

References

[1] Baron, J. A. and Ying, S. P., "Acoustic emission source location", in *Non-destructive Testing Handbook*, Vol. 5, Sect. 6 (ASNT, Columbus, OH, 1987), pp. 136-154.

[2] Sachse W. and Sancar, S., "Acoustic emission source location on plate-like structures using a small array of transducers", U. S. Patent No. 4,592,034 (1986).

[3] Castagnede B., Sachse, W. and Kim, K. Y., *J. Acoust. Soc. Am.*, **86**(3), 1161-1171 (1989).

[4] Kelly, A. and Tyson, W. R., *J. Mech. Phys. Solids*, **13**, 329-350 (1965).

[5] Drzal, L. T., Rich, M. J. and Lloyd, P. F., *J. Adhesion*, **16**, 1-30 (1982).

[6] Drzal, L. T., Rich, M. J., Koenig, M. F. and Lloyd, P. F., *J. Adhesion*, **16**, 133-152 (1983).

[7] Bascom W. D. and Jensen, R. M., *J. Adhesion*, **19**, 219-239 (1986).

[8] Netravali, A. N., Henstenburg, R. B., Phoenix, S. L. and Schwartz, P., "Interfacial Shear Strength Studies using the Single-Filament-Composite Test I. Experiments on Graphite Fibers in Epoxy". In Press: *Polymer Composites.*

[9] Petrina, P. and Phoenix, S. L., "Single-Fiber-Composite Studies: Graphite Fibers in Epoxy". In Preparation.

[10] Netravali, A. N., Topoleski, L. T. T., Sachse, W. and Phoenix, S. L., *Composites Science and Technology*, **35**, 13-29 (1989).

[11] Netravali, A. N., Li, Z.-F., Sachse, W. and Wu, H. F., "Interfacial Shear Strength of Epoxy and High-strength Fibers Using Acoustic Emission Technique", Materials Science Center Report #6746, Ithaca, NY (August 1989). Submitted for publication.

[12] Li, Z.-F., Netravali, A. N. and Sachse, W. "Effect of Ammonia Plasma Treatment on Ultra-high Strength Polyethylene and Kevlar Fibers and their Interfaces with Epoxies", Materials Science Center Report #6750, Ithaca, NY (August 1989). Submitted for publication.

[13] Scruby, C. B., "Quantitative Acoustic Emission Techniques", Chapt. 4, in *Research Techniques in Nondestructive Testing*, Vol. VIII, R. S. Sharpe, (ed.), (Academic Press, London, 1985), pp. 141-210.

[14] Aki, K. and Richards, P. G., *Quantitative Seismology: Theory and Methods*, Vol. 1, Freeman, San Francisco (1980), Chapt. 3.

[15] Ceranoglu, A. N. and Pao, Y. H., *ASME J. Appl. Mech.* **48**, 125-147 (1981).

[16] Michaels, J. E., Michaels, T. E. and Sachse, W., *Materials Evaluation*, **39**(11), 1032-1037 (1981).

[17] Sachse, W., "The Processing of AE Signals", in: *Progress in Acoustic Emission IV*, I. Kimpara, Ed., (Japanese Society of NDI, Tokyo, 1988), pp. 26-38.

[18] Hsieh, P. N., *Quantitative Acoustic Emission Source Characterization in an Aluminum Specimen*, Ph. D. Dissertation, Cornell University, Ithaca, NY (August 1987).

[19] Kim, K. Y. and Sachse, W., *J. Appl. Phys.*, **65**(11), 4234-4244 (1989).

[20] Chang, C. and Sachse, W., *J. Acoust. Soc. Am.*, **77**(4), 1335-1341 (1985).

[21] Chang, C. and Sachse, W., "Separation of spatial and temporal effects in an ultrasonic transducer", in *Review of Quantitative Non-destructive Evaluation*, **5A**, D. O. Thompson and D. E. Chimenti, (eds.), (Plenum Press, New York, 1985), pp. 139-143.

[22] Sachse, W. and Kim, K. Y., *Ultrasonics*, **25**, 195-203 (1987).

[23] Weaver, R. L., Sachse, W. and Niu, L., *Journal of the Acoustical Society of America*, **85**(6), 2255-2261; 2262-2267 (1989).

[24] Castagnede, B. and Sachse, W., "Optimized determination of elastic constants of anisotropic solids from wavespeed measurements", in *Review of Progress in Quantitative Nondestructive Evaluation*, **8B**, D. O. Thompson and D. E. Chimenti, (eds.), (Plenum Press, New York, 1988), pp. 1855-1862.

[25] Castagnede, B., Jenkins, J. T., Sachse, W. and Baste, J.-S., "Optimal Determination of the elastic constants of composite materials from ultrasonic wavespeed measurements", Materials Science Center Report #6688, (May 1989). In Press: *Journal of Applied Physics.*

[26] Castagnede, B., Sachse, W. and Thompson, M. O., "Determination of the elastic constants of anisotropic materials via laser-generated ultrasound", in *UI'89 Conference Proceedings*, (Butterworth Scientific Ltd., Guildford, Surrey, UK, 1989).

[27] Grabec, I. and Sachse, W., *J. Appl. Phys.*, **66**, - (1989).

[28] Grabec, I. and Sachse, W., *J. Acoust. Soc. Am.*, **85**(3), 1226-1235 (1989).

[29] Sachse, W. and Grabec, I., "Experimental Characterization of Ultrasonic Phenomena by a Neural-like Learning System", in *Review of Progress in Quantitative Nondestructive Evaluation*, **8A**, D. O. Thompson and D. E. Chimenti, (eds.), (Plenum Press, New York, 1988), pp. 649-656; and "Application of an Intelligent Signal Processing System to Acoustic Emission Analysis", in: *Progress in Acoustic Emission IV*, I. Kimpara, (ed.), (Japanese Society of NDI, Tokyo, 1988), pp. 75-80.

[30] Kohonen, T., *Proc. Intl. Conf. Neural Networks*, San Diego, CA (1987), pp. 1-77.

[31] Kohonen, T., *Neural Networks*, **1**, 3-16 (1988).

[32] Kohonen, T., *Self-organization and Associative Memory*, (Springer Verlag, New York, 1984).

Elastic Waves and Ultrasonic Nondestructive Evaluation
S.K. Datta, J.D. Achenbach and Y.S. Rajapakse (Editors)

STRAIN-INDUCED ANISOTROPY IN ELASTIC WAVE PROPAGATION

Manfred BRAUN

Department of Mechanical Engineering
University of Duisburg
Duisburg, Federal Republic of Germany

An isotropic elastic material, which is exposed to a static deformation, loses its isotropy in the sense that superimposed elastic waves exhibit anisotropic behavior. The transition from isotropy to anisotropy due to a small triaxial predeformation is analyzed. Special attention is payed on how the polarization of quasitransverse waves depends on the direction of the wave normal.

1. INTRODUCTION

In an isotropic elastic material small disturbances of the stress-free state are propagated as longitudinal and transverse waves. The propagation speed of the transverse waves does not depend on their polarization, i. e. disturbances of arbitrary transverse directions are propagated at the same speed. When the material is deformed, the propagation speeds of superimposed waves are changed and the amplitudes deviate from strictly longitudinal and transverse directions, in general. Moreover, the common speed of transverse waves is split in two distinct propagation speeds, the corresponding waves having distinct polarizations, which emerge from the manifold of arbitrary transverse directions.

Both the propagation speeds and the corresponding polarization directions depend on the direction of the wave normal, thus indicating anisotropic behavior. In contrast to *material anisotropy*, which is a material property, the *strain-induced anisotropy* is generated by the deformation to which an inherently isotropic material is exposed.

The object of this paper is to analyze the transition from the isotropic behavior of the material in its natural state to the anisotropic behavior in its deformed state. To this end the acoustic tensor of the deformed material is expanded with respect to the strain tensor. For a given wave normal the propagation speeds and the corresponding polarization directions are determined by the eigenvalues and eigenvectors of the acoustic tensor. A perturbation procedure applied to the eigenvalue problem yields the change of the propagation speeds due to a small elastic predeformation.

2. NOTATIONS AND GOVERNING EQUATIONS

The time-dependent deformation of an elastic body is decribed by functions $x^k = x^k(\xi^\alpha, t)$, which give the coordinates x^k of the actual position at time t of a material particle, which had been located at coordinates ξ^α in the stress-free state. The *local* deformation is described by the deformation gradient $\mathbf{F}$, a tensor with components

$$F^k{}_\alpha = \frac{\partial x^k}{\partial \xi^\alpha}. \tag{2.1}$$

The material properties of the elastic body are characterized by the mass density ϱ and the strain energy density function

$$W = W(\mathbf{F}). \tag{2.2}$$

Both densities are understood per unit volume of the *undeformed* material. If the material under consideration is isotropic, the strain energy density is expressible as a function

$$W = W(I_1, I_2, I_3), \tag{2.3}$$

where I_1, I_2, I_3 are the principal invariants of the right CAUCHY-GREEN tensor

$$\mathbf{C} = \mathbf{F}^\mathsf{T}\mathbf{F}. \tag{2.4}$$

In the equations of motion the strain energy function W appears in form of its second derivative, a fourth-order tensor with components

$$A_k{}^\alpha{}_l{}^\beta = \frac{\partial^2 W}{\partial F^k{}_\alpha \partial F^l{}_\beta}, \tag{2.5}$$

which still depend on the deformation gradient $\mathbf{F}$.

The theory of waves propagating in a deformed elastic material is well established [6]. In the present paper the *material* description is preferred, which means that all quantities are referred to the undeformed material. Especially, the normal vector $\mathbf{n}$ and the speed of propagation U refer to the image of the wave in the undeformed configuration rather than to the actual wave propagating in the deformed material. The propagation speed U and the amplitude or polarization vector $\mathbf{d}$ of a wave are governed by the propagation condition

$$[\mathbf{Q}(\mathbf{n}) - \varrho U^2 \mathbf{1}]\,\mathbf{d} = 0, \tag{2.6}$$

where the acoustic tensor $\mathbf{Q}$ has the components

$$Q_{kl} = A_k{}^\alpha{}_l{}^\beta n_\alpha n_\beta. \tag{2.7}$$

The acoustic tensor depends on the unit normal vector $\mathbf{n}$ of the wave and, in virtue of the elasticity tensor (2.5), also on the static deformation, onto which the wave is superimposed.

3. SERIES EXPANSION OF THE ACOUSTIC TENSOR

The polar decomposition theorem states that the deformation gradient can be written as a product of an orthogonal rotation tensor $\mathbf{R}$ and a symmetric stretch tensor. Since, in the stress-free state, the stretch tensor is unity, its deviation from the unity tensor is introduced as strain tensor $\mathbf{E}$. This correponds to BIOT's definition of finite strain. The polar decomposition of the deformation gradient then reads

$$\mathbf{F} = \mathbf{R}(\mathbf{1} + \mathbf{E}). \tag{3.1}$$

Without loss of generality it can be assumed that there is locally no rotation, viz. $\mathbf{R} = \mathbf{1}$.

Through the equations (3.1), (2.2), (2.5) and (2.7) the acoustic tensor $\mathbf{Q}$ depends on the strain tensor $\mathbf{E}$. Since the strains are always much less than unity, the acoustic tensor can be expanded with respect to the strain tensor. A lengthy but straightforward calculation yields a series expansion of the form

$$\begin{aligned} \mathbf{Q}(\mathbf{n}) \approx \; & \mu\mathbf{1} + (\lambda + \mu)\,\mathbf{n} \otimes \mathbf{n} + \\ & + [(\lambda + \nu_2)\,\mathrm{tr}\,\mathbf{E} + 2\,(\mu + \nu_3)\,\mathbf{n} \cdot \mathbf{E}\mathbf{n}]\;\mathbf{1} + (\nu_1 + \nu_2)(\mathrm{tr}\,\mathbf{E})\,\mathbf{n} \otimes \mathbf{n} + \\ & + 2\,(\mu + \nu_3)\,\mathbf{E} + (\lambda + \mu + 2\nu_2 + 2\nu_3)(\mathbf{E}\mathbf{n} \otimes \mathbf{n} + \mathbf{n} \otimes \mathbf{n}\mathbf{E}) + \dots \,. \end{aligned} \tag{3.2}$$

The first two terms of this expansion do not depend on the strain and correspond to the stress-free state. The remaining terms are linear in the strain tensor, terms of higher order are neglected. The coefficients λ, μ, ν_1, ν_2, ν_3 are certain combinations of the derivatives of the strain-energy function (2.3) taken in the stress-free reference state, i. e. at $I_1 = I_2 = 3$, $I_3 = 1$. Regardless of their explicit representation in terms of the strain energy function (2.3), the coefficients can be interpreted as material parameters characterizing the elastic properties of the material: λ and μ are LAMÉ's constants, the third-order elastic moduli ν_1, ν_2, ν_3 are those introduced by TOUPIN and BERNSTEIN [5].

4. PRINCIPAL WAVES

In a first step the eigenvalue problem (2.6) will be solved for a special choice of the normal vector $\mathbf{n}$. The strain tensor can be represented as

$$\mathbf{E} = \varepsilon_1\,\mathbf{e}_1 \otimes \mathbf{e}_1 + \varepsilon_2\,\mathbf{e}_2 \otimes \mathbf{e}_2 + \varepsilon_3\,\mathbf{e}_3 \otimes \mathbf{e}_3 \tag{4.1}$$

in terms of the principal strains ε_k and the unit vectors $\mathbf{e}_k$ of the corresponding principal directions. By inspection of (3.2) it can be seen that, if the wave normal happens to coincide with one of the principal directions of strain, the three direction vectors $\mathbf{e}_1$, $\mathbf{e}_2$, $\mathbf{e}_3$ are eigenvectors of the acoustic tensor.

A principal wave is characterized by its wave normal $\mathbf{n}$ and its polarization direction $\mathbf{d}$, both of which are aligned with one of the principal directions of strain. If the wave normal and the polarization are specified by

$$\mathbf{n} = \mathbf{e}_n, \quad \mathbf{d} = \mathbf{e}_m, \tag{4.2}$$

the corresponding propagation speed U will be denoted by c_{nm}. From the propagation condition (2.6), with (3.2) inserted for the acoustic tensor, one obtains

$$\varrho c_{nn}^2 = \lambda + 2\mu + (\lambda + \nu_1 + 2\nu_2)\,\mathrm{tr}\,\mathbf{E} + 2\,(\lambda + 3\mu + 2\nu_2 + 4\nu_3)\,\varepsilon_n, \tag{4.3}$$

$$\varrho c_{nm}^2 = \mu + (\lambda + \nu_2)\,\mathrm{tr}\,\mathbf{E} + 2\,(\mu + \nu_3)(\varepsilon_n + \varepsilon_m), \quad n \neq m. \tag{4.4}$$

The propagation speeds of the transverse principal waves satisfy the symmetry relation $c_{nm} = c_{mn}$, as can be seen from (4.4). Unlike the transverse waves in the undeformed material, which admit arbitrary transverse polarization, the transverse principal waves in the deformed medium have distinct polarizations, in general.

The propagation speeds (4.3) and (4.4) seem to disagree with the results of TOUPIN and BERNSTEIN [5]. The disparity is due to different descriptions of the propagating wave: TOUPIN and BERNSTEIN refer the propagation speeds to the actual wave front in the deformed material, while those presented here refer to the image of the wave front in the undeformed configuration. If the propagation speeds (4.3), (4.4) are transformed to the actual configuration, they coincide in fact with those given by TOUPIN and BERNSTEIN.

5. PROPAGATION SPEEDS

The wave normal $\mathbf{n}$ is now allowed to assume arbitrary directions. As before, the propagation speeds are determined by the eigenvalue problem (2.6) for the acoustic tensor $\mathbf{Q}(\mathbf{n})$. Since the expansion (3.2) of the acoustic tensor is valid only up to linear terms in $\mathbf{E}$, it is adequate to keep this level of accuracy and approximate the eigenvalues also only up to linear terms in $\mathbf{E}$.

In the undeformed medium ($\mathbf{E} = \mathbf{0}$) longitudinal waves correspond to the *simple* eigenvalue $\lambda + 2\mu$ of the acoustic tensor with $\mathbf{n}$ as eigenvector. According to the perturbation theory of symmetric operators [2] the perturbed eigenvalue, due to a strain tensor $\mathbf{E} \neq \mathbf{0}$, is $\mathbf{n} \cdot \mathbf{Qn}$ in the first approximation. With (3.2) inserted for $\mathbf{Q}$ one obtains the explicit formula

$$\varrho U^2 = \lambda + 2\mu + (\lambda + \nu_1 + 2\nu_2) \operatorname{tr} \mathbf{E} + 2(\lambda + 3\mu + 2\nu_2 + 4\nu_3)\, \mathbf{n} \cdot \mathbf{En}, \tag{5.1}$$

which approximates, up to linear terms in $\mathbf{E}$, the propagation speed U of quasilongitudinal waves in the deformed material. By use of (4.3) this expression can be reduced to

$$U^2 = c_{11}^2 n_1^2 + c_{22}^2 n_2^2 + c_{33}^2 n_3^2, \tag{5.2}$$

where c_{11}, c_{22}, c_{33} denote the propagation speeds of the principal longitudinal waves and n_1, n_2, n_3 are the components of the normal vector with respect to the triad of principal directions of strain.

Transverse waves in the stress-free medium correspond to the *double* eigenvalue μ of the unperturbed acoustic tensor. Every vector orthogonal to the wave normal $\mathbf{n}$ is an eigenvector in this case. Here the perturbed eigenvalues of $\mathbf{Q}$ are, in the first approximation, identical with the eigenvalues of the second-rank tensor $(\mathbf{1} - \mathbf{n} \otimes \mathbf{n})\mathbf{Q}(\mathbf{1} - \mathbf{n} \otimes \mathbf{n})$. Again the result can be simplified by introducing the principal wave speeds c_{nm}. The common speed of transverse waves in the stress-free medium is split up by the deformation in two distinct propagation speeds of quasitransverse waves. In a first approximation they are given by the roots of the quadratic equation

$$\begin{aligned} U^4 - \left[(c_{12}^2 + c_{13}^2)\, n_1^2 + (c_{23}^2 + c_{12}^2)\, n_2^2 + (c_{13}^2 + c_{23}^2)\, n_3^2\right] U^2 + & \\ + \left(c_{12}^2 c_{13}^2 n_1^2 + c_{23}^2 c_{12}^2 n_2^2 + c_{13}^2 c_{23}^2 n_3^2\right) &= 0, \end{aligned} \tag{5.3}$$

where c_{12}, c_{13}, c_{23} denote the propagation speeds (4.4) of the principal transverse waves in the deformed medium and, as before, n_1, n_2, n_3 are the components of the normal vector of the wave with respect to the triad of principal directions of strain.

The dependence of propagation speeds on the direction of the wave normal is usually represented by means of a slowness surface, which is generated by all possible slowness vectors $\mathbf{s} = (1/U)\mathbf{n}$. Slowness surfaces have been analyzed for various kinds of *material* anisotropy [4]. Here the geometrical shape of the slowness surfaces resulting from *strain-induced* anisotropy is described.

If equation (5.2) is divided by U^2, it can be restated in terms of the slowness vector $\mathbf{s}$ as

$$c_{11}^2 s_1^2 + c_{22}^2 s_2^2 + c_{33}^2 s_3^2 = 1. \tag{5.4}$$

The slowness surface corresponding to the quasilongitudinal wave is an ellipsoid, its axes being aligned with the principal directions of strain.

Equation (5.3), after division by U^4, yields

$$\begin{aligned} 1 - \left[(c_{12}^2 + c_{13}^2)\, s_1^2 + (c_{23}^2 + c_{12}^2)\, s_2^2 + (c_{13}^2 + c_{23}^2)\, s_3^2\right] + \\ + (s_1^2 + s_2^2 + s_3^2)(c_{12}^2 c_{13}^2 s_1^2 + c_{23}^2 c_{12}^2 s_2^2 + c_{13}^2 c_{23}^2 s_3^2) = 0. \end{aligned} \tag{5.5}$$

This algebraic equation can be transformed to the simpler form

$$\frac{s_1^2}{1 - c_{23}^2 s^2} + \frac{s_2^2}{1 - c_{13}^2 s^2} + \frac{s_3^2}{1 - c_{12}^2 s^2} = 0, \tag{5.6}$$

where $s = |\mathbf{s}|$ denotes the magnitude of the slowness vector. The algebraic surface of fourth order described by (5.5) or (5.6) is the FRESNEL surface, which is well known from geometrical optics [3].

6. POLARIZATION OF QUASITRANSVERSE WAVES

The waves propagating in the deformed material are neither longitudinal nor transverse, in general. For small strains the deviation of the amplitude vector from strictly longitudinal and transverse directions may be neglected. More pronounced is the following effect: Unlike transverse waves in the stress-free medium, which admit arbitrary transverse polarization, each of the two quasitransverse waves in the deformed material has its specific polarization, unless the roots of equation (5.3) coincide.

Again the perturbation theory of linear operators can be used to determine the polarization directions. In a zeroth approximation, where the deviation from strictly transverse direction is neglected, the polarization vectors $\mathbf{d}$ of quasitransverse waves are eigenvectors of the tensor $(\mathbf{1} - \mathbf{n} \otimes \mathbf{n})\mathbf{E}(\mathbf{1} - \mathbf{n} \otimes \mathbf{n})$. The dependence of these directions on the wave normal admitts the following graphical representation: Take any unit normal vector $\mathbf{n}$ and attach to its tip the corresponding polarization vectors of quasitransverse waves. If the normal vector is allowed to assume arbitrary positions in space, the corresponding polarization vectors will provide the unit sphere with two mutually orthogonal vector fields.

Figure 1 shows the polarization directions for the strain ratio $\varepsilon_1 : \varepsilon_2 : \varepsilon_3 = 0 : 3 : 4$. In the plane of maximum and minimum strain there are four singular points, which allow arbitrary transverse polarization. Such singular points must exist, since, according to a mathematical theorem, one cannot provide a sphere with a smooth vector field. V. I. ARNOLD calls this the "hedgehog theorem", since it states that one cannot comb a hedgehog [1].

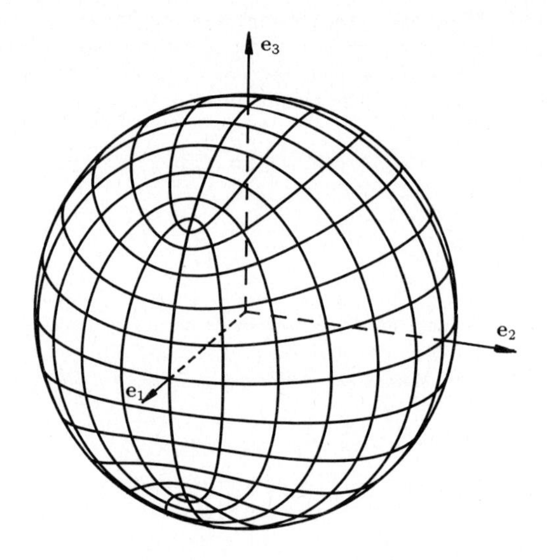

FIGURE 1
Dependence of the polarization directions on the wave normal

7. CONCLUSION

The slowness surface of waves propagating in a stress-free, isotropic elastic material consists of a simple and a double sphere, corresponding to the longitudinal and transverse waves, respectively. Any small predeformation, establishing a triaxial state of strain in the material, transforms the simple sphere to a triaxial ellipsoid and unfolds the double sphere to a fourth-order surface of FRESNEL type. The shape of the complete slowness surface is determined by the six propagation speeds of the principal waves.

In the stress-free medium transverse waves of arbitrary polarization are propagated at the same speed. Any triaxial deformation, however small, causes two *distinct* polarization directions to emerge from the manifold of arbitrary transverse directions. These directions can be represented by two mutually orthogonal vector fields on the unit sphere.

REFERENCES

[1] ARNOLD, V. I., Ordinary Differential Equations (The MIT Press, Cambridge, 1978)

[2] KATO, T., Perturbation Theory for Linear Operators (Springer-Verlag, Berlin 1976)

[3] KLINE, M. and KAY, I. W., Electromagnetic Theory of Geometrical Optics (Interscience Publishers, New York, 1965)

[4] MUSGRAVE, M. J. P., Crystal Acoustics (Holden-Day, San Francisco, 1970)

[5] TOUPIN, R. A. and BERNSTEIN, B., Journal of the Acoustical Society of America **33** (1961) 216–225

[6] TRUESDELL, C., Archive of Rational Mechanics and Analysis **8** (1961) 263–296

Elastic Waves and Ultrasonic Nondestructive Evaluation
S.K. Datta, J.D. Achenbach and Y.S. Rajapakse (Editors)

AN ACOUSTOELASTIC TECHNIQUE FOR THE COMPLETE EVALUATION OF PLANE STATES OF RESIDUAL STRESS

Jay J. DIKE and George C. JOHNSON

Department of Mechanical Engineering
University of California
Berkeley, CA 94720

A new acoustoelastic technique for the complete evaluation of plane states of residual stress is presented. This particular technique uses longitudinal waves propagating normal to the plane of stress. The two-dimensional equilibrium equations are differentiated and added, yielding a Poisson's equation for the in-plane shear stress in which the forcing function is obtained from the measured velocity changes. The basic theory is presented along with experimental results for residual stress determination in an aluminum plate.

1. INTRODUCTION

Acoustoelasticity is a nondestructive technique proposed for the evaluation of residual stress which involves the measurement of the variation in the phase velocities of ultrasonic waves caused by the presence of the stress field. Within the broad heading of acoustoelasticity, there are a range of different methods that have been considered, all of which are limited to the evaluation of plane states of stress. The most common acoustoelastic technique is called the birefringence technique (Pao, et al., 1984). This technique is based on the fact that the difference in the speeds at which two shear waves propagating normal to the plane of stress, but polarized in the principal stress directions, is proportional to the difference in the principal stresses. Another technique which is currently receiving considerable attention involves the difference in the speeds of two SH waves propagating in one principal direction and polarized in the other (King and Fortunko, 1983; Man and Lu, 1987). In this case, the difference in the square of the SH wave speeds is equal to the difference in principal stresses divided by the material's mass density. There is no acoustoelastic constant for this technique. A third technique, called the longitudinal wave technique, involves the change in the speed of a longitudinal wave travelling normal to the plane of the stress (Kino, et al., 1979).

Each of the three acoustoelastic techniques discussed in the previous paragraph has certain advantages and disadvantages. A clear advantage of the SH wave technique is the absence of an acoustoelastic constant whose uncertainty affects the precision of the resulting stresses. The birefringence technique often has an advantage in that there is a relatively larger velocity variation per unit stress than in either of the other techniques. Advantages of the longitudinal wave technique are the ease with which measurements can be made over a large region of a sample, and the spatial resolution which can be achieved. All of these techniques use relative, as opposed to absolute, measurements (Dike and Johnson, 1989).

The remainder of this paper addresses the application of the longitudinal wave technique to evaluating the residual stress state throughout a sample. A recent analytical development has made it possible to estimate the complete residual stress state everywhere in a planar structure (Johnson and Dike, 1988;

Dike and Johnson, 1989). A summary of the theoretical basis and experimental results for an aluminum ring are given here. For more details on the numerical and experimental aspects, see Dike and Johnson (1989).

2. THEORETICAL BASIS

Consider a body with traction-free boundaries subject to a plane state of residual stress, with cartesian components σ_{xx}, σ_{yy}, and σ_{xy}, and a longitudinal wave propagating in the direction normal to the plane. The material in question is taken to be initially homogeneous and acoustoelastically orthotropic, so that the shift in the speed V of this longitudinal wave from the speed V_o in the unstressed material is (King and Fortunko, 1983; Johnson and Mase, 1984)

$$\frac{V - V_o}{V_o} = A_x\sigma_{xx} + A_y\sigma_{yy} = A_+(\sigma_{xx} + \sigma_{yy}) + A_-(\sigma_{xx} - \sigma_{yy}), \tag{1}$$

where A_x and A_y are acoustoelastic constants which may be different for an anisotropic material, and

$$A_+ = \tfrac{1}{2}(A_x + A_y), \qquad A_- = \tfrac{1}{2}(A_x - A_y). \tag{2}$$

Note that for a material which is acoustoelastically isotropic, $A_- = 0$ and the change in wave speed from the unstressed state is proportional to the sum of the in-plane normal stresses.

The equilibrium equations for the stresses in the absence of body forces are

$$\sigma_{xx,x} + \sigma_{xy,y} = 0, \qquad \sigma_{xy,x} + \sigma_{yy,y} = 0, \tag{3}$$

where comma denotes partial differentiation with respect to the indicated coordinate. A Poisson's equation for the shear stress σ_{xy} in terms of the normal stresses may be obtained by differentiating Eq. $(3)_1$ with respect to y and Eq. $(3)_2$ with respect to x, and adding. Thus,

$$\nabla^2\sigma_{xy} = -(\sigma_{xx} + \sigma_{yy})_{,xy}, \tag{4}$$

where ∇^2 is the two-dimensional Laplace operator.

If an accurate estimate of the sum of the normal stresses can be obtained from acoustoelastic measurements, then Eqs. (3) and (4) can be solved for the entire stress field in the body. If $A_- = 0$ (acoustoelastic isotropy), the right-hand side of Eq. (4) is directly related to the acoustoelastic measurements. Let us now focus attention to the problem posed by a material which is acoustoelastically anisotropic.

In order to find a solution for the shear stress by integration of Eq. (4), the values of σ_{xy} along the boundary must be known. Because the stresses are residual, the boundaries are taken to be traction free. Consider a point on the boundary whose outward normal vector makes an angle Θ with the x-axis. The stress tensor at this point may be expressed either in terms of the cartesian (x-y) components used above, or in terms of normal-tangential (n-t) components σ_{nn}, σ_{tt}, and σ_{nt}, which are related to the cartesian components as

$$\sigma_{nn} = \sigma_{xx}\cos^2\Theta + \sigma_{yy}\sin^2\Theta + \sigma_{xy}\sin 2\Theta$$
$$\sigma_{tt} = \sigma_{xx}\sin^2\Theta + \sigma_{yy}\cos^2\Theta - \sigma_{xy}\sin 2\Theta \tag{5}$$
$$\sigma_{nt} = \tfrac{1}{2}(\sigma_{yy} - \sigma_{xx})\sin 2\Theta + \sigma_{xy}\cos 2\Theta$$

In the case considered here, the only nonvanishing stress component is σ_{tt}. Thus, the cartesian components along the boundary may be written

$$\sigma_{xx} = \sigma_{tt}\sin^2\Theta, \qquad \sigma_{yy} = \sigma_{tt}\cos^2\Theta, \qquad \sigma_{xy} = -\tfrac{1}{2}\sigma_{tt}\sin 2\Theta. \tag{6}$$

In light of Eq. (1), the velocity change from the unstressed state is related to the tangential component of stress as

$$\frac{V - V_o}{V_o} = \sigma_{tt}(A_x\sin^2\Theta + A_y\cos^2\Theta). \tag{7}$$

Assuming that measurements of velocity change can be made along the boundary and that the geometry of the boundary (Θ) is known, the shear stress σ_{xy} can be determined through Eqs. $(6)_3$ and (7).

Unfortunately, because the material is not isotropic, we cannot obtain the right-hand side of Eq. (4) directly from the measurements. Instead, we propose to use an iterative scheme in which the velocity data is used to provide an estimate of $\sigma_{xx} + \sigma_{yy}$ which is updated at the end of each step of the iteration. An initial estimate of the sum of the stresses is obtained by letting $A_-=0$. The boundary values for σ_{xy} and this initial guess are used to solve for the stresses throughout the sample. At this point, we have estimates of σ_{xx}, σ_{yy}, and σ_{xy} which are not consistent with Eq. (1). However, by using these estimates and the actual value of A_-, we obtain at the end of each step of the iteration, a new estimate of the sum of the stresses through the equation

$$(\sigma_{xx} + \sigma_{yy})_{n+1} = \frac{1}{A_+}\left[\frac{V - V_o}{V_o} - A_-(\sigma_{xx} - \sigma_{yy})_n\right], \tag{9}$$

where the subscripts "n" and "n+1" refer the the iteration steps involved.

To demonstrate that this scheme is effective, we consider the stress state generated by the far-field tension of an infinite plate of elastically isotropic material containing a circular hole. We know the exact stress state in terms of components expressed in either polar or cartesian coordinates. Our approach is to use the known normal stresses in Eq. (1) to generate the synthetic velocity variations given various choices of acoustoelastic constants. The known shear stresses σ_{xy} along the edges of an annular region are used with the velocity variations to estimate the stress state in the interior of the annulus through numerical integration of Eqs. (3) and (4).

In the example considered, synthetic velocities are provided at discrete polar grid points on one quadrant of the annulus, with 9 radial locations between the inner and outer boundaries, and a circumferential spacing of 5°. This is a rather coarse grid (only 153 data points in the interior of the region), but it serves to show that the finite difference algorithm used provides stress values which, for the case of a 100 MPa far-field tension, are everywhere within 5 MPa of the exact values.

When the material is taken to be acoustoelastically anisotropic, the same stress state leads to a different velocity variation. However, the same stress pattern emerges after the iterative process described above. Table I gives the number of iterations required to reduce the maximum stress difference between subsequent iterations to within 0.1% for a range of different anisotropies. We find that the technique converges for all cases. In particular, the last case shown is an extreme case corresponding to $A_x = 20\ TPa^{-1}$, $A_y = 0\ TPa^{-1}$.

Table I. Number of iterations required for convergence in acoustoelastically anisotropic materials. In all cases, $A_+ = 10\ TPa^{-1}$.

A_-	Iterations
1	3
2	4
4	5
6	7
8	9
9	11
10	14

3. EXPERIMENTAL RESULTS

An annulus of 6061-T6 aluminum, with nominal thickness of 12.7 mm, inside diameter of 38.1 mm, and outside diameter of 63.5 mm, was loaded in diametral compression until permanently deformed and then completely unloaded. Relative measurements of spatial variation in velocity were made over one quadrant of the specimen on a 1 mm radial, 2.5° circumferential grid. The absolute velocity variations necessary in Eq. (1) were determined from the relative measurements through the procedure described by Dike and Johnson (1989). Because measurements cannot be made at the very edge of the sample, the interior measurements are extrapolated to obtain the boundary values of the stresses. The extrapolation procedure used for the results shown below involved a linear least-squares fit to the points near the boundary. The interior data is used where it is available, with extrapolated data used only where necessary. Experimentally determined stress contours are compared with those estimated by the NIKE2D finite element code (Hallquist, 1986).

The contours of the sum of the residual normal stresses given in Fig. 1 show generally good agreement between the experimental and numerical estimates. Note in particular the agreement of the two approaches for the zero contour (C), indicating that the use of the relative measurements is acceptable for evaluating the stress. The fact that the experimental contours are noisier than the numerical contours is to be expected due to the intrinsic uncertainty in the measurements. This noise is especially noticeable in the low-stress regions of the annulus. We also note that the contours generally have the correct shape and are properly located spatially.

Experimental and numerical estimates of the shear stress σ_{xy} are presented in Fig. 2. Again, the zero-stress contour (E) has the same basic pattern throughout the region and is noisier in the experimental plot. The regions of positive and negative shear are in uniform agreement, though there are again certain regions within which the magnitudes are somewhat different. Under the corner of the flat at the top, for example, the experimental contours accurately denote the stress concentration at the edge of the flat, but overpredict the magnitude of the shear stress at this point.

Contours for the normal stress σ_{xx}, which is the hoop stress at the top of the sample are given in Fig. 3. The region in which this stress component is small is accurately delineated and, as in the previous plots, the zero-stress contours agree reasonably well. The regions of tension and compression are in spatial agreement, although the magnitudes of the experimental estimates are slightly higher at the boundaries than are the numerical estimates.

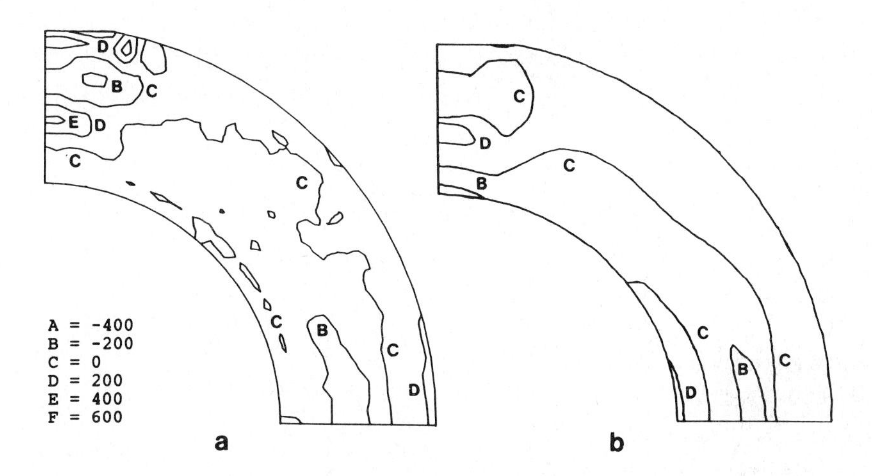

Figure 1. Residual stress sum $\sigma_{xx} + \sigma_{yy}$ obtained (a) experimentally and (b) numerically for an aluminum ring after diametral compression and unloading.

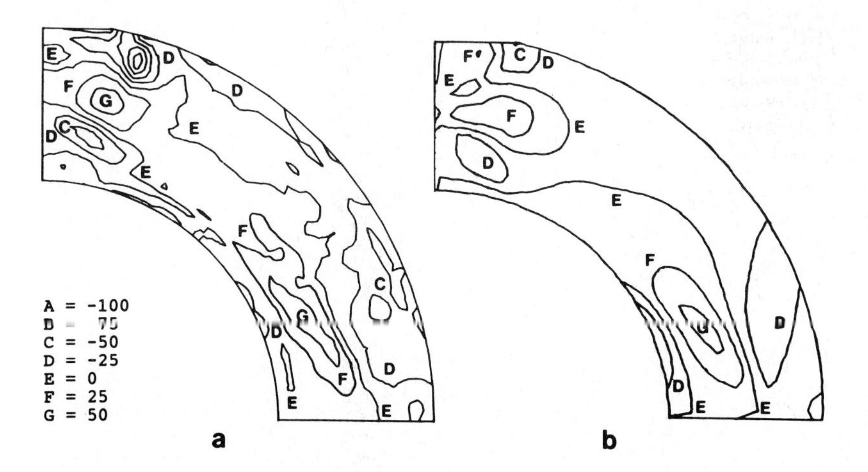

Figure 2. Residual shear stress σ_{xy} obtained (a) experimentally and (b) numerically for an aluminum ring subject to diametral compression and unloaded.

ACKNOWLEDGEMENTS

This work was supported by the Lawrence Livermore National Laboratory through the Engineering Research and Development Program, and by IBM Almaden Research Center. The authors would like to thank Elane Flower of LLNL for her input during the many stimulating discussions of this problem and for her assistance in performing the numerical calculations.

REFERENCES

Dike, J.J and Johnson, G.C., 1989, "Residual Stress Determination using Acoustoelasticity," to appear in Journal of Applied Mechanics.

Hallquist, J.O., 1979, NIKE2D - A Vectorized Implicit, Finite Deformation, Finite Element Code, Lawrence Livermore National Laboratory UCID No. 19677, Livermore, CA.

Johnson, G.C. and Dike, J.J., 1988, "Complete Evaluation of Residual Stress States using Acoustoelasticity," in Review of Progress in Quantitative NDE, D.O. Thompson and D.E. Chimenti, ed., Plenum Press, New York, Vol. 7B, pp. 1391-1398.

Johnson, G.C. and Mase, G.T., 1984, "Acoustoelasticity in Transversely Isotropic Materials,", Journal of the Acoustical Society of America, Vol. 75, pp. 1741-1747.

King R.B. and Fortunko, C.M., 1983, "Determination of in-plane Stress States in Plates Using Horizontally Polarized Shear Waves," Journal of Applied Physics, Vol. 54, pp. 3027-3035.

Kino, G.S., Hunter, J.B., Johnson, G.C., Selfridge, A.R., Barnett, D.M., Herrmann, G., and Steele, C.R., 1979, "Acoustoelastic Imaging of Stress Fields," Journal of Applied Physics, Vol. 50, pp. 2607-2613.

Man, C.-S. and Lu, W.Y., 1987, Towards an Acoustoelastic Theory for Measurement of Residual Stress", Journal of Elasticity, Vol. 17, pp. 159-182.

Pao, Y.-H., Sachse, W., and Fukuoka, H., 1984, "Acoustoelasticity and Ultrasonic Measurements of Residual Stress," in Physical Acoustics, W.P. Mason and R.N. Thurston, ed., Academic Press, New York, Vol. XVII, Chap. 2, pp. 61-143.

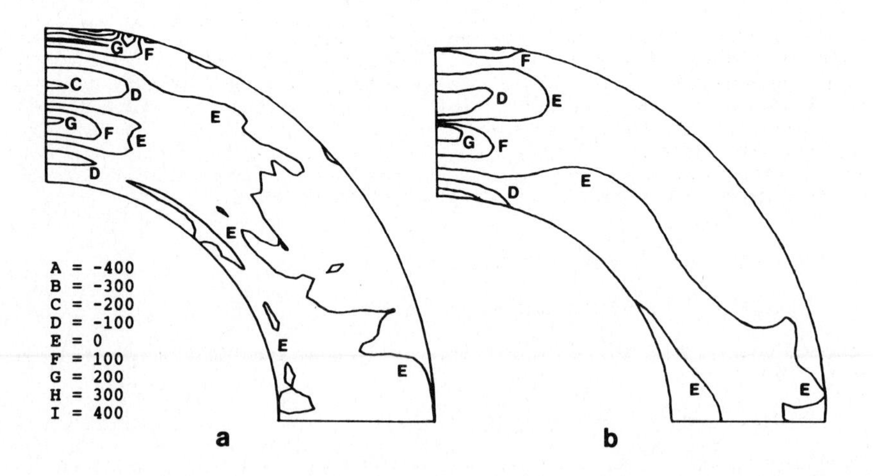

Figure 3. Residual normal stress σ_{xx} obtained (a) experimentally and (b) numerically for an aluminum ring subject to diametral compression and unloaded.

Elastic Waves and Ultrasonic Nondestructive Evaluation
S.K. Datta, J.D. Achenbach and Y.S. Rajapakse (Editors)

NUMERICAL SIMULATIONS OF ACOUSTIC EMISSION AND INVERSION OF MICROCRACKING

Manabu ENOKI and Teruo KISHI

Research Center for Advanced Science and Technology
The University of Tokyo
Tokyo, Japan

Numerical simulation is indispensable for obtaining more accurate Green's functions of finite media including a macroscopic crack. The effort of the current work is directed at obtaining a three-dimensional finite difference method (FDM) representation to simulate a Green's function. Then Green's functions for infinite plates are obtained using the derived relationship and these results compared with the available theoretical solutions. The Green's functions of a precracked one inch compact tension specimen is obtained using the three-dimensional FDM formulation. And then the advanced analysis system has been developed to evaluate AE signals quantitatively.

1. INTRODUCTION

The generation of displacement discontinuities caused by deformation or fracture in a solid has been investigated in the fields of dislocation theory and seismology, where it has been found that the generation of a microcrack can be represented by the seismic moment tensor[1, 2]. This moment tensor can be obtained by computing the inverse deconvolution integral of detected acoustic emission (AE) signals using dynamic Green's functions of the media. So the moment tensor is termed the source function in AE technology. In any case, accurate evaluation of Green's function is a critical factor in obtaining an accurate moment tensor. The dynamic Green's function can be evaluated by analytical[3, 4], experimental and numerical methods. Numerical simulation is indispensable for obtaining more accurate Green's functions of finite media including a macroscopic crack. Thus, the effort of the current work is directed at obtaining a new three-dimensional finite difference method (FDM) representation to simulate a Green's function. As mentioned below, the AE waveform with more than six channels must be recorded and the multiple deconvolution must be carried out in multiple convolution equation to determine the deformation moment tensor and characterize the AE sources.

2. NUMERICAL SIMULATIONS OF ACOUSTIC EMISSION

2.1. Formulation of Acoustic Emission Wave

The equations of wave motion in an elastic, isotropic and homogeneous medium in vector form are:

$$\rho\partial^2\mathbf{U}/\partial t^2 = \mu\nabla^2\mathbf{U} + (\lambda+\mu)\nabla(\nabla\cdot\mathbf{U}), \tag{1}$$

where the vector **U** consists of the displacements, λ and μ are the Lamè constants and ρ is the mass density of the material.

The boundary conditions are that all tractions on the free surfaces are zero. On the upper surface shown by the shaded portion in Figure 1, for example, we get the following formulations as boundary conditions:

$$\begin{aligned}\sigma_{zz} &= (\lambda+2\mu)U_{z,z} + \lambda(U_{x,x} + U_{y,y}) = 0,\\ \sigma_{zx} &= \mu(U_{x,z} + U_{z,x}) = 0,\\ \sigma_{zy} &= \mu(U_{z,y} + U_{y,z}) = 0.\end{aligned} \quad (2)$$

The initial conditions are

$$\mathbf{U} = \partial\mathbf{U}/\partial t = 0 \quad \text{for} \quad t \leq 0. \quad (3)$$

Let us consider the source for a monopole force, $F_i(\mathbf{x})r(t)$, where $F_i(\mathbf{x})$ is the monopole force in the i-direction on a point $\mathbf{x}$, and $r(t)$ is a non-dimensional time function. In a very small square Δl^2 including the point $\mathbf{x}$, the surface stress p_i at the point $\mathbf{x}$ is given by the following equation:

$$p_i(\mathbf{x},t) = F_i(\mathbf{x})r(t)/\Delta l^2. \quad (4)$$

Let us consider the input function for moment tensors, $D_{jk}(\mathbf{x})r(t)$, at an inner point $\mathbf{x}$, where $D_{jk}(\mathbf{x})$ is the moment resulting from the product of the force in the j-direction with the arm in the k-direction, and $r(t)$ is the same time function as was used in the case of a monopole force. By considering the equation of motion containing the body forces at two points, $\mathbf{x}^{k+}$ and $\mathbf{x}^{k-}$, the following equations result:

$$\begin{aligned}\rho\partial^2U_j(\mathbf{x}^{k+})/\partial t^2 &= \mu\nabla^2U_j(\mathbf{x}^{k+}) + (\lambda+\mu)U_{j,j}(\mathbf{x}^{k+}) + F\cdot r(t)/\Delta l^3,\\ \rho\partial^2U_j(\mathbf{x}^{k-})/\partial t^2 &= \mu\nabla^2U_j(\mathbf{x}^{k-}) + (\lambda+\mu)U_{j,j}(\mathbf{x}^{k-}) - F\cdot r(t)/\Delta l^3.\end{aligned} \quad (5)$$

2.2. Finite Difference Method

Let us take the same increments Δl in the x, y and z-directions in Equations (4) and (5), and take an increment Δt in time. $\mathbf{U}(i,j,k,p)$ represents the approximate components of displacement at a grid point $(i\Delta l, j\Delta l, k\Delta l)$ at time $p\Delta t$. For simplicity we denote the point by (i, j, k). Here i, j, k and p are integers $0 \leq i \leq IE$, $0 \leq j \leq JE$, $0 \leq k \leq KE$, where IE, JE, and KE represent the total numbers of the mesh points in x, y and z-directions, respectively. For an inner point, the Equation (1) is discretized. Using a central difference

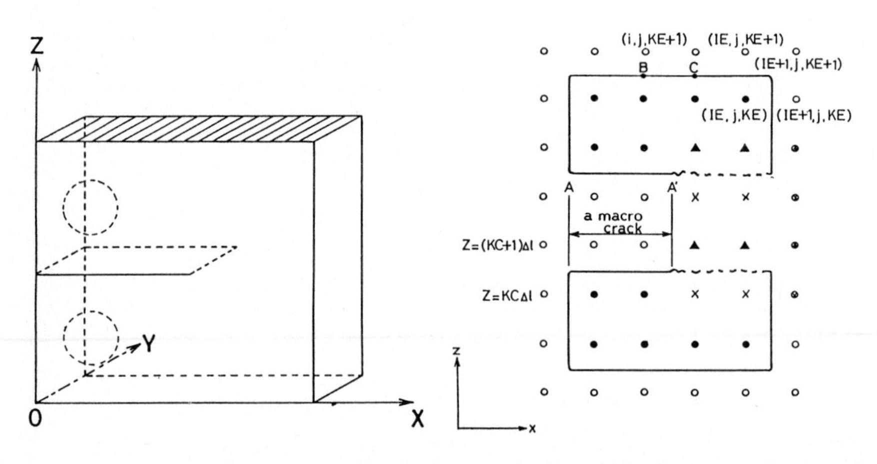

FIGURE 1
A model of 1TCT specimen.

FIGURE 2
Finite difference mesh around the free surfaces and crack surfaces.

approximation for Equation (1) leads to the following discretized formulation:

$$\begin{aligned} U_x(i,j,k,p+1) = &-U_x(i,j,k,p-1) + (2 - 2\Delta t^2(\lambda+4\mu)/\rho\Delta l^2)\, U_x(i,j,k,p) \\ &+ (\Delta t^2(\lambda+\mu)/4\Delta l^2)\, (U_y(i+1,j+1,k,p)+U_y(i-1,j+1,k,p) \\ &\quad -U_y(i+1,j-1,k,p)-U_y(i-1,j-1,k,p) \\ &\quad +U_z(i+1,j,k+1,p)+U_z(i-1,j,k-1,p) \\ &\quad -U_z(i+1,j,k-1,p)-U_z(i-1,j,k+1,p)) \\ &+ (\Delta t^2(\lambda+2\mu)/\rho\Delta l^2)\, (U_x(i+1,j,k,p)+U_x(i-1,j,k,p)) \\ &+ (\Delta t^2\mu/\rho\Delta l^2)\, (U_x(i,j+1,k,p)+U_x(i,j-1,k,p) \\ &\quad +U_x(i,j,k+1,p)+U_x(i,j,k-1,p)). \end{aligned} \quad (6)$$

The stability condition of these equations is

$$L = \Delta t/\Delta l \leq (\alpha^2 + 2\beta^2)^{-1/2}, \quad (7)$$

where $\alpha^2=(\lambda+2\mu)/\rho$ and $\beta^2=\mu/\rho$. It is well known that Equation (6) generally converges to the exact solution of Equation (1), as Δl is decreased while keeping the value of L constant. Let us consider the fictitious points around the free surfaces and the continuous plane in front of a macrocrack as shown in Figure 2. Using the central difference approximation, the finite difference formulation at the boundary points is obtained by substituting these approximation into Equation (2). A different treatment is necessary at the corners. The normal and shear stresses at these point, σ_{nn} and σ_{nt}, are given by

$$\begin{aligned} \sigma_{nn} &= \sigma_{xx}/2 + \sigma_{xz} + \sigma_{zz}/2 = 0, \\ \sigma_{nt} &= (\sigma_{zz} - \sigma_{xx})/2 = 0. \end{aligned} \quad (8)$$

2.3. Numerical Results

The first simulation is a check of the accuracy of the present method when applied to the case of an Al-plate with a monopole force at the center of a lower surface. The results for the epicentral displacements are shown in Figure

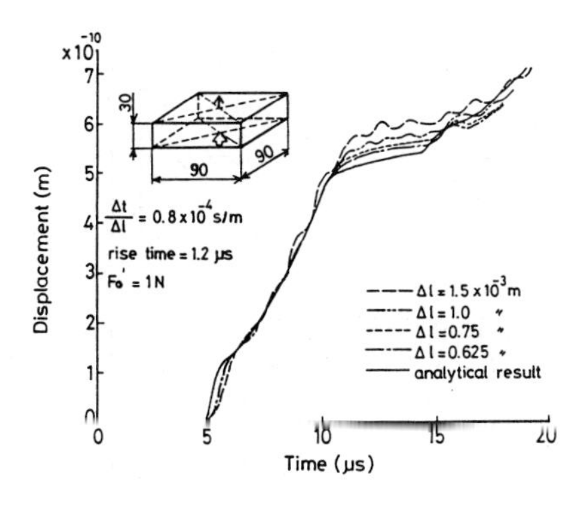

FIGURE 3
The effect of mesh size on the epicentral response (a monopole).

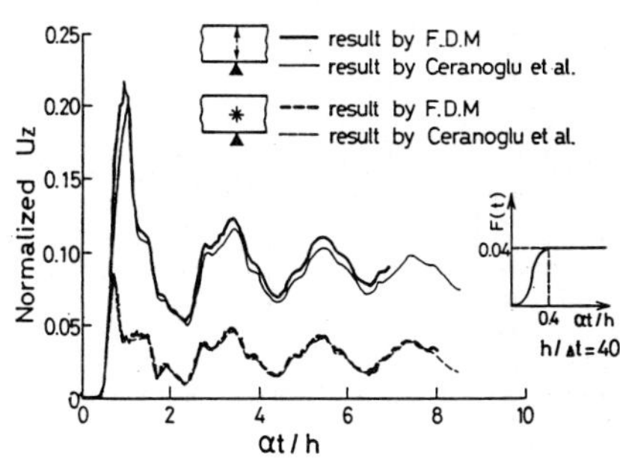

FIGURE 4
Epicentral response for a double force and a center of explosion.

3. When $\Delta l<1.0$ mm, the oscillations of the displacement with respect to time vanishes. The results of this simulation approach Knopoff's analytical solution as Δl is decreased.

Figure 4 shows comparisons between the present simulation and the analytical results by Ceranoglu and Pao[4] for the response at the epicenter. The solid lines represent the solutions to a vertical double force response, and the broken lines, the solutions to a center of explosion response. The displacement is normalized by $\pi\mu h^2 U_z/D_0$. The response for a center of explosion is obtained from an input of three mutually orthogonal double forces. There is good agreement between the finite difference solution and the analytical one. From these results, it can be concluded that the present approach is of sufficient accuracy to obtain a good solution of the response function.

2.4. Application to a 1TCT Specimen

The Green's functions of a one inch compact tension (1TCT) type specimen shown in Figure 1 are simulated using the present FDM. The material is steel with the material constants, $\lambda=1.07\times10^{11}$ Pa, $\mu=7.67\times10^{10}$ Pa, and $\rho=7.87\times10^{3}$ Kg/m^3. In order to obtain the Green's functions, the nine couples of moment are applied to the three-dimensional localized points. The response functions at the six points of transducers are simulated. An example of the results at a transducer point on the surface is given in Figure 5.

3. INVERSION OF MICROCRACKING

3.1. Theory of Acoustic Emission

Denoting the response function in the i-direction at the position **x** on the

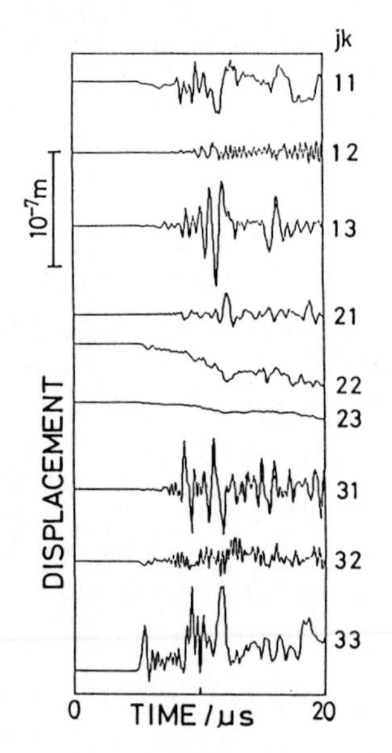

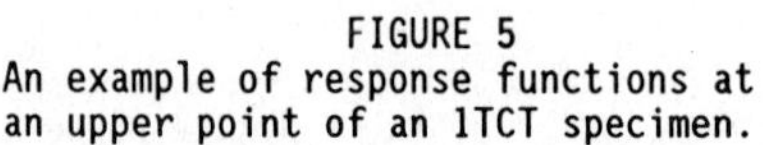

FIGURE 5
An example of response functions at an upper point of an 1TCT specimen.

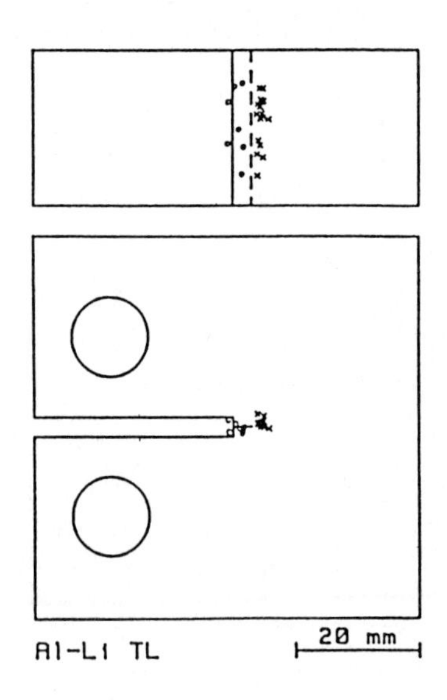

FIGURE 6
The results of location of A sources detected during fracture toughness test in Al-Li alloy.

surface by $S_i(x, t)$, the detected signals $V(x', t)$ due to moment tensor $D_{jk}(x, t)$ can be expressed as[5],

$$V(x', t) = S_i(x', t) * G_{ij,k}(x', x, t) * D_{jk}(x, t). \tag{9}$$

where $G_{ij}(x', x, t)$ is the displacement field in the direction x_i at position x' at the time t due to an impulsive force in the direction x_j at **x** at time 0, and the comma indicates a differentiation and * means a convolution integral with respect to time. And deformation moment tensor $D_{jk}(x, t)$, which is the quantity to represent a microcracking, is defined as, for an isotropic material,

$$D_{jk}(x, t) = \{\lambda\ [u_m]\ \nu_m\ \delta_{jk} + \mu\ ([u_j]\ \nu_k + [u_k]\ \nu_j)\}\ \Delta A, \tag{10}$$

where λ and μ are the Lamè's constants, ΔA is the area of surface A.

Assuming that a microcrack is a disc-like crack subject to a normally applied stress σ, the crack radius a and the angle between $[u_i]$ and ν_i by θ can be represented as respectively

$$a = \{3\ (1 - 2\gamma)\ D_{ii}\ /\ 16\ (1 - \gamma^2)\ \sigma\}^{1/3}. \tag{11}$$

$$\cos\theta = [u_m]\ \nu_m\ /\ ([u_k]\ [u_k])^{1/2}. \tag{12}$$

where E is the Young's modulus, γ is Poisson's ratio.

3.2. Analysis System and Results

The location of each source event is determined by measuring the differences in the wave arrival time between two transducers[6]. A nonlinear least-square method can be used to solve the equation for source location **r**. Figure 6 shows the location data of Al-Li alloy in fracture toughness test with 1TCT specimen

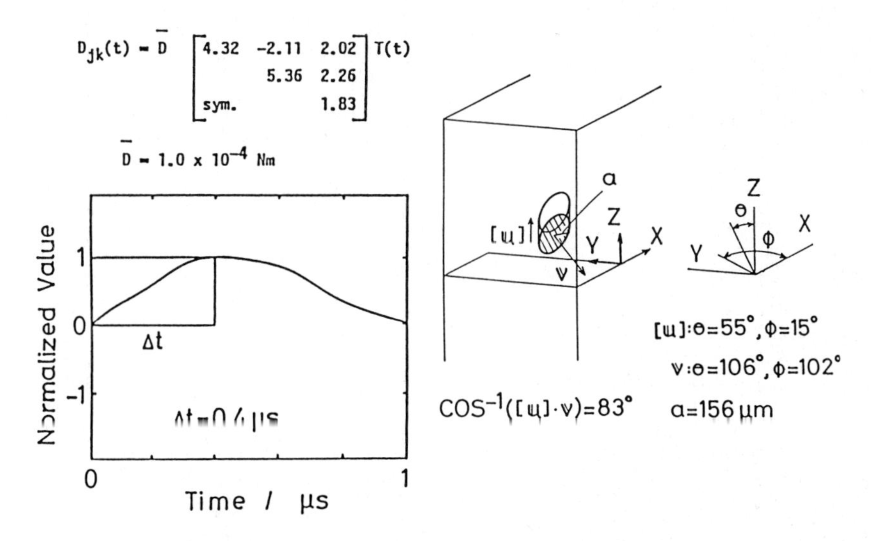

FIGURE 7
The obtained result of moment tensor due to a microcracking in Al-Li alloy.

FIGURE 8
The results of the orientation, fracture mode and size of the microcrack in Al-Li alloy.

of TL direction. AE events can be classified two groups. Figure 6 shows that source events under the low applied load are generated at the back of the fatigue precrack tip, while those under the high applied load are generated in front.

The response function of measuring system is obtained from the result of the deconvolution with the known source function of the a pencil breaking lead [7] and the above Green's function. The deformation moment tensor has to be determined to obtain the mode and orientation of microcracking. We have presented the multiple deconvolution method to determine the moment tensor[5, 8]. Figure 7 shows an example of moment tensor D_{jk} due to microcracking in Al-Li alloy. As shown in Figure 7, a microcrack is generated with a rise time of about 0.4 μs. Applying the nonlinear least-square method to Equation (10), the displacement discontinuity $[u_i]$ and the normal ν_i are obtained from the determined moment tensor D_{jk}. Figure 8 indicates the inclination of the microcrack plane to the main crack surface and the inclination of the microcrack plane to the direction of the displacement discontinuity. This result has demonstrated that a microcracking occurs in mixed mode of tensile and shear, but shear component is more stronger. The crack radius a is estimated as 156 μm from Equation (11) by assuming $\sigma = 3\ \sigma_{ys}$. The estimated value of radius a agrees well with the size of the separation crack which is observed at the location of source event in front of the precrack tip.

4. CONCLUSIONS

To obtain the Green's functions of a finite medium including a macrocrack, which is out of the main problems for AE source characterization, a program has been developed which simulates three-dimensional wave propagation. The conclusions may be summarized as follows:

1. For the continuous plane in front of a macrocrack, fictitious points may be introduced in a manner similar to that used for free surfaces, which makes the treatment of a macrocrack much simpler.
2. The present results for an infinite plate show good agreement with the analytical solutions both for monopole and dipole forces. The maximum error is several percent under the condition of appropriate mesh size and time increment.
3. With the present method, the Green's function of a finite medium with a macrocrack can be obtained with a good accuracy, taking account of the reflection from boundary surfaces and a macrocrack.
4. From the obtained result of moment tensor by the multiple deconvolution method, size, inclination of microcrack surface and fracture mode of microcrack were quantitatively evaluated. The observation by scanning electron microscope verified that these AE sources were identified as the separation crack in front of the precrack in Al-Li alloy.

REFERENCES

[1] Mura, T., Micromechanics of Defects in Solids (Martinus Nijhoff Publishers, The Hague, 1982)
[2] Aki, K. and Richards, P. G., Quantitative Seismologiy Vol.1 (W. H. Freeman and Company, San Francisco, 1980)
[3] Knopoff, L., J. Appl. Phys., 29 (1958) 661.
[4] Ceranoglu, A. N. and Pao, Y. H., J. Appl. Mech., 48 (1981) 125.
[5] Enoki, M. and Kishi, T., Int. J. Fracture, 38 (1988) 295.
[6] Scruby, C. B. and Baldwin, G. R., J. Acoust. Emiss., 3 (1984) 182.
[7] Hsu, N. N., Simmons, J. A. and Hardy, S. C, Mater. Eval., 35 (1977) 100.
[8] Enoki, M. and Kishi, T., Progress in Acoustic Emission IV (The Japanese Society for Non-Destructive Inspection, 1988) pp. 140-147.

Elastic Waves and Ultrasonic Nondestructive Evaluation
S.K. Datta, J.D. Achenbach and Y.S. Rajapakse (Editors)

ACOUSTOELASTIC STRESS MEASUREMENT WITH POWER-CEPSTRUM ANALYSIS

H. FUKUOKA, M. HIRAO, Y. MURAKAMI, M. YOKOYAMA, and A. TANAKA

Faculty of Engineering Science
Osaka University
Toyonaka, Osaka 560, Japan

Power cepstrum analysis is applied to decompose two polarized shear-wave signals and thereby calculate their arrival-time difference for the birefringent acoustoelastic stress analysis. The technique is first tested as to the texture-induced anisotropy and the acousto-elastic constant in the uniaxial stress state and then used to measure the principal stress difference and direction in a biaxially stressed specimen. The results show a good agreement with those by the strain gauge measurements.

1. INTRODUCTION

The phenomenon that a velocity of elastic wave varies with stress in material is known as acoustoelastic effect and it enables a nondestructive evaluation of residual stress in structural elements. In elastic solids, the shear and longitudinal waves are fundamental modes and, if we combine proper boundary conditions with these fundamental waves, we moreover have such modes as surface wave and plate waves. Therefore, there are various aspects in the acoustoelastic effect. One of the most practical techniques is the birefringent acoustoelasticity, in which the velocity difference of two shear waves polarized in two principal stress directions is proportional to the principal stress difference. For the measurement of velocity difference, there are many techniques such as sing-around, phase interference, time-averaging, and spectrum analysis methods etc. These techniqes have their own problems. For example, in the sing-around method [1], the measurement should be done by placing a shear-wave transducer in two principal directions which are unknown in advance; the transducer rotation may cause a change in the coupling condition between transducer and specimen, leading to an erroneous result. The spectrum analysis method [2] suffers from the finite band width of transducers, because it detects the difference of frequencies that give the minima in the amplitude spectrum.

In this study, we apply a power cepstrum method to the birefringent acousto-elastic stress measurement to overcome some of these problems. With the power cepstrum method, we can measure the acoustic birefringence even for the cases of lower SN ratio and of small difference in arrival times between two shear waves; the minimum difference measurable is smaller than the spectrum method. Moreover, since only one measurement is required for each scanning point, the stress measurement can be made quickly and it is free from the change of coupling condition. We measure the uniaxial and biaxial stresses with the power cepstrum method. The results are compared with those obtained by the sing-around method and the strain-gauge measurement, respectively.

2. BIREFRINGENT ACOUSTOELASTIC LAW

According to the nonlinear theory of anisotropic elasticity, the velocity difference, $V_{T1}-V_{T2}$, of two shear waves propagating in the thickness direction and polarized in the principal directions is proportional to the principal stress difference, $\sigma_1-\sigma_2$, as shown in Eq. (1) for a plane stress state;

$$B \equiv (V_{T1}-V_{T2})/V_{T0} = \alpha + C_A(\sigma_1-\sigma_2), \quad (1)$$

We have assumed that the principal stress directions coincide with the textural axes. In Eq. (1), V_{T0} is the velocity of shear wave in the isotropic stress-free state. Acoustic birefringence, B, is the sum of the texture anisotropy, α, and the stress effect, $C_A(\sigma_1-\sigma_2)$; C_A is the acoustoelastic constant and it approximately takes the values of -0.7×10^{-5} /MPa for steels and -4.0×10^{-5} /MPa for aluminum alloys.

Because the propagation path is the same for two shear waves, the acoustic birefringence is obtained by measuring the time-of-flight alone and Eq. (1) is replaced by

$$B \equiv (T_2-T_1)/T_0 = \alpha + C_A(\sigma_1-\sigma_2), \quad (2)$$

where T_1 and T_2 are the times-of-flight of shear waves polarized in the first and second principal stress directions; T_0 is the mean value of T_1 and T_2.

When the principal stress directions do not coincide with the textural axes, the birefringent acoustoelasticity law is modified to [3]

$$B = (T_{II}-T_I)/T_0 = [\alpha^2 + 2C_A(\sigma_1-\sigma_2)\cos2\theta + C_A{}^2(\sigma_1-\sigma_2)^2]^{\frac{1}{2}}, \quad (3)$$

$$\tan2\phi = [C_A(\sigma_1-\sigma_2)\sin2\theta]/[\alpha + C_A(\sigma_1-\sigma_2)\cos2\theta], \quad (4)$$

where T_I and T_{II} are the times-of-flight of polarized shear waves, T_0 is their average value, θ is the angle between the principal stress direction and the texture axis, and ϕ is the angle between the polarization direction and the texture axis. In this case, we need B and ϕ to determine $\sigma_1-\sigma_2$ and θ for given α and C_A.

3. POWER CEPSTRUM ANALYSIS

A cepstrum technique is often used to decompose a particular signal within multiple wavelets overlapping each other, and it has been proved to be useful in many applications [4]. There are two kinds of cepstra, one being a power cepstrum and the other a complex cepstrum. The power cepstrum is used to detect the time lags between component wavelets and their relative amplitudes, while the complex cepstrum is used to reconstruct the signal waveform. Independent variable of the cepstrum is called "quefrency," which has the dimension of time. In this paper, we use the power cepstrum that is defined by the power spectrum of the logarithm of the power spectrum of a signal.

Suppose that a time signal x(t) is composed of s(t) and as(t - τ), the latter of which has a time lag, τ, with respect to the first signal s(t);

$$x(t) = s(t) + as(t - \tau), \quad (5)$$

where a is the amplitude ratio. The power spectrum of x(t) is $F_x(\omega)$ and is given by

$$F_x(\omega) = F_s(\omega)(1+a^2+2a\cos\omega\tau), \quad (6)$$

where $F_s(\omega)$ is the power spectrum of s(t) alone. Taking the logarithm of Eq. (6), we have

$$\log F_x(\omega) = \log F_s(\omega) + \log(1+a^2+2a\cos\omega\tau). \quad (7)$$

The first term of the right hand side can be eliminated by making use of a power spectrum of the reference signal obtained separately. The second term can be expanded into a convergent infinite series as

$$\log(1+a^2+2a\cos\omega\tau) = 2\sum_{n=1}^{\infty}(1/n)(-1)^{n+1}a^n\cos(n\omega\tau) \qquad \text{when } |a|<1, \quad (8a)$$

or $$\log(1+a^2+2a\cos\omega\tau) = \log a^2 + 2\sum_{n=1}^{\infty}(1/n)(-1)^{n+1}a^{-n}\cos(n\omega\tau) \qquad \text{when } |a|>1. \quad (8b)$$

The power spectrum of Eq. (8) is the power cepstrum, $C_x(t)$, and it has a series of peaks with magnitudes of $2(1/n)a^n$(or $2(1/n)a^{-n}$) at quefrency $t = n\tau$ for $|a|<1$ (or $|a|>1$). The maximum peak appears at quefrency $t = \tau$ and this quefrency gives the difference of arrival times of two polarized shear waves, τ.

The auto-correlation function defined by the inverse Fourier-transformation of the power spectrum of a signal is commonly used for the same purpose. This method, however, has a disadvantage that a waveform distortion, i.e., the difference between spectra of wavelets, leads to noticeable errors in the results. In case of the power cepstrum analysis, the effect of waveform distortion is well suppressed by taking the logarithm of the power spectrum. The power cepstrum yields the best indication of time interval between wavelets even in the present of waveform distortion and noise [4].

4. EXPERIMENTS

Figure 1 shows dimensions of the biaxial test specimen with a rectangular coordinate system used for the identification of the measuring points. The biaxial stress is generated by tightening a screw at the ends of bifurcated part. This specimen as well as uniaxial compression and tension specimens are cut out from the rolled plate of 7075-T651 aluminum alloy, which is 25 mm thick. Uniaxial stress directions are parallel to the rolling direction.

The block diagram of our experimental setup is shown in Fig. 2. A PZT shear wave transducer of 10 MHz resonance frequency is placed at nearly 45 deg to the principal stress direction for giving optimum efficiency. A pulser/receiver sends a wide-band pulse to the transducer and it emits the shear wave into the specimen. As the shear wave travels, it splits into two polarized waves having the different velocities. After reflections, the echoes are received by the same transducer. A pair of echoes is selected by the gate circuit and is sent into an A-D converter, where the signal is digitized and averaged over 128 repetitions to increase the SN ratio. The waveform data of 1024 points are then transferred to a personal computer to perform the power cepstrum analysis.

The flow chart of power cepstrum analysis is shown in Fig.3. After the first preprocessing of extending the data length to 4096 points and removing the dc

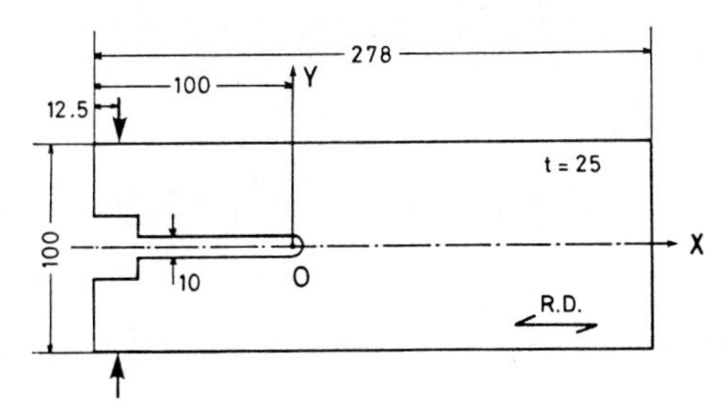

FIGURE 1
Specimen for biaxial acoustoelastic stress mesaurement.

component, the power spectrum is calculated with an FFT program. We then obtain the difference of the logarithm of the power spectrum and the logarithmic spectrum of a reference waveform. For the reference wave form, the range of 2.2 to 17.4 MHz is taken considering frequency characteristics of the transducer. After the second preprocessing of extending the data from 128 to 1024 points, the power cepstrum is calculated again with the FFT program. These operations are done for each step of loading (or scanning) and for the first and the second echoes.

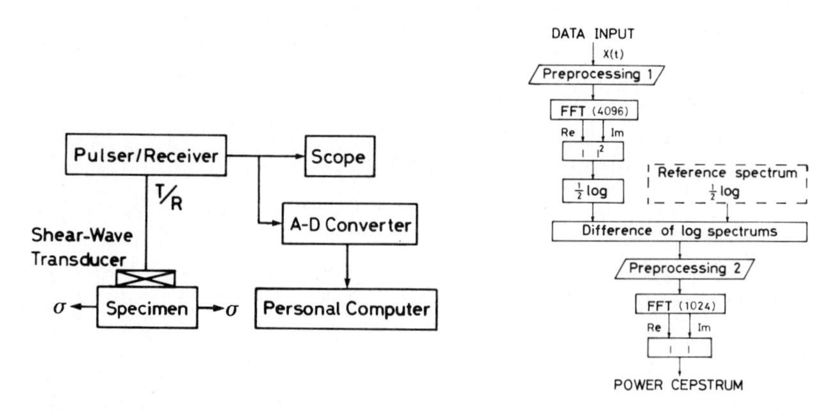

FIGURE 2
Block diagram of experimental setup.

FIGURE 3
Flow chart of power cepstrum analysis.

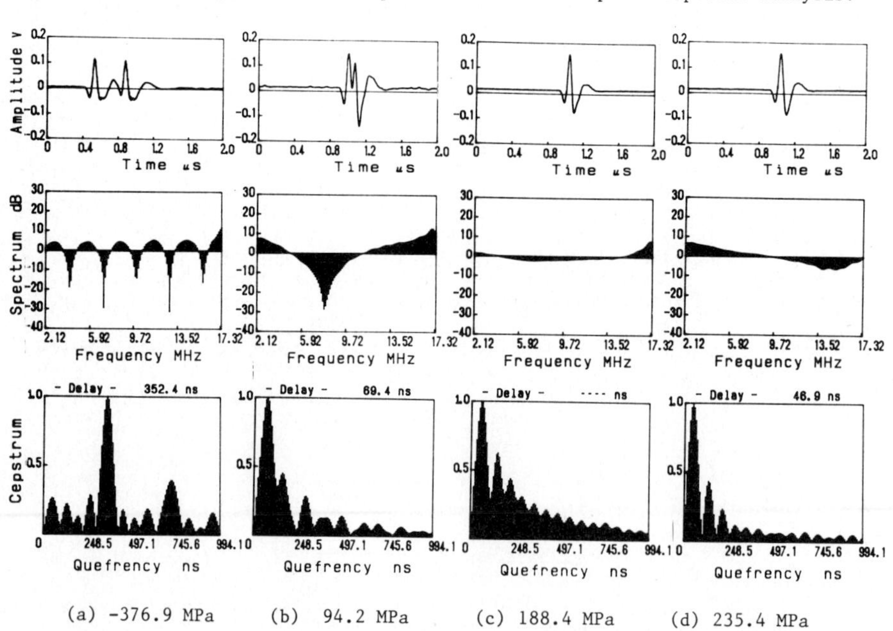

FIGURE 4
Waveforms, logarithmic spectra, and power cepstra of the first echoes for four steps of uniaxial loading.

5. RESULTS AND DISCUSSIONS

Figure 4 shows the waveforms, their logarithmic spectra, and the power cepstra of the first echoes for uniaxial tests. In these figures, the logarithmic spectrum has been subtracted by the logarithmic spectrum of reference waveform. The power cepstrum is normalized so that the maximum value becomes unity. Those values above the figures of the power cepstra are the quefrencies at which the power cepstra take the maximum peak values, that is, the differences of the arrival times of two polarized shear waves. For Fig. 4(c), we cannot obtain the difference of the arrival time, because the two polarized shear waves overlap each other almost completely. Although the overlapping of two polarized shear waves is rather severe in Fig. 4(d), the power cepstrum method is capable of measuring their arrival-time difference.

Figure 5 presents the variation of acoustic birefringence, B, due to uniaxial stresses obtained from the first echoes. Acoustic birefringence at zero stress gives the value of texture anisotropy, α, and the slope of the fitting line gives the acoustoelastic constant, C_A. Table 1 presents a comparison of the values for α and C_A obtained with the power cepstrum method to those obtained with the sing-around method; we see a good agreement between them. In the power cepstrum method, the minimum resolution of time lags depends mainly on the band width of the transducer and the SN ratio of the signal. For the present case, the observable minimum time lag was about 40 ns. The uncertainty of the measurement has come in during the ensemble averaging of the waveform data, the first and second preprocessings, and the FFT calculations. It is very difficult to evaluate the accuracies for these series of calculations. Maximum scatters from the fitting lines of stress-acoustic birefringence were 0.074 % and 0.044 % for the first and second echoes, respectively, in compression test; they were 0.088 % and 0.078 % for the first and second echoes, respectively, in tension test. The maximum scatter of 0.088% corresponds to the error in stress of about 20 MPa.

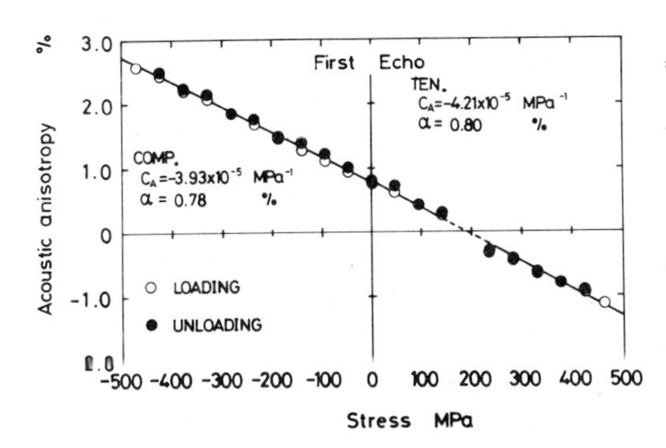

FIGURE 5
Variations of acoustic anisotropy for uniaxial loading and unloading (first echoes). A part of data unavailable is indicated by the broken line.

TABLE 1
Comparison of α and C_A measured by the power cepstrum method and the sing-around method.

	α %	C_A $\times 10^{-5}MPa^{-1}$
Compression Test, 1st Echo	0.78	-3.93
Compression Test, 2nd Echo	0.78	-3.95
Compression Test, Sing Around	0.77 (Ave.)	-3.97 (Ave.)
Tension Test, 1st Echo	0.80	-4.21
Tension Test, 2nd Echo	0.74	-4.05
Tension Test, Sing Around	0.75 (Ave.)	-4.14 (Ave.)

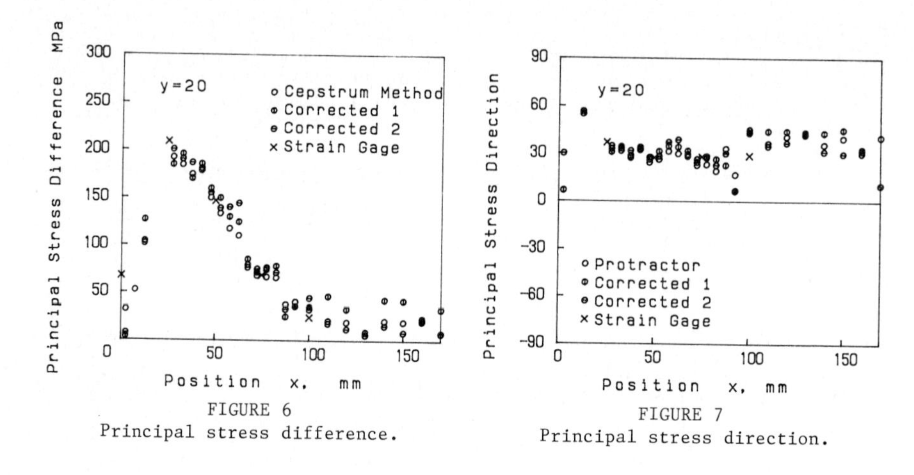

FIGURE 6
Principal stress difference.

FIGURE 7
Principal stress direction.

Typical results of the biaxial test are shown in Figs. 6 and 7. Figure 6 shows a comparison of the principal stress difference along an off-axis scanning line (y = 20 mm), measured by the present cepstrum method, and using the strain gauges. Figure 7 shows a comparison of the principal stress direction obtained by the two methods. The angle ϕ + 45 deg was first measured by reading a protractor attached to the transducer, when two signals took the equal amplitudes on the viewing scope. It was then corrected using the relative magnitudes of the first two or three peaks of the power cepstrum. They are plotted as "correction 1" and "correction 2". Equation (8a) (when $|a|<1$) shows that the amplitude ratio a is theoretically determind from the magnitudes of first three peaks of power cepstrum as 2a, a^2, and $(2/3)a^3$. The angle between the direction of vibration of the shear-wave transducer and the polarization direction is given by $\tan^{-1}\sqrt{a}$.

6. CONCLUSIONS

Values of texture anisotropy α and acoustoelastic constant C_A obtained by the power cepstrum method in uniaxial tests have shown good agreement with those obtained by the conventional sing-around method. The power cepstrum method proposed in this paper is capable of detecting the difference of arrival times of two wavelets as small as 40 ns. However, it cannot exactly measure "zero" difference of them. The maximum scatter from the fitting line of stress-acoustic birefringence will give an uncertainty of about 20 MPa in stresses.

Principal stress differences and principal stress directions in the plane stress state obtained by the power cepstrum method and those by the strain gauge measurement agree with each other very well. We then conclude that the power cepstrum method can be used for a practical birefringent acoustoelastic stress measurement. A potential advantage is that it can be easily incorporated in an automatic stress measurement.

REFERENCES

[1] Crecraft, D.I., J. Sound Vib. 5 (1967) 173.
[2] Blinka, J. and Sachse, W., Exp. Mech. 16 (1976) 448.
[3] Iwashimizu, Y. and Kubomura, K., Int. J. Solid Struct. 9 (1973) 99.
[4] Kemerait, R. and Childers, D., IEEE Trans. Inform. Theory, 18 (1972) 745.

Elastic Waves and Ultrasonic Nondestructive Evaluation
S.K. Datta, J.D. Achenbach and Y.S. Rajapakse (Editors)

ULTRASONIC CHARACTERIZATION OF TEXTURE IN ZINC-COATED STEEL SHEETS

M. Hirao and H. Fukuoka

Faculty of Engineering Science
Osaka University
Toyonaka, Osaka 560, Japan

K. Fujisawa and R. Murayama

System Engineering Division
Sumitomo Metal Industries, Ltd.
Amagasaki, Hyogo 660, Japan

Application of the ultrasonic texture measurement to monitoring the sheet formability is surveyed. In cold-rolled steel sheets, an orientation distribution coefficient, W_{400}, dominates others in magnitude and it is the key parameter determining the $\bar{r}$ value. This paper proposes to use the planar average of the transit times in the S_0 (fundamental symmetric) mode, measured with electromagnetic acoustic transducers (EMATs), to extract the information of W_{400}. As for zinc-coated sheets, a simple formula is available, with which the velocity decrease due to the coating can be easily compensated.

1. INTRODUCTION

An ultrasonic technique is currently under development for the nondestructive characterization of formability in metal sheets such as steel and aluminum. The technique makes use of the dependence of ultrasonic velocities on texture, that is, the nonrandom orientation of crystallites making up the polycrystal. Because texture is the common source of formability (plastic anisotropy) and the anisotropy in elastic wave velocities, it is natural to monitor the formability from the velocities. The Voigt-Reuss-Hill averaging scheme gives the aggregate elastic constants in textured polycrystals in terms of the fourth-order orientation distribution coefficients (ODCs), W_{4m0} ($m = 0,2,4$), which quantify the texture. The tensor of elastic constants shows weakly orthorhombic symmetry depending on W_{4m0}.

This work briefly reviews the dispersion relation for the S_0 (fundamental symmetric) modes guided in the plane of textured sheet. It then presents the experimental results obtained with low carbon steel sheets and electromagnetic acoustic transducers (EMATs), which show the relations of the transit time with the plastic strain ratio ($\bar{r}$ value) and the ND//<111> axis density.

Zinc-coated steel sheet is also of industrial interest, which shows resistance for a corrosive environment. Because zinc is a "soft" material, the coating induces a decrease in the S_0 mode velocity. Final result is an approximate formula, which enables the correction for the effect of zinc coating from known coating/substrate thickness ratio. As a result, the formability estimation of the underlying steel is made possible.

2. ULTRASONIC-FORMABILITY RELATION

2.1. Phase velocity of S_0 mode in textured polycrystalline sheets

Texture in polycrystalline metals modifies elastic constants and then ultrasonic velocities, making them anisotropic. Texture is a nonrandom distribution of crystallite orientation, on the basis of which the single-crystal elastic constants are weighted to derive the aggregate elastic behavior in an appropriate averaging. The Voigt-Reuss-Hill average shows that three ODCs, W_{4m0} (m

= 0,2,4), enter in the result.

The S_0 plate mode has the phase velocity $V_{so}(\gamma)$ in the low frequency range;

$$V_{so}(\gamma) = V_0(1 - \tfrac{1}{2}\Delta) + (c/\rho V_0)(s_0 W_{400} + s_2 W_{420}\cos 2\gamma + s_4 W_{440}\cos 4\gamma). \tag{1}$$

When the sheet is isotropic, the S_0 mode with limiting low frequency is propagated at $V_0 = [4\mu(\lambda+\mu)/(\lambda+2\mu)\rho]^{\frac{1}{2}}$. This basic velocity is superimposed by two perturbations, being weak in magnitude but indispensable. One is the dispersion effect due to a small but finite thickness-to-wavelength ratio and in Eq. (1) it is represented by $\Delta = [\lambda/(\lambda+2\mu)]^2(kd)^2/3$, where k is the wavenumber and 2d is the sheet thickness. Another is the texture term, which makes V_{so} depend on the angle γ measured from the rolling direction. The coefficients s_ℓ's are constant. The W_{400} term does not contribute to the velocity anisotropy but to the shift in its planar average. Since we have such an explicit form for the texture dependence of the S_0 mode velocity, it is possible to evaluate the ODCs from the ultrasonic experiments.

2.2. Experimental observations

Figure 1 gives elementary observations regarding the formability-ultrasonic relation. The specimens are low carbon steel sheets. The thicknesses range from 0.667 to 0.901 mm. The abscissa is the planar average of the transit times in the S_0 mode, $\langle T_c \rangle$, measured with meander-coil EMATs and then corrected for the dispersion with known thickness and frequency (0.7 MHz). Using Eq. (1), $\langle T_c \rangle$ is found to depend only on W_{400},

$$\langle T_c \rangle = (L/V_0)[1 - (c/\rho V_0^2)s_0 W_{400}] + \tau, \tag{2}$$

where L is the path length and τ is the delay within the measuring devices. In Fig. 1 (a), the data appear to exhibit linear behavior just as Eq. (2) predicts. The W_{400} is measured from the ratio of the through-thickness longitudinal wave velocity to the average velocity of shear waves polarized in the plane of sheet. This velocity ratio is also proportional only to W_{400}.

Figure 1 (b) shows a correlation between $\langle T_c \rangle$ and $\bar{r}$ which was measured by conventional tensile test; $\bar{r} = (r_0+2r_{45}+r_{90})/4$, where r_γ is the ratio of plastic strains in the width and thickness directions of a tensile specimen cut at an angle γ from the rolling direction. The $\bar{r}$ value is an important parameter in sheet industries, since it gives the limiting drawing ratio. Large $\bar{r}$ values indicate good formability. Isotropic sheet has $\bar{r} = 1$. To date, the $\bar{r}$ value has been inspected destructively after sampling small pieces. We now have a nondestructive means for its evaluation. In Fig. 1 (b), the data scattering within ±0.1 is observed from the quadratic regression curve.

Such a close correlation between $\langle T_c \rangle$ (or W_{400}) and $\bar{r}$ occurs, because texture is the main source for both of the elastic and plastic anisotropy. These steel sheets have been manufactured by controlling the chemical composition and the conditions of hot rolling, cold rolling, and annealing so that the {111} crystallographic planes tend to lie in the sheet plane. The volume fraction of the crystallites having this orientation roughly determines the $\bar{r}$ value. At the same time, this orientation has the negative largest W_{400}. Hence, there approximately exists an inverse proportionality between W_{400} and $\bar{r}$. Figure 1 (c) supports this point, where the ND//<111> axis density has been measured with x-ray diffraction method and then normalized by the background intensity of a standard specimen without texture (ND: normal direction to the sheet plane).

The above is the outline of the ultrasonic-formability relation. Detailed discussions can be found elsewhere [1-4]. The S_0 mode velocity measurement with EMATs seems most promising, partly because the EMATs enable the measurements on moving sheets.

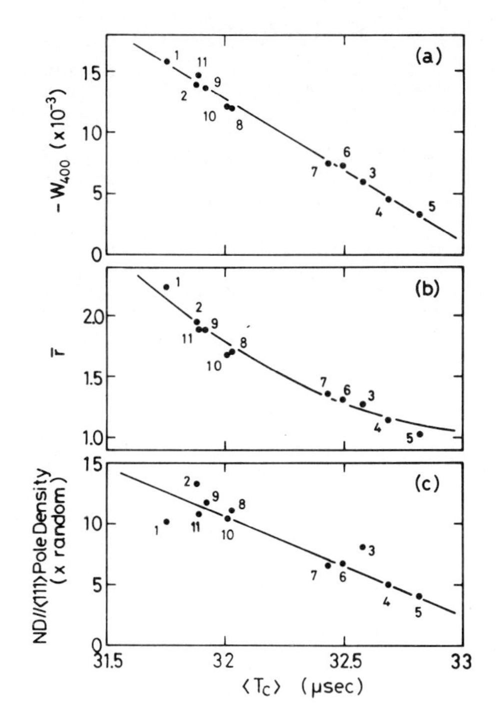

FIGURE 1
Dependence of $\langle T_c \rangle$ on W_{400} and its relation to the $\bar{r}$ value and the ND//<111> pole density for eleven low carbon steel sheets.

3. APPROXIMATE FORMULA FOR ZINC-COATING EFFECT

Steel sheets are often given zinc-coating to avoid rusting for the use in the corrosive environments. The ultrasonic-formability relation is desired to include such cases, not only uncoated sheets. There are two processes of coating; one is the hot-dipping and the other the electroplating. Although the coating is usually thin, as small as 10 μm, it may influence the ultrasonic-formability interrelation. First, during the production process, especially in case of hot-dipping, the atomic diffusion occurs across the interface between the steel substrate and the zinc coating. The effects of metallurgical changes, including this alloying layer, on the formability have not been investigated thoroughly. Second, the tensile test is usually made for having the r value. The zinc-coated sheet may show different r value from the uncoated sheet having originally the same r value. The third point is that the zinc coating lowers the propagation velocities of plate modes, because zinc is an "elastically soft" material. In this paper, we derive an approximate formula to predict how much the S_0 mode velocity decreases with the thickness ratio of coating to matrix.

We use three assumptions for simplification:

(1) The zinc coating is isotropic in itself.
(2) The coating is much thinner than the substrate, as is actually the case.
(3) The sheet is thin enough to consider a uniform strain through the entire thickness.

Zinc is a typical metal that shows strong elastic anisotropy [5,6] and moreover the zinc coatings have relatively sharp texture [7]. But, for the very thin coatings, its effect would be less important. In this situation, the coating will not produce any anisotropy in the S_0 mode velocity. Instead, it will change the overall velocity level, thus affecting the evaluation of W_{400}.

Suppose that an uncoated isotropic sheet is elastically stretched in the x_1 direction in its plane; ND = x_3. Under the plane stress condition that $\sigma_{33} = \varepsilon_{22} = 0$, the resulting stress and strain satisfy

$$\sigma_{11} = M\varepsilon_{11}, \tag{3}$$

where $M \equiv 4\mu(\lambda+\mu)/(\lambda+2\mu)$. It is this combination of elastic constants that gives the limiting velocity of the S_0 mode, V_0, through $V_0 = (M/\rho)^{\frac{1}{2}}$ (see Eq. (1)).

We now extend this idea to find the expression of $\bar{M}$ for the coated sheet, where the substrate thickness is 2d and the total coating thickness is 2Δd. On the basis of assumption (3) above, we can apply the Voigt averaging procedure for $\bar{M}$. By weighting M^{Fe} and M^{Zn}, for steel and zinc, respectively, with their thicknesses, we obtain

$$\bar{M} = (dM^{Fe} + \Delta dM^{Zn})/(d + \Delta d). \tag{4}$$

The same expression holds for the average density $\bar{\rho}$. We then have the S_0 mode velocity in the coated sheet, V'_{so}; that is, $V'_{so} = (\bar{M}/\bar{\rho})^{\frac{1}{2}}$. Following assumption (2), V'_{so} is further approximated to be

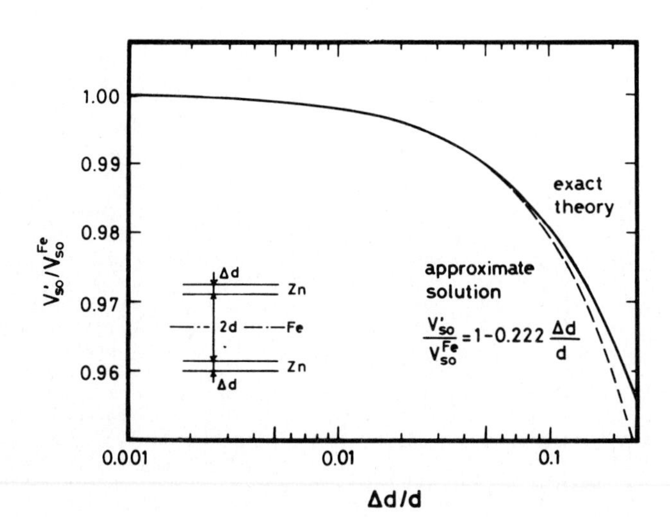

FIGURE 2
Comparison of approximate formula to the exact solution for the velocity decrease of S_0 mode due to the zinc coating. For steel, $\lambda+2\mu$ = 277.0 GPa, μ = 81.7 GPa, ρ = 7850 kg/m^3 [3]; for zinc, $\lambda+2\mu$ = 121.3 GPa, μ = 39.6 GPa, ρ = 7120 kg/m^3 [5,6].

$$V'_{so} = \sqrt{\frac{\bar{M}}{\bar{\rho}}} = \sqrt{\frac{M^{Fe}}{\rho^{Fe}}} \cdot [1 + \tfrac{1}{2}(\frac{\Delta d}{d})(\frac{M^{Zn}}{M^{Fe}} - \frac{\rho^{Zn}}{\rho^{Fe}})], \tag{5}$$

where only the first-order terms in the thickness ratio are retained. Within this range of approximation, V'_{so} varies linearly with $\Delta d/d$.

Figure 2 compares Eq. (5) with the exact solution. For the latter, a set of six simultaneous equations is obtained by solving the equations of motion in the base sheet and the coatings on both sides, being symmetric about the central plane, with the boundary conditions on the free surfaces and the interfaces. The solution V'_{so} is numerically found so that it makes the coefficient determinant zero. The polycrystalline elastic constants derived from single-crystal moduli by the Hill average are used for iron and zinc. We see no difference between two solutions up to $\Delta d/d = 0.05$. As $\Delta d/d$ increases, the approximate solution overestimates the velocity decrease increasingly.

Next, Eq. (5) was tested experimentally using six electroplated sheets, whose substrates were originally the same sheet of $\bar{r} = 1.65$. The same EMATs were used as in the case of uncoated sheets (Fig. 1). They were separated 145 mm apart by a rigid fixture. It is observed that $\langle T_c \rangle$ is proportional to Δd over far wider range than $\Delta d/d < 0.05$. The slope of the best fitting line yields the coefficient 0.184, which is slightly lower than the theoretical result, 0.222. After being corrected (solid circles), six sheets exhibit nearly the same $\langle T_c \rangle$, verifying that the formula has compensated for the coating effect. A small difference from the original $\langle T_c \rangle$ could be due to the texture in coating, which has been excluded.

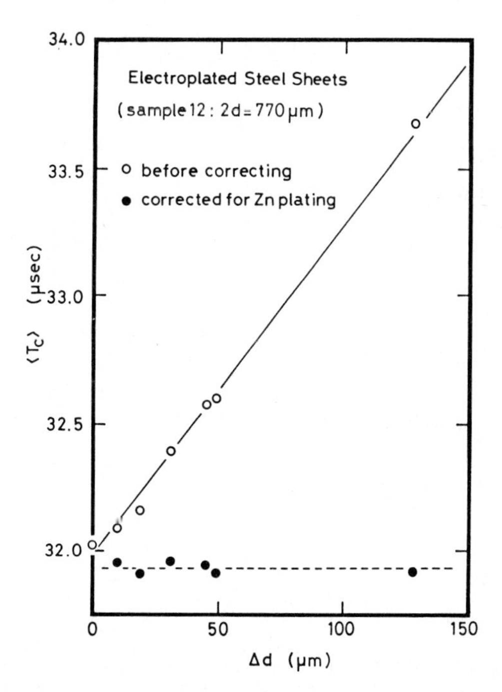

FIGURE 3
Correction for the electroplated coatings.

4. CONCLUSIONS

To summarize, the S_0 mode phase velocity in a coated sheet can be written as

$$V'_{so}(\gamma) = V_0 + \underset{(kd)^2}{(\text{Dispersion Term})} + \underset{\Delta d/d,\ C^{Zn}/C^{Fe}}{(\text{Zn-Coating Term})} + \underset{W_{4m0},\ \gamma}{(\text{Texture Term})}. \quad (6)$$

Under each term, operational parameters are given. For evaluating W_{400} and then $\bar{r}$, the dispersion and zinc-coating terms must be considered for correction. Terms in W_{420} and W_{440} can be canceled by suitable averaging over different γ. Comparing Fig. 1 (b) and Fig. 3, we observe that the ignorance of the zinc-coating term may disturb the correlation between $\langle T_c \rangle$ and $\bar{r}$ to a considerable extent. If the quadratic curve of Fig. 1 (b) is used for a coated sheet, an underestimated $\bar{r}$ results. Equation (6) implies that three small effects are just additive to V_0. The mutual interactions among them are in the second-order magnitude, and for a first-order approximation, they may be neglected.

We have presented a simple formula and found it useful for calculating the decrease of the S_0 mode velocity due to the thin zinc coatings. This is one of problems to be solved before getting a reliable formability prediction for coated sheets. Unsolved problems are:

(1) the influences of zinc coating on the sheet texture and on the plastic deformation during the tensile test for the r value,
(2) the effect of texture in the coatings, which may depend on the process, hot-dipping or electroplating, as well as on the texture of the base steel sheet [7], and
(3) the transduction mechanism of the EMATs for steel sheets coated with nonferrous metals.

ACKNOWLEDGMENTS

The authors are grateful to Dr. A. V. Clark, Jr. and Dr. H. M. Ledbetter for many helpful discussions; and to S. Toyosima for the assistance with the ultrasonic experiments. The sample sheets were supplied by Sumitomo Metal Industries, Ltd.

REFERENCES

[1] Hirao, M., Hara, N., Fukuoka, H., and Fujisawa, K., J. Acoust. Soc. Am. 84 (1988) 667.
[2] Clark, Jr., A.V., Thompson, R.B., Reno, R.C., Blessing, G.V., and Matlock, D., SAE Technical Paper Series #890505 (Intern. Congress and Exposition, Detroit, Feb. 27 - Mar. 3, 1989) 89.
[3] Hirao, M., Fukuoka, H., Fujisawa, K., and Murayama, R., Metall. Trans. in press.
[4] Thompson, R.B., Smith, J.F., Lee, S.S., and Johnson, G.C., Metall. Trans. in press.
[5] Alers, G.A. and Neighbors, J.R., J. Phys. Chem. Solids 7 (1958) 58.
[6] Ledbetter, H.M., J. Phys. Chem. Ref. Data 6 (1977) 1181.
[7] Takechi, H., Matsuo, M., Kawasaki, K., and Tamura, T., in Proc. of 6th Intern. Confer. on Texture of Mater. (Tokyo, 1981), The Iron and Steel Institute of Japan.

Elastic Waves and Ultrasonic Nondestructive Evaluation
S.K. Datta, J.D. Achenbach and Y.S. Rajapakse (Editors)

NONDESTRUCTIVE STRESS AND MICROSTRUCTURE ANALYSIS BY ULTRASONICS

P. Höller, E. Schneider, H. Willems
Fraunhofer-Institut für zerstörungsfreie Prüfverfahren
Saarbrücken, Federal Republic of Germany

I. INTRODUCTION

Nondestructive testing technology can be divided into three major categories:
- NDT for defects, NDTd;
- NDT for stresses, NDTs;
- NDT for properties, NDTp; respectively NDT for structure, especially microstructure, NDTst and materials characterization, NDTmc.

NTDd is the most developed and applied NDT technology. For NDTs, X-ray diffraction and relaxation methods are the most used techniques. By X-ray diffraction, elastic strains induced by stress states are measured. This technique suffers from low penetration ($\lesssim$ 50μm). Relaxation techniques are applied for residual stresses only. By drilling holes or machining notches, stresses to be measured are released by elastic strains which are measured by rosettes of small strain gauges. This technique is nondestructive only as far as the machined notches or holes do not prohibit the use of the component under investigation. NDTs by ultrasonics is based on nonlinear elastic behaviour which is described by second and third order elastic constants. This technique was and still is a subject of intensive research and development but also approved and applied for many years. Measured quantities in NTDd and NDTs are directly related to locus and geometrical structures of defects respectively elastic strains related to stresses.

As far as physical properties like electrical or thermal conductivity, magnetic permeability and elasto-mechanic properties are considered, the situation is similar to NDTd and NDTs or even more fortunate; but the most important mechanical properties are those characterizing the behaviour of materials under loads inducing stresses up to and beyond the yield strength. Direct determination of these properties is only possible by destructive testing. Nevertheless, there is access to these properties by NDT via measurements of composition, microstructure and microstresses. These parameters are strongly correlated to the most important mechanical properties, also to the nonelastic ones. Therefore, most NDTp techniques are based on measurements and analysis of microstructures.

In this paper, we give an overview on ultrasonic NDT technologies for stresses and properties respectively structure considering only those techniques which are already transferred into industrial application or are ready to be transferred and applied.

II. QUANTITIES TO BE MEASURED, WAVE MODES, PROBES AND SPECIAL INSTRUMENTATIONS

Microstructure, texture and residual stress states influence the elastic behaviour of a material and hence the propagation velocities of ultrasonic waves. In order to evaluate one of the mentioned states, particular preassumptions have to be fulfilled, but in each case, a very precise time-of-flight measurement is needed. Depending on the materials parameter to be analyzed and on the geometry of the component, the appropriate ultrasonic modes and probes must be selected.

II.1 NDT for stresses

The combined use of two ultrasonic modes yield to evaluation equations for stress states, in which the **ultrasonic time-of-flight** is the only measured quantity. For cases of application the following **mode-combinations** have been used successfully:

* Longitudinal and radially polarized shear waves for stress/strain evaluations in screws and bolts /1/;
* Longitudinal and linearly polarized shear waves for evaluation of two-axial bulk stresses and for characterization of three-axial bulk stress states /2/;
* Skimming longitudinal and/or Rayleigh- and/or SH-waves for evaluation of surface stress states /3/.

Ultrasonic probes for the mentioned modes are either available on the market or, for some cases of application, special probes have been developed and optimized /4/:

* Probes for simultaneous excitation of a longitudinal and a radially polarized shear wave;
* Electro-magnetic probe for simultaneous excitation of two shear waves, polarized perpendicular to each other;
* Electro-magnetic probes with permanent magnetization for linearely polarized shear waves;
* Fixed-distance-probes with one transmitter and two receivers for skimming longitudinal-, Rayleigh-, and SH-waves;

Special Instrumentations have been developed and are available to drive the ultrasonic probes and to perform the high precision time-of-flight measurement:

* Transmitter/Receiver units for EMUS with permanent magnetization and center frequencies of 0.5 - 2.5 MHz;
* Transmitter/Receiver units for EMUS with pulsed magnetization and 1.5 MHz center frequency;
* Transmitter/Receiver unit for piezoelectric probes with center frequencies between 20 and 50 MHz /5/;
* Set-up for the automated time-of-flight measurement of ultrasonic waves with 1 till 10 MHz center frequency /6/.

II.2 NDT for structure

II.2.1 Texture

Ultrasonic techniques enable the characterization of orthorhombic textures in materials with cubic symmetry in terms of the three fourth order expansion coefficients of the orientation distribution function. Of particular interest are only two of these expansion coefficients, because of correlations between these coefficients and the two parameters r_m and Δr, characterizing the deep drawing behaviour of rolled sheets.

* The application of SH_0 wave modes has advantages, because the two coefficients of the ODF (Orientation Distribution Function) can be evaluated from the same measured data /7/. The suitable probe is mentioned under II.1, as well as the time-of-flight measuring unit.

II.2.2 Microstructure

The nondestructive characterization of microstructure, especially of polycrystalline materials, is an important task in order to improve quality assurance during fabrication. Either **ultrasonic attenuation** or **velocity** measurements can be used to evaluate microstructural characteristics such as grain size, hardening depth, porosity, etc. Velocity mesurements are mainly used for detection and analysis of porosity; the same modes and hardware are used as in case of a stress analysis.

Ultrasonic attenuation measurements are applied in order to evaluate grain sizes and to detect structural inhomogeneities. The **measured quantities** are:

* attenuation coefficient α, evaluated using transmission or backscattering technique;
* scattering cofficient $\alpha_s = S \cdot d^3 \cdot f^4$ (d - grain size, f - frequency, S - scattering parameter);
* absorption coefficient α_A, evaluated using reverberation technique /8/ α_A = f (dislocations, magnetostriction).

The suitable **ultrasonic modes** and frequencies are:

* normal incident longitudinal waves;
* 45° shear waves and surface waves;
* in both cases, frequencies used are such that wavelength do not exceed five times the grain size.

Instrumentation includes

* Broadband ultrasound generated by laser pulses and received by a heterodyne interferometer (OPUS = photo-optical ultrasonic device);
* Narrowband piezoelectric device and high speed averager for signal averaging of ultrasonic backscattered signals.

III. APPLICATIONS

III.1 NDT for stresses

Measuring the time-of-flight of a longitudinal and a radially polarized shear wave, propagating along the axes of bolts and screws, their strain and stress states were evaluated. The most important advantage is that there is no need to know the initial length. The piezoelectric-EMUS-combination probe was applied on screws with temperatures up to 120°C; excellent agreement was found with the actual strain state in the screws /1,4/.
Skimming longitudinal waves were applied to evaluate the surface stress states in a welded plate. The principle stresses parallel and perpendicular to the weld were found in good agreement with those, evaluated destructively in the plate, in the heat affected zone and the seam /3/.
The same technique was applied to evaluate axial and hoop stresses in rolls after different heat teatments /3/.
Bulk stress states in turbine rotors are characterized by using shear waves, polarized perpendicular to each other, in order to control heat treatments.
Longitudinal and linearly polarized shear waves are applied to evaluate the profiles of the three principle stresses in a plate of sintered metal /2/.

III.2 NDT for structure

III.2.1 Texture

The nondestructive characterization of the deep-drawing behaviour is essential for the process control and quality assurance of rolled sheets and strips. Complementary to X-ray diffraction techniques, ultrasonic techniques are developed to characterize the deep drawing behavior in terms of the parameters Δr and r_m. Measuring the times-of-flight of ultrasonic waves, two fourth-order expansion coefficients of the orientation distribution function are evaluated. Published empirical and theoretical results are used to link the planar anisotropy parameter Δr with the expansion coefficient C_4^{13} and the normal anisotropy parameter r_m with the coefficient C_4^{11}. In different sets of cold and hot rolled ferritic steel sheets linear correlations are found between the two deep drawing parameters and these expansion coefficients. Based on these correlations an alogrithm is developed and applied to avaluate Δr- and r_m-values using ultrasonic techniques. The excellent correlation (Fig. 1,2) between these data and the Δr- and r_m-values given by the manufacturer of the sheets show that ultrasonic techniques hold high promise to characterize the deep-drawing behavior of sheets /7/.

III.2.2 Grain size

The nondestructive determination of grain size in metals is based on the dependence of the ultrasonic attenuation coefficient α on the grain diameter d assuming Rayleigh scattering, i.e., $d/\lambda \ll 1$ (λ-ultrasonic wavelength). Several techniques can be used to measure α. Usually, α is determined from the amplitude decay of a backwall echo sequence with the restriction that parallel surfaces are required. Independent of this restriction, α can be measured by means of backscattering techniques if multiple scattering is negligible /9/.

Backscattering techniques for grain size measurements have been developed, for example, by BAM (FRG) /10/, BNF (UK) /11/ and IzfP (FRG) /12/. Both, the BAM-system and the BNF-system operate in immersion technique. The ultrasonic pulses are insonified at the critical angle. A leaky Rayleigh wave is excited, and the amplitude of the backscattered volume waves is measured within a certain time window. After a correction for sheet thickness, the grain size is obtained from the averaged amplitude using calibration curves. The BAM-system was installed at Krupp Steel Company, FRG, and is used for on-line grain size inspection of austenitic strips during continuous annealing /13/. Here the grain sizes are in the range 10-30μm. The BNF-technique /11/ is especially designed for grain size measurements in copper alloys with grain sizes in the range 10-80μm.

Whereas the above described techniqes are single frequency techniques which require calibration, the IzfP-technique developed by Goebbels /12/ involves backscattering measurements at two different frequencies simultaneously. This allows the separation of the measured attenuation coefficient into absorption coefficient and scattering coefficient. From the scattering coefficient the grain size can be evaluated directly. The technique can be used for bulk measurements applying 45°-shear waves as well as for surface measurements applying surface waves /14/. Using appropriate frequency combinations the covered grain size range is about 10-200 μm (Fig. 3).

In case of sheets and strips with parallel surfaces, transmission techniques too can be applied for grain size measurements. Broadband ultrasonic pulses are used and the ultrasonic attenuation coefficient α (f) is determined from the spectral analysis of the backwall echoes. The grain size is then determined by fitting the Rayleigh approximation to the measured α(f)-curve with d as fitting parameter. Current work includes the application of the OPUS-system (see II.2.2), /15/.

III.2.3 Hardening depth

Backscattering measurements are also appropriate for detecting structural inhomogeneities. In the case of induction hardening, for example, the hardened near-surface layer with a martensitic structure is rather transparent to ultrasound. At the transition from the martensite to the ferritic base material, scattering is enhanced due to the increase in grain size. The hardening depth is calculated from the measured time-of-flight between the surface echo and the start of the enhanced scattering signal from the base material. Using frequencies of around 20 MHz, the backscattering technique can be applied for measuring hardening depths above 2 mm /16/. It should be emphasized that the backscattering technique is sensitive to microstructural changes, whereas the determination of hardening depth, in practice, is related to a fixed hardness value. However, a direct correlation between both techniques can only be expected in the cases where sharp microstructural transitions are present. Otherwise, a more detailed evaluation is needed.

III.2.4 Creep damage

For components operating under creep conditions, it is of fundamental importance to detect nondestructively the progress of creep damage, especially in the transition phase from creep stage II to creep stage III. In today's practice, metallographic replication techniques are used. This technique is only sensitive to porosity very near to the surface. In /17/, it was shown that high precision ultrasonic velocity measurements are strongly related to volume percentage of porosity. This technique is well supported by theory /18, 19/. A volume fraction of about 10^{-3} is the present limit of detection. Nevertheless, a further enhancement of detectability is requested from the practical point of view. The main advantage of the ultrasonic velocity technique is its capability for fast screening of large areas in big components. When surface waves are used, the analyzed depth is given by penetration depth of the surface wave which is about one wavelength (1 mm at 3 MHz in steel).

References

/1/ Schneider, E; Repplinger, W.: Bestimmung der Lastspannungen in Schrauben mittels Ultraschallverfahren; wird veröffentlicht.

/2/ Schneider, E.; Pitsch, H.; Hirsekorn, S.; Goebbels, K.: Nondestructive Detection and Analysis of Stress States with Polarized Ultrasonic Waves; in Review of Progress in Quantitative NDE; Eds. D.O. Thompson and D.E. Chimenti; Plenum Press (1985), 4 B, 1079-1088.

Mohrbacher, H.; Bruche, D.; Schneider, E.: Bestimmung von Eigenspannungen in Hartmetallen mittels Ultraschallverfahren. DGZfP Jahrestagung 1989; wird veröffentlicht.

/3/ Schneider, E.; Herzer, H.; Bruche, D.: Bestimmung oberflächennaher Spannungszustände in Walzen mittels Ultraschallverfahren. DGZfP Jahrestagung 1989; wird veröffentlicht.

Chu, S.L.; Peukert, H.; Schneider, E.: Evaluation of Residual Stress States in Welded Plates using Ultrasonic Techniques. In Residual Stresses in Science and Technology; Eds: E. Macherauch, V. Hauk; DGM 1987, 1, 334-341.

/4/ Wilbrand, A; Repplinger, W.; Hübschen, G.; Salzburger, H.-J.: Emus Systems for Stress and Texture Evaluation by Ultrasound. 3rd International Symposium on ND Characterization of Materials, 1988; will be published by Springer.

/5/ Pangraz, S.; Simon, H.; Herzer, H.; Arnold, W.: ND Evaluation of engineering ceramics by high frequency acoustic techniques. Proceedings 18th Intern. Symp. on Acoustical Imaging 1989, to be published.

/6/ Herzer, H.; Schneider, E.: Instrument for the Automated Ultrasonic Time-of-Flight Measurement - A Tool for Materials Characterization-. 3rd International Symposium on ND Characterization of Materials, 1988; will be published by Springer.

/7/ Spies, M.; Schneider, E.: ND Analysis of Textures in Rolled Sheets by Ultrasonic Techniques submitted to be published in "Textures and Microstructures"; Ed. H.J. Bunge.

Spies, M.; Schneider, E.: ND Analysis of the Deep-Drawing Behaviour of Rooled Sheets with Ultrasonic Techniques. 3rd International Symposium on ND Characterization of Materials, 1988; will be published by Springer.

/8/ Willems, H: A new method for the measurement of Ultrasonic absorption in polycristalline materials; in Review of Progress in Quantitative NDE; Eds. D.O. Thompson and D.E. Chimenti; Plenum Press (1987), 6 A, 473-481.

/9/ H. Willems, K. Goebbels: Ultrasonic attenuation measurement using back-scattering technique, Review of progress in Quantitative NDE, 1988, to be published.

/10/ A. Hecht, R. Thiel, E. Neumann, E. Mundry: Nondestructive Determination of Grain Size in Austenitic Sheet by Ultrasonic Backscattering. Mat. Eval. 39, Sept. 1981, 934-938.

/11/ D. Boxall: Measuring grain size, Sheet Metal Industries, Vol. 43, No. 471, 1966, 549-553.

/12/ K. Goebbels: in: Research Techniques in Nondestructive Testing; Ed. R.S. Sharpe, Academic Press, Vol. IV, 1980, 87-154.

/13/ K.H. Michel, H. Mülders, A. Hecht: Stahl und Eisen 106 (1986) Nr. 8, 377-382.

/14/ H. Willems, K. Goebbels: Characterization of microstructure by backscattered ultrasonic waves. Metal Science 15 (1981) 549-553.

/15/ S. Faßbender, M. Kulakov, B. Hoffmann, M. Paul, H. Peukert, W. Arnold: Non-contact and nondestructive evaluation of grain sizes in thin metal sheets. 3rd International Symposium on ND Characterization of Materials, 1988, will be published by Springer.

/16/ Bruche, D.; Herzer, H.; Schneider, E.: Bestimmung von Härtetiefen mittels Ultraschall-Rückstreuverfahren. DGZfP-Jahrestagung 1989; wird veröffentlicht.

/17/ H. Willems, W. Bendick, H. Weber: ND Characterization of Materials, Eds. J.F.Bussière, J.P.Monchalin, C.O.Ruud, R.E.Green, Plenum Press, New York 1987, 471-480.

/18/ S. Hirsekorn, IzfP-Report # 790218-TW, Saarbrücken 1979.

/19/ C.M. Sayers, R.L. Smith: The propagation of ultrasound in porous media, Ultrasonics, Sept. 1982, 201-205.

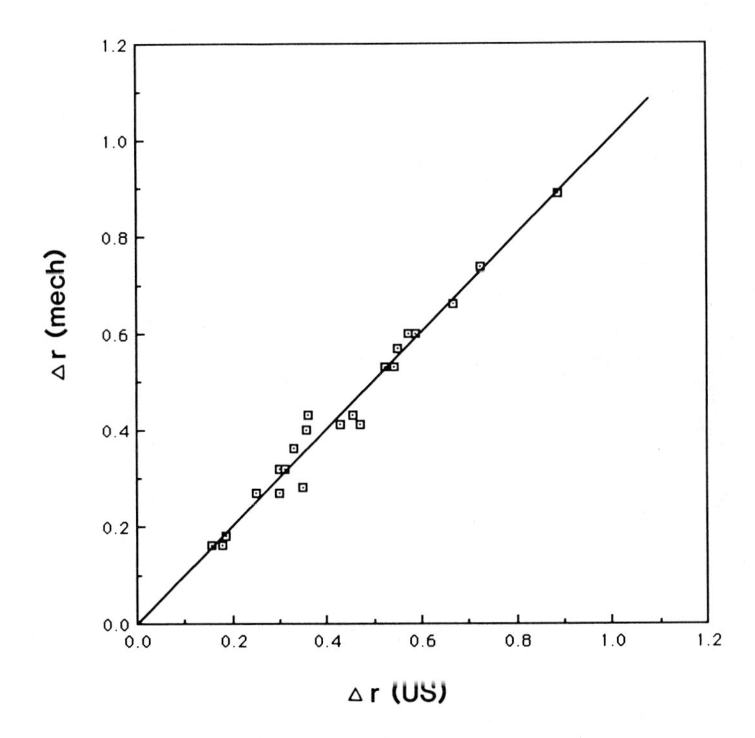

Fig. 1 Correlation between the mechanically determined Δ r and Δ r determined by ultrasonics

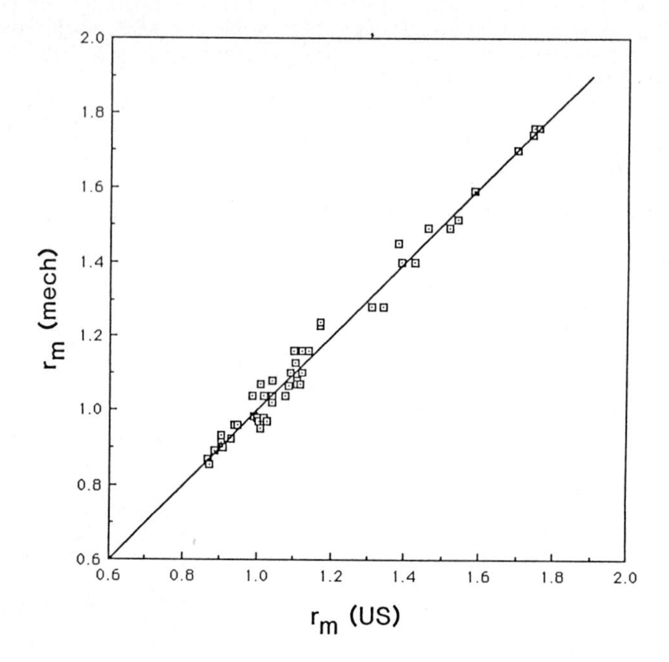

Fig. 2 Correlation between the mechanically determined r_m and r_m determined by ultrasonics

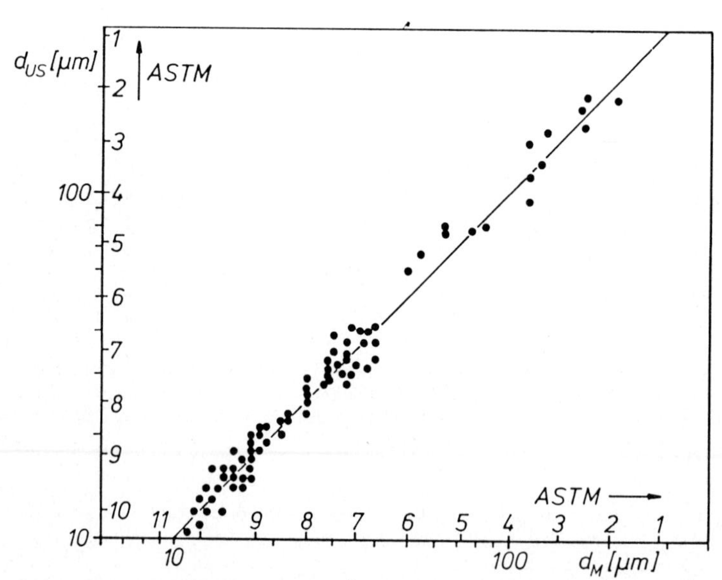

Fig. 3 Comparison between ultrasonic grain size determination (d_{us}) and metallographic grain size determination (d_M)

Elastic Waves and Ultrasonic Nondestructive Evaluation
S.K. Datta, J.D. Achenbach and Y.S. Rajapakse (Editors)
© Elsevier Science Publishers B.V. (North-Holland), 1990

UNIAXIAL DISSIPATIVE ELASTIC WAVES DUE TO HIGH VELOCITY IMPACT

Hans IRSCHIK and Franz ZIEGLER

Department of Civil Engineering, E201
Technical University of Vienna, Austria, A 1040

Wave propagation in physically dissipative rods,e.g. of elastic-viscoplastic type, is considered applying an exact elastic-inelastic analogy to a numerical time-stepping procedure.Treating the incompatible parts of strain as additional sources of selfstress in the non-dissipative elastic structure, linear methods can be used in their most powerful form. Semi-infinite rods with dynamic and kinematic impact conditions are treated to demonstrate the method for different types of constitutive equations.

1. INTRODUCTION AND GENERAL ELASTIC-INELASTIC ANALOGY.

This paper is concerned with wave propagation in physically dissipative elastic bodies, using linear elastic solutions. As is well known, a linear elastic structure with some added dissipative mechanisms, e.g of viscoplastic type, is completely analogous to the structure without those mechanisms, but with an additional distribution of sources of selfstresses.This is immediately seen from the constitutive relations,which in both cases can be put into the quasilinear rate form $\sigma_{ij,t} = E_{ijkl}(\varepsilon_{kl,t} - \bar{\varepsilon}_{kl,t})$, with stresses σ, strains ε, stiffness E and time t. (Geometric relations, conservation of momentum, as well as boundary conditions are coinciding.) Above, the sources $\bar{\varepsilon}$ denote the incompatible, possibly nonlinear parts of strain, while important examples for self- or eigenstress loadings are the thermal parts of strain, $\bar{\varepsilon}_{ij} = \alpha_{ij}\Theta$ (see e.g. [1],p.181.), or the problem of imposed dislocation fields. Of course, seen from the point of material sciences, dissipative defects in linear elastic bodies correspond to dislocations (Mura [2],p.308). Therefore, the above analogy may be considered to be of a tautological type.From a computational point of view, however, it allows a consistent use of well established linear methods in inelastic problems. Dating back to Reißner [3], first numerical applications of the analogy are due to Mendelson et.al.[4], where the evolution of the unknown dislocation distributions has been calculated from the constitutive equations in a time stepping procedure.An extension of this formulation, which basically is related to the "initial strain and stress" techniques of FEM (e.g.Zienkiewicz,Cormeau[5]),to vibration problems of inelastic structures can be found in [6] to [9],see [10] for a review. Thereby methods for thermal shock problems of linear elastic bodies (Boley,Barber [11]) have been used, where the total response is split into a quasistatic and into a dynamic part, see [12]. Applying the method of influence functions and modal analysis (stiffness of the structure is fixed, only the fictitious sources are updated), compatibility and dynamic equilibrium are satisfied exactly throughout the numerical procedure. Below, problems of inelastic wave propagation are treated as a new application. The dynamic domain integral formulation for linear elastic bodies under the action of time dependent dislocations is used as an analytic starting point:

$$u_i^*(\underline{x},t) = -\int_0^t \int_V E_{jlmn}\, \bar{\varepsilon}_{mn}(\underline{\xi},\tau)\, G_{ij',l}(\underline{\xi},\underline{x};t,\tau)\, dV(\underline{\xi})\, d\tau, \qquad (1)$$

where u^* is the deformation, G denotes the linear elastic Green's tensor due to a single, unit, instantaneous force applied in $\underline{x}$ at $t=\tau$, and the derivative has to be taken with respect to $\underline{\xi}$; see Mura [2],p.47 for the infinite domain. The geometrical relations are assumed to be linearized.Total solution is the sum $u_i=u_i^*+u_i^0$, where u_i^0 denotes the linear elastic part due to the imposed external force loading. Considering only uniaxial waves and neglecting any thermal coupling and the temperature field (which ,however,could be incorporated), semi-infinite rods under longitudinal impact (Heaviside-type kinematic and dynamic boundary conditions) are treated by applying symmetric or antimetric source distributions $\bar{\varepsilon}$ to the infinite domain (imaging technique). An explicit time-integration scheme is developed to calculate the evolution of $\bar{\varepsilon}$ for viscoelastic and elasto-viscoplastic materials, using the dynamic influence function G. Also in case of a rather rough incrementation, satisfactory numerical accuracy is achieved in comparison to results published in the literature,and no numerical instabilities,well-known from standard Finite Difference Methods (e.g. Lorimer,Haddow [13], Bodner, Aboudi [14]), are found to occur. Even for elastoplastic unloading waves, the explicit scheme, refering to the state at the beginning of the time interval, yields quite satisfactory results.This favorable numerical behaviour is a result of the underlying consistent elastic-inelastic analogy.

2.TIME-STEPPING PROCEDURE FOR UNIAXIAL WAVE PROPAGATION.

Green's function G for longitudinal wave propagation in an infinite rod is a box-type pulse with intensity (c/2E), the wave fronts travelling with the elastic wave speed c in opposite directions, conf. e.g. Graff [15], p.25. E is Young's modulus.Accordingly,the kernel in equ.(1) is

$$G^*(x,\xi;t,\tau) = -E\, G_{,\xi} = \frac{c}{2}\left(\delta[x-\xi-c(t-\tau)]-\delta[x-\xi+c(t-\tau)]\right) H(t-\tau), \qquad (2)$$

corresponding to a unit source of selfstress applied in ξ at $t=\tau$. δ is Dirac's delta, H is Heaviside's step function. Time convolution and volume integration in equ.(1) is easily performed for arbitrary distributions of $\bar{\varepsilon}$. Especially, subdividing the rod's axis into cells, and time into intervals, e.g. assuming the increments of the fictitious additional sources $\Delta_{ij}\bar{\varepsilon}$ to be switched on at the beginning t_{j-1} of the time interval j according to a Heaviside function, and to be constant within the i-th cell, the corresponding increments of selfstress follow as box-type stress pulses travelling with speed c:

$$\Delta_{ij}\sigma^*(x,t)= E\ (\Delta_{ij}u^*_{,x}-\Delta_{ij}\bar{\varepsilon})= -E\ \Delta_{ij}\bar{\varepsilon}\ [H(x-x_{ij1})-H(x-x_{ij2})]/2, \tag{3}$$

$$\text{where } x_{ij2}=x_i+\Delta x_i/2+c(t-t_{j-1}),\ x_{ij1}=x_{ij2}-\Delta x_i.$$

Δx_i denotes the length of the cell and x_i its midpoint. Equ.(3) applies for $t\geq t_{ji}=t_{j-1}+\Delta x_i/c$, $x>x_i+\Delta x_i/2$. For $x<x_i-\Delta x_i/2$ a stress wave of identical shape travels in the opposite direction. Within the i-th interval the increment of stress is zero for $t\geq t_{ji}$. Δu^*_i itself is antimetric with respect to $x=x_i$.Therefore,according to the method of imaging , a symmetric distribution of $\Delta\bar{\varepsilon}$ with respect to $x=0$ yields the condition of a clamped end at this cross-section , while an antimetric distribution gives a free end.The former case is used for the simulation of an imposed velocity at the end of a semi-infinite rod , and the latter for an imposed stress. Assuming impact conditions, those inhomogeneous boundary conditions will be satisfied by the linear solution part u^0 itself. During the time-stepping, the total solution is found by adding to this associated linear elastic solution a series of incremental linear "thermal shock" solutions , where the summation is performed in Mach's (x,t) plane according to equ.(3) -following a time optimal characteristic path and using the proper imaging conditions. In order to do so, the increments $\Delta\bar{\varepsilon}$ have to be calculated from the material's law, which is assumed in the evolutionary form of a first order differential equation in time of the type $\bar{\varepsilon}_{,t}= f(\sigma,q,t)$, f being a possibly nonlinear function and q an internal variable, often defined by an evolutionary relation itself.This formulation is appropriate for a broad class of viscoelastic and elasto-viscoplastic materials, see e.g. Mukherjee [19],where various time-integration methods and their error measures for a time-step control are discussed.Taking the simple one-step Euler method, frequently sufficient in dynamics, where time steps have to be chosen small enough to account for the rapid change of inertial forces,

$$\Delta_{ij}\bar{\varepsilon} = \Delta t_j\ f|_{t=t_{j-1}}, \tag{4}$$

the value of f being taken at the beginning of the time interval j, immediately before $\Delta_{ij}\bar{\varepsilon}$ is switched on in $x=x_i$. Thus the source increments are refered to a state, which is already known. Δt_j is the j-th time-step. For computational convenience, Mach's plane is subdivided into equal intervals $\Delta t=\Delta x/c$, Δx denoting the constant cell length. Of course, other integration methods instead of the one-step Euler's could be used. It is emphasized, however, that even in implicit , iterative schemes only the constitutive equation - i. e. the increment of a fictitious selfstress-loading - is approximated, while the increments of stress, deformation and strain in each iterative step keep their proper physical meaning due to the use of the dynamic influence function G^* of equ.(2). Mach's plane and the propagation of an incremental stress pulse according to equ.(3) is sketched in fig.(1).

3.EXAMPLES.

As a first example, a viscoelastic rod of Maxwell type is considered, where a constant velocity v is suddenly applied at the end $x=0$ of the rod. Using the dimensionless quantities $\Sigma=\sigma/E$, $T=tE/\eta$, $\Phi=v/c$ $\Gamma=(E/\eta)(x/c)$, equ.(4) becomes $\Delta_{ij}\bar{\varepsilon} = \Sigma_{ij-1}\Delta T$. η denotes the viscosity parameter, and Σ_{ij-1} is the total stress in Γ_i at the beginning of the j-th time interval.For the purely elastic rod, $\Sigma^0(\Gamma,T)= -\Phi[1-H(\Gamma-T)]$.
The numerical procedure of sec.2 is applied using equ.(3) with $\Delta T = \Delta\Gamma$. Satisfactory coincidence with the analytical result of Lee,Kanter [16] is found: Using the rough increment $\Delta T=0.1$, the error of the stresses for an observation period of $T\leq 6$ is smaller than 4.0%; in case of $\Delta T = 0.01$ it is smaller than 0.9% . This largest errors occur at the wave front,where the stresses are known to decay expoonentially. It can be shown that the presented method here leads to a truncated Taylor-expansion of the exponential function (canceled after the linear term.) The smallest errors occur in x=0, where the decay corresponds to the product of a Bessel function and an exponential function. These errors are smaller than 0.15% for $\Delta T=0.01$ and $T\leq 6$.The numerical solution turns out to be smooth with a sharp wave front; no smearing of the wave front or numerical oscillations -well known from standard Finite Difference Methods, comp.[13]- are present. To demonstrate this, fig.(2) gives the stress wave propagation in a Maxwell rod with dynamic boundary condition $\Sigma(0,T)=\Sigma_s H(T)$ as a function of Γ for various values of T.Using $\Delta T=0.01$,practically coincidence with the semi-analytical results of Flügge [17] is obtained.
The second example corresponds to an elasto-viscoplastic rod, obeying the nonlinear law of Bodner,Partom [18], appropriate for strongly rate sensitive materials, e.g. for Titanium. A Finite Difference solution for the wave propagation in such a rod with a suddenly aplied constant velocity at x=0 has been given by Bodner,Aboudi [14] .Using the material and loading parameters as well as the size of time stepping of ref. [14], the explicit scheme of sec 2 is applied to the Bodner Partom-law,which assumes the Prandtl-Reuss flow equations formally to be valid. The proportionality factor,however, corresponding to the ratio of the second invariants of plastic strain rates and deviatoric stress, contains an inner variable , which is given in evolutionary form,see fig.(3) for basic relations and parameters. For the results of fig.(3) ,the explicit scheme of sec. (2), refering to the strain at the beginning of the time interval, is extended with repect to an inner loop, which calculates the current value of the inner variable Z in an iterative manner. Results agree well with those of [14], but show no unphysical oscillations, which are due to the standard Finite Difference formulation in [14].
The illustration of fig.(4) shows ~~an example~~ of an unloading wave in an elastoplastic, hardening material. The analytic solution of this problem is due to White [20]. The increment of the nonlinear part of strain is $\Delta\bar{\varepsilon}= \Delta\Sigma(E/E_1-1)$, $E_1>0$,comp.fig (4). This case of a finite nonlinear material formulation obviously is not suitable for an explicit strategy , contrary to evolutionary laws . In order to show the versatility of the underlying method, however,an explicit scheme is applied to this problem, neglecting the change in stress due to the change of $\Delta\bar{\varepsilon}$ during the current time-step. The result, which of course should be improved by an implicit iteration, turns out to be surprisingly satisfactory: Elastic precurser as

well as the unloading wave are given almost exactly, only the plastic wave front is smeared out. Fig. (4) is for $t/T=1.5$, T denoteing the loading period at $x=0$, where a given stress is suddenly aplied.

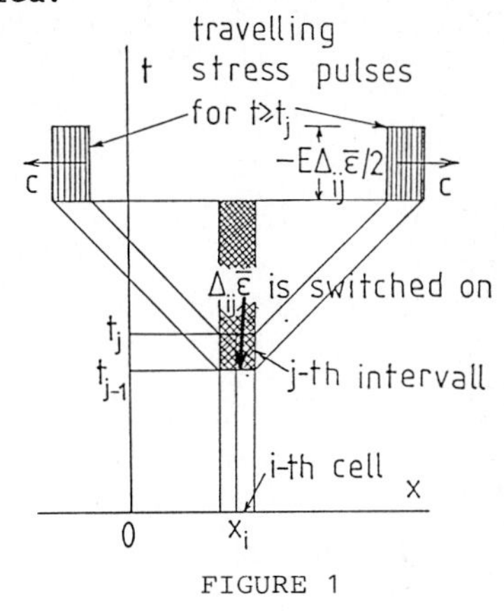

FIGURE 1

Stress wave propagation due to $\Delta_{ij}\bar{\varepsilon}$ for $t>t_j$, shown in the (x,t)-plane.

FIGURE 2

Stress wave propagation in a Maxwell rod with dynamic boundary condition $\Sigma(0,T)=\Sigma_s H(T)$.

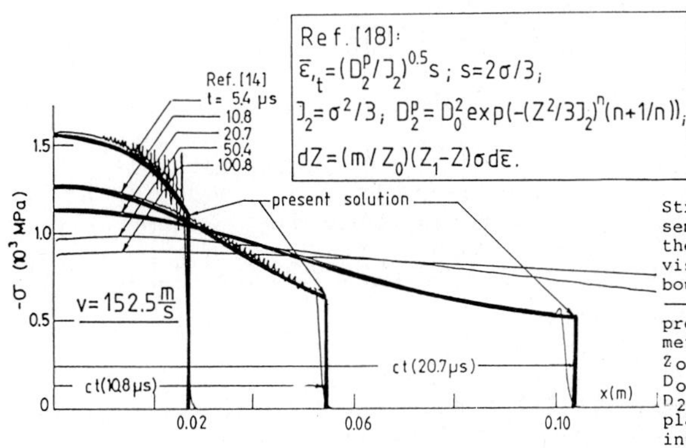

FIGURE 3

Stress wave propagation in a semi-infinite rod according to the Bodner, Partom [18] elastic-viscoplastic law; kinematic boundary condition $u_{,t}(0,t)=vH(t)$. ——— Result of [14], ▬▬▬ present results. Material parameters: $E = 1.18\ 10^5$ MPa, $Z_0 = 1150$ MPa, $Z_1 = 1400$ MPa, $D_0 = 10^4$/sec, $n = 1$, $m = 100$. D_{2p} denotes second invariant of plastic strain rate, Z is the internal variable.

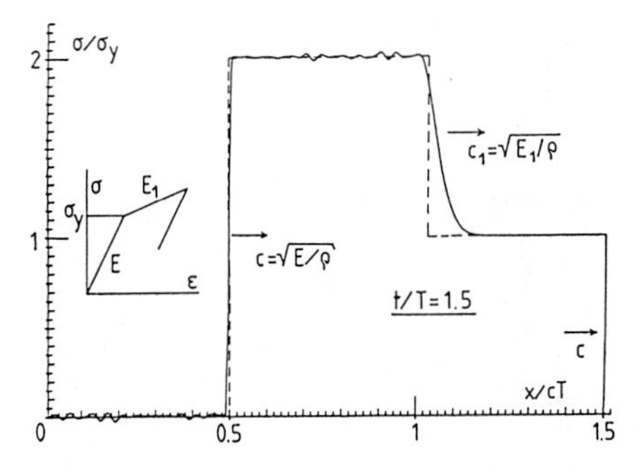

FIGURE 4

Stress wave propagation in a semi-infinite rod of elastoplastic, hardening material. Dynamic boundary condition $\sigma(0,t)=\sigma_s(H(t)-H(t-T))$; T denotes time of unloading, $\sigma_s > \sigma_y$. Fig. 4 shows the result for $E_1/E=0.5$, $\sigma_s/\sigma_y = 2$, $t/T = 1.5$: – - – - Ref. [20], ——— present result.

CONCLUSIONS

The numerical solution procedure with fictitious sources of selfstress, being based on a consistent elastic-inelastic analogy, turned out to be stable, time efficient and accurate in case of various problems of uniaxial, dissipative wave propagation. Extensions to multi-dimensional problems and implicit schemes will be reported elsewhere.

REFERENCES

[1] Ziegler, F. and Irschik, H., Thermal Stress Analysis Based on Maysel's Formula, in: Hetnarski, R. B., (ed.), Thermal Stresses II (North- Holland, Amsterdam, 1987)

[2] Mura, T., Micromechanics of Solids (Mart. Nijhoff Pub., The Hague, 1982)

[3] Reißner, H., Zeitschr. Angew. Math. Mech. 11 (1931) 1-8.

[4] Mendelson, A., Hirschberg, M. H. and Manson, S. S., J. Basic Eng., Trans. Am. Soc. Mech. Eng. 81D (1959) 585-589.

[5] Zienkiewicz, O. C. and Cormeau, I. C., Int. J. Num. Meth. Eng. 8 (1974) 821-845.

[6] Ziegler, F. and Irschik, H., Int. J. Solids Struct. 21 (1985) 819-829.

[7] Irschik, H., Acta Mechanica 62 (1986) 332-342.

[8] Fotiu, P., Irschik, H. and Ziegler, F., Acta Mechanica 69 (1987) 193-203.

[9] Fotiu, P., Irschik, H. and Ziegler, F., Dynamic Plasticity: Structural Drift and Modal Projection, in: W. Schiehlen, (ed.), Proc. IUTAM-Symp. Nonlinear Dynamics in Engng. Systems, Stuttgart 1989 (Springer Verlag, Berlin), in print.

[10] Irschik, H. and Ziegler, F., Zeitschr. Angew. Math. Mech. 68 (1988) 199-205

[11] Boley, B. and Barber, A. D., J. Appl. Mech. 24 (1957) 413-420.

[12] Irschik, H. and Ziegler, F., J. Thermal Stresses 8 (1985) 53-69.

[13] Lorimer, S. A. and Haddow, J. B., Comm. Appl. Num. Meth. 2 (1986) 563-569.

[14] Bodner, S. R. and Aboudi, J., Int. J. Solids Struct. 19 (1983) 305-319.

[15] Graff, K. F., Wave Motion in Elastic Solids (Clarendon Press, Oxford, 1975)

[16] Lee, E. H. and Kanter, I., J. Appl. Physics 24 (1953) 1115-1122.

[17] Flügge, W., Viscoelasticity (Blaisdell Publ., Waltham, Mass., 1967)

[18] Bodner, S. R., Partom, Y., J. Appl. Mech. 42 (1975) 385-389.

[19] Mukherjee, S., Boundary Element Methods in Creep and Fracture (Appl. Sc. Pub., London, 1982)

[20] White, M. P., J. Appl. Mech. 16 (1949) 39-45.

Elastic Waves and Ultrasonic Nondestructive Evaluation
S.K. Datta, J.D. Achenbach and Y.S. Rajapakse (Editors)
Elsevier Science Publishers B.V. (North-Holland), 1990

EFFECTS OF STRESS AND TEMPERATURE ON ULTRASONIC VELOCITY

OSAMI KOBORI

Department of Mechanical Engineering
Osaka Sangyo University, Osaka, Japan

YUKIO IWASHIMIZU

Department of Mechanical Engineering
Ritsumeikan University, Kyoto, Japan

The effects of stress and temperature on ultrasonic velocities were studied experimentally to extend the acoustoelastic law for thermo-elastic materials. The temperature dependence of acoustoelastic constants, which is the most important point in this study, was confirmed by two kinds of experiment I and II on mild steel. Following this, the experiment II was also performed on aluminium alloy, and the coefficients in modified acoustoelastic law were determined for both materials.

1. INTRODUCTION

In most previous studies on acoustoelasticity, the effect of temperature was not treated in detail. This effect, however, seems important, not only in practical acoustoelastic measurements where some temperature change will occur, but also in extending the applicability of acoustoelasticity to thermal stress measurements.
In the present experimental study, though restricted to uniaxial stress and small temperature change, we have confirmed that the velocity change is proportional to the temperature change, and that the proportionality factor is also linearly dependent on the stress. This means that the acoustoelastic constant is varied directly with the temperature change.
The effects of temperature and stress on ultrasonic velocity were first studied by Salama et al.[1][2][3]. They found the equivalent result to the linear dependence of the acoustoelastic constant, and applied this to residual stress measurements.

2. MODIFIED ACOUSTOELASTIC RELATIONS

We consider the case where an isotropic thermoelastic material is initially in a stress-free state with a temperature T_0, and then subjected to a uniaxil stress σ along the X_1-axis and a small temperature change θ (Fig.1). In this state, there are three velocities of waves propagated along the X_3-axis. one is longitudinal velocity V_L and the other two are transverse ones V_1 and V_2 with respective polarizations along the X_1- and X_2-axes. Here the velocity change due to σ and θ are assumed to be expressed by

$$\Delta V_L/V_{L0} = (V_L-V_{L0})/V_{L0} = (K_L+ K'_L\theta)\sigma + L_L\theta + M_L\theta^2, \quad (1)$$

$$\Delta V_A/V_{T0} = (V_A-V_{T0})/V_{T0} = (K_A+ K'_A\theta)\sigma + L_A\theta + M_A\theta^2 \quad (A=1,2), \quad (2)$$

where V_{L0} and V_{T0} are, respectively, the longitudinal and transverse velocity in the reference (initial) state. Since the temperature is expected to change the velocity more than the stress, only the terms of θ^2 and $\sigma\theta$ are taken into account in the modified acoustoelastic relations (1) and (2).

By differentiation of these equations, we have

$$\partial(\Delta V/V_0)/\partial\sigma = K + K'\theta, \quad (3)$$

$$\partial(\Delta V/V_0)/\partial\theta = L + K'\sigma + 2M\theta. \quad (4)$$

$\partial(\Delta V/V_0)/\partial\sigma$ is the acoustoelastic constant, and K' is its temperature coefficient. According to Eq.(3), K and K' can be determined by the velocity measurements where the stress is varied under some constant temperatures (Experiment II). According to Eq.(4), L, K' and M can be determined by the velocity measurements where the temperature is varied under some constant stresses (Experiment I). In the present study, K, K', L and M were experimentally determined for mild steel and aluminium alloy. For mild steel, in particular, both Experiments I and II [4] were performed to find the same K'. For aluminium alloy, however, Experiment II was soley adopted.

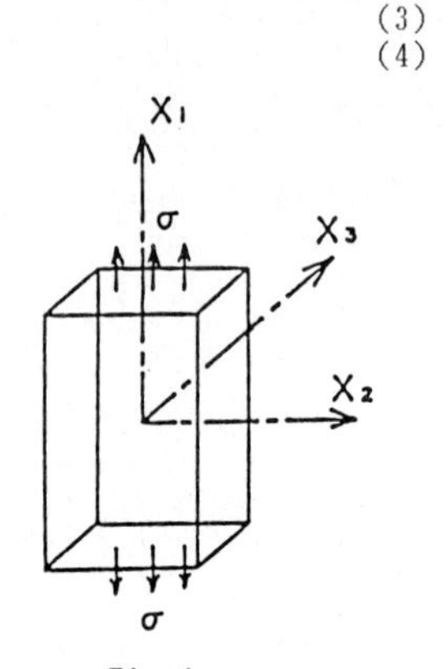

Fig.1
X_k-axis and uniaxial stress

3. EXPERIMENT

Specimens for uniaxial tensile test were made from rolled plates of S45C steel and aluminium alloy (Al:97.5, Fe:0.23, Si:0.015, Cu:0.011 %). The thickness and width of the specimens were, respectively, 8.00 mm and 30 mm, and the longitudinal direction of them was along the rolling direction of the plates. Ultrasonic transducers used were 5mm square PZT with a resonance frequency 5MHz, and the relative time delay of two successive echoes was measured by the sing around method. The temperature was measured at 3 points on a specimen surface, and well controlled with an adiabatic box illustrated in Fig.2. The tested range of temperature was 5～28℃ for steel, and 0～20℃ for aluminium.

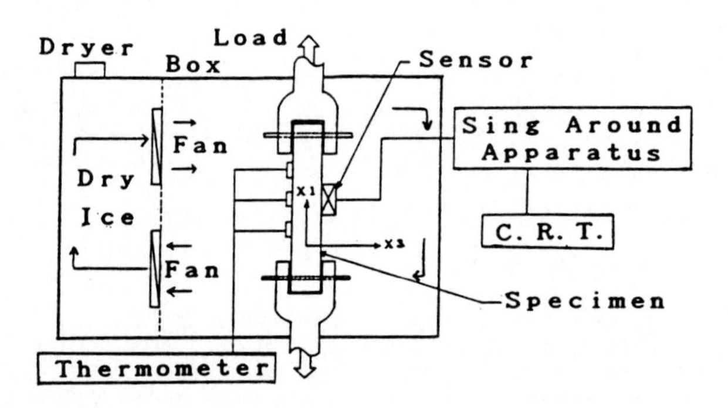

Fig.2 Schematic set up in experiment

4. EXPERIMENTAL RESULTS

In illustrating the experimental results with figures, use is made of the change in the relative time delay, $\Delta\tau^* = \tau - \tau^*$, where τ^* is the relative time delay in some reference state chosen conveniently in each case. $(-\Delta\tau^*)$ may be understood to be proportional to the velocity change.

Table 1 Temperature coefficients of velocity (steel,Al)

	Wave		Velocity m/s (20 °C)	Temp. Coef. L 10^{-4}/ °C	Kobori [5] V_0 m/s	Kobori [5] L 10^{-4}/ °C
Steel	Longitudinal		6066	−1.16	5911	−1.49
	Transverse	σ//	3230	−1.54	3167	−1.27
		σ⊥	3202	−1.38	3170	−1.22
Al	Longitudinal		6119	−2.58	5995	−2.97
	Transverse	σ//	3185	−2.87	3131	−3.11
		σ⊥	3183	−2.68	3129	−3.05

4.1 Velocity Change due only to Temperature

$\Delta\tau^*$ in the unstressed aluminium specimen was varied with temperature as shown in Fig.3. (The reference value τ^* was chosen at 0°C.) Since $\Delta\tau^*-\theta$ relation was linear in all waves, the coefficients M in Eqs(1) and (2) were found to be zero. This was also confirmed in steel. Table 1 lists the coefficients L obtained from the slopes of these linear relations. It is to be noted that there was some difference between L_1 and L_2 for two transverse waves.

Fig.3
Variations in relative time delay with temperature

4.2 Temperature Coefficient $\partial(\Delta V/V_0)/\partial\theta$ of Velocity (Experiment I)

Fig.4 shows an example of $\Delta\tau^*$ measured in steel by varying the temperature under some constant stresses. From the slope of each linear relation, the temperature coefficient of velocity $\partial(\Delta V/V_0)/\partial\theta$ was obtained, and plotted in Fig.5. This shows the linear dependence of $\partial(\Delta V/V_0)/\partial\theta$ on the stress σ, that is, the relation expressed by Eq.(4). The coefficients K' obtained from this were listed in Table 3.

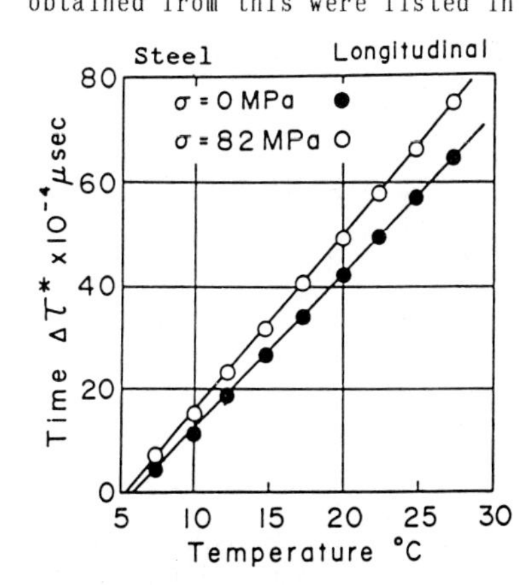

Fig.4
Variations in relative time delay with temperature

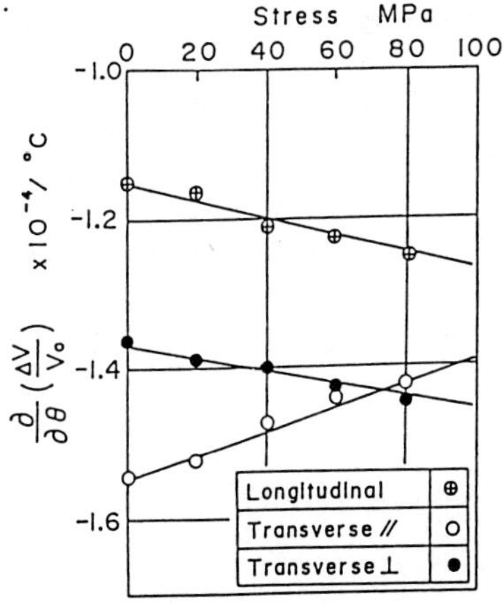

Fig.5
Temperature coefficient of velocity

4.3 Acoustoelastic Constant $\partial(\Delta V/V_0)/\partial\sigma$ (Experiment II)

Fig.6 shows an example of $\Delta\tau^*$ measured in aluminium by varying stress under several constant temperatures. From the slope of the linear relation between $\Delta\tau^*$ and σ, the acoustoelastic constant $\partial(\Delta V/V_0)/\partial\sigma$ was obtained. This was listed in Table 2, and plotted in Fig.7(b). In Table 2 and Fig.7(a), the similar result for steel was given. Figs 7(a) and (b) clearly show that the acoustoelastic constant was dependent linearly on the temperature. Hence, the coefficient K' in Eq.(3) was given by this slope, and listed in Table 3. As for steel, K' obtained by Experiment I and II differes to some degree. However, the extent of errors in the present sing around measurements may be estimated corresponding to the uncertainty of ±12, ±4 and ±8 % in K_L, K_1 and K_2, respectively. Taking account for this, the above difference of K' may be thought immaterial.

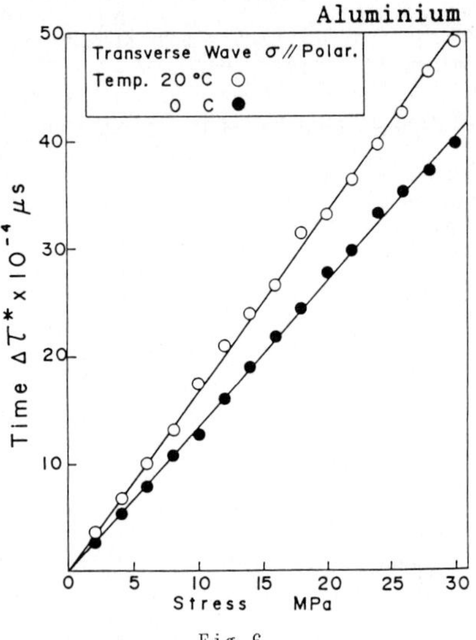

Fig.6
Variations in relative time delay with stress

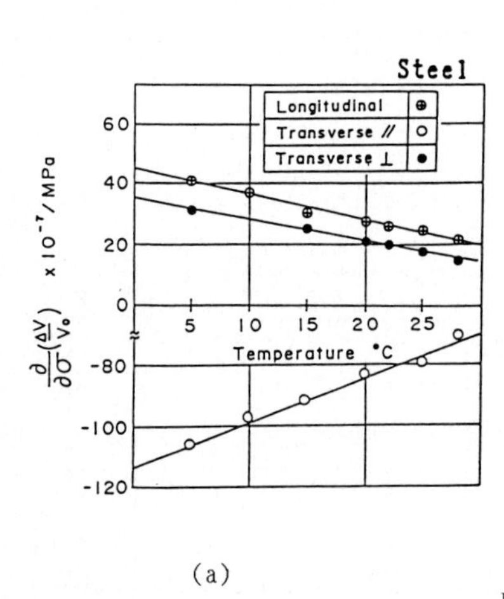

(a)

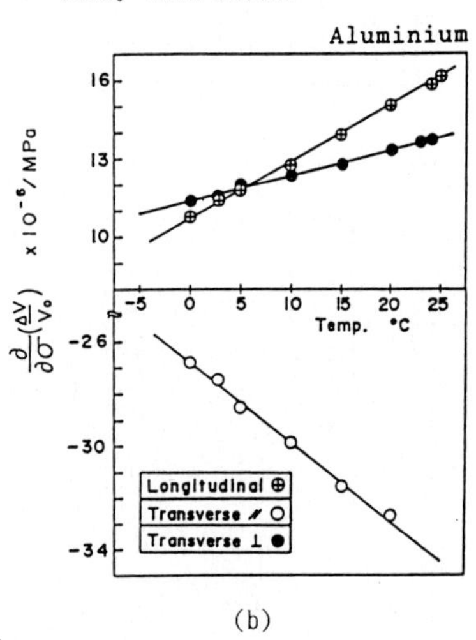

(b)

Fig.7
Temperature dependence of acoustoelastic constants

5. CONCLUSIONS

The results of the present study are summarized as follows:

① In unstressed steel and aluminium, the velocity was varied directly with temperature change. Therefore, the coefficients M in Eqs (1) and (2) were found to be zero, and the coefficients L were determined.

② In steel, the temperature coefficients of velocity change, $\partial(\Delta V/V_0)/\partial\theta$, were confirmed linearly dependent on the stress σ, and the coefficients K' were determined.

③In steel and aluminium, the acoustoelastic constants $\partial(\Delta V/V_0)/\partial\sigma$ were confirmed linearly dependent on the temperature change θ, and the coefficients K' were determined. As for steel, these values were shown to coincide with those mentioned above to within the measurement errors.

These show the validity of the modified acoustoelastic relations, Eqs (1) and (2) with the coefficients M = 0.

Salama et al.[2] performed the experiments similar to Experiment I in the present study, and obtained the ratio Δ/σ = K'/L for steel and aluminium. Their ratio for the transverse waves in steel is different even in sign from the present one. Though the sign of K' is responsible for this, the relative change of the acoustoelastic constant K'/K has the equal sign. Therefore, this difference seems not essential, but due to some different definition of the polarization directions or specimen direction.

Table 2 Acoustoelastic constant (Steel,Al)

	Wave			Acoustoelastic Constant x 10^{-7} /MPa		
				0℃	5℃	20℃
Steel	Longitudinal		K_L		41	28
	Transverse	σ//	K_1		-107	-83
		$\sigma\perp$	K_2		31	21
Al	Longitudinal		K_L	108		152
	Transverse	σ//	K_1	-269		-328
		$\sigma\perp$	K_2	114		134

Table 3 Stress-temperature coefficient K' (Steel,Al)

				Stress-temp. Coef. K' (x10^{-7}/MPa·°C)		
				Exp. I	Exp. II	Mean
Steel	Longitudinal		K'_L	-0.87	-1.04	-1.0 ±0.1
	Transverse	σ//	K'_1	1.49	1.43	1.46 ±0.06
		$\sigma\perp$	K'_2	-0.71	-0.81	-0.76 ±0.06
Al	Longitudinal		K'_L	2.2 ±0.1		
	Transverse	σ//	K'_1	-3.1 ±0.1		
		$\sigma\perp$	K'_2	1.1 ±0.1		

REFERENCES

[1] Salama, K. and Ling, C. K., J. Appl. Phys., vol.51, p.1503 (1980)
[2] Chandrasekaran, N. and Salama, K., Nondestructive Methods for Material Property Determination (ed. Ruud, C. O. and Green, R. E.), p.399 (1983), Plenum
[3] Salama, K., et al., Rev. Prog. QNDE, vol.3B, p.1355 (1984)
[4] Kobori, O. and Iwashimizu, Y., Trans. Jpn Soc. Mech. Eng., vol.54, p.245 (1988)
[5] Kobori, O., Report for Grant No.58350005 for Scientific Research from the Ministry of Education of Jpn, p.61 (1986)

Elastic Waves and Ultrasonic Nondestructive Evaluation
S.K. Datta, J.D. Achenbach and Y.S. Rajapakse (Editors)

ULTRASONIC-WAVE STUDIES OF THE NEW HIGH-T_C OXIDE SUPERCONDUCTORS

Hassel LEDBETTER

Materials Science and Engineering Laboratory
National Institute of Standards and Technology
Boulder, Colorado 80303, USA

I highlight studies that began in spring 1987 on high-critical-temperature oxide superconductors. The superconducting systems include La–Sr–Cu–O, Y–Ba–Cu–O, and Bi–Sr–Ca–Cu–O. In the latter system, T_c approaches 110 K. Most existing studies focus on sound velocities and elastic constants between ambient temperature and 4 K. Because almost all studies focus on polycrystalline sintered ceramics, I consider effects of voids on sound velocity. Some specimens contain more than fifty-percent voids. Thus, large effective-physical-property changes occur, especially if the voids assume nonspherical shapes. The relatively low Poisson ratios (near 0.21) shown by the superconductors suggest that their interatomic bonding resembles that in perovskites such as $BaTiO_3$ and $SrTiO_3$ and in some, but not all, other metal oxides. Temperature hysteresis of sound velocities suggests a phase transition similar to that occurring in relaxor ferroelectrics. Sound velocity provides an especially sensitive probe of this transition, more sensitive than other physical properties such as electrical resistivity or specific heat. Because many believe that strong anharmonic effects occur in these materials, I consider the Grüneisen parameter, which one can obtain either from sound velocities or their pressure derivatives.

1. INTRODUCTION

In spring 1986, Bednorz and Müller [1] redirected research on superconductors. They showed that not only do oxides superconduct, but they superconduct at surprisingly high temperatures. Their compound — Ba–La–Cu–O — showed a T_c near 30 K. Follow-on studies at other laboratories yielded higher transition temperatures: Y–Ba–Cu–O, $T_c \simeq 90$ K [2]; Bi–Sr–Ca–Cu–O, $T_c \simeq 105$ K [3]; Tl–Ba–Ca–Cu–O, $T_c \simeq 125$ K [4].

Virtually every imaginable probe, both experimental and theoretical, has been applied to these oxides with hopes of understanding, improving, and applying these oxides [5–8]. Among these probes is ultrasound.

2. ULTRASOUND

For probing a solid's bulk properties, ultrasonic wave velocity, or ultrasound, provides four common possibilities: velocity and attenuation of both longitudinal and transverse waves. If we use other waves — extensional, flexural, torsional — then other probes emerge. If we excite the normal-vibration frequencies of a regular solid (sphere, cube, and so on), then a spectrum of probes emerges. And, we should remember the distinction between phase and group velocities.

Ultrasound reflects long-wavelength acoustic phonons. Sometimes, optical phonons couple with acoustic and affect the ultrasound. The cubic–tetragonal phase transition near 110 K in $SrTiO_3$ provides a good example of this coupling. Upon cooling through the transition, the R-corner <u>optical</u> mode softens. Yet, the elastic constants, which reflect <u>acoustical</u> modes change drastically: the elastic compliance S_{11} increases enormously.

Ultrasound responds to changes in various material defects. These include lattice defects such as vacancies, dislocations, stacking faults, and twin boundaries. These include macroscopic defects such as cracks and voids. Ultrasound also responds to internal strain (residual stress).

Ultrasound velocity combines with mass density to give all the usual elastic constants. In turn, these relate to other physical properties: heat capacity, thermal expansivity, atomic volume, and so on.

Ultrasound provides a sensitive probe of phase transitions. It couples with various order parameters: strain, dielectric constant, magnetic ordering, and so on. Most lattice instabilities appear as ultrasound-velocity changes.

Ultrasound represents a fundamental physical property. We can show that the mean ultrasound velocity v_m depends only on atomic volume V_a and the Debye characteristic temperature Θ_D:

$$v_m \sim \Theta_D V_a^{1/3}. \tag{1}$$

For conventional superconductors, ultrasound provides an invaluable probe. For example, in vanadium [9] the elastic stiffnesses C_{ij} increase during cooling above T_c; below T_c they decrease continuously; the (100) [100] shear modulus C_{44} shows the largest decrease. The BCS theory predicts that the ultrasonic attenuation α should decrease steadily below T_c according to the following relationship [10]:

$$\alpha_s/\alpha_n = 2f(\Delta) = 2/[1 + \exp(\Delta/kT_c)]. \tag{2}$$

Here, f denotes the Fermi function and $\Delta(T)$ the energy-gap width, which vanishes at T_c. Thus, we see an explicit, simple relationship to electron-pairing mechanisms. Surprisingly, no one yet connected elastic constants, say the bulk modulus, to BCS theory. This should be possible; the electrons contribute to the bulk modulus. For a free-electron gas, in atomic units,

$$B_e = (1/V_0)(22.1/9)/r_0^2. \tag{3}$$

Here, V_0 denotes atomic volume: $(4/3)\,\pi r_0^3$.

Because of Eq. (1), the ultrasound velocity appears implicitly in the BCS expression for the superconducting-transition temperature:

$$T_c = 1.14\ \Theta_D \exp(-1/\lambda). \tag{4}$$

The electron–phonon parameter λ also depends on Θ_D, therefore on v_m. Using the BCS theory and taking $\lambda \sim \Theta^{-2}$ leads to a maximum T_c at $\lambda = 2$. This represents an artificial result of the weak-coupling approximation. Strong-coupling theories predict that T_c increases monotonically with increasing λ.

For the new oxide superconductors, ultrasound provides further opportunities. It couples with most known quasiparticles, many of which appear in various suggested models of a high T_c.

We can see the principal features of ultrasonic-wave propagation in the following well-known relationship:

$$u = A \exp(-\alpha x) \exp[ik(vt-x)]. \tag{5}$$

Here u denotes displacement, α attenuation, k wave number $(2\pi/\lambda)$, v velocity, t time, and x propagation direction.

The present report highlights the many wave-propagation studies. Limited space prohibits a comprehensive review. (Our files already contain about two hundred reprints–preprints on sound-velocity, elastic-constant studies on oxide superconductors.) On velocities and elastic constants, there exist at least three reviews [11–13]. We ignore completely the interesting, essential question of ultrasonic attenuation.

3. LOW-TEMPERATURE STUDIES

3.1. La–Sr–Cu–O

Probably, the first sound-velocity, elastic-constant study on high-T_c oxides arose from AT&T Laboratories. Bishop and coworkers [14] reported on La–Sr–Cu–O. They reported the 10–250-K relative longitudinal-velocity change. Figure 1 shows a subsequent, more-thorough study reported by Ledbetter and coworkers [15] for La_2CuO_4 (a semiconductor) and $La_{1.85}Sr_{0.15}CuO_4$ (a superconductor with $T_c \simeq 35$ K).

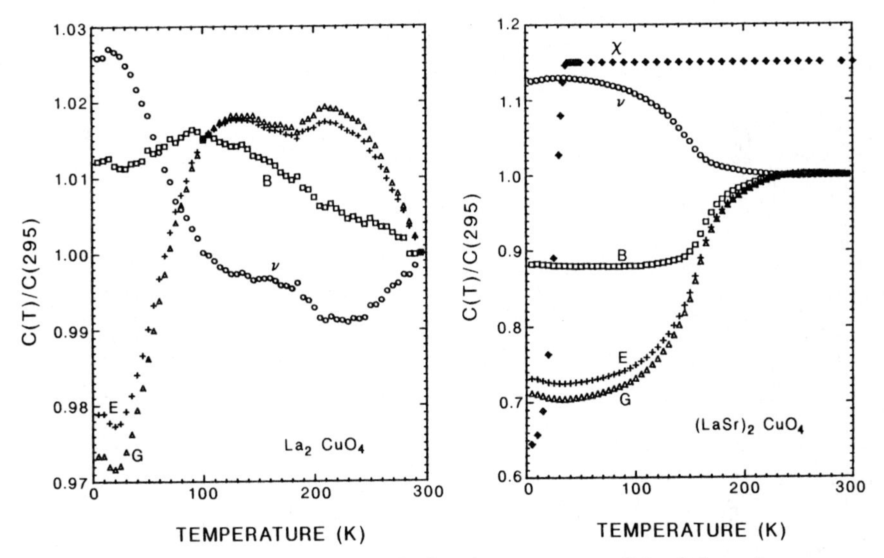

FIGURE 1. Temperature dependence of elastic constants of La_2CuO_4 and $La_{1.85}Sr_{0.15}CuO_4$. Note scale difference. For the superconductors, we show dc magnetic susceptibility, χ. B denotes bulk modulus, E Young modulus, G shear modulus, and ν Poisson ratio.

In their elastic-constant behavior, La–Cu–O and La–Sr–Cu–O show two similarities. First, the ambient-temperature elastic constants are quite similar; La–Sr–Cu–O is roughly ten percent stiffer. They show nearly identical Poisson ratios, which indicates similar interatomic bonding. At low temperatures, ~20 K for La–Cu–O, ~40 K for La–Sr–Cu–O, they show an elastic-stiffness minimum. This minimum may correspond to thermal-expansion anomalies, where in La–Sr–Cu–O the transition is normal–superconductive. The largest difference in the two materials appears in their temperature dependence. Upon cooling, La–Cu–O softens about three percent. La–Sr–Cu–O softens about thirty percent. Both softenings suggest irregular behavior during cooling. The La–Sr–Cu–O reversible, nearly hysteresis-free softening transition over a wide temperature interval suggests a second-order magnetic transition. The recent inelastic-neutron-scattering study by Shirane and coworkers [16] supports this view. They found that a spin-excitation intensity "approximately constant between 300 and 150 K, drops precipitously for temperatures down to about 50 K, and is then approximately constant down to 5 K."

3.2. Y-Ba-Cu-O

This oxide was the first found with a T_c exceeding liquid-nitrogen temperature (77 K). Probably, more Y-Ba-Cu-O studies exist than for all other combined oxide superconductors. Again, the first sound-velocity, elastic-constant report on this material arose from AT&T Laboratories. Bishop and coworkers [17] reported changes in longitudinal velocity between 50 and 150 K. Figure 2 shows a complete 5-295-K set of elastic constants reported by Ledbetter and Kim [18], who describe large thermal-hysteresis effects.

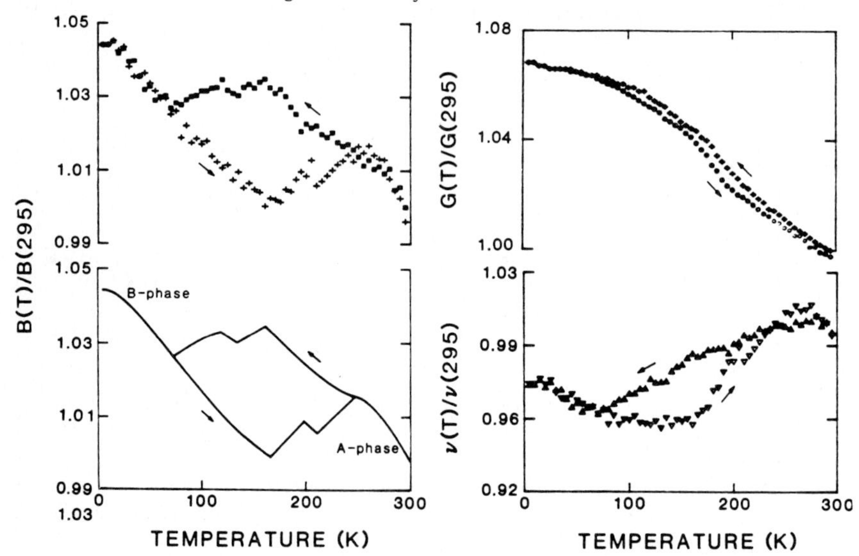

FIGURE 2. For $Y_1Ba_2Cu_3O_{7-x}$, temperature variation of B = bulk modulus, G = shear modulus, ν = Poisson ratio. Especially the dilatation modes show a large thermal hysteresis, which seems to consist of a two-step transition during both cooling and heating.

In another study, Ledbetter [19] summarized his group's findings on Y-Ba-Cu-O:

1. On different specimens, we detected no systematic dependence of elastic constants on oxygen content. We believe these effects are masked by material defects: voids and cracks, twins, texture.

2. The highest elastic constants yield Debye temperatures that agree with those derived from thermal-expansion and specific-heat studies.

3. Among all usual elastic constants, Poisson ratio shows the smallest variation: 0.21 ± 0.02, close to values for $BaTiO_3$ and $SrTiO_3$.

4. Measured versus temperature, sound velocities and elastic constants provide a sensitive, useful probe of changes occurring within the $Y_1Ba_2Cu_3O_{7-x}$ superconductor.

5. Elastic constants show irregularities below and above T_c. During

cooling, all the elastic constants show an irregularity near 200 K.

6. Within measurement error (about 1 part in 1000), none of the elastic constants show an abrupt change at T_c, 91 K.

7. Below T_c, all the elastic constants show regular behavior — except ν. This indicates continued changes in interatomic forces. This may support Geballe's view that the large elastic-constant change near T_c suggests that many electrons enter into Cooper pairs.

8. During cooling, between 160 and 70 K, the material behaves like it undergoes a sluggish phase transformation.

9. The large (4-percent) decrease in Poisson ratio is unexpected. It suggests large interatomic-force changes.

10. The unusual flatness of $\nu(T)$ near 295 K suggests unusual material-property changes above ambient temperatures.

3.3. Bi–Sr–Ca–Cu–O

Perhaps, Ledbetter and coworkers [20] first reported on this oxide. They reported results for the compound $(Bi,Pb)_2Sr_2Ca_2Cu_3O_{10}$. They found that Bi–Pb–Sr–Ca–Cu–O shows almost the same temperature behavior as Y–Ba–Cu–O. But it is much softer elastically, showing an elastic Debye temperature of 312 K versus 437 K for Y–Ba–Cu–O. Like Y–Ba–Cu–O, it shows hysteresis, bulk-modulus softening during cooling, and a large Poisson-ratio change. Suggesting similar interatomic bonding, the Poisson ratios show similar values.

3.4. Tl–Ba–Ca–Cu–O

So far, no one reported sound velocities of this oxide, where the $Tl_2Ba_2Ca_2Cu_3O_{10}$ compound shows the highest-known T_c: 125 K. Partly, this lack of information arises from thallium's toxicity and the difficulty of preparing large specimens. One expects this situation to change soon: probably a report will emerge before the present paper is published.

3.5. Other Oxides

Although we lack space to describe them, we mention studies on other oxides: Nd–Ce–Cu–O [21,22]; Ba–Pb–Bi–O [23].

4. CORRECTION TO THE VOID-FREE STATE

Most bulk oxide-superconductor specimens are sintered ceramics. Thus, they contain porosity, from a few percent up to fifty percent. Ledbetter and coworkers [24] gave an analytical procedure for correcting for spherical voids. For nonspherical voids, the problem is much more difficult, and we use a model by Ledbetter and Datta [25]. For Y–Ba–Cu–O, Fig. 3 shows the effects of both spherical and nonspherical voids.

5. GRÜNEISEN PARAMETER

From either the sound velocities or their pressure derivatives one can get the Grüneisen parameter γ, the best single parameter for characterizing anharmonic phenomena. Gamma enjoys general importance, especially for the superconductors where many believe the interatomic potential is anharmonic. For Y–Ba–Cu–O, some studies report $\gamma \simeq 3$. However, Ledbetter [26] argued that $\gamma \simeq 1.5$, a value consistent with most oxides.

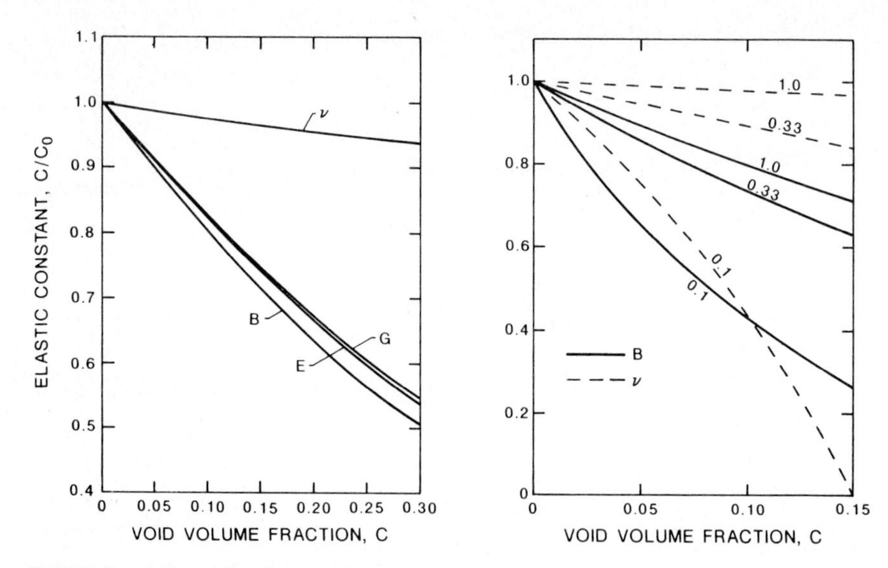

FIGURE 3. Left side shows effect of spherical voids on elastic constants ν = Poisson ratio, E = Young modulus, G = shear modulus. For B and ν, right side shows effect of voids with three aspect ratios: 1.0, 0.33, 0.1 (oblate disc).

REFERENCES

[1] J. G. Bednorz and K. A. Müller, Z. Phys. B (1986) 189.
[2] M. K. Wu et al., Phys. Rev. Lett. 58 (1987) 908.
[3] H. Maeda et al., Jap. J. Appl. Phys. 27 (1988) L209.
[4] Z. Z. Sheng and A. M. Hermann, Nature 332 (1988) 55.
[5] D. M. Ginsberg (ed.), Physical Properties of High-Temperature Superconductors I (World Scientific, Singapore, 1989).
[6] R. M. Metzger (ed.), High-Temperature Superconductivity: The First Two Years (Gordon and Breach, New York, 1988).
[7] J. C. Phillips, Physics of High-T_c Superconductors (Academic, San Diego, 1989).
[8] C. P. Poole, T. Datta, and H. A. Farach, Copper-Oxide Superconductors (Wiley, New York, 1988).
[9] G. A. Alers and D. L. Waldorf, IBM J. Res. Dev. 6 (1962) 89.
[10] J. Schrieffer, Theory of Superconductivity (Benjamin, N.Y., 1964) p. 107.
[11] H. Ledbetter, J. Metals 40 (1988, Jan.) 24.
[12] P. B. Allen, Z. Fisk, and A. Migliori, in Ref. [5], p. 213.
[13] J. Dominec, Supercond. Sci. Technol. 2 (1989) 91.
[14] D. J. Bishop et al., Phys. Rev. B 35 (1987) 8788.
[15] H. Ledbetter et al., Physica C (forthcoming).
[16] G. Shirane et al., BNL preprint (1989).
[17] D. J. Bishop et al., Phys. Rev. B 36 (1987) 2408.
[18] H. Ledbetter and S. Kim, Phys. Rev. B 38 (1988) 11857.
[19] H. Ledbetter, Proc. MRS Int. Meet. on Adv. Mater. (Tokyo, 1988).
[20] H. Ledbetter et al., Phys. Rev. B 39 (1989) 9689.
[21] A. Al-Kheffaji et al., submitted to J. Phys.: Cond. Matter.
[22] H. Ledbetter et al., nonpublished research (NIST, Boulder, 1989).
[23] T. Hikata et al., Phys. Rev. B 36 (1987) 5578.
[24] H. Ledbetter et al., J. Mater. Res. 2 (1987) 786.
[25] H. Ledbetter and S. K. Datta, J. Acoust. Soc. Amer. 79 (1986) 239.
[26] H. Ledbetter, Physica C 159 (1989) 488.

Elastic Waves and Ultrasonic Nondestructive Evaluation
S.K. Datta, J.D. Achenbach and Y.S. Rajapakse (Editors)

ACOUSTOELASTICITY OF MAGNETOELASTIC MATERIALS WITH ORTHOTROPIC SYMMETRY

Eiji MATSUMOTO* and Shinya MOTOGI**

* Department of Mechanical Engineering, Kyoto University, Kyoto 606, Japan
** Department of Mechanical Engineering, Osaka City University, Osaka 558, Japan

Acoustoelastic formulae for a magnetoelastic material with orthotropic symmetry are derived. The formulae enable us to separate the effects of the texture- and the stress-induced anisotropies on the speeds or the birefringences of ultrasonic waves. It is shown that the plane stress and the unknown anisotropic contribution to the birefringence can be determined from the birefringence of a normal incident wave by applying several kinds of magnetic fields.

1. INTRODUCTION

Many kinds of structural materials such as Fe-Ni-Co alloys are ferromagnetic materials. Some experimental studies have been done on the nondestructive stress evaluation by means of the magnetic fields. Utrata and Namkung [1,2] and Kwun [3,4] measured uniaxial stresses and Kwun [5] measured a hydrostatic stress from the speed shifts of ultrasonic waves due to applied magnetic fields. Recently, the authors [6] obtained the explicit expressions of the speeds of the waves in an isotropic deformed and magnetized material. They [7] showed for the isotropic material that the principal values and directions of a plane stress can be determined by measuring the acoustic birefringence under several kinds of magnetic fields.
In many cases, however, structural materials have weak orthotropic anisotropies induced by working, which influence the speeds of ultrasonic waves. Separation of the two effects on the speeds is an important subject of nondestructive stress evaluation based on acoustoelasticity. In this paper, we generalize our results to a material with orthotropic symmetry and propose a method for stress evaluation available in the case where weak unknown anisotropies are included.

2. GENERAL BASIC EQUATIONS

In order to derive acoustoelastic formulae for magnetoelastic materials, we should obtain the speeds of waves propagating in predeformed states under magnetic fields. Similarly to the case of (nonmagnetic) elastic materials, we define three states of deformations. That is, the original state is undeformed and nonmagnetized, the initial state is statistically and homogeneously deformed and magnetized, and the final state is achieved by superposing a small dynamic motion on the initial state. Let ξ_α , X_I and x_i denote the common Cartesian coordinate systems in the three states. Variables evaluated in each state are distinguished by superscripts r, i, f, respectively. When the applied magnetic field is weak or when a weak oscillating magnetic field is superposed, the relation of the magnetization and the magnetic field is reversible and linear. Then the material can regarded as a soft ferromagnetic material. According to Pao and Yeh [8], such soft ferromagnetic materials are governed by the balance equations:

Mass: $\dot{\rho} + \rho\, \nabla \cdot \mathbf{v} = 0$, (1)

Linear Momentum: $\nabla \cdot \mathbf{t} + \mu_0(\mathbf{M} \cdot \nabla)\mathbf{H} = \rho\dot{\mathbf{v}}$, (2)

Angular Momentum: $\mathbf{t} + \mu_0 \mathbf{H} \otimes \mathbf{M}$ = symmetric , (3)

Energy: $\rho\dot{U} = \mathrm{tr}[\mathbf{t}(\nabla \mathbf{v})] + \mu_0 \rho \mathbf{H} \cdot \dot{\overline{(\mathbf{M}/\rho)}}$ (4)

and the quasi-magnetostatic field equations:

Definition: $\mathbf{B} = \mu_0(\mathbf{H} + \mathrm{M})$, (5)

Gauss' Law: $\nabla \cdot \mathbf{B} = 0$, (6)

Ampere's Law: $\nabla \times \mathbf{H} = \mathbf{0}$. (7)

Here μ_0 denotes the magnetic permeability, ρ the mass density, $\mathbf{v}$ the velocity, t the Cauchy stress, $\mathbf{M}$ the magnetization, $\mathbf{H}$ the magnetic field, $\mathbf{B}$ the magnetic induction, and U the internal energy per unit mass. If the exchange energy is not taken into account, U can be expressed in terms of the Lagrangian strain $E_{\alpha\beta}$ and the quantity $N_\alpha = (M_I/\rho)(\partial X_I/\partial \xi_\alpha)$ in the original coordinate system:

$$2U = \rho^r \mu_0 A_{\alpha\beta} N_\alpha N_\beta + \rho^r \mu_0 B_{\alpha\beta\gamma\delta} E_{\alpha\beta} N_\gamma N_\delta + \frac{1}{\rho^r} C_{\alpha\beta\gamma\delta} E_{\alpha\beta} E_{\gamma\delta} + \frac{1}{3\rho^r} C_{\alpha\beta\gamma\delta\epsilon\eta} E_{\alpha\beta} E_{\gamma\delta} E_{\epsilon\eta}, \quad (8)$$

where we retain only the terms up to the order required in later discussions and $A_{\alpha\beta}$ is the inverse of the susceptibility tensor, $B_{\alpha\beta\gamma\delta}$ the magnetoelastic constants, $C_{\alpha\beta\gamma\delta}$ and $C_{\alpha\beta\gamma\delta\epsilon\eta}$ the second and the third order elastic constants.
From the fact that U is a function of $\mathbf{E}$ and $\mathbf{N}$, (4) implies that

$$T_{\alpha\beta} = \rho^r \frac{\partial U}{\partial E_{\alpha\beta}} + M_\alpha \frac{\partial U}{\partial N_\beta}, \quad (9)$$

$$\mu_0 H_\alpha = J \frac{\partial U}{\partial N_\alpha}, \quad (10)$$

where J is the Jacobian and $\mathbf{T}$ the Kirchhoff stress.
Define the following deviations from the initial to the final states

$$\mathbf{t} = \mathbf{T}^f - \mathbf{t}^i,\ \mathbf{m} = \mathbf{M}^f - \mathbf{M}^i,\ \mathbf{h} = \mathbf{H}^f - \mathbf{H}^i,\ \mathbf{b} = \mathbf{B}^f - \mathbf{B}^i . \quad (11)$$

Suppose that the initial deformation is small and the deviations are of a small order compared with the initial deformation. In the following calculations, we neglect higher order terms than the first order for the initial deformation and the second order for the initial magnetization. Substituting (8) into (9) and (10) and taking the difference of the results corresponding to the final and the initial states, we obtain the constitutive equations for the deviations:

$$h_I = \mathcal{A}_{IJ} m_J + \mathcal{B}_{IJK} \frac{\partial u_J}{\partial X_K}, \quad (12)$$

$$t_{IJ} = \mathcal{C}_{IJKL} \frac{\partial u_K}{\partial X_L} + \mathcal{D}_{IJK} m_K, \quad (13)$$

where

$$\mathcal{A}_{IJ} = A_{IJ}(1+e^i_{NN}) + A_{IK}\frac{\partial u^i_J}{\partial X_K} + A_{KJ}\frac{\partial u^i_I}{\partial X_K} + B_{KLIJ} e^i_{KL}, \quad (14)$$

$$\mathcal{B}_{IJK} = 2\delta_{JK}\mathcal{A}_{IL}M^i_L + \bar{B}_{JKIL}M^i_L \ , \tag{15}$$

$$\mathcal{C}_{IJKL} = \bar{C}_{IJKL} + \mu_0\delta_{KL}(\mathcal{A}_{JM}M^i_I + \bar{B}_{IJMN}M^i_N)M^i_M + \mu_0\bar{B}_{KLJM}M^i_M M^i_I \ , \tag{16}$$

$$\mathcal{D}_{IJK} = \mu_0\mathcal{A}_{IK}M^i_J + \mu_0\delta_{JK}\mathcal{A}_{IL}M^i_L + \mu_0\bar{B}_{IJLK}M^i_L \ , \tag{17}$$

$$\bar{B}_{IJKL} = B_{IJKL}(1+e^i_{NN}) + B_{IJKM}\frac{\partial u^i_L}{\partial X_M} + B_{IJML}\frac{\partial u^i_K}{\partial X_M} + B_{IMKL}\frac{\partial u^i_J}{\partial X_M} + B_{JMKL}\frac{\partial u^i_I}{\partial X_M} \ , \tag{18}$$

$$\bar{C}_{IJKL} = C_{IJKL}(1-e^i_{NN}) + C_{MJKL}\frac{\partial u^i_I}{\partial X_M} + C_{IMKL}\frac{\partial u^i_J}{\partial X_M} + C_{IJML}\frac{\partial u^i_K}{\partial X_M} + C_{IJKM}\frac{\partial u^i_L}{\partial X_M} + C_{IJKLMN}e^i_{MN} . \tag{19}$$

Subtracting the basic equations (1)-(7) in the initial state from those in the final state and substituting (12) and (13) into the result, we finally obtain the basic equations for the deviations:

$$\mathcal{G}_{IJKL}\frac{\partial^2 u_K}{\partial X_J \partial X_L} + \mathcal{K}_{IJK}\frac{\partial m_K}{\partial X_J} = \rho^i \frac{\partial^2 u_I}{\partial t^2} \ , \tag{20}$$

$$(\mathcal{A}_{IJ} + \delta_{IJ})\frac{\partial m_J}{\partial X_I} + \mathcal{B}_{IJK}\frac{\partial^2 u_J}{\partial X_I \partial X_K} = 0 \ , \tag{21}$$

$$\varepsilon_{IJK}\left(\mathcal{A}_{KL}\frac{\partial m_L}{\partial X_J} + \mathcal{L}_{KLM}\frac{\partial^2 u_L}{\partial X_J \partial X_M}\right) = 0 \ , \tag{22}$$

where

$$\mathcal{G}_{IJKL} = \bar{C}_{IJKL} + t^i_{JL}\delta_{IK} + \mu_0(\bar{B}_{IJMN}\delta_{KL} + \bar{B}_{KLIM}\delta_{JN} + \bar{B}_{KLJM}\delta_{IN} + \mathcal{A}_{LM}\delta_{IK}\delta_{JN} + \mathcal{A}_{IM}\delta_{KL}\delta_{JN} + \mathcal{A}_{JM}\delta_{KL}\delta_{IN})M^i_M M^i_N \ , \tag{23}$$

$$\mathcal{K}_{IJK} = \mu_0(\bar{B}_{IJKL} + \mathcal{A}_{JL}\delta_{IK} + \mathcal{A}_{IK}\delta_{JL} + \mathcal{A}_{JK}\delta_{IL})M^i_L \ , \tag{24}$$

$$\mathcal{L}_{KLM} = (\mathcal{A}_{KN}\delta_{LM} + \mathcal{A}_{MN}\delta_{KL} + \bar{B}_{LMKN})M^i_N \ . \tag{25}$$

Here the balance equations of mass, angular momentum and energy are automatically satisfied and we have used the condition that the initial deformation and the initial magnetization are static and homogeneous. Note that the basic equations are valid for a material with arbitrary anisotropy.

3. PLANE HARMONIC WAVES AND ORTHOTROPIC SYMMETRY

3.1. Propagation Conditions

Let us consider a plane harmonic wave with a small amplitude superposed on the initial state:

$$u_I = \bar{u}_I \exp[i(k\, N_J X_J - \omega t)] \ , \tag{26}$$

$$m_I = \bar{m}_I \exp[i(k\, N_J X_J - \omega t)] \ . \tag{27}$$

Substituting (26) and (27) into the basic equations (20)-(22), we obtain the system of equations for $\bar{\mathbf{u}}$ and $\bar{\mathbf{m}}' = i\bar{\mathbf{m}}$:

$$(k^2\Lambda_{IK} - \rho^i\omega^2\delta_{IK})\,\bar{u}_K - k\,\Gamma_{IK}\,\bar{m}'_K = 0 \ , \tag{28}$$

$$k\,\Phi_{IK}\,\bar{u}_K - \Psi_{IK}\,\bar{m}'_K = 0\,, \tag{29}$$

where

$$\Lambda_{IK} = \mathcal{G}_{IJKL}N_J N_L\,, \quad \Gamma_{IK} = \mathcal{K}_{IJK}N_J\,,$$
$$\Phi_{IK} = (\mathcal{B}_{JKL}N_L N_I + \mathcal{L}_{IKJ} - \mathcal{L}_{JKL}N_L N_I)N_J\,, \quad \Psi_{IK} = \mathcal{A}_{IK} + N_I N_K\,. \tag{30}$$

The speed and the coupled fields of a wave are given by a proper value and a proper vector of (28) and (29). In general, the wave has the coupled fields of the mechanical deformation and the magnetization, i.e., a magnetoelastic wave.

3.2. Orthotropic Symmetry Induced by Working

In many cases, a weak orthotropic anisotropy is induced by working like rolling, pulling etc. For single crystals, it is known that a strong magnetic uniaxial anisotropy is induced by rolling. The direction of the symmetry depends on the ratio of rolling and the rolling direction relative to the crystal symmetry. For polycrystals, from Namkung [9] and Kwun [10], the effects of the induced magnetoelastic anisotropy on the wave speeds are of the same order as those of the applied stresses. Since the stress effects are the same order as those of the mechanical anisotropy induced by working, we may assume that the magnetic properties have also orthotropic anisotropy of the same order as the mechanical anisotropy and their axes of orthotropy are common. If we take the axes of orthotropy as the coordinate axes, the material constants in (8) take the forms.

$$A_{IJ}=\begin{vmatrix} a_1 & 0 & 0 \\ 0 & a_2 & 0 \\ 0 & 0 & a_3 \end{vmatrix},\ B_{IJKL}=\begin{vmatrix} b_{11} & b_{12} & b_{13} & 0 & 0 & 0 \\ b_{21} & b_{22} & b_{23} & 0 & 0 & 0 \\ b_{31} & b_{32} & b_{33} & 0 & 0 & 0 \\ 0 & 0 & 0 & b_{44} & 0 & 0 \\ 0 & 0 & 0 & 0 & b_{55} & 0 \\ 0 & 0 & 0 & 0 & 0 & b_{66} \end{vmatrix},\ C_{IJKL}=\begin{vmatrix} c_{11} & c_{12} & c_{13} & 0 & 0 & 0 \\ c_{12} & c_{22} & c_{23} & 0 & 0 & 0 \\ c_{13} & c_{23} & c_{33} & 0 & 0 & 0 \\ 0 & 0 & 0 & c_{44} & 0 & 0 \\ 0 & 0 & 0 & 0 & c_{55} & 0 \\ 0 & 0 & 0 & 0 & 0 & c_{66} \end{vmatrix}. \tag{31}$$

For the third order elastic constants C_{IJKLMN}, the independent nonvanishing components are given in Voigt notation by

$$\begin{aligned} &c_{111},\ c_{112},\ c_{113},\ c_{122},\ c_{123},\ c_{133},\ c_{144},\ c_{155},\ c_{166},\ c_{222} \\ &c_{223},\ c_{233},\ c_{244},\ c_{255},\ c_{266},\ c_{333},\ c_{344},\ c_{355},\ c_{366},\ c_{456} \neq 0\,. \end{aligned} \tag{32}$$

4. SPEEDS OF WAVES AND ACOUSTIC BIREFRINGENCES

For simplicity, we assume that one of the principle axes of the strain coincides with one of the axes of orthotropy, i.e., X_3 axis and the wave propagates along X_3 axis. We restrict ourselves to the special cases where the initial magnetization is perpendicular to the propagation direction. Even in the special cases, it needs enormous analytical calculations to obtain the speeds of the waves, so that we use symbolic manipulation system REDUCE for the derivation. When the direction of the initial magnetization coincides with the X_1 axis, the speeds of the longitudinal and the two transverse waves are expressed, respectively, as

$$\begin{aligned} \rho^i v_1{}^2 =\ & e^i{}_{11}\mu_0 M_1{}^2 b_{55}(b_{13}b_{55}-b_{13}-6b_{55}+3)+e^i{}_{11}(c_{55}+c_{155})+e^i{}_{22}\mu_0 M_1{}^2 b_{55}(b_{23}b_{55} \\ & -b_{23}-2b_{55}+1)+e^i{}_{22}(-c_{55}+c_{255})+e^i{}_{33}\mu_0 M_1{}^2 b_{55}(b_{33}b_{55}-b_{33}-6b_{55}+3)+e^i{}_{33} \\ & (c_{55}+c_{355})+t^i{}_{33}+\mu_0 M_1{}^2 b_{55}(-b_{55}+1)+c_{55}\,, \end{aligned} \tag{33}$$

$$\rho^i v_2{}^2 = e^i{}_{11}(-c_{44}+c_{144})+e^i{}_{22}(c_{44}+c_{244})+e^i{}_{33}(c_{44}+c_{344})+t^i{}_{33}+c_{44}\,, \tag{34}$$

$$\rho^i v_3{}^2 = [e^i{}_{11}\mu_0 M_1{}^2 b_{31}{}^2(-3a_1+b_{11}) + e^i{}_{11}a_1{}^2(-c_{33}+c_{133}) + e^i{}_{22}\mu_0 M_1{}^2 b_{31}{}^2(-a_1+b_{21})$$
$$+ e^i{}_{22}a_1{}^2(-c_{33}+c_{233}) + e^i{}_{33}\mu_0 M_1{}^2 b_{31}{}^2(-5a_1+b_{31}) + e^i{}_{33}a_1{}^2(3c_{33}+c_{333})$$
$$+ t^i{}_{33}a_1{}^2 - \mu_0 M_1{}^2 a_1 b_{31}{}^2 + a_1{}^2 c_{33}]/a_1{}^2 \ . \qquad (35)$$

The acoustic birefringence is then given from (34) and (35) by

$$B = \rho^i v_1{}^2 - \rho^i v_2{}^2 = e^i{}_{11}\mu_0 M_1{}^2 b_{55}(b_{13}b_{55} - b_{13} - 6b_{55} + 3) + e^i{}_{11}(c_{44} + c_{55} - c_{144} + c_{155})$$
$$+ e^i{}_{22}\mu_0 M_1{}^2 b_{55}(b_{23}b_{55} - b_{23} - 2b_{55} + 1) + e^i{}_{22}(-c_{44} - c_{55} - c_{244} + c_{255}) + e^i{}_{33}\mu_0 M_1{}^2 b_{55}$$
$$(b_{33}b_{55} - b_{33} - 6b_{55} + 3) + e^i{}_{33}(-c_{44} + c_{55} - c_{344} + c_{355}) + \mu_0 M_1{}^2 b_{55}(-b_{55} + 1) - c_{44} + c_{55} \ . \qquad (36)$$

We see that the speeds and the acoustic birefringence are always of the second order in the initial magnetization, which agrees with the experimental results.

5. SEPARATION OF STRESS- AND TEXTURE-INDUCED ANISOTROPIES

5.1. Acoustoelastic Formulae Expressed in Terms of $\mathbf{e}^i$ and $\mathbf{M}^i$

The acoustic birefringences given in the preceding section are expressed in terms of the initial strain and the initial magnetization in the form

$$B = k_1 + k_2 M^2 + (k_3 + k_4 M^2)e^i{}_{11} + (k_5 + k_6 M^2)e^i{}_{22} + (k_7 + k_8 M^2)e^i{}_{33} \ , \qquad (37)$$

where M is the amplitude of the magnetization and the coefficients k_1-k_8 are determined by the direction of the magnetization and the material constants. In order to apply the birefringence formulae to stress evaluation, the initial strain should be replaced by the initial stress. The initial magnetization should be also replaced by the initial magnetic field, because the magnetic field is easily controlled rather than the magnetization. To do it, we must solve equations (9) and (10) for the initial strain and the initial magnetization. By use of REDUCE, the derivations are possible only in special cases such that uniaxial and plane stresses.

5.2. Plane Stress

Consider the case where the principal directions of a plane stress coincide with X_1 and X_2 axes and the wave propagates along the X_3 axis (a normal incident wave). The direction of the initial magnetic field is assigned to coincide with X_1 or X_2 axis. Then the acoustic birefringence takes the form

$$B = k_9 + (k_{10} + k_{11}\cos 2\theta)H^2 + [k_{12} + (k_{13} + k_{14}\cos 2\theta)H^2]\sigma_1 + [k_{15} + (k_{16} + k_{17}\cos 2\theta)H^2]\sigma_2 \ , \qquad (38)$$

where σ_1 and σ_2 are the principal stresses, H the amplitude of the magnetic field and θ the angle between x_1 axis and the direction of the initial magnetic field. The above coefficients k_9-k_{17} are expressed in terms of the material constants. The explicit expressions of the coefficients are too long to lay down here, but they can be determined by simple experiments from (38) by applying specified stresses and magnetic fields. When the material is isotropic, we have

$$k_9 = k_{10} = 0, \ k_{11} = 0, \ k_{12} = -k_{15}, \ k_{13} = -k_{16}, \ k_{14} = k_{17} \ . \qquad (39)$$

In view of (38) and $(39)_1$, k_9 indicates the effect of the texture-induced anisotropy on the acoustic birefringence. We can also show that the term including k_{10} in (38) is sufficiently small compared with other terms. Since the induced anisotropies are small, we can approximate the coefficients in the orthotropic case by those in the isotropic case, when they are multiplied by other small quantities. Thus, we may employ the data on an isotropic virgin material for the coefficients except k_9 and k_{10}. To determine the principal values of the stress and the unknown anisotropic contribution k_9, we first measure the birefringence B_0 without applying the magnetic field and we obtain from (38)

$$k_9 + k_{12}\sigma_1 + k_{15}\sigma_2 = B_0 . \tag{40}$$

We next apply two kinds of magnetic fields in the directions of X_1 and X_2 axes, and write the two acoustic birefringences as B_{H1} and B_{H2}, respectively. Putting $\mathbf{H} = (H, 0, 0)$ and $\mathbf{H} = (0, H, 0)$ in (38), we have

$$k_9 + (k_{10}+k_{11})H^2 + [k_{12}+(k_{13}+k_{14})H^2]\sigma_1 + [k_{15}+(k_{16}+k_{17})H^2]\sigma_2 = B_{H1} , \tag{41}$$

$$k_9 + (k_{10}-k_{11})H^2 + [k_{12}+(k_{13}-k_{14})H^2]\sigma_1 + [k_{15}+(k_{16}-k_{17})H^2]\sigma_2 = B_{H2} . \tag{42}$$

To obtain more accurate values of the right hand sides of (40)-(42), we may draw two curves, $B_{H1}-H^2$ and $B_{H2}-H^2$, from the acoustic birefringences by changing the amplitudes of the magnetic fields. Let ψ_1 be the angle between the horizontal line (H^2 axis) and the bisector of B_{H1} and B_{H2} curves, and ψ_2 the angle between the two curves. Although the two curves may not be straight for large H, we need only their initial slopes at $H = 0$. As a result, the three equations (40)-(42) can be solved for the principal values of the stress and k_9 as

$$\begin{aligned}
\sigma_1 &= (k_{17}\tan\psi_1 - k_{16}\tan\psi_2 + k_{11}k_{16} - k_{10}k_{17}) / \Delta , \\
\sigma_2 &= (-k_{14}\tan\psi_1 + k_{13}\tan\psi_2 - k_{11}k_{13} + k_{10}k_{14}) / \Delta , \\
k_9 &= B_0 - [(k_{12}k_{17} - k_{14}k_{15})\tan\psi_1 + (k_{13}k_{15} - k_{12}k_{16})\tan\psi_2 - k_{11}k_{12}k_{16} \\
&\quad + k_{10}k_{16}k_{17} - k_{11}k_{13}k_{15} + k_{10}k_{14}k_{15}] / \Delta ,
\end{aligned} \tag{43}$$

where $\Delta = k_{13}k_{17} - k_{14}k_{16}$. It is worth noting that the speed shift of a wave due to the stress and the magnetic field is expressed in a similar form to (38). Thus if the speed shift can be measured in a sufficient accuracy, a similar evaluation of the stress is possible instead of measuring the birefringences. The case of uniaxial stresses is a special case of the above discussions where $\sigma_1 = 0$ or $\sigma_2 = 0$.

REFERENCES

[1] Namkung, M., Utrata, D., Allison, S.G. and Heyman, J.S., Quantitative Nondestructive Evaluation, 5B (1986) 1481.

[2] Utrata, D. and Namkung, M., Quantitative Nondestructive Evaluation, 6B (1987) 1585.

[3] Kwun, H. and Teller, C.M., J. Appl. Phys., 54 (1983) 4856.

[4] Kwun, H., J. Appl. Phys., 57 (1985) 1555.

[5] Kwun, H., Mater. Eval., 44 (1986) 1560.

[6] Matsumoto, E., and Motogi,S., Acoustoelastic Effects in Soft Ferromagnets, in: Yamamoto, Y. and Miya, K.,(eds.), Electromagnetomechanical Interactions in Deformable Solids and Structures (North-Holland, 1986) pp. 303-308.

[7] Motogi, S. and Matsumoto, E., Proceedings of the Japan Congress on Materials Research (1989) 110.

[8] Pao, Y.-H. and Yeh, C.-S., Int. J. Engng Sci., 11 (1973) 415.

[9] Namkung, M., Utrata, D., Allison, S.G. and Heyman, J.S., Quantitative Nondestructive Evaluation, 5B (1986) 1489.

[10] Kwun, H., J. Appl. Phys. 57 (1985) 1397..

Elastic Waves and Ultrasonic Nondestructive Evaluation
S.K. Datta, J.D. Achenbach and Y.S. Rajapakse (Editors)

THEORETICAL REPRESENTATION OF AE AND SOURCE INVERSION PROCEDURE

Masayasu OHTSU

Department of Civil and Environmental Engineering
Kumamoto University, Kumamoto 860, Japan

Acoustic Emission (AE) is defined as elastic waves emitted due to microfracturing. To confirm the applicability of elastodynamics to AE waveforms, a simulation analysis of a detected waveform in the debonding process is performed, based on the integral representation with a dynamic dislocation model. By utilizing a moment tensor representation, a source inversion procedure is proposed for classifying crack types and determining crack orientations. Source characteristics determined in a pull-out test are in remarkable agreement with experimental findings.

1. INTRODUCTION

Elastic waves generated by dislocation, transformation, and microfracturing in a solid are defined as acoustic emission (AE). The generation of elastic waves due to fracturing has long been investigated in seismology. Therefore, theoretical treatment of AE is closely associated with quantitative seismology [1]. Lamb's problem of determining transient elastic disturbance was first applied by Breckenridge et al. [2] to AE research. They showed experimental AE waveforms similar to Lamb's solutions. Their success bred the study of AE waveforms, based on a dislocations model and Green's functions [3],[4],[5].

The present paper summarizes theoretical treatments on AE and proposes a source inversion procedure based on a moment tensor representation. A SiGMA (Simplified Green's function for Moment tensor Analysis) inversion to determine moment tensor components and a unified decomposition of eigenvalues to determine crack types and crack orientations are discussed. The proposed procedure is applied to a simulation analysis of AE waveforms in the debonding process of stainless steel from base metal and a source inversion in a pull-out test of anchor-bolt from concrete block.

2. THEORETICAL REPRESENTATION OF ACOUSTIC EMISSION

2.1 Integral Representation

Dynamic displacement component, $u_i(\mathbf{x},t)$, is represented by the integration on boundary surface S in elastodynamics,

$$u_i(\mathbf{x},t) = \int_S [G_{ik}(\mathbf{x},\mathbf{x'},t)*t_k(\mathbf{x'},t) - T_{ik}(\mathbf{x},\mathbf{x'},t)*u_k(\mathbf{x'},t)]dS. \quad (1)$$

Where G_{ik} is Green's function and T_{ik} is the associated traction,

$$T_{ik} = C_{kjpq}G_{ip,q}n_j, \quad (2)$$

where C_{kjpq} shows components of elastic constants and n_j is the unit normal vector to the boundary surface. $G_{ip,q}$ implies the spatial derivative of G_{ip}

in respect to y_q.

The effect of crack formation is introduced, replacing the surface integration on boundary S by crack surface F. The surface, F, is assumed to consist of two surfaces, which are originally in contact and in motion together. Due to the crack nucleation, the discontinuity of displacement components b_k is generated on crack surfaces F^+ and F^-, as follows;

$$b_k(\mathbf{x'},t) = u_k^+(\mathbf{x'},t) - u_k^-(\mathbf{x'},t). \tag{3}$$

While traction t_k is kept continuous on the crack surface. Substituting these conditions on crack surface F, eq. 1 becomes [5],

$$u_i(\mathbf{x},t) = \int_F T_{ik}(\mathbf{x},\mathbf{x'},t) * b_k(\mathbf{x'},t)dS. \tag{4}$$

In the case that the crack size is small compared with the distance from AE source $\mathbf{x'}$ to sensor location $\mathbf{x}$, the term, T_{ik}, can be taken outside the integration. In addition, separating spatial and time dependences of the crack function, $b_k(\mathbf{x},t)$, the integration becomes,

$$\int_F b_k(\mathbf{x'},t)dS = bl_k S(t). \tag{5}$$

In eq. 5, l_k indicates the direction vector of crack motion and b represents the crack volume. S(t) is the source-time function, presenting the time dependence of crack formation. The determination of the source-time function is well known as the deconvolution analysis [6].

2.2 SiGMA Inversion

Not only traction Green's function, T_{ik}, contains the unit normal to the crack surface, but also crack vector b_k exists in eq. 4. It implies that both two directions must be determined in a source inversion procedure. Therefore, eq. 4 is modified and thus a moment tensor is introduced, taking into account eqs. 2 and 5,

$$\begin{aligned} u_i(\mathbf{x},t) &= C_{kjpq} G_{ip,q}(\mathbf{x},\mathbf{x'},t) n_j b l_k * S(t) \\ &= G_{ip,q} m_{pq} * S(t). \end{aligned} \tag{6}$$

m_{pq} is called the moment tensor. In the case of isotropic materials, the moment tensor is represented, by employing Lame constants λ and μ,

$$m_{pq} = b(\lambda l_k n_k \delta_{pq} + \mu l_p n_q + \mu l_q n_p). \tag{7}$$

To determine moment tensor components from P-wave amplitudes of AE waveforms, a SiGMA (Simplified Green's function for Moment tensor Analysis) inversion is developed. Selecting the P-wave term in an infinite space, eq. 6 becomes,

$$u_i = r_i r_p r_q m_{pq} / (4 \pi \rho v_p^3 R), \tag{8}$$

where ρ is the mass density and v_p is P wave velocity. R is the distance between sensor location $\mathbf{x}$ and AE source $\mathbf{x'}$, and r_i is its direction cosine. Because only the amplitude of P wave is taken into account, the time dependence is omitted in eq. 8.

Provided that a source location analysis is carried out, based on the travel time differences of P waves, information on the distance, R, and the direction cosine, $\mathbf{r}$ is obtained. Consequently, a set of eqs. 8 at all sensor locations constitutes linear algebraic equations with unknown m_{pq}, substituting each P-wave amplitude into the left-hand-side of eq. 8. Because relative values of tensor components are necessary, the minimum number of sensor locations is six.

In this case, the requirement for monitoring AE waveforms is the equivalent sensitivity at all detecting devices. This is the attainable condition, because the conventional standards for commercial AE devices demands for the sensitivity variation within 3 dB.

2.3 Unified Decomposition

The determination of crack type and crack orientation is derived from an eigenvalue analysis of the moment tensor [7]. Because direction vector **l** is parallel to normal vector **n** for a pure tensile crack, three eigenvalues are easily expected from eq. 7, as follows;

$$E_1 = b(\lambda + 2\mu),\ E_2 = b\lambda \text{ and } E_3 = b\lambda. \tag{9}$$

These eigenvalues are further decomposed into compensated linear vector dipoles (CLVD) [8] of $4\mu b/3$, $-2\mu b/3$, and $-2\mu b/3$, and an isotropic component of $b(3\lambda + 2\mu)/3$.

Results of the eigenvalue analysis provides three corresponding eigenvectors,

$$\mathbf{l} + \mathbf{n},\ \mathbf{l} \times \mathbf{n} \text{ and } \mathbf{l} - \mathbf{n}. \tag{10}$$

It shows that the direction of crack opening (**l** or **n**) is identical to the direction of the first eigenvector.

For a pure shear crack, crack vector **l** is vertical to normal vector **n** and the eigenvalue analysis provides the following results,

$$E_1 = \mu b,\ E_2 = 0 \text{ and } E_3 = -\mu b. \tag{11}$$

In this case, either the direction of crack motion **l** or unit normal **n** is determined from the sum of the first and the third eigenvectors, as can be seen in eq. 10.

In a practical situation, AE sources are reasonably assumed to have the combined effects of tensile and shear cracks. On the basis of the eigenvalue analysis, the eigenvalues are uniquely decomposed as shown in Fig. 1. From eq. 11, the contribution of shear crack is referred to as (X, 0, -X). The CLVD components correspond to (Y, -0.5Y, -0.5Y), and the isotropic component is represented as Z. These are mathematically expressed, as follows;

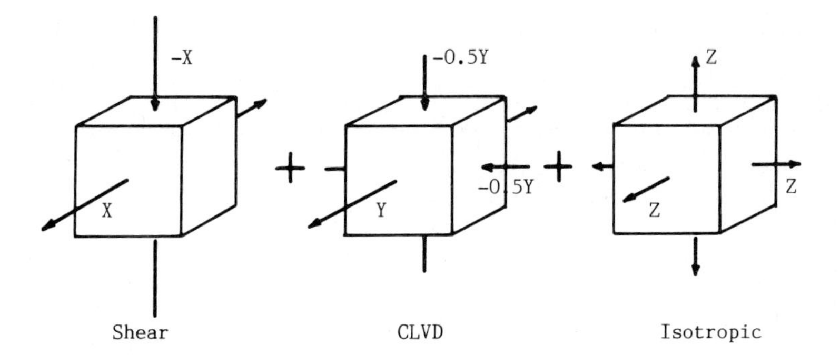

Fig. 1 Unified decomposition of eigenvalues into shear, CLVD, and isotropic models.

$$\begin{aligned} 1.0 &= X + Y + Z \\ E_2/E_1 &= 0 - 0.5Y + Z \\ E_3/E_1 &= -X - 0.5Y + Z \end{aligned} \qquad (12)$$

This unified decomposition provides a rational basis for the crack classification. Taking into account the results of error estimation [7], the following criterion is developed. In the case that the ratio X (%) is greater than 55 %, AE source is classified into a shear crack and the crack orientation of shear type is determined. In the reverse case (the ratio X (%) is less than 45 %), AE source is classified into a tensile crack and the crack orientation is determined as the direction of crack opening.

3. SIMULATION ANALYSIS

Fig. 2 (a) shows an AE waveform detected during the debonding process of stainless steel from base metal. According to the micrgraph observation, a tensile crack of approximately penny-shape was found at the interface between stainless steel and base metal. To simulate the AE waveform, a penny-shaped crack is assumed,

$$b(\mathbf{x'}) = 2h\sqrt{1 - (r/a)^2}, \qquad r = |\mathbf{x'} - \mathbf{xc}|$$

where 2h is the crack opening displacement and a is the radius of the crack, of which center is located at **xc**. Thus, eq. 5 becomes,

$$\int_F b(\mathbf{x'})dS = 4\pi a^2 h/3. \qquad (13)$$

The micrograph observation informed that $h = 2\,\mu m$ and $a = 20\,\mu m$. The source-time function previously employed [5] was selected to implement the time dependence of the debonding source,

$$\begin{aligned} S(t) &= t/Tr - 2\sin(2\pi t/Tr)/(3\pi) + \sin(4\pi t/Tr)/(12\pi) \quad &\text{for } 0 < t < Tr, \\ &= 1 \quad &\text{for } t > Tr, \end{aligned}$$

where Tr is the rise time of crack nucleation and was chosen as 20 μs in the simulation analysis.

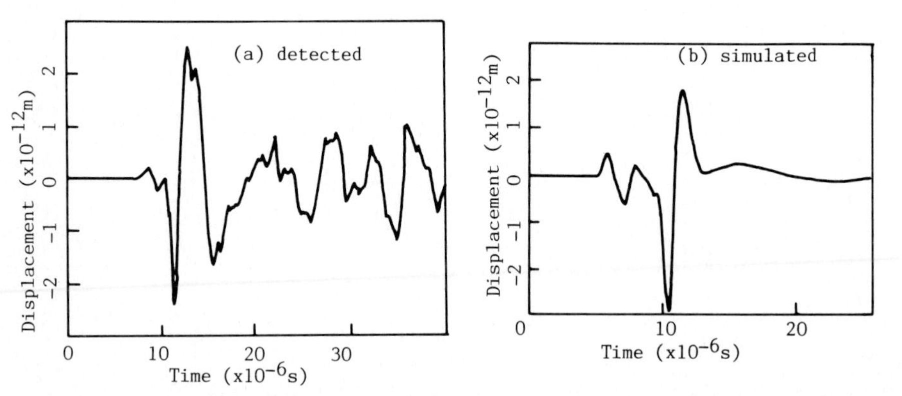

Fig. 2 (a) AE waveform detected during the debonding process and (b) the simulated one in a half space.

The waveform in Fig. 2 (a) was detected on the stainless layer of 5 mm thickness bonded to base metal. By the source location analysis [9], the epicenter of the source in Fig. 2 (a) was located at 27 mm apart from the displacement sensor. Because the waveform was detected at the surface of the stainless layer, Green's function in a half space [10] was applied. A result of the symulation analysis by eqs. 4 and 13 is shown in Fig. 2 (b). The duration preceding the first motion in Fig. 2 (a) is meaningless because it depends on the triggering time of the recorder, while that of Fig. 2 (b) is exactly identical to the travel time of P wave. Note that the time scale of the horizontal axsis in Fig. 2 (a) is different from that of Fig. 2 (b), because the detected waveform was recorded in analog form. Taking it into consideration, it is known that the beginning portion is in good agreement with that of the simulated one. The latter portion of the detected wave is smeared due to reflections which are not observed in the half-space solution. The comparison of the beginning portions in Figs. 2 (a) and (b) shows that the essential feature of the detected wave is recovered by the simulation analysis, including the amplitudes. This confirms the applicability of elastodynamics to the quantitative analysis of AE waveforms.

4. SOURCE INVERSION

The source inversion procedure of the SiGMA inversion and the unified decomposition were applied to AE waveforms in a pull-out test of hooked anchor-bolt from concrete block. A concrete block was of dimension 0.6 m x 0.6 m x 0.3 m, of which concrete strength was 45.8 Mpa under uni-axial compression and P wave velocity was 4070 m/s. An anchor plate with a nut was embedded in the block and was pulled out by a center-hole jack until the a frustum of concrete cone was spalled off. Six AE sensors of 150 kHz resonant frequency were attached to the concrete block. Total amplification of AE signals was 60 dB and the frequency range was 30 kHz to 80 kHz.

Reading P-wave travel time, locations of AE sources were determined by the 3D source location procedure developed [10]. Moment tensor components were obtained, reading the amplitudes of first motions and substituting into eq. 8. Following the eigenvalue analysis, the unified decomposition was performed and all AE events were classified into either tensile cracks or shear cracks, on

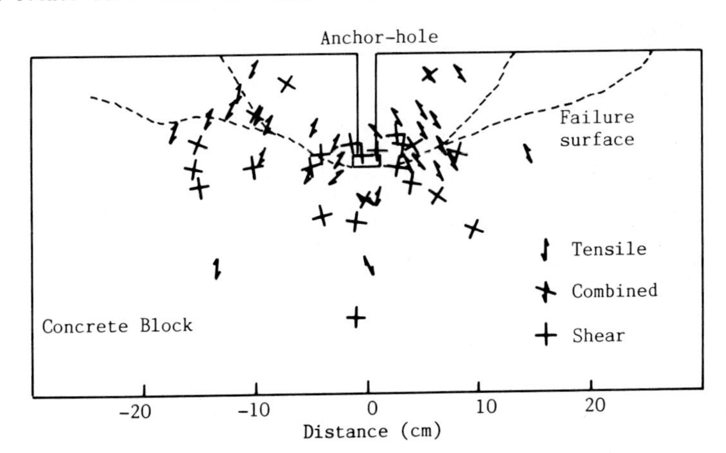

Fig. 3 A plane view of locations, types, and orientations of AE sources analyzed in the pull-out test of anchor-bolt, based on the source inversion procedure.

the basis of the ratio X. In some AE events, however, X ratios were determined between 45% and 55%. For these events, both directions of a tensile case and a shear case were plotted as "combined".

A plane view of AE sources located, classified, and oriented is shown in Fig. 3. AE sources are plotted at their locations with their types and orientations. In the figure, spalled failure surfaces are also drawn by broken lines. It is observed that the directions of tensile cracks are in remarkable agreement with opening directions of the failure surface. In addition, AE events of shear type are oriented parallel to the failure surafces. These results agree well with experimental findings.

5. CONCLUSIONS

Theoretical treatment of AE is summarized, based on the integral representation and the moment tensor representation. Both time and spatial dependences of the crack function are prescribed for the AE event detected in the debonding process of stainless steel from base metal. On the basis of micrograph observation data, the beginning portion of AE waveform is reasonably simulated as the half-space solution, including the amplitudes.

The source inversion procedure of the SiGMA inversion and the unified decomposition was applied to AE waveforms detected during the pull-out test of anchor-bolt from concrete block. Crack locations, types, and orientations determined by the source inversion procedure are in good agreement with experimental findings. It shows a great promise of the SiGMA inversion and the unified decomposition for clarifying the mechanisms of crack generation.

ACKNOWLEDGEMENTS

The author wishes to thank Dr. S. Yuyama in Physical Acoustics Corp. Japan, Mr. M. Shigeishi in Kumamoto University, and Mr. H. Iwase in Gifu University for their valuable efforts in performing this research.

REFERENCES

[1] Aki, K. and Richards, P. G., Quantitative Seismoligy Theory and Methods, Vol. I and Vol. II (W. H. Freeman and Company, San Francisco, 1980).
[2] Brechenridge, F. R., Tschiegg, C. E. and Greenspan, M., J. Acoust. Soc. Am, 57(3) (1975) 626.
[3] Pao, Y. H., Theory of Acoustic Emission: Elastic Waves and Nondestructive Testing of Materials, AMD Vol. 29 (ASME, New York, 1978), pp. 107-128.
[4] Wadley, H. N. G., Scruby, C. B. and Shrimpton, G., AERE-R9644, AERE Harwell (1980).
[5] Ohtsu, M., J. Acoustic Emission, 1(2) (1982) 103.
[6] Hsu, N. N. and Hardy, S. C., Experiments in Acoustic Waveform Analysis for Characterization of AE Sources, Sensors and Structures: ibid. [3], pp. 85-106.
[7] Ohtsu, M., Materials Evaluation, 45(9) (1987) 1070.
[8] Knopoff, L. and Randall, M. J., J. Geophys. Res., 75(26) (1970) 4957.
[9] Yuyama, S., Imanaka, T. and Ohtsu, M., J. Acoust. Soc. Am., 83(3) (1988) 976.
[10] Ohtsu, M. and Ono, K., J. Acoustic Emission, 3(1) (1984) 27.
[11] Ohtsu, M. and Ono, K., NDT International, 21(3) (1988) 143.

Elastic Waves and Ultrasonic Nondestructive Evaluation
S.K. Datta, J.D. Achenbach and Y.S. Rajapakse (Editors)

ULTRASONIC STUDIES OF DEFORMATION AND FAILURE OF ROCKS

Colin M. SAYERS

Koninklijke/Shell Exploratie & Produktie Laboratorium
Volmerlaan 6
2288 GD Rijswijk ZH
The Netherlands

The failure of brittle rocks is preceded by the formation of microcracks at defects which enable the microcracks to grow under a local tensile stress, even when the externally applied stress is compressive. Ultrasonic wave velocities in rock are reduced in the presence of open microcracks and fractures and may therefore be used to monitor the progressive damage of the rock. In general, these microcracks are not randomly orientated and the rock displays an elastic anisotropy determined by the shape and content of the cracks and by the crack orientation distribution function. This function gives the probability of a crack having a given orientation with respect to a set of axes fixed in the rock, and is used to calculate the variation of elastic wave velocity with propagation direction. The coefficients $W_{\ell mn}$ of a series expansion of the crack orientation distribution function in generalised spherical harmonics can be obtained to order $\ell = 4$ from the angular variation of the ultrasonic wave velocity. This allows construction of microfracture pole figures, which may be compared with those obtained by petrofabric examination. Since the anisotropy in strength properties originates from the preferred orientation of microstructural defects, the prediction of the microfracture orientation distribution will have an important application in the field of rock fracture.

1. INTRODUCTION

The failure of brittle rocks during compression is thought to be preceded by the formation, growth and coalescence of microcracks [1, 2]. Tensile stresses necessary for microcrack growth originate from shear along pre-existing microcracks and stress concentrations around inhomogeneities. Laboratory measurements of ultrasonic wave velocities in rock samples as a function of confining pressure have demonstrated that the presence of microcracks significantly decreases the elastic wave velocities in the rock [3, 4]. As the pressure is increased the measured velocities increase markedly. This behaviour is attributed to the closure of microcracks with increasing pressure until they have no effect on the elastic properties of the rock. Ultrasonic wave velocities may therefore be used to monitor the progressive damage of the specimen.

Recently, an inversion technique for obtaining the orientation distribution of microcracks from measured ultrasonic wave velocities has been developed [5] and was applied [6] to determine the orientation distribution of open microcracks in Barre granite under uniaxial compressive stress from the measurements of Nur and Simmons [7]. The stresses applied during this experiment, however, were rather low and crack propagation did not occur. In the present work ultrasonic velocity measurements in three orthogonal directions in a Berea sandstone cube stressed to peak in a true triaxial loading apparatus are discussed and application of the theory to the determination of in-situ stress is made.

2. THEORY

It is assumed that the mechanical behaviour of rock is determined by the formation, growth and coalescence of microcracks. As a result of the anisotropy of the stress field, these microcracks show some degree of preferred orientation. To model the effect of this on ultrasonic velocities these cracks are approximated by ellipsoidal voids embedded in an isotropic background medium. The orientation of an ellipsoidal crack with principal axes $OX_1X_2X_3$ with respect to a set of axes $Ox_1x_2x_3$ fixed in the rock may be specified by three Euler angles ψ, θ and ϕ. The orientation distribution of cracks is then given by the crack orientation distribution function $W(\xi,\psi,\phi)$ where $\xi = \cos\theta$. $W(\xi,\psi,\phi)d\xi d\psi d\phi$ gives the fraction of cracks between ξ and $\xi + d\xi$, ϕ and $\phi + d\phi$ and ψ and $\psi + d\psi$.

The elastic stiffnesses of the rock may be calculated in terms of the coefficients $W_{\ell mn}$ in an expansion of the crack orientation distribution function in generalised Legendre functions [5, 6]. Since the elastic stiffness tensor is of fourth rank it depends only on the coefficients $W_{\ell mn}$ of the expansion of $W(\xi,\psi,\phi)$ for $\ell \le 4$. It is assumed that the orientation distribution of cracks is orthotropic with symmetry axes coincident with the reference axes Ox_1, Ox_2 and Ox_3. For orthotropic material symmetry and ellipsoidal cracks, the non-zero $W_{\ell mn}$ are all real and are restricted to even values of ℓ, m and n. In the following, circular cracks are assumed with a=b>>c, for which $W_{\ell mn} = 0$ unless n = 0. The elastic stiffnesses are determined in this case by W_{200}, W_{220}, W_{400}, W_{420}, W_{440} and three anisotropy factors a_1, a_2 and a_3 [5, 6]. The ultrasonic wave velocities in the rock may then be obtained as solutions of the Christoffel equations [8]. In principle, the velocities enable the determination of the $W_{\ell m0}$ for $\ell \le 4$ provided that the anisotropy factors a_i can be determined. For dry, penny-shaped cracks the a_i may be calculated from the work of Hudson [9]. a_1 is found to be much smaller than a_2 and a_3 [5, 6]. As a result, the coefficients W_{4m0} make only a small contribution to the measured ultrasonic wave velocities and may be neglected, the resultant equations for the velocities allowing W_{200} and W_{220} to be determined.

3. STRESS-INDUCED ANISOTROPY OF BEREA SANDSTONE

Velocity measurements made on a 50 mm cube of Berea sandstone cut parallel to the bedding plane are shown in Figures 1 and 2 [10]. The initial velocity measurement was made at a hydrostatic compressive stress of 4.1 MPa, after which the stress perpendicular to the bedding plane was raised in steps of 4.1 MPa to 120 MPa, keeping the other two stress components fixed. The apparatus illustrated in Figure 3 was used [10]. A reference set of axes is chosen with Ox_3 perpendicular to the bedding plane and parallel to the maximum stress direction and x_1 and x_2 perpendicular to the remaining cube edges. The velocities measured were v_{11}, v_{22} and v_{33} (P-waves) and v_{12}, v_{23} and v_{31} (S-waves). v_{33} increases monotonically with stress, v_{23} and v_{31} rise monotonically throughout most of the test, but with a small drop immediately prior to the peak stress. v_{11}, v_{22} and v_{12} rise at first, but then drop after the first third of the test.

Below about 30 MPa all velocities plotted in Figures 1 and 2 rise with increasing compressive stress applied along Ox_3. At these low stress levels no additional cracks are expected to be introduced into the rock and this behaviour is due to closure of pre-existing microcracks and open grain boundaries in the rock. As the stress increases still further, new microcracks form and pre-existing ones grow and coalesce. The parameters W_{200} and W_{220} may be obtained independently from the P and S-wave velocities and are plotted in Figure 4. The anisotropy parameters W_{200} and W_{220} derived from the S-wave measurements in the high stress region are larger in magnitude than those derived from the P-wave measurements. This difference may result from the neglect of any contact along part of the crack length since the theory treats the cracks as ellipsoidal voids. It is interesting to compare the values of W_{200} and W_{220} plotted in Figure 4 with those corresponding to complete crack alignment. For complete alignment of crack normals along Ox_3, $W_{200} = 0.04005$, $W_{220} = 0$, whilst cracks parallel to Ox_3 with normals randomly oriented in the x_1x_2 plane correspond to $W_{200} = -0.02003$, $W_{220} = 0$. Perfect alignment of cracks with normals along Ox_1 corresponds to $W_{200} = -0.02003$, $W_{220} = 0.02453$ whilst for complete alignment along Ox_2, $W_{200} = -0.02003$, $W_{220} = -0.02453$. It is seen, by comparing these values with Figure 4, that at 120 MPa compressive stress applied perpendicular to the bedding plane most cracks are orientated with their normals lying randomly in the bedding plane (W_{200} negative, $W_{220} \simeq 0$), any preexisting cracks with normals lying along the maximum stress direction having been closed.

4. APPLICATION TO IN-SITU STRESS MEASUREMENT

The removal of rock by coring is accompanied by strain relaxation and generally involves both an instantaneous component and a time-dependent component [11]. This relaxation is found to be greatest in the maximum in-situ stress direction and least in the minimum in-situ stress direction. Teufel [11] argued that an important mechanism in the strain recovery process is the formation and growth of microcracks in the core. Ultrasonic P-wave velocities measured for propagation in the three principal directions of strain relaxation in two orientated cores of sandstone as a function of compressive hydrostatic stress are plotted in Figure 5 [11]. The cores were obtained from the MWX-2 well on the U.S. Department of Energy's multi-well experimental site near Rifle, Colorado. This well is vertical and the sandstone cores came from the Mesa Verde formation which consists of nearly flat-lying sandstones, mudstones and siltstones. It is seen in Figure 5 that at atmospheric pressure the P-wave velocity is anisotropic with the slowest velocity in the direction of maximum strain recovery (maximum in-situ principal stress). All velocities increase with increasing confining stress, the greatest increase occurring in the direction of maximum strain recovery. As a result the velocity anisotropy decreases with increasing confining stress. These results were interpreted by Teufel as indicating that the strain recovery process is a direct result of the formation and opening of microcracks and that the cores were isotropic at depth. Since the MWX-2 well is vertical and the beds are flat-lying it is convenient to choose the reference axes $Ox_1x_2x_3$ with x_3 vertical and x_1 and x_2 parallel to the minimum and maximum horizontal stress directions. Table 1 gives the values of W_{200} and W_{220} obtained [12] and the principal stress magnitudes calculated by Teufel [11] using the anelastic strain relaxation method.

Table 1. W_{200}, W_{220} and in-situ stresses (MPa) determined from P-wave velocity and anelastic strain recovery measurements [11] for orientated sandstone core [12].

Depth	W_{200}	W_{220}	σ_{11}	σ_{22}	σ_{33}
1995.5 m	0.00393	-0.00228	34.73	43.05	46.83
2404.3 m	0.00117	-0.00332	46.70	58.00	56.43

The values of W_{200} and W_{220} derived above can be used to plot the orientation distribution of crack normals as described elsewhere [5, 6]. Figure 6 shows equal area projections of the distribution of crack normals using the values of W_{200} and W_{220} given in Table 1. The x_3 (vertical) direction lies at the centre of these figures. For random orientation of the cracks these contours would have the value one everywhere. The peaks therefore represent a preferential orientation of crack normals. Anelastic strain recovery measurements on 26 orientated cores from the MWX-2 well showed the horizontal stresses to vary in a consistent way, with the ratio of horizontal to vertical stresses increasing with depth [11]. Above 2 km the overburden is the maximum principal stress, but below this depth the maximum horizontal stress is the maximum principal stress. The in-situ stress magnitudes determined by Teufel [11] for the two sandstone cores for which P-wave velocity measurements were made are listed in Table 1. From a comparison of these values with the crack orientation distributions plotted in Figure 6, it can be seen that the maximum density of crack normals lies along the maximum in-situ stress direction whilst the minimum density of crack normals lies along the minimum in-situ stress direction. The change in maximum stress direction from the Ox_3 direction (vertical) to the Ox_2 direction is seen to be reflected by the change in the microcrack orientation distribution. It appears that during coring and the subsequent strain relaxation, a rock will expand most by crack-opening in the direction of maximum principal in-situ stress and least in the direction of minimum principal in-situ stress. The use of ultrasonics to obtain the microcrack orientation distribution function therefore appears promising for obtaining in-situ stress information.

5. CONCLUSIONS

In this paper measurements of P and S-wave velocities in a cubic sample of Berea sandstone stressed to peak were inverted to obtain the crack orientation distribution. Large changes in ultrasonic wave velocities consistent with crack growth occur in Berea sandstone as it is stressed towards failure. Application of the theory to the determination of in-situ stress was discussed. It was found that the maximum density of crack normals obtained from an inversion of the measured ultrasonic velocities lies along the maximum in-situ stress direction whilst the minimum density of crack normals lies along the minimum in-situ stress direction. The use of ultrasonics therefore appears promising for obtaining in-situ stress information.

ACKNOWLEDGEMENTS

This paper is published by permission of Shell Internationale Research Maatschappij. I wish to thank J.G. van Munster for making the measurements plotted in Figures 1 and 2 and M.S. King for helpful discussions.

REFERENCES

[1] Paterson, M.S., Experimental Rock Deformation - The Brittle Field (Springer, Berlin, 1978)
[2] Kranz, R.L., Tectonophysics 100 (1983) 449.
[3] Birch, F., J. Geophys. Res. 65 (1960) 1083.
[4] Birch, F., J. Geophys. Res. 66 (1961) 2199.
[5] Sayers, C.M., Ultrasonics 26 (1988) 73.
[6] Sayers, C.M., Ultrasonics 26 (1988) 311.
[7] Nur, A. and Simmons, G., J. Geophys. Res. 74 (1969) 6667.
[8] Musgrave, M.J.P., Crystal Acoustics (Holden-Day, San Fransisco, 1970)
[9] Hudson, J.A., Geophys. J. Roy. astr. Soc. 64 (1981) 133.
[10] Sayers, C.M., van Munster, J.G. and King, M.S., submitted to the International Journal of Rock Mechanics (1989).
[11] Teufel, L.W., SPE 11649, presented at the SPE/DOE Symposium on Low Permeability Reservoirs, Denver, Colorado, March 14-16, 1983.
[12] Sayers, C.M., submitted to the International Journal of Rock Mechanics (1989).

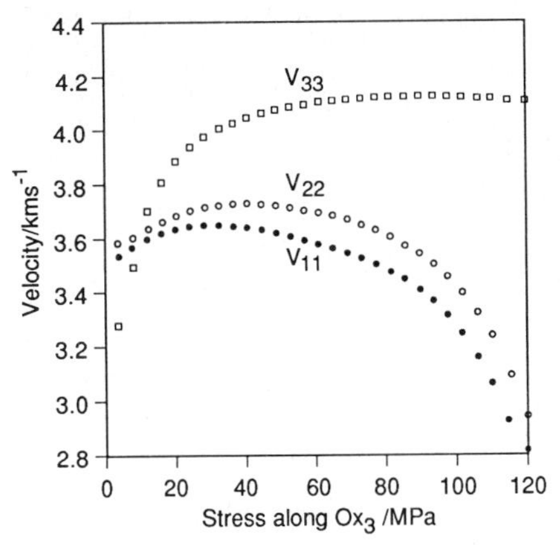

Figure 1
Compressional-wave velocities in dry Berea sandstone as a function of major principal stress.

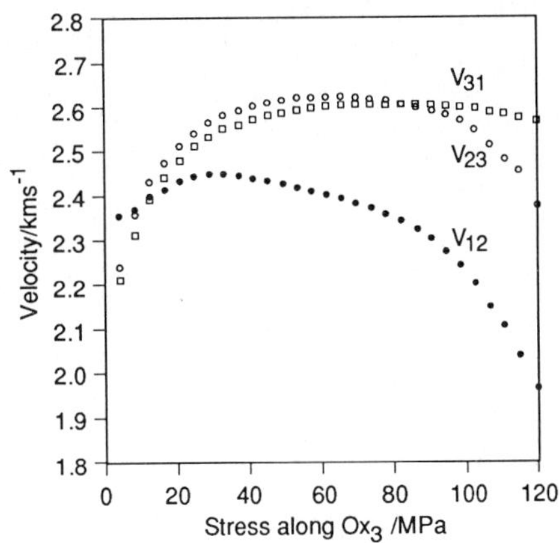

Figure 2
Shear-wave velocities in dry Berea sandstone as a function of major principal stress.

Figure 3
Experimental test set-up showing the loading system within the triaxial frame and a rock sample in the centre.

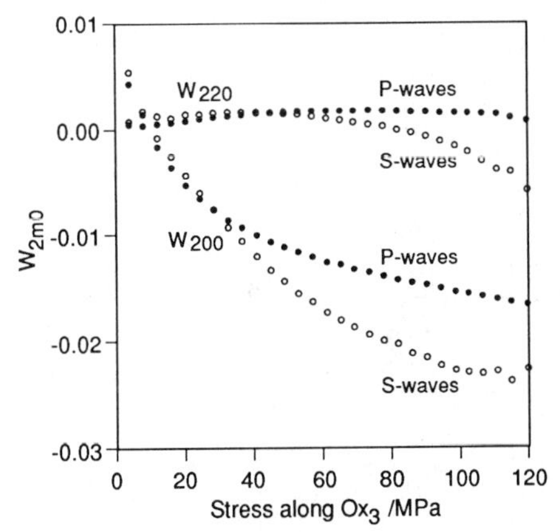

Figure 4
Crack orientation parameters W_{200} and W_{220} obtained from the measurements of Figures 1 and 2.

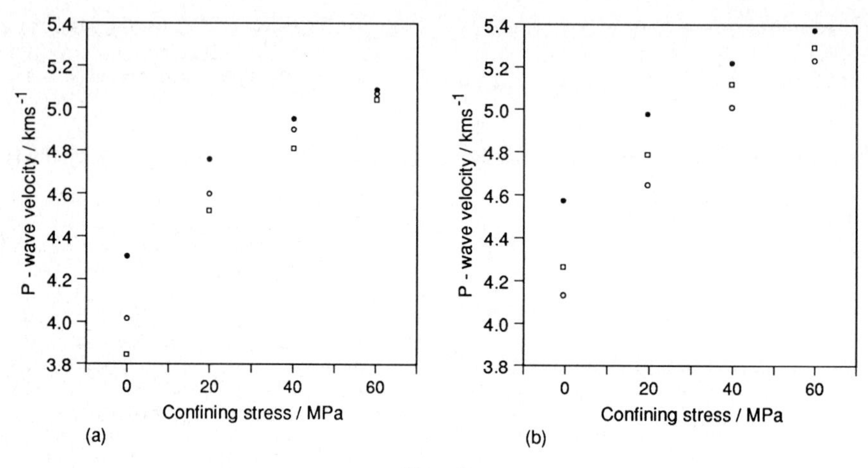

Figure 5

P-wave velocities measured on orientated core as a function of confining pressure. (a) Sandstone from 1995.5m, (b) Sandstone from 2404.3m. Measurements along Ox_1, Ox_2 and Ox_3 are denoted by •, ○ and □ respectively. The reference axes $Ox_1x_2x_3$ are chosen with x_3 vertical and x_1 and x_2 parallel to the minimum and maximum horizontal stress directions. The maximum principal stress (Table 1) is along Ox_3 for (a) and along Ox_2 for (b).

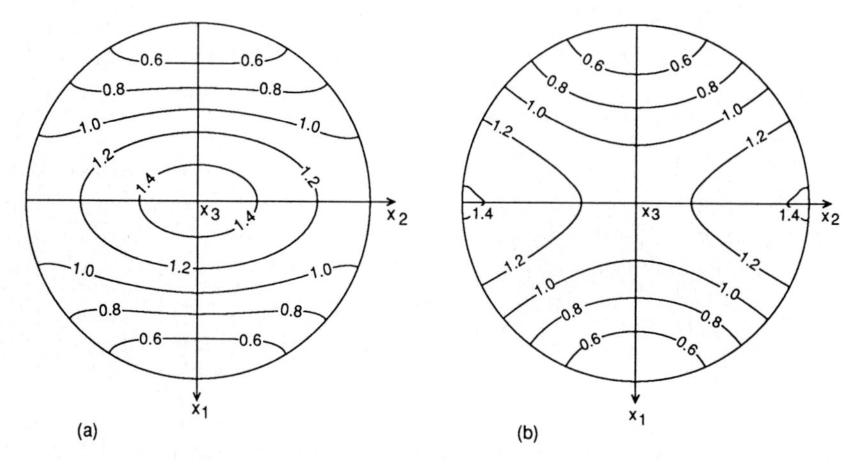

Figure 6

Contoured equal area projection of the orientation distribution of crack normals calculated from the values of W_{200} and W_{220} given in Table 1. (a) Sandstone from 1995.5m, (b) Sandstone from 2404.3m. The x_3 (vertical) direction lies at the centre of these figures. The maximum principal stress (Table 2) is along Ox_3 for (a) and along Ox_2 for (b).

Elastic Waves and Ultrasonic Nondestructive Evaluation
S.K. Datta, J.D. Achenbach and Y.S. Rajapakse (Editors)

CRACK GROWTH IN WELD REGION: CHARACTERIZATION BY ACOUSTIC EMISSION

M K VIJAYENDRA, C R L MURTHY, A K RAO

Department of Aerospace Engineering,
Indian Institute of Science,
Bangalore 560 012, India.

Joining by welding is an important and extensively used fabrication process in engineering. The process involves intense heat inputs, due to which the material at the joint undergoes significant variations in chemical and physical properties and associated changes in the microstructure and consequently the mechanical properties. These changes occur in a comparatively small region in the vicinity of the weld bead. For convenience, the weld region can be split into three zones: the Weld bead (WB), the Heat affected zone (HAZ), and the Parent metal (PM). The strength and toughness properties of the three zones are different. Therefore, under identical environment and stress conditions, similar defects tend to grow differently in each of these zones. The HAZ is the interface between WB and PM. The rate of crack growth is highest in the HAZ [1]. Due to either configuration of the defects or the inadquate sensitivity of the conventional NDT methods, some defects always go undetected, particularly close to the Weld bead. Further, there is a need for a Non-destructive evaluation (NDE) method which can discriminate in-service growth of cracks in the three zones to help monitor structural integrity and ensure safety. In this context, Acoustic Emission (AE) technique is suitable and is most promising. However, the classical source location technique is inadequate for the purpose because the weld region is very narrow. Hence, characterization and location by signal analysis become necessary. Methods and procedures for characterizing AE signals to extract this required information are not readily available.

An attempt is made to understand and characterize data obtained through controlled experiments specifically designed for evaluation of the three distinct zones in the weld region by Acoustic emission. The experiments and the results are presented in this paper.

1. INTRODUCTION

Welding is an important and extensively used fabricating process in engineering [2]. It is found from experience that most of the defects found in pressurized components are associated with weld or weld repaired areas. A crack in weld region propagates rapidly and it is very difficult to arrest it. This can lead to a catastrophic failure [3]. In this region, there are three distinct zones which have different metallurgical structures and mechanical and fracture properties. These are: Weld bead (WB), Heat affected zone (HAZ) and Parent metal (PM). The HAZ being brittle is the most vulnerable zone and calls for special attention.

Conventional NDT techniques have limitations (of size, shape and nature of defects) in detecting defects. Always some of the defects present are difficult to detect and may consistently go undetected [4]. These defects, under certain conditions may become dynamic and result in failure of the structure. Thus there is a need for a NDE method which can discriminate in-service growth of crack in the three zones to help monitor structural integrity and ensure safety.

1.1. Acoustic Emission in Relation to Weld Examination

Acoustic emission is a transient elastic wave generated by rapid release of energy within a material [5]. The phenomenon appears to be most suitable and promising for NDE of weld region in terms of discrimination of zones, as the emission generaterd from growing cracks in materials with different microstructures is different. However as these differences are not explicitly apparent systematic investigations through controlled experiments are necessary to bring them out. A program was undertaken towards this end. A commonly used pressure vessel Steel (ASTM 516 Gr.70) was chosen for the experiments. The plate was welded by 'Metal Inert Gas welding' process with carbon di-oxide as the shielding gas. Twelve tensile specimens were prepared from the plate transverse to the weld bead. Macro etching was done to delineate the different zones. Subsequently, for characterizing each zone, four specimens were tested each with an edge notch in that zone. Each specimen was fatigue cycled to initiate a crack and was then subjected to a uniformly increasing load till fracture. Acoustic emission data was acquired and analysed. There are distinct differences in the AE characteristics of the three zones. These could be utilized for incipient failure detection and to take further course of action to confirm the existence of the flaw in that zone by conventional NDT methods and to repair the area.

The advantages of AE techniques over other techniques are global coverage and high sensitivity to crack growth. A common method adopted for evaluating the integrity of a large structure using acoustic emission is to utilize source location with a multi-sensor configuration. But, by this method source location is very difficult in the weld region due to the limitations associated with location accuracy. In such cases signal analysis can better serve the purpose.

2. EXPERIMENTAL PROGRAMME

2.1 Material

Pressure vessel steel ASTM 516 Gr.70 which is also commonly used for a variety of other applications was considered appropriate for the current set of investigations. Some of its Mechanical properties and Chemical compositions are presented in Table 1.

Table 1

Mechanical properties		Chemical composition		
Yield Strength	: 26.676 Kg/mm^2	Carbon	: 0.28%	
Tensile Strength	: 48.31 Kg/mm^2	Manganese	: 0.65%	Heat analysis
Elongation in 200 mm	: 17%	Phosphorus	: 0.2%	
		Sulphur	: 0.045%	
		Silicon	: 0.035%	

2.2 Specimen preparation

A plate of 300 mm x 300 mm x 30 mm was cut into two pieces of 150 mm x 300 mm x 30 mm with assymetric double 'V' groove edge preparation. The plate was welded by MIG (Carbon di-oxide) process. The plate was flush ground and ultrasonically examined for of defects. Defective regions were marked and eliminated in the preparation of specimens. Tensile specimens (Fig.1)were prepared by cutting the plate transverse to the weld bead with guage length as per ASTM standards while other dimensions were suitably scaled to suit the mounting fixture of the testing machine. A total of 12 specimens were selected for the experimental program.

All specimens were macro-etched by Nital solution to delineate the three zones in the weld region and subsequently three sets of four specimens each were prepared. Edge notches were introduced in Parent metal in set 1, in Heat affected zone in set 2 and for set 3 in the Weld bead. All specimens were precracked to a length of 3 mm under identical fatigue cycling with a stress ratio of 0.3 and a frequency of 20 cycles/sec.

2.3 Experimentation

A schematic of the experimental set up is shown in fig.2. The specimen surfaces to be gripped were knurled in advance. An MTS 810 system with wedge grips is used in the 10% range (5,000 kg). An AET 5000 system [6] was used for acquiring data. Twelve specimens were tested successively.

A 375 kHz sensor was mounted close to the weld region by applying silicone grease and held firmly by a clamp.The sensor was connected to a pre-amplifier of 250 - 500 kHz band width with a gain of 60 dB. The total system gain was set to 94 dB with a threshold of 1 volt automatic. A tensile load is applied to the specimen at a uniform rate of 1.08 kg/sec till fracture. Simultaneously AE was continuously monitored and data recorded.

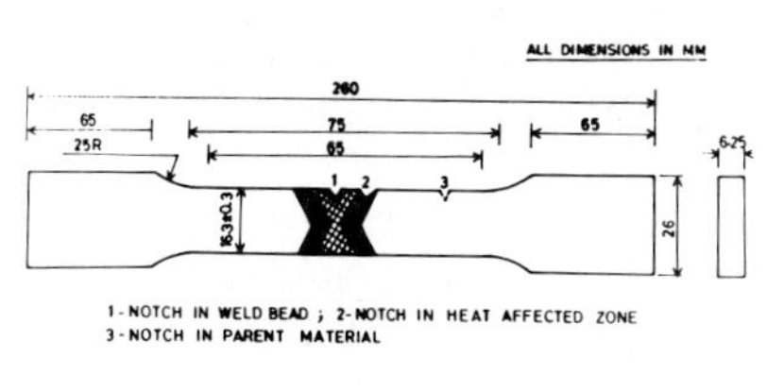

Fig.1

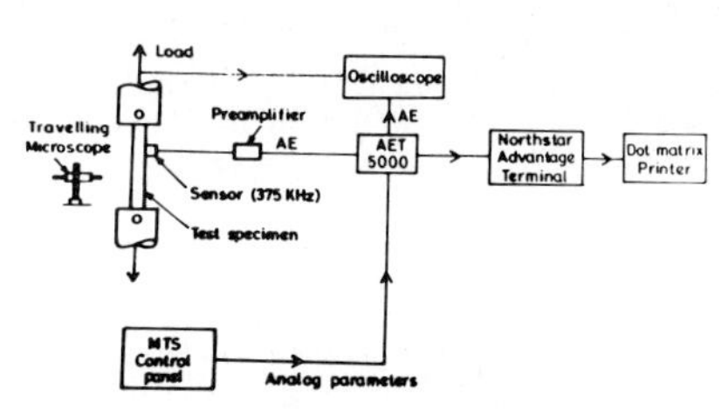

Fig.2

3.RESULTS AND DISCUSSION

The data set consisting of four AE parameters Ringdown counts (RDC), Event duration (ED), Peak Amplitude (PA) and Rise time (RT) along with the load, were stored in different files in the computer in the AET 5000 system. This data was post processed and a number of AE parametric and distribution plots were generated. Of these, three plots viz., 1.Total events .vs.time, 2. Total ringdown counts .vs.time and 3. distribution of events by peak amplitude appear to contain characteristics discriminating the three zones. The trends observed in each of these plots are now discussed. For convenience, typical figures from the specimens are shown in Figs.3 through 6 and Table 2.

Figs.3 and 4 show cumulative AE activity in terms of total events for cracks in WB and HAZ. PM and WB show similar features; WB being more critical. It is seen that in both cases AE activity increases with respect to time i.e., with load. But there are some distinct differences. In HAZ the activity grows uniformly. In WB, there is slow growth (upto approximately 50% of fracture load), a sudden jump and then again slow growth upto fracture. This sudden activity at approximately 50% of fracture load corresponds to a phenomenon like 'pop-in'. Cumulative RDC plots also show similar trends.

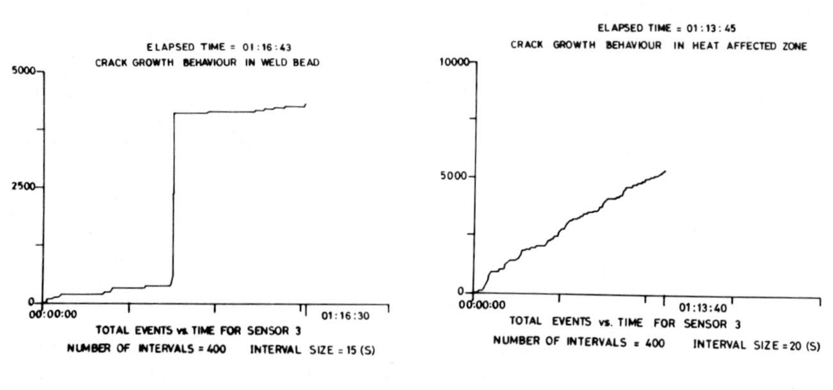

Fig.3 Fig.4

The distribution plots yield a different picture. Figs.5 and 6 show amplitude distribution plots for WB and HAZ. The distributional shapes are different. To bring out the salient points which are not apparent in these plots, statistical analysis was carried out and the results are summarized in Table 2. It can be observed that the discrimination shown in the total events plots is also shown in the distributions. However an important point to be mentioned is that the PM and WB show negative skewness in the distribution plots. This is possibly because of the similarity in microstructure. The positive skewness shown in the HAZ can be attributed to the differences in microstructure in terms of the presence of additional phases.

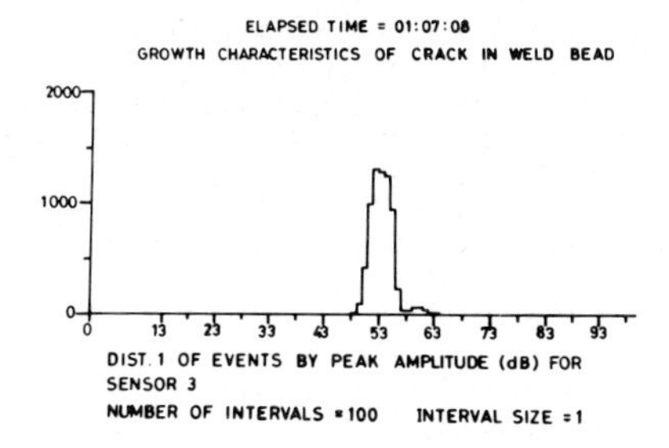

Fig.5

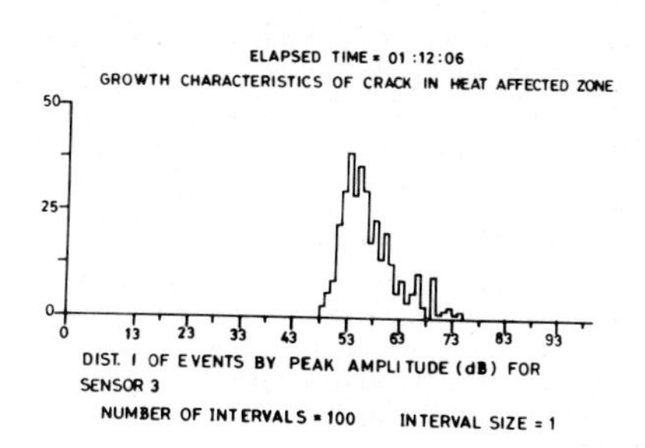

Fig.6

Table 2

NOTCH IN	PM		WB		HAZ	
SPECIMEN NO.	1	2	1	2	1	2
Pop-in	YES	YES	YES	YES	NO	NO
Event Rate/min	12	15	35	41	70	80
RDC Rate/min	950	970	196	220	2200	1950
Mean Peak Amp.	53.43	52.9	52.59	51	56.42	62.71
Std.Deviation(PA)	3.159	3.2	2.03	2.5	4.24	4.63
Co.eff.of Variation (PA)	0.059	0.060	0.038	0.04975	0.075	0.0738
Skewness (PA)	LEFT -2.915	LEFT -2.5	LEFT -2.822	LEFT -2.9	RIGHT +1.075	RIGHT +0.135

The differences in AE characteristics shown in the three zones, can be understood and verified by considering the differences in crack growth behaviour in terms of fracture mechanics theory and microstructural differences. For this purpose, a piece of 50 mm x 20 mm was cut, polished, etched and observed under metallurgical microscope at 1500 magnification. The parent metal showed alternate layers of Pearlite and Ferrite and weld bead showed uniformly distributed Ferritic and pearlitic regions, whereas heat affected zone showed in addition to pearlite, martensite with certain degree of spherodization. Since martensite has a non-cubic structure and since carbon is still present in the lattice, slip does not occur readily and therefore martensite is very hard, strong and brittle [7]. Therefore, crack growth is much faster and energy liberated in the process is much higher. Hence event rate is higher with a larger slope (fig.4). Since martensite formation from austenite is accompanied by an increase in specific volume (about 3%); this may be a reason why large stresses are set up in hardening that distort the article being hardeneed and result in cracks. The lattice distortion in martensite is reflected in high tensile strength and hardness and low ductility.

The Parent metal and Weld bead show 'pop-in' type of behaviour (HAZ is distinguished from PM and WB by the absence of such a phenomenon). As such a phenomenon is associated with liberation of bursts of energy, the plot of cumulative events .vs. time (Fig.3) indicates a sharp jump. Otherwise, the event rate in PM and WB is much lower than in HAZ. Large number of events are observed in the AE data. which is evident from the plot of cumulative events vs. time (figures 3 and 4). Otherwise the crack growth rate in parent metal as well as weld bead is lower while in the heat afected zone it is higher.

4.CONCLUSION

Non-destructive evaluation of weld region is important from the point of view of structural integrity. Heat affected zone being the most vulnerable area for failure due to growing cracks [8]. It is desirable to discriminate crack growth in this zone. This is difficult to achieve either by conventional methods or by source location using AE. Thus a feasibility study has been undertaken to characterize crack growth in different zones by AE signal analysis. The results obtained show that the cumulative plots as well as distribution plots aid in identifying the HAZ.

The present study confirms that signals emitted by defects in the three zones of a weld region have distinctive features which provide means for discrimination. The signals from HAZ which is the most critical zone are clearly identifiable by the positive skewness of Peak amplitude distribution. As such, it is possible to evolve a suitable algorithm for separating the critical defects in the HAZ and other zones.

ACKNOWLEDGEMENTS

The help rendered by Prof.Kishore, Department of Metallurgy, I.I.Sc, Bangalore, India, for extending facilities for carrying out metallurgical examination is gratefully acknowledged.

REFERENCES

1. "Rational Welding Design" , Gray T G F and Spence J , II edition, Butterworths, 1982. pp.73 - 119

2. "Welding methods and Metallurgy" , Jackson M D.

3. "Analysis of Welded Structures", Koichi Masubuchi, MIT, USA. edn.1980

4. "Special Technical Publication", 505 and 697.

5. "Metals Handbook" Volume 11, 8th edn. 1988. pp 243-243.

6. "AET 5000 Manual", AET Corporation, Sacremento, USA.

7. "A text book of Materials science and Metallurgy", Khanna O P, Dhanpat Rai and sons, New Delhi. pp.251 - 287.

8. "Investigation of Hydrotest failure of Thiokol Chemical Corporation - 260" dia. SL 1 motor case" NASA TMX 119, Lewis Research centre, Cleveland, Ohio, Jan 1966.

Elastic Waves and Ultrasonic Nondestructive Evaluation
S.K. Datta, J.D. Achenbach and Y.S. Rajapakse (Editors)

Radiative Transfer and Diffusivity of Ultrasound in Polycrystals

Richard L. Weaver
Department of Theoretical and Applied Mechanics
University of Illinois
Urbana, Illinois 61801

Inasmuch as microstructure is a major determinant of the mechanical properties of materials and inasmuch as standard optical methods for assessment of microstructures are time consuming, destructive and expensive, there has developed a substantial interest in using 10 MHz range ultrasound as a nondestructive probe of such microstructures. That such ultrasound, with inverse wavenumbers of order 100 microns would be sensitive to heterogeneity on the length scales of importance in metallic microstructures has long been appreciated. Conventional procedures, which scrutinize the scattering-based attenuation of coherent ultrasonic beams suffer, however, from complex systematic errors which become prohibitive outside a well controlled laboratory setting. For this reason researchers are increasingly studying the incoherent randomly scattered part of the field rather than the coherent part. The present paper is concerned with the theory of that incoherent field.

The theory of the attenuation of coherent ultrasound in materials with random microstructures in general, and in polycrystals in particular, is well developed. It dates to the 1940's with the work of Mason and McSkimm [1] and continues to the present in the work of Stanke and Kino[2] and Hirsekorn[3]. The theory adresses the calculation of the ensemble average response of a statistically homogeneous random medium and is, in the first-order-smoothing-approximation[9] formulation used by Stanke and Kino[2], usually thought to be accurate to leading order in the strength of the heterogeneity, that is, in the anisotropy of the constituent crystallites. The formulation used by Hirsekorn[3] and others, it should be noted however, is thought to be accurate in a rather different limit. Questions regarding theoretical accuracy aside, there remain major issues regarding the application of these theories to laboratory measurements. In a typical ultrasonic measurement of attenuation corrections must be introduced for diffraction, for non parallel specimen faces, for absorption, for reflection coefficients less than unity and for rough surfaces. Often such considerations will be beyond ready implementation in a field application.

The theory of the mean square ultrasonic response of a random medium , on the other hand, is almost nonexistent. This mean square response is the more relevant when experiments such as the backscattering work pioneered by Fay[5] or the fully diffuse work of Guo, Holler and Goebbels[6] or Weaver[7,8] are contemplated. Such experiments study an ultrasonic field which had the appearance of great incoherence, *ie*, the signals look stochastic. Signals which vary strongly with slight variations in transducer position cannot be studied by means of a theory of mean responses such as that of Stanke and Kino[2]. An ensemble or spatial average of such signals merely incurs phase cancellations which lead in turn to a vanishing mean response.

There is some debate as to whether *any* low order ensemble moment of the field such as a mean or a mean square will capture the salient behavior of the ultrasonic field in a typical realization of the ensemble, *i.e.*, in a typical laboratory specimen. There are some questions as to the ergodicity of acoustic multiple scattering in random media. The experiments such as those of references [5-8], however, are often conducted by averaging the square of the response in frequency and/or space. There is therefore, in any comparison of theory and experiment, an implicit ergodic hypothesis which merely requires the equivalence of frequency and space averages with ensemble averages. This hypothesis seems less unlikely than the usual ergodic presumption.

While the backscattering configuration of Fay[5] and others has long resisted thorough quantitative analysis, the fully developed multiply scattered diffuse field studied by Guo *et al*[6] and Weaver[7,8] is potentially more tractable. These works, which focussed on the ultrasonic field on time scales long compared to the mean time between successive scatterings of an ultrasonic ray off the random microstructure, found that the evolution of the ultrasonic energy was well modeled by a heat equation with an additional term to represent the internal friction.

$$D \nabla^2 E - \sigma E + I = E_{,t} \tag{1}$$

where D is the (possibly frequency dependent) diffusion constant and σ is the dissipation rate due to internal friction. I is the source of ultrasonic energy and E is the ultrasonic energy density. The laboratory work reported in references [6-8] successfully extracted a best fit value for the diffusivity D and then made some attempts to relate it to the microstructure.

Guo *et al* argued that the diffusivity should be one third the product of the wave speed and a mean free path. They then identified that mean free path with the inverse of twice the attenuation. This argument may be modified to account for the presence of two wave modes in the medium, each with its own speed and attenuation. One concludes

$$D = f_T * c_T / 6 \alpha_T + f_L * c_L / 6 \alpha_L \tag{2}$$

where the f's are the relative fractions with which the transverse(T) and longitudinal(L) wave modes contribute to the transport of ultrasonic energy. The values for the f's are not clear *a priori*. Regardless, however, of the values of the f's, the above formula has clear implications for the asymptotic behavior of D. In the low frequency, Rayleigh, regime where wavelength is much greater than polycrystalline grain sizes, the attenuations scale with the fourth power of frequency; hence the diffusivity should scale with the inverse of the fourth power of frequency. On the high frequency "stochastic" asymptote, the attenuations scale with the second power of frequency; correspondingly, the diffusivity should scale inversely with that power. (In the extremely high frequency geometrical optics limit where the attenuations becomes frequency independent, one would expect D to become independent of frequency also. Such high frequencies are,though, not relevant in the work of Guo *et al.*) Guo *et al* observed, however, a frequency dependence which was very clearly an inverse *first* power.

While Guo *et al* ascribed the disagreement to a failure of the weak scattering approximation implicit in the calculations of the attenuations, it will be seen in the present work that the error is in the identification of "mean free path" with the inverse of twice the attenuation. Such an identification will be seen to have been correct in the Rayleigh limit where scattering is isotropic. At higher frequencies, however, where the scattering is increasingly biased towards the forward direction, a typical ray must scatter several times before it has significantly changed its direction. A "transport mean free path" is therefore somewhat greater than the mean free path against any scattering at all. Correspondingly, the above formula increasingly underestimates the diffusivity at the higher frequencies. The remainder of this communication will report on a more formal derivation for the diffusivity.

The theory of the mean response, which is employed for the prediction of wave speeds and attenuations, is reviewed here before the theory of mean square responses is presented. The theory is nicely presented by Frisch[9] in abstract diagrammatic form but can be found in earlier work by Karal and Keller[10] where examples are given from acoustic, electromagnetic and elastic waves in piecewise constant isotropic media. Because the diagrammatic form is more compact, because it demonstrates the essential features of the integral equations, and because the present application is towards a piecewise constant anisotropic medium, we will confine the present exposition to the diagrammatic form. A more mathematically explicit exposition can be found in reference [11].

The mean Green's function (actually a second rank tensor) in a random medium is given by [9]

$$\langle G \rangle = \quad = \quad + \qquad (3)$$

which is an integral equation for < **G** >. Angle brackets represent ensemble averages. This equation is the Dyson equation and is exact. A heavy dark horizontal line represents the average Green's function, and a lighter horizontal line represents the Green's function of the ensemble average (in this case Voigt average) medium and is very readily calculated if the medium is statistically homogeneous and isotropic. The heavy black circle is the self-energy operator (also a second rank tensor) and is given formally by an expansion in powers of the strength of the inhomogeneity of the medium. In the case of a polycrystal this is equivalent to an expansion in powers of the anisotropy of the constituent crystallites.

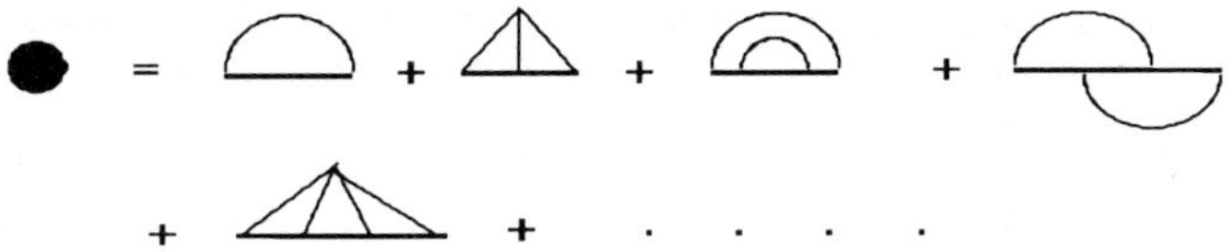

$$= \quad + \quad + \quad + \quad + \quad + \; \cdot \; \cdot \; \cdot \; \cdot \qquad (4)$$

Each horizontal line is again the bare Green's function. Each connection represents a scattering off two or more correlated points of the medium. Hence the first term is the leading term (second order) in powers of the strength of the scattering; the second term is of third order and the remaining illustrated terms are of fourth order. The "Keller" approximation[2,10] or "first-order-smoothing" approximation[9] expresses the self energy by its first term only. It is thought to be valid for sufficiently weak microstructural scattering and is in any case the only approximation that calls for only the lowest order statistics, *i.e.*, the mean and variance, of the moduli of the medium. Higher order statistics are generally unknown. There is a self-consistent-like approximation which sums all the rainbow graphs like the first and third above and only calls for the second order statistics, but it is difficult to claim that it should be any more valid than the simple Keller approximation; it too ignores terms like the second one above which are third order in the strength of the scattering.

Upon making the Keller approximation the mean Green's function as expressed by the Dyson equation is seen to be represented by a succession of forward scatterings by pair-correlated inhomogeneities.

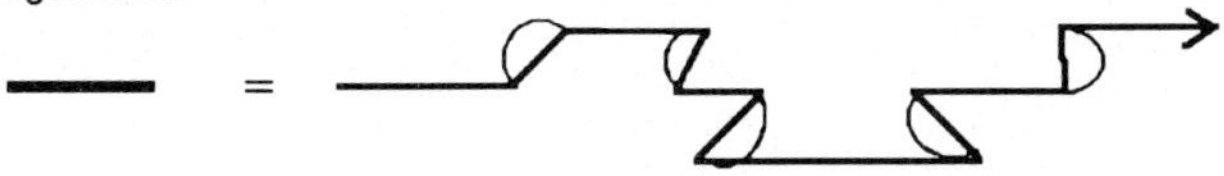

$$=$$

The pair-correlation operator  involves the co-variance of the moduli of the medium and therefore is given in terms of an eighth rank tensor function of position.

$$= \Xi \, v^2 \, \eta(\mathbf{r}_1 - \mathbf{r}_2) \qquad (5)$$

where η gives the probability that two points $\mathbf{r}_1$ and $\mathbf{r}_2$ are in the same crystallite. This form for the variance, where Ξ is a constant eighth rank tensor, follows from an assumption that the medium is statistically homogeneous and that neighboring crystallites are have uncorrelated orientations. This form was also assumed by Stanke and Kino[2] who also assumed that η is given by

$$\eta(\mathbf{r}_1 - \mathbf{r}_2) = \exp(-\beta r) \qquad (6)$$

where r is the distance between the points and β is an inverse measure of the microscale length, related to a typical crystallite radius. ν measures the degree of anisotropy in the individual crystallites. After making the Keller approximation the Dyson equation is readily solved in the domain of spatial Fourier transform. The results for attenuation by this formal diagrammatic means are found in reference [11] to be identical to those of Stanke and Kino[2].

There also exists, in Frisch's treatise[9], a diagrammatic formalism for mean square responses.

= + (7)

The above equation is the Bethe-Salpeter equation, an integral equation for the covariance of the response of the random medium. It is given in terms of the "intensity operator"[9] which, like the self-energy operator, has a formal expansion in powers of the strength of the scattering. The "ladder" approximation, so-called because of the shapes of the associated diagrams, approximates the intensity operator by its leading term in that expansion. The response covariance is, in that approximation, and written out in a summable infinite series,

= + + + . . . (8)

Equivalently, the mean square response is given as the sum of processes like the one pictured below wherein the field propagates coherently (and suffers attenuation) by each of the heavy dark mean field Greens functions between successive scatterings off pair-correlated points of the medium.

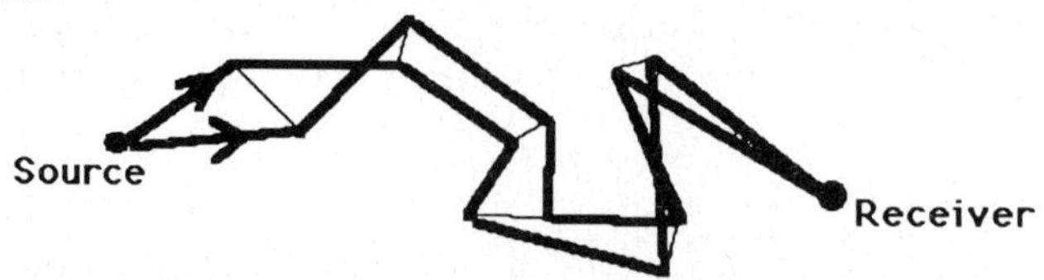

The integral Bethe-Salpeter equation of which the above is merely the iterated expansion, takes the form of an equation for a tensor quantity defined on a space of two spatial Fourier transform variables Δ and p. In the limit that the attenuation is weak over the distance of a wavelength, $\alpha\lambda << 1$ (which assures that energy is propagated coherently "on shell") and in the limit that the spatial scale of the diffuse field measurement is much larger than a wavelength, $\Delta\lambda << 1$, neither limit posing additional practical difficulty, the integral equation may be reduced somewhat and an equation of *radiative transfer* derived.

$$S^m_n(\mathbf{p}, \boldsymbol{\Delta}) = \delta^m_n + \int d^2\hat{\mathbf{s}}$$

$$\Big[{}^m_n K^l_i(\mathbf{p}, \mathbf{p}; \hat{\mathbf{s}}\omega/c_L, \hat{\mathbf{s}}\omega/c_L)\hat{s}_l \hat{s}_i \hat{s}_j \hat{s}_k R^L(\hat{\mathbf{s}}, \boldsymbol{\Delta}) S^j_k(\hat{\mathbf{s}}\omega/c_L, \boldsymbol{\Delta}$$

$$+ \; {}_n^m K_i^l \,(\mathbf{p}, \mathbf{p}; \hat{\mathbf{s}}\omega/c_T, \hat{\mathbf{s}}\omega/c_T)\,\{\delta_l^i - \hat{s}_l \hat{s}_i\}\{\delta_j^k - \hat{s}_j \hat{s}_k\}$$

$$R^T(\hat{\mathbf{s}}, \Delta) S_k^j \,(\hat{\mathbf{s}}\omega/c_T, \Delta)\Big] \qquad (9)$$

At this point the equations can be made to admit of simple physical interpretation. The tensor source-of-flux **S** is seen to be made up of two contributions. The first (δ) is due to the original deposition of energy by the source transducer. The second is in two parts. In the first term in the integral one can see, reading from right to left, a source of flux in a longitudinal mode in the $\hat{\mathbf{s}}$ direction being propagated by the longitudinal coherent energy propagator R and then scattered by the operator Kssss from the longitudinal mode into the mode **p**. The second term in the integral accounts for the energy scattered into mode **p** from a transverse mode propagating in the $\hat{\mathbf{s}}$ direction. The scattering operator **K** is related to the covariance of the medium moduli Ξ.

$$\mathbf{K}(\mathbf{p},\mathbf{p},\mathbf{q},\mathbf{q}) \;=\; \eta(\mathbf{p}-\mathbf{q})\, v^2\, \mathbf{ppqq}\, \Xi \qquad (10)$$

where η is the spatial Fourier transform of the two point correlation function and where the eighth rank tensor Ξ has been contracted with four wavevectors to result in a fourth rank tensor **K**.

The radiative transfer equations above are similar to the more usual equations of electromagnetic radiative transfer. The major difference, other than that they are here presented in a Fourier transform domain, is that we have here an apparent total of nine "Stokes parameters", the components of the tensor **S**, unlike the case of only four for electromagnetic waves. The extra parameters are due to the presence of three modes of propagation of elastic waves unlike the two of electromagnetic waves. In the "diffusion" limit, however, where Δ/α is taken as small -- corresponding to a limit such that the length scale over which one is studying the multiply scattered field is much greater than a mean free path, *i.e.* on a scale such that the field has scattered many times[11]-- these nine parameters reduce to only two independent non trivial parameters representing the energy flux in the longitudinal mode and in each of two identical transverse modes.

$$\mathbf{S}(\mathbf{p},\Delta) \;=\; \hat{\mathbf{p}}\hat{\mathbf{p}}\; A(\mathbf{p},\Delta) \;+\; \{\,\mathbf{I} - \hat{\mathbf{p}}\hat{\mathbf{p}}\,\}\; B(\mathbf{p},\Delta) \qquad (11)$$

The equations of radiative transfer then become, upon substituting this form for **S**, a pair of coupled integral equations for A and B defined on the unit sphere. A and B represent respectively the energy fluxes in the longitudinal and transverse modes.

The equations are readily solved by means of an expansion of A and B in a series of spherical harmonics, or alternatively and equivalently, a series in powers of Δ/α. As described in [11], the diffusivity can be extracted from that solution. It is given in terms of various spherical-harmonic weighted integrals of **K** over the unit sphere. These integrals are similar to integrals found in the derivation of attenuations from the mean Green's function calculations. The resulting diffusivity is then written in terms of several attenuation like quantities.

$$D = \frac{1}{6\left(\frac{1}{c_L^3} + \frac{2}{c_T^3}\right)} \; \frac{(\alpha_T - \alpha'_{TT} + \alpha'_{LT})/c_L^2 \;+\; 2(\alpha_L - \alpha'_{LL} + \alpha'_{TL})/c_T^2}{(\alpha_L - \alpha'_{LL})(\alpha_T - \alpha'_{TT}) - \alpha'_{LT}\alpha'_{TL}} \qquad (12)$$

where α_T and α_L are the usual transverse and longitudinal coherent wave attenuation rates. The primed quantities are angle weighted attenuations equivalent to the unprimed quantities except for a weighting of the cosine of the scattering angle in the integral definition. A primed attenuation in fact vanishes in the limit that scattering is isotropic. The double subscripted quantities, *e.g.* α_{TL}, are the mode-conversion partial attenuation rates, *e.g.*, the transverse-to-longitudinal part. While the form (12) was derived[11] within the specific context of a polycrystalline aggregate of cubic crystallites, the form is one which suggests itself as being more general.

As anticipated in the opening paragraphs, the diffusivity is given in terms of the attenuations and wave speeds. But because of the need for transport mean free paths rather than the more simple scattering mean free paths, it is the angle-weighted attenuations that appear. In the low frequency Rayleigh limit where the scattering is isotropic, the primed quantities vanish and the diffusivity simplifies. The result is identical to that anticipated in equation (2) where values of the f's are taken which correspond to simple equipartition between transverse and longitudinal energy densities.

$$f_T / f_L = 2 c_L^3 / c_T^3 \tag{13}$$

as deduced by mode counting techniques [12].

$$D_{Rayleigh} = \frac{1}{6\left(\frac{1}{c_L^3} + \frac{2}{c_T^3}\right)} \left(\frac{c_L / \alpha_L}{c_L^3} + \frac{2 c_T / \alpha_T}{c_T^3} \right) \tag{14}$$

In the case of an aggregate of cubic crystallites with pair correlation function as in equation (6) the Rayleigh limit diffusivity may be written explicitly as

$$D_{Rayleigh} = (c_T/6\beta)\ x^{-4}\ (c_T^4/v^2)\ 125\ (\ 2/3+K/4\ )/[\ (2+K^{-3})(1+2K^{-5/3})\ ] \tag{15}$$

where K is the ratio of longitudinal to transverse wave speeds and x is dimensionless frequency, $x = \omega / (\beta c_T)$. v is the ratio of the third elastic constant of an individual crystallite to the mass density of the medium. The inverse fourth power dependence is apparent.

On the high frequency stochastic asymptote, the diffusivity takes a similarly simple form, which to leading order scales inversely with the logarithm of frequency. This result has been derived only for an aggregate of cubic crystallites but one conjectures that the logarithmic dependence is universal.

$$D_{stochastic\ asymptote} = 350(c_T/\beta)K^3(c_T^4/v^2)[K^2/20 + 1/7]/[(1+2K^3)\mathrm{Log}_e x] \tag{16}$$

In the first correction to this there is a term independent of frequency in the denominator. Its dependence on K is very complicated and it has not been calculated.

The diffusivity as calculated from equation(12) is plotted in Figure 1. Diffusivity is nondimensionalized by a factor of 4200 $c_T^5/v^2\beta$. The low frequency fourth power dependence is observable below x equal to about 0.3 Above x equal to about 3, one observes the onset of a relatively frequency independent regime where the diffusivity scales more slowly than an inverse power of frequency. Four data points taken from the measurements of Guo *et al* are also plotted here. It can be seen that their observed inverse first power dependence on frequency is well described by the present theory. In order to plot their experimental data on the dimensionless plot here, it was necessary to adjust one unknown parameter, the magnitude of the microstructural length scale β^{-1}. A value of 100 microns was chosen in order to obtain the observed good agreement. As discussed in [11], this value is consistent with the micrograph published by Guo[6].

This work was supported by the National Science Foundation Solid Mechanics Program through grant number MSM 87-22413.

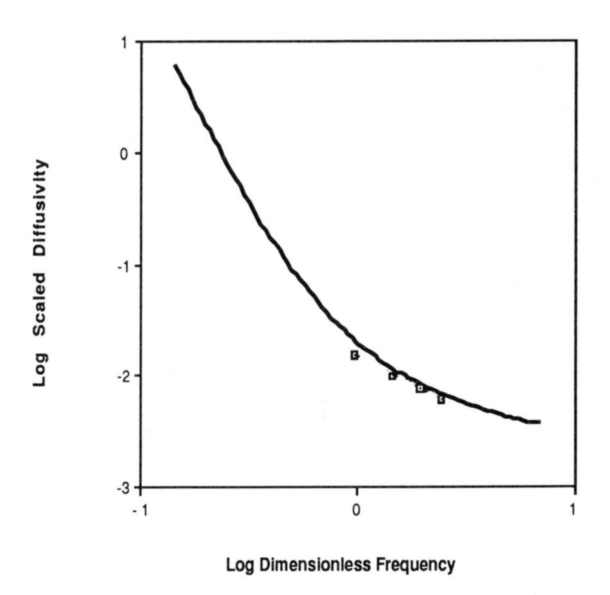

Figure 1. The ultrasonic diffusivity of a polycrystalline aggregate of cubic crystallites is plotted versus dimensionless frequency $\omega/\beta c_T$.

References

1] W.P. Mason and H.J. McSkimm, "Attenuation and Scattering of High Frequency Sound Waves in Metals and Glasses" J. Acoust Soc. Am. **19**, 464 (1947).
2] F.E. Stanke and G. Kino, "A Unified Theory for Elastic Wave Propagation in Polycrystalline Materials" J. Acoust. Soc. Am **75**, 665 (1984).
3] S.Hirsekorn, "The Scattering of Ultrasonic Waves by Polycrystals" J. Acoust. Soc. Am **72**, 1021 (1982).
4] S.Hirsekorn, "The Scattering of Ultrasonic Waves by Polycrystals II: Shear Waves" J. Acoust. Soc. Am **73**, 1160-63 (1983).
5] B. Fay, "Theoretical Considerations of Ultrasound Backscatter" Acustica **28**, 354 (1973).
6] C.B. Guo, P. Holler, and K. Goebbels, "Scattering of Ultrasonic Waves in Anisotropic Polycrystalline Metals" Acustica **59**, 112 (1985)
7] R.L. Weaver "Anderson Localization of Ultrasound" Wave Motion, in press (1989)
8] R.L. Weaver "Diffuse Field Measurements of Grain Size" to appear in *Nondestructive Testing and Evaluation for Manufacturing and Construction*, Proceedings of the Conference held in Urbana, Illinois August 1988 (edited by H. Reis) Hemisphere, NY.
9] U. Frisch, "Wave Propagation in Random Media" in *Probabalistic Methods in Applied Mathematics*, **1**, A.T. Bharucha-Reid, ed. (Academic Press, N.Y.1968)
10] F.C. Karal and J.B. Keller, "Elastic, Electromagnetic and Other Waves in a Random Medium" J. Math. Phys. **5**, 537 (1964).
11] R.L. Weaver "Diffusivity of Ultrasound in Polycrystals" in press J.Mechs.Phys.Sols.(1989)
12] R.L. Weaver " On Diffuse Waves in Solid Media" J. Acoust. Soc. Am. **71**, 1608 (1982)

D

POSTER PRESENTATIONS

Elastic Waves and Ultrasonic Nondestructive Evaluation
S.K. Datta, J.D. Achenbach and Y.S. Rajapakse (Editors)

Lacunas in welds

I. D. Abrahams [1] & G. R. Wickham [2]

1 Introduction

This study is concerned with the propagation of ultrasound through a composite body consisting of two semi–infinite elastic solids welded together at a plane interface $z = 0$, where (x, y, z) are rectangular cartesian coordinates. In $z < 0$, the material is supposed to be homogeneous and isotropic, in contrast with that in $z > 0$ which will be taken to have a particular inhomogeneous and anisotropic structure. Specifically, we assume that the material in $z > 0$ is composed of transversely isotropic crystals whose zonal axes (axes of symmetry) all lie parallel to the (x, z)-plane and that the material is homogeneous in the y-direction; that is, all cross-sections perpendicular to the y-axis are identical. This configuration is motivated by applications in the non–destructive evaluation of austenitic steel welds. Such welds have a grain structure large compared with the ultrasonic wavelength. Further, the orientation of the zonal axes of the grains is largely determined by the flow of heat during the formation of the weldment and, in particular, varies rapidly near the fusion interface. The model considered here assumes that the orientation of the zonal axis of each crystal or grain is a function of its distance from the fusion interface ($z = 0$). Thus if θ is the angle that the zonal axis makes with the z-axis at a particular point in the material, then

$$\theta = \theta(z).$$

In this case the general stress–strain law relative to cartesian coordinates (x, y, z) is of the form

$$\tau_{ij} = C_{ijkl}(\theta)\varepsilon_{kl}.$$

Using the orthogonal transformation which expresses $C_{ijkl}(\theta)$ in terms of $\bar{C}_{ijkl}$, the constant stiffness tensor relative to the zonal axis of a single crystal in the anisotropic material, we obtain the equations of motion for plane strain and anti–plane strain. Here, for the sake of brevity, we describe the latter case corresponding to an incident time harmonic SH line source situated in the isotropic phase at a distance z' from the interface. The governing differential equation in $z > 0$ is

$$(\bar{c}_{44} - \bar{c}_{66}) \sin\theta \cos\theta \frac{\partial^2 v}{\partial x \partial z} + (\bar{c}_{66} \cos^2\theta + \bar{c}_{44} \sin^2\theta) \frac{\partial^2 v}{\partial x^2} + (\bar{c}_{44} - \bar{c}_{66}) \frac{\partial}{\partial z} \left\{ \sin\theta \cos\theta \frac{\partial v}{\partial z} \right\}$$

$$+ \frac{\partial}{\partial z} \left\{ (\bar{c}_{66} \sin^2\theta + \bar{c}_{44} \cos^2\theta) \frac{\partial v}{\partial z} \right\} + \rho\omega^2 v = 0, \qquad z > 0, \tag{1}$$

where $v(x, z)$ is the elastic displacement, ρ is the density and ω is the frequency of the incident waves. At the interface between the two materials we require continuity of the stress and displacement fields.

2 Solution

Taking the Fourier transform of v with respect to x, $V(\alpha, z)$ say, the differential equation (1) has a solution of the form,

$$V(\alpha, z) = e^{iN\alpha\psi(z)} \left\{ \mathrm{Ai}(N^{2/3}\zeta) \sum_{n=0}^{\infty} \frac{a_n(\zeta)}{N^{2n}} + \frac{\mathrm{Ai}'(N^{2/3}\zeta)}{N^{4/3}} \sum_{n=0}^{\infty} \frac{b_n(\zeta)}{N^{2n}} \right\}, \tag{2}$$

[1]Department of Mathematics, University of Newcastle, Newcastle upon Tyne, NE1 7RU, U.K.
[2]Department of Mathematics, University of Manchester, Manchester, M13 9PL, U.K.

where $\zeta = \zeta(\alpha, z)$, $N = \omega a\sqrt{\rho/\bar{c}_{44}}$ and a is a typical length scale in the inhomogeneous material. Equation (2) is known to be convergent for N sufficiently large. Away from the turning point $\zeta = 0$ we find, after Fourier inversion and a stationary phase analysis, that

$$v(x,z) = \frac{ae^{-3i\pi/4}}{\sqrt{2\pi N}} \frac{v_0(x.z;z')}{|q(x,z;z')|^{1/2}} e^{iN\phi_0(x,z;z')} + O(1/N^{3/2}), \qquad z > 0, \tag{3}$$

where

$$\begin{aligned}
v_0(x,z;z') &= \left(\frac{p(0)-\eta\alpha_0^2}{p(z)-\eta\alpha_0^2}\right)^{1/4} \frac{1}{[(p(0)-\eta\alpha_0^2)^{1/2}+(\nu^2-\alpha_0^2)^{1/2}]}, \\
\phi_0(x,z;z') &= \alpha_0(\psi(z)-x)+\chi(\alpha_0,z)-(\nu^2-\alpha_0^2)^{1/2}z', \\
q(x,z;z') &= -\eta\int_0^z \frac{dt}{(p(t)-\eta\alpha_0^2)^{3/2}} + \frac{\nu^2 z'}{(\nu^2-\alpha_0^2)^{3/2}}, \qquad p(z) = 1-(1-\eta)\sin^2\theta(z), \\
\psi(z) &= \int_0^z \frac{(1-\eta)\sin\theta\cos\theta}{p(z)}\,dz, \qquad \chi(\alpha,z) = \int_0^z \frac{(p(z)-\eta\alpha^2)^{1/2}}{p(z)}\,dz,
\end{aligned}$$

$\eta = \bar{c}_{66}/\bar{c}_{44}$, $\nu^2 = \bar{c}_{44}/\mu$ and μ is the shear modulus in the isotropic material. The stationary phase point α_0 for a given x, z and z' is the solution of the transcendental equation

$$\psi(z) - x = \eta\alpha_0 \int_0^z \frac{dt}{p(t)\sqrt{p(t)-\eta\alpha_0^2}} - \frac{\alpha_0 z'}{\sqrt{\nu^2-\alpha_0^2}} = \alpha_0 F(\alpha_0^2, z; z'). \tag{4}$$

3 Discussion

It can easily be shown that there exist regions of the (x,z)–plane where there are no solutions of (4) for α_0, i.e. there is no stationary phase point. These regions correspond to quiet zones or "lacunas"; their boundaries are given by $x = \psi(z) \mp \sqrt{p(z)/\eta}F(p(z)/\eta, z; z')$ and are shown in the following figures computed using data for austenitic steel with the parabolic grain structure $\theta(z) = \arctan(z/a)$.

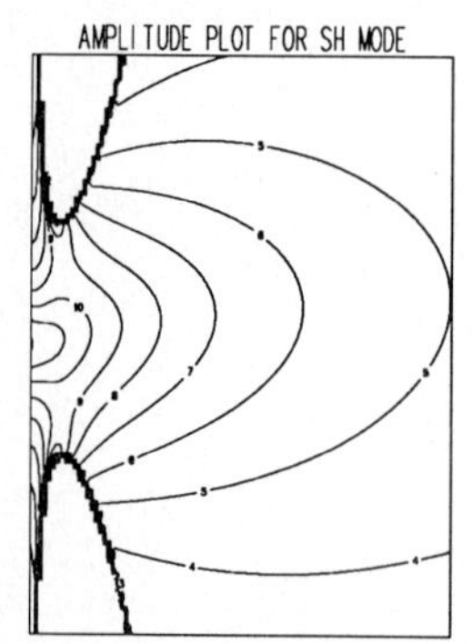

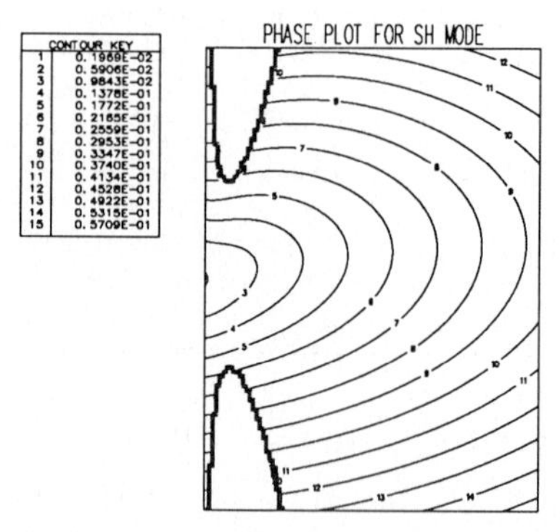

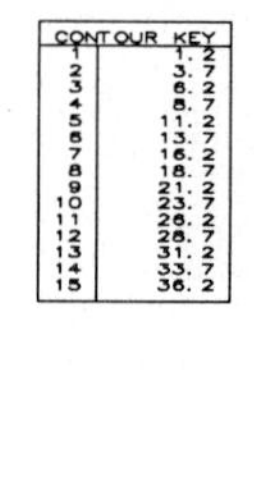

This example shows how the inhomogeneous material distorts the radiation from an isotropic source. A particularly notable feature, apart from the lacuna formation, is the tendency for the energy to follow the general direction of the grain axes. In the case of plane strain, it is possible to develop a related vector theory, though this is considerably more complicated. A full discussion of the work outlined here and the latter case will be described in future articles. The authors gratefully acknowledge the support of the Central Electricity Generating Board (UK).

Elastic Waves and Ultrasonic Nondestructive Evaluation
S.K. Datta, J.D. Achenbach and Y.S. Rajapakse (Editors)

Scattering by a small surface breaking crack

I. D. Abrahams[1] & G. R. Wickham [2]

1 Introduction

In this paper we consider the scattering of elastic waves by a compact surface defect of general shape in an elastic half space. As an example we consider an incident Rayleigh surface wave so that the motion in the material is one of plane strain. Previous studies have been concerned mainly with the *scalar* problem for horizontally polarized shear waves, or, when the general elastic wave equation has been considered, attention has focussed on submerged scatterers or slowly varying surface deviations; see Datta & Sabina[1]. Our general approach to the surface defect problem is the method of matched asymptotic expansions. This involves expanding the solution in an inner region in terms of the non-dimensional wave-number $2\pi a/\lambda$, where λ is the wavelength and a is a typical dimension of the defect. Each term in this expansion is posed as a problem in elastostatics which belong to a small class admitting explicit solution.

In this article we confine our attention to the example of scattering by an edge crack. This leads to a matrix Wiener-Hopf problem in the inner region, first investigated by Khrapkov[2]. The outer expansion in our matching scheme is furnished by fundamental sources placed in the surface of the elastic material; the generalised Lamb's problem. Here the difficulty is to expand these outer potentials about the singularities and in the present study this is acheived by expressing the elastodynamic field in terms of cylindrical waves (Brind & Wickham[3]).

2 The outer expansion

The configuration and notation is described in the following figure:

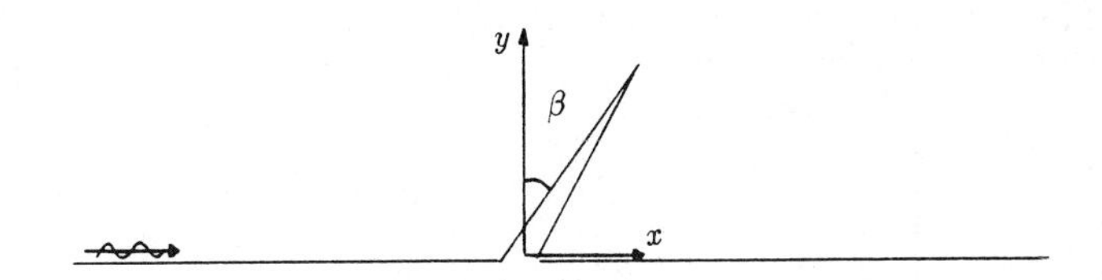

The total elastic displacement in the material satisfies the equation

$$-K^2 \text{curl curl}\mathbf{u} + k^2 \text{grad div}\mathbf{u} + \mathbf{u} = \mathbf{0} \tag{1}$$

and hence, expressing $\mathbf{u}$ in terms of incident and scattered Lamé potentials and applying Green's theorem we obtain

$$\mu K^2 \phi^{(s)}(\mathbf{r}') = \int_{\partial C} u_i(\mathbf{r}) \Sigma_{ij}^P(\mathbf{r}, \mathbf{r}') \nu_j \, |d\mathbf{r}|, \quad \mathbf{r}' \in D, \tag{2}$$

where μ is the shear modulus, ν_j is the j'th component of the outward normal vector on the crack face ∂C, Σ_{ij}^P is the stress tensor corresponding to an isotropic compressional line source situated

[1]Department of Mathematics, University of Newcastle, Newcastle upon Tyne, NE1 7RU, U.K.
[2]Department of Mathematics, University of Manchester, Manchester, M13 9PL, U.K.

at $\mathbf{r}'$ in the elastic half space whose surface is supposed traction free. A similar expression may be obtained for the scattered shear wave potential $\psi^{(s)}(\mathbf{r})$. We now define inner and outer coordinates according to

$$(X,Y) = (x,y)/a, \qquad (\bar{x},\bar{y}) = k(x,y) \tag{3}$$

respectively, and expanding $\Sigma^P_{ij}(\mathbf{r},\mathbf{r}')$ in (2) in a Taylor series about $\mathbf{r} = \mathbf{0}$ with $|\mathbf{r}'| > a$ we obtain

$$\phi^{(s)}(\bar{x},\bar{y}) = 2aU_0(1-\frac{k^2}{K^2})\left(-\frac{i}{4}H_0^{(1)}(\bar{\mathbf{r}}) + \frac{1}{4\pi}\int_{-\infty}^{\infty} e^{i\alpha\bar{x}-\gamma(\alpha)\bar{y}}\frac{S(\alpha)}{\gamma(\alpha)R(\alpha)}\,d\alpha\right)C \tag{4}$$

where

$$\left.\begin{matrix} R(\alpha) \\ S(\alpha) \end{matrix}\right\} = (2\alpha^2 - K^2)^2 \mp 4\alpha^2\gamma(\alpha)\delta(\alpha), \quad \gamma(k_0) = \sqrt{k_0^2 - k^2}, \quad \delta(k_0) = \sqrt{k_0^2 - K^2}$$

and k_0 is the positive real zero of $R(\alpha) = 0$. The constant

$$C = \int_{\partial C} \bar{u}_1\nu_1\,|d\mathbf{R}| \tag{5}$$

must be determined by solving the inner problem.

3 The inner expansion

Substituting the inner variables (3) into (1), and expanding for small ka, gives the elastostatic boundary value problem near the crack. This may be solved explicitly by first applying Mellin transforms to the Airy stress function which results in the matrix Wiener–Hopf equation

$$\frac{\sin p\pi}{\Delta(p)}\mathbf{Q}(p)\left\{\begin{pmatrix} A^- \\ B^- \end{pmatrix} + \frac{1}{p}\begin{pmatrix} l_1 \\ l_2 \end{pmatrix}\right\} = E^*\begin{pmatrix} A^+ \\ B^+ \end{pmatrix}, \tag{6}$$

defined in the strip of the complex transform p plane $1-\epsilon < \Re(p) < 1, \epsilon > 0$. In (6) $A^-(p), B^-(p)$ are Mellin transforms of the stress on the continuation of the crack from $R = 1$ to ∞ and are consequently analytic in the left half plane $\Re(p) < 1$. Similarly $A^+(p)$,$B^+(p)$ are the transforms of the dislocation density from 0 to 1 and are analytic in $\Re(p) > 1-\epsilon$. The vector (l_1, l_2) is known from the forcing and

$$\begin{aligned}
\mathbf{Q}(p) &= n(p)\mathbf{I} + m(p)(p-1)\mathbf{J}(p), \\
\Delta(p) &= \det(\mathbf{Q}(p)) = n^2(p) - m^2(p)(p-1)^2(1-(p-1)^2\cos^2\beta), \\
\mathbf{J}(p) &= \begin{pmatrix} \sin\beta & (p-2)\cos\beta \\ -p\cos\beta & -\sin\beta \end{pmatrix}, \quad E^* = \frac{u_0 E}{4(1-\nu^2)a}, \\
n(p) &= \cos 2(p-1)\beta + \cos p\pi - 2(p-1)^2\cos^2\beta\cos 2(p-1)\beta, \quad m(p) = 2\cos\beta\sin 2(p-1)\beta.
\end{aligned}$$

To solve (6) we must write $\sin p\pi\mathbf{Q}(p)/\Delta(p)$ as $\mathbf{Q}_+(p)\mathbf{Q}_-(p)$ which is explicitly possible in this case as the factorization is commutative.

4 Solution and conclusions

The leading order inner solution may be found for the dilatation, or trace of the stress tensor σ_{ii} as $R \to \infty$. This can be directly related to the dilatation of the outer field as $\bar{r} \to 0$ and hence the unknown constant (5) can be determined. Alternatively the integral in (5) may be evaluated directly from $A^+(2)$, $B^+(2)$ and both methods yield

$$C = \frac{1}{4E^*}(1 + \cos 2\beta, \sin 2\beta)\mathbf{Q}_+(2)\mathbf{Q}(0)\begin{pmatrix} l_1 \\ l_2 \end{pmatrix}, \tag{7}$$

which gives the Rayleigh wave transmission and reflection coefficients as

$$A_\pm^R = \frac{8ikaU_0(1 - k^2/K^2)\delta(k_0)k_0^2 C}{kR'(-k_0)}(1 + O(ka)). \tag{8}$$

The higher order terms in the expansion in powers of ka show the asymmetry in the outgoing Rayleigh wave amplitudes.

A systematic scheme can be devised for obtaining the asymptotic expansions to any order, and the technique is applicable to any near surface or surface breaking imperfections. The defects presently under investigation by the authors include a circular bite and mound, and a rigidly–bonded strip.

References

[1] S. K. Datta & F. Sabina. 1986 *Matched asymptotic expansions applied to diffraction and high frequency asymptotics* ed. V. K. Varadan & V. V. Varadan. North Holland.

[2] A. A. Khrapkov. 1971 *Prikl. Mat. Mekh.* **35**, pp. 677–689.

[3] R. J. Brind & G. R. Wickham. 1989 *Proc. Roy. Soc. Lond. A* (submitted).

Elastic Waves and Ultrasonic Nondestructive Evaluation
S.K. Datta, J.D. Achenbach and Y.S. Rajapakse (Editors)

Three-dimensional numerical simulation of elastic body waves scattered from a half-space with hemispherical or shallow surface indentations

Peter H Albach and Leonard J Bond

NDE Centre
Department of Mechanical Engineering
University College London
LONDON WC1E 7JE
United Kingdom

This paper examines the scattering of time harmonic elastic waves from a three-dimensional axisymmetric surface indentation in a half-space, utilizing a boundary method. The surface indentation represents an isolated corrosion pit in a thick aluminium plate, and the incident plane wave is an idealised ultrasound wave generated by a compressional or shear wave transducer

1. INTRODUCTION

The area of detection and characterisation of corrosion has been neglected for a long time in favour of methods for the detection of cracks. This study looks into ways of detecting pitting corrosion located on the remote side of thick aluminium plates using ultrasound. A single isolated pit is represented as a three-dimensional axisymmetric indentation in the stress free surface of an elastic half-space. An incoming ultrasound plane wave is scattered from the surface obstacle, and the scattered field should provide information about the size and the shape of the surface perturbation and therefore of the extent of corrosion present on the surface.

2. NUMERICAL METHOD

The method used here is a boundary method which originated in geophysics [1], where the scattering of earthquake waves from surface indentations is of interest for the determination of surface movements. The co-ordinate system most appropriate to the given geometry is the spherical co-ordinate system. A vectorial plane wave can be expanded in spherical co-ordinates into an infinite series of spherical vector functions. The scattered field can be described as an infinite series of outgoing spherical waves with unknown coefficients. The stress components due to the scattered field are obtained by differentiating the spherical vector functions accordingly. For numerical purposes the series are truncated to a finite number of terms. Applying the boundary conditions - vanishing normal stress components along the free boundary - using a least squares matching method gives the coefficients for the outward travelling waves. Due to the axisymmetric nature of the surface obstacle it is possible to reduce the three-dimensional scattering problem into a set of two-dimensional subproblems with a given angular dependence. Each of these subproblems is solved separately, and the results are superimposed to give the full three-dimensional solution. The accuracy of the computer program that was developed was checked against published results for the near-field displacements [1, 2]. Details of the numerical method can be found in [3, 4].

3. RESULTS

The method is applied here to a plane wave incident at an angle α on to a half-space made of aluminium ($\nu = 0.34$). The wave vector is located in the x-z plane and the surface disturbance is the section of a hemisphere, see Fig 1. The surface diameter to incident wavelength ratio was chosen as $2a/\lambda = 1.0$, and the depth of the indentation was set to $d = 0.5a$.

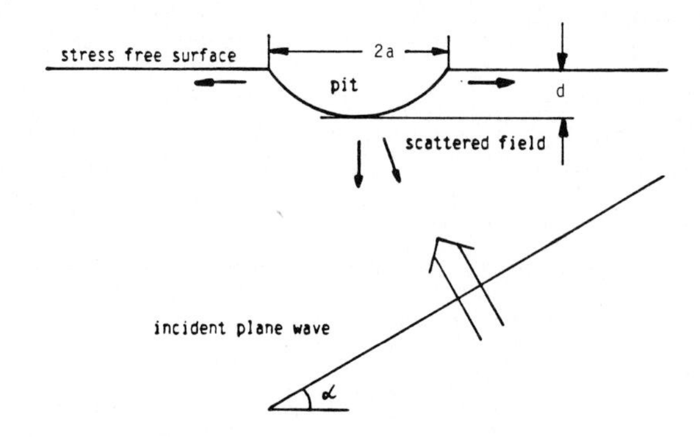

Fig 1: Time harmonic plane wave incident at an angle α on to an axisymmetric surface indentation in a half-space.

Fig. 2 shows the surface displacements in the vicinity of the surface indentation for a plane compressional wave incident at 60 degrees. An interesting result is that the horizontal surface movements are stronger in the plane perpendicular to the incident wave vector (y-z plane) than in the plane of the incident wave vector. Fig. 3 shows the scattered far-field for the same case. It can be seen that in the plane of the incident wave vector the scattered field consists of two major beams, one is reflected back into the bulk material, the other is oriented along the surface in the direction of the incident wave. This surface component is a Rayleigh wave, which is approximated by the spherical vector functions (i.e. body waves) chosen to represent the scattered field. A more accurate representation of this surface wave could be obtained by including surface wave terms explicitly in the expansion of the scattered field. For the plane perpendicular to the incident wave vector it can be seen that the scattered surface component is not very pronounced, but a strong shear wave (Phi component) is scattered back into the bulk material. This component is responsible for the strong horizontal surface movements in that plane.

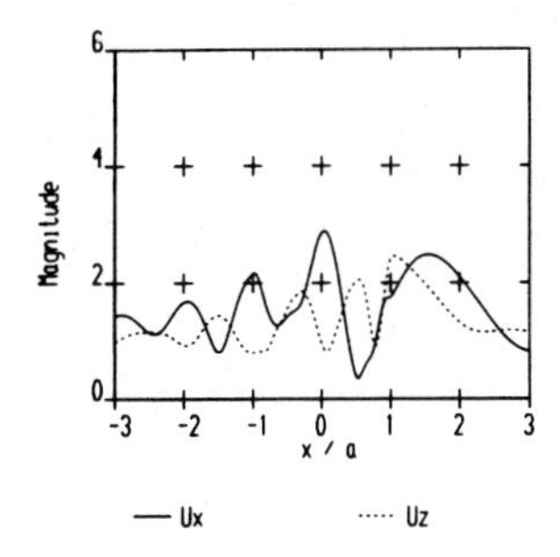

— Ux ····· Uz

— Ux ····· Uz
- - Uy

Surface displacements in x-z plane
Type of surface obstacle : Segment of a hemisphere
Order of approximation : 13
Incident wavetype : P
Incident angle : 60.00 deg
Poisson's ratio : 0.34
2a / lambda incident : 1.00
d / a : 0.50

Surface displacements in y-z plane
Type of surface obstacle : Segment of a hemisphere
Order of approximation : 13
Incident wavetype : P
Incident angle : 60.00 deg
Poisson's ratio : 0.34
2a / lambda incident : 1.00
d / a : 0.50

Fig. 2: Surface displacements for a P wave incident under 60 degrees on to a surface indentation with the shape of a segment of a hemisphere.

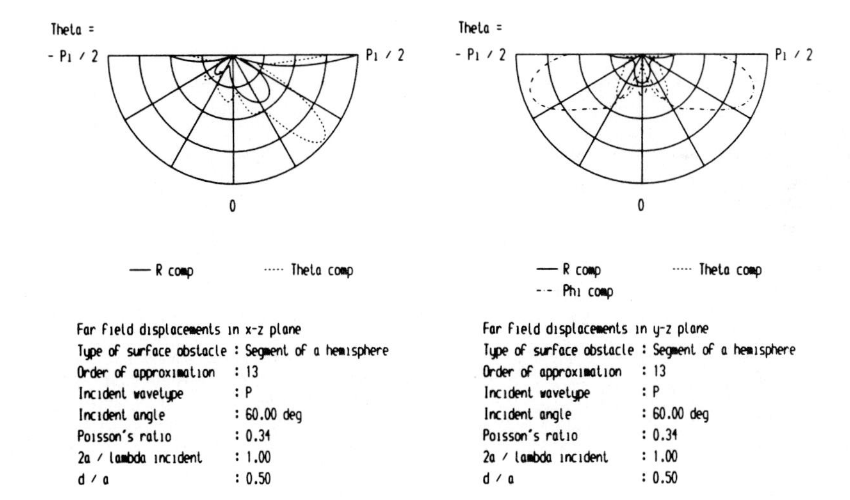

Fig. 3: Far-field displacements for a P wave incident under 60 degrees on to a surface indentation with the shape of a segment of a hemisphere.

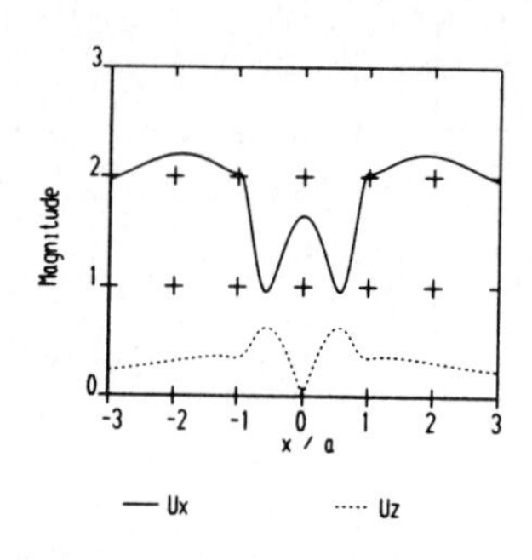

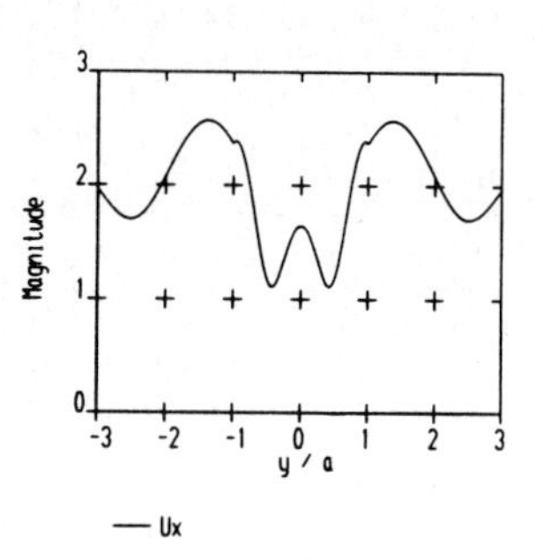

Surface displacements in x-z plane
Type of surface obstacle : Segment of a hemisphere
Order of approximation : 11
Incident wavetype : SV
Incident angle : 0.00 deg
Poisson's ratio : 0.34
2a / lambda incident : 1.00
d / a : 0.50

Surface displacements in y-z plane
Type of surface obstacle : Segment of a hemisphere
Order of approximation : 11
Incident wavetype : SV
Incident angle : 0.00 deg
Poisson's ratio : 0.34
2a / lambda incident : 1.00
d / a : 0.50

Fig. 4: Surface displacements for a SV wave normally incident on to a surface indentation with the shape of a segment of a hemisphere.

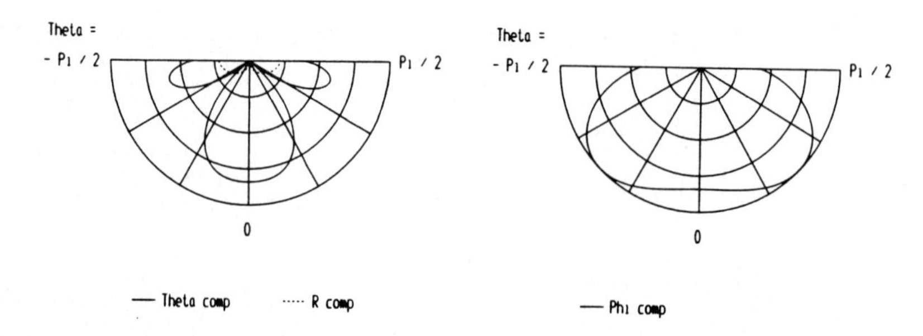

Far field displacements in x-z plane
Type of surface obstacle : Segment of a hemisphere
Order of approximation : 11
Incident wavetype : SV
Incident angle : 0.00 deg
Poisson's ratio : 0.34
2a / lambda incident : 1.00
d / a : 0.50

Far field displacements in y-z plane
Type of surface obstacle : Segment of a hemisphere
Order of approximation : 11
Incident wavetype : SV
Incident angle : 0.00 deg
Poisson's ratio : 0.34
2a / lambda incident : 1.00
d / a : 0.50

Fig. 5: Far-field displacements for a SV wave normally incident on to a surface indentation with the shape of a segment of a hemisphere.

Fig. 4 and Fig. 5 show the surface displacements and the scattered far-field for the same geometry and a normally incident shear wave. Due to the normalisation with respect to the incident shear wavelength (which, in aluminium, is approximately half of the compressional wavelength) the scattering is not as strong as in the previous case. The scattered surface wave components are negligible, but again the scattered shear wave component in the y-z plane is much stronger than either the scattered compressional or shear waves in the x-z plane. It is also interesting to note that the shear wave components scatter more strongly than the compressional wave components, independent of whether the incident wave is of compressional or shear wave type. Further results for various axisymmetric surface indentations can be found in [3, 4].

ACKNOWLEDGEMENTS

This work was performed with the support of the Procurement Executive of the Ministry of Defence and the Department of Trade and Industry, UK. Peter Albach would like to thank Prof J. Brian Davies from University College London and Dr David Bruce from RAE Farnborough, UK, for their help and advice.

REFERENCES

[1] Francisco J Sanchez-Sesma, Diffraction of Elastic Waves by Three-Dimensional Surface Irregularities, Bull. Seism. Soc. Am., 73, pp 1621 - 1636, December 1983

[2] Francisco J Sanchez-Sesma, Site effects on strong ground motion, Soil Dynamics and Earthquake Engineering, 6, pp 124 - 132, 1987

[3] Peter H Albach and Leonard J Bond, Numerical Simulation of Elastic Wave Scattering from Three-Dimensional Axisymmetric Surface Features, Conference Proceedings of the Review of Progess in Quantitative Nondestructive Evaluation, Vol 9, Plenum Press, in print

[4] Peter H. Albach and Leonard J Bond, 3-D Elastic Wave Scattering from Axisymmetric Surface Indentations, to be published.

Elastic Waves and Ultrasonic Nondestructive Evaluation
S.K. Datta, J.D. Achenbach and Y.S. Rajapakse (Editors)

MEASUREMENT OF WELDING RESIDUAL STRESSES BY ACOUSTOELASTIC TECHNIQUE USING LONGITUDINAL AND TRANSVERSE WAVES

Yoshio ARAI and Hideo KOBAYASHI

Department of Mechanical Engineering Science
Faculty of Engineering, Tokyo Institute of Technology
Ohokayama 2-12-1, Meguro-ku, Tokyo, JAPAN

An acoustoelastic technique using ultrasonics is a promising method to measure stresses nondestructively based on the same principle as the photoelasticity [1-3]. In this study a new method using longitudinal and transverse waves to adjust the isotropic component of the elastic constant of anisotropic materials is presented. The new method is applied to the measurement of welding residual stresses of a welded joint of carbon steel.

1. NONDESTRUCTIVE EVALUATION OF STRESS AND ANISOTROPY

In a textured material, following relations between ultrasonic velocities and an texture coefficient W_{400} can stand [4].

$$R_L = \frac{V_{yy}^2}{\Sigma V^2} = \frac{C_{11} - (2/5)C + (2/15)\pi^2 C W_{400}}{(C_{11} + 2C_{44})} \tag{1}$$

$$\Delta R_T = \frac{V_{yz}^2 - V_{yx}^2}{\Sigma V^2} = \frac{2\pi^2 C W_{400}}{9(C_{11} + 2C_{44})} \tag{2}$$

where V_{ij} is the ultrasonic velocity, T_{ij} is the traveling time, the first index means the propagation direction, the second one means the vibration direction, $\Sigma V^2 = V_{yy}^2 + V_{yx}^2 + V_{yz}^2$, C_{ij} is the elastic constant of the single crystal, and $C = C_{11} - C_{12} - 2C_{44}$. The isotropic component of elastic constant is $C_{11}-(2/5)C$ and the other part is the anisotropic one in Eq. (1). A correction factor, K, is introduced to the isotropic component of elastic constant. The texture coefficient W_{400} corrected by K is defined as follows:

$$W_{400}(R_L, K) = \frac{15[R_L(C_{11} + 2C_{44}) - \{C_{11} - (2/5)C\}K]}{2\pi^2 C} \tag{3}$$

where K can be determined by a next relation.

$$W_{400}(R_L, K) = W_{400}(\Delta R_T) \tag{4}$$

If an error of measurement is 0, K becomes 1. An uniaxial stress, σ, along the weld line and W_{400} can be evaluated by following relations.

$$\sigma = \frac{(C_{11} + 2C_{44})(5R_L + 3\Delta R_T) - 5K\{C_{11} - (2/5)C\}}{(C_{11} + 2C_{44})(5A + 3B)} \tag{5}$$

$$W_{400} = \frac{45[(C_{11} + 2C_{44})(A\Delta R_T - BR_L) + KB\{C_{11} - (2/5)C\}]}{2\pi^2 C(5A + 3B)} \tag{6}$$

where A and B are material constants depending stress.

2. EXPERIMENTAL PROCEDURE

A material used in this study was a carbon steel (JIS-STS42) for nuclear piping. Dimensions of the specimen were 125mm wide, 120mm long and 7mm thick. A sing-around method was used to measure the traveling time of ultrasonic waves. An ultrasonic frequency used was 5MHz. dimensions of ultrasonic sensor were ϕ5mm (longitudinal wave) and ϕ7mm(transverse wave). Measurements were performed along the center line perpendicular to the weld line of the specimen.

3. RESULTS AND DISCUSSION

W_{400} are evaluated by the ultrasonic velocities in the unstressed state. The relation between W_{400} and ΔR_T is shown in Fig. 1. An average value of K which is calculated by Eq. (4) is 1.00563. The solid symbols in Fig. 1 is $W_{400}(R_L, K)$ using calculated K value. $W_{400}(R_L, K)$ and $W_{400}(\Delta R_T)$ are coincide each other. σ and W_{400} in welded specimen are evaluated from the ultrasonic velocities using calculated K value. The distribution of σ is shown in Fig. 2. The results of the present method show a good agreement with the results obtained by the strain gauge method.

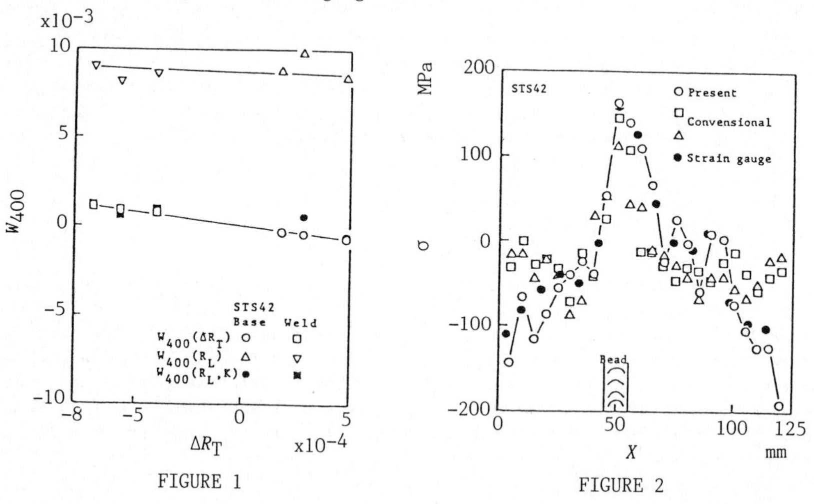

FIGURE 1

Relation between texture coefficient and difference in transverse wave velocities

FIGURE 2

Distribution of residual stresses

4. CONCLUSIONS

The method using longitudinal and transverse waves is applied successfully to the nondestructive measurement of welding residual stresses of a welded joint of carbon steel.

REFERENCES

[1] Arai, Y. and Kobayashi, H., Proc. of RFMMT, (1987) 709.
[2] Arai, Y., Kobayashi, H. and Suzuki, H., Proc. ICEM, Vol. I, (1988) 624.
[3] Kobayashi, H., et. al., Proc. ICRS-2, to be published.
[4] Sayers, C. M., "Solid mechanics research for QNDE", (1987) 320.

Elastic Waves and Ultrasonic Nondestructive Evaluation
S.K. Datta, J.D. Achenbach and Y.S. Rajapakse (Editors)

WHAT WE KNOW ABOUT FIRST TRANSONIC STATES IN ANISOTROPIC LINEAR ELASTIC SOLIDS

David M. BARNETT* and Jens LOTHE**

*Department of Materials Science and Engineering
Stanford University
Stanford, CA. 94305-2205, U.S.A.

**Institute of Physics
University of Oslo
P.O. Box 1048
Blindern, Oslo 3, Norway

There exist six possible types of ordinary first transonic states in stable anisotropic linear elastic solids. Four of these (Types 1, 2, 4, and 6) have been shown to occur in physically realizable crystals of cubic symmetry. This work deduces the orientations and elastic constant restrictions admitting a Type 3 transonic state in crystals of hexagonal and tetragonal symmetry. We prove that Type 3 states can never occur in stable crystals of isotropic, cubic, and trigonal symmetry.

1. INTRODUCTION

It is now well-established that the existence or non-existence of subsonic free surface (Rayleigh) waves in anisotropic linear elastic half-spaces is determined by the behavior of the hermitian impedance matrix at the so-called limiting speed $\hat{v}$, which is the smallest speed at which the equations of motion for plane steady disturbances lose ellipticity [1]. If $\mathbf{m}$ and $\mathbf{n}$ are two real orthogonal unit vectors along the surface wave propagation direction in the half-space boundary and normal to the half-space boundary, respectively, $\hat{v}^{-1}$ is the maximum projection along $\mathbf{m}$ of the centered outer slowness branch in the $\mathbf{m}$-$\mathbf{n}$ plane. Alternatively, a straight line L in the $\mathbf{m}$-$\mathbf{n}$ plane, parallel to $\mathbf{n}$ and displaced from the origin by $v^{-1}\mathbf{m}$, first contacts the outer slowness branch tangentially at $v = \hat{v}$ as the speed v is increased from zero. The state of affairs at $\hat{v}$ is called the first transonic state.

There exist only six possible types (Types 1-6) of ordinary first transonic states in stable linear elastic solids [2], depending on the number of first contacts and the number of slowness branches participating in the contacts. Transonic states of Types 1, 2, 4, and 6 have been found in many physically realizable crystals of cubic symmetry [3]. The Type 5 state has not yet been found in physically realizable or artificially constructed crystals. This work focuses on determining the conditions (orientations of $\mathbf{m}$ and $\mathbf{n}$ and restrictions on the elastic constants) admitting the existence of a Type 3 transonic state (when L contacts all three slowness branches tangentially at a single point) in media of isotropic, cubic, hexagonal, trigonal, and tetragonal symmetry.

2. THE TYPE 3 TRANSONIC STATE

It is easy to deduce that the necessary and sufficient conditions for a Type 3 transonic state to occur are:

$$(1) \qquad C_{ijkl}m_i m_l \;=\; \rho\hat{v}^2\delta_{jk} \qquad (\rho \text{ is mass density})$$
$$(2) \qquad C_{ijkl}(m_i n_l + m_l n_i) \;=\; 0.$$

A systematic study of the application of these conditions to the five crystal classes mentioned above yields the following results ($\mathbf{e}_I$, $\mathbf{e}_{II}$, and $\mathbf{e}_{III}$ are the natural crystal base vectors, e.g., see Nye [4]).

(1) Crystals of isotropic, cubic, and trigonal symmetry *do not admit the existence* of Type 3 states.

(2) Hexagonal symmetry admits a Type 3 state if and only if :
$\mathbf{m} = \mathbf{e}_{III}$; $\mathbf{n}$ = any vector in the basal plane and
$C_{11} > C_{33}$; $C_{33} = C_{44} = -C_{13}$.

(3) Tetragonal 6 or 7 symmetries admit a Type 3 state if either:
(a) The same conditions for hexagonal symmetry apply, or
(b) $\mathbf{n} = \mathbf{e}_{III}$; $\mathbf{m} = \mathbf{e}_I \cos\theta + \mathbf{e}_{II} \sin\theta$; $C_{13} = -C_{44}$ and
(i) Tetragonal 7
$\tan 2\theta = -(C_{11} - C_{66})/2C_{16} = -2C_{16}/(C_{12} + C_{66})$ and
$C_{11} - C_{66} > 0$; $C_{12} + C_{66} > 0$; $C_{11} + C_{66} = 2C_{44}$;
$(C_{11} - C_{66})(C_{12} + C_{66}) = 4C_{16}^2 > 0$.
(ii) Tetragonal 6. Either
(α) $\theta = 0, \pi/2$; $C_{11} = C_{66} = C_{44}$; $C_{12} + C_{66} > 0$ or
(β) $\theta = \pi/4, 3\pi/4$; $C_{11} - C_{66} > 0$; $C_{12} = -C_{66}$; $C_{11} + C_{66} = 2C_{44}$,

or

(c) $C_{11} + C_{66} = C_{33} + C_{44}$; $\mathbf{m}$ and $\mathbf{n}$ are any two orthogonal unit vectors normal to $\mathbf{t} = \mathbf{e}_I \cos\theta - \mathbf{e}_{II} \sin\theta$ with
(i) Tetragonal 7
$\tan 2\theta = (C_{11} - C_{66})/2C_{16} = 2C_{16}/(C_{12} + C_{66})$;
$(C_{11} - C_{66})(C_{12} + C_{66}) = 4C_{16}^2$; $C_{33} = C_{44} = -C_{13}$.
(ii) Tetragonal 6
$\theta = \pi/4, 3\pi/4$; $C_{11} > C_{33}$; $C_{12} + C_{66} = 0$; $C_{33} = C_{44} = -C_{13}$;
$C_{11} - C_{66} > 0$; $C_{11} + C_{66} = 2C_{44}$.

No known physically realizable crystals have elastic constants satisfying the above restrictions.

REFERENCES

[1] Barnett, D.M. and Lothe, J., Proc. Roy. Soc. Lond. A 402 (1985), 135-152.

[2] Chadwick, P. and Smith, G.D., Adv. in Appl. Mech. 17 (1977), 303-376.

[3] Barnett, D.M. and Lothe, J., Proc. Roy. Soc. Lond. A 411 (1987), 251-263.

[4] Nye, J.F., Physical Properties of Crystals (Oxford at the Clarendon Press, 1972), pp. 282-283.

Elastic Waves and Ultrasonic Nondestructive Evaluation
S.K. Datta, J.D. Achenbach and Y.S. Rajapakse (Editors)

TRANSIENT ELASTIC AND VISCOELASTIC RESPONSE OF A HALF-SPACE

Piotr BOREJKO * and Franz ZIEGLER *

Civil Engineering Department
Technical University of Vienna
A-1040 Wien, Austria

The transient elastic and viscoelastic response of a uniform half-space, $-\infty<x<\infty$ and $z\geq0$, produced by a line of vertical surface traction is investigated. At first the elastic response is evaluated by the method of generalized rays [1]. Subsequently, the application of the Fast Fourier Transformation (FFT) and of the numerical version of the correspondence principle [2,3] renders the viscoelastic response. A three parameter standard linear solid is considered.

In Figs. 1 (a) and 1 (b) the time records of the elastic and viscoelastic displacements at the surface receiver (1.5,0) are given, respectively, when the initial pulse varies with time like the triangular pulse. Results presented in Fig. 1 (a) may be compared with the analytical ones of Pekeris [4] and with the experiment given in [5, p. 65]. The predominant pulses in Figs. 1 (a) and 1 (b) represent the elastic and viscoelastic Rayleigh waves, respectively. In Fig. 1(b) two pulses following the Rayleigh wave are presumably the additional surface waves which are possible in the viscoelastic half-space [3].

Figures 2 (a) and 2 (b) show the Fourier spectra of the elastic and viscoelastic displacements at the surface receiver (1.5,0), respectively, when the initial pulse varies with time like two opposite triangular pulses (see Fig. 2 (c) for its frequency spectrum). For low frequencies, the viscoleastic amplitudes are smaller than the elastic ones, while, for high frequencies, the opposite is true.

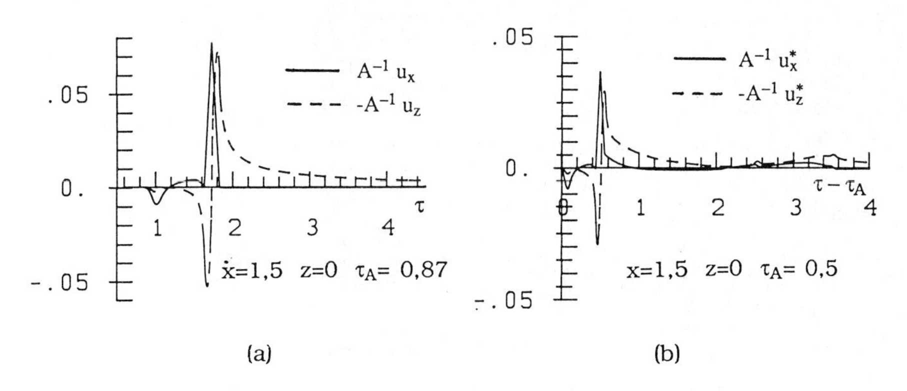

Fig. 1. (a) Elastic horizontal u_x and vertical u_z displacements as functions of time.
(b) Viscoelastic horizontal u_x^* and vertical u_z^* displacements as functions of time. Here $A=(2\pi\mu)^{-1}$, μ is the shear modulus, τ is the normalized time and τ_A denotes the arrival time of the P-wave.

* Research sponsored by the Austrian FWF undr Project S30-01.

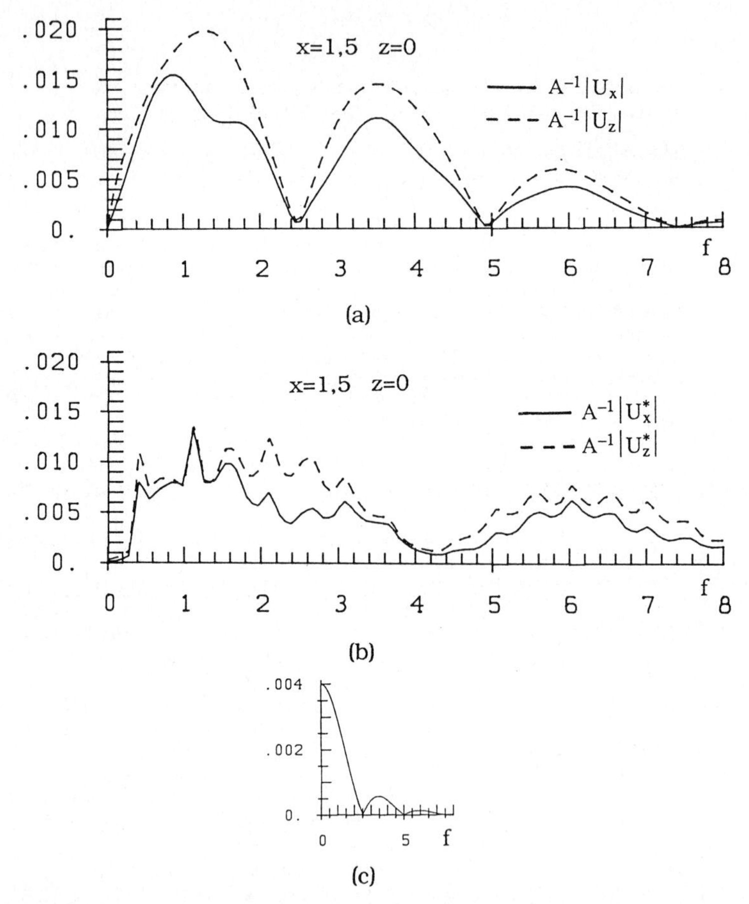

Fig. 2. (a) Fourier amplitude spectrum of the elastic displacements. (b) Fourier amplitude spectrum of the viscoelastic displacements. (c) Fourier spectrum of the initial pulse. Here f is the normalized frequency and for A see Fig. 1.

References

[1] Pao, Y.H. and Gajewski, R.R., The generalized ray theory an transient responses of layered elastic solids, in: Mason, W.P. and Thurston, R.N., (eds.), Physical Acoustics 13 (Academic Press, New York, 1977) pp. 183-265.

[2] Dasgupta, G. and Sackman, J.L., ASME J. Appl. Mechanics 44 (1977) 57.

[3] Borejko, P. and Ziegler, F., Surface waves on an isotropic viscoelastic half-space: The method of generalized rays, in: Parker, D.F. and Maugin, G.A., (eds.), Recent Developments in Surface Acoustic Waves (Springer Series on Wave Phenomena, Springer Verlag, Berlin, 1988) pp. 299-308.

[4] Pekeris, C.L., Proc. Natl. Acad. Sci, U.S.A. 41 (1955) 469.

[5] Ewing, W.M., Jardetzky, W.S. and Press, F., Elastic Waves in Layered Media (McGraw-Hill, New York, 1957)

Elastic Waves and Ultrasonic Nondestructive Evaluation
S.K. Datta, J.D. Achenbach and Y.S. Rajapakse (Editors)

AN APPROXIMATION METHOD FOR HIGH FREQUENCY DIFFRACTION OF ELASTIC WAVES BY A CRACK: THE SH-WAVE CASE

S.K.BOSE
Regional Engineering College, Durgapur (W.B.), India.

An analog of the "slender body approximation" of fluid mechanics is proposed here to approximately obtain the high frequency diffraction of elastic waves by cracks which may have small curvatures. Considering SH-waves it is postulated that, mainly the contribution to the displacement and stress at a point P near the crack comes from c.o.d nearest to P. (For theory and applications of "slender body approximation" in wave propagation problems, see Gear [1], [2] and Bose [3]). It is possible to generalize the method to deal with cases other than SH-waves.

INTEGRAL REPRESENTATION OF DIFFRACTED FIELD AND "SLENDER BODY" ANALOG FOR NEAR-FIELD

Referring to Fig.1, the total transverse displacement can be written as $v(z,x) = v_i(z,x) + v_d(z,x)$, where by an application of Green's theorem

$$v_d(z,x) = -\frac{1}{2\pi}\int_0^1 V(\sigma)\frac{\partial G}{\partial \nu}\, d\sigma \qquad (1)$$

Suppressing $\exp(i\omega t)$ in the above, $V(\sigma)$ is the c.o.d and G the Green's function given by

$$G = \frac{i\pi}{2} H_0^{(2)}[k\{(\zeta-z)^2 + (\xi - x)^2\}^{1/2}]$$

$$= -\frac{1}{2}\int_{-\infty}^{\infty} e^{-s_1|\zeta-z|} e^{-i\xi_1|\xi-x|} \frac{d\xi_1}{s_1} \qquad (2)$$

$s_1^2 = \xi_1^2 - k^2.$

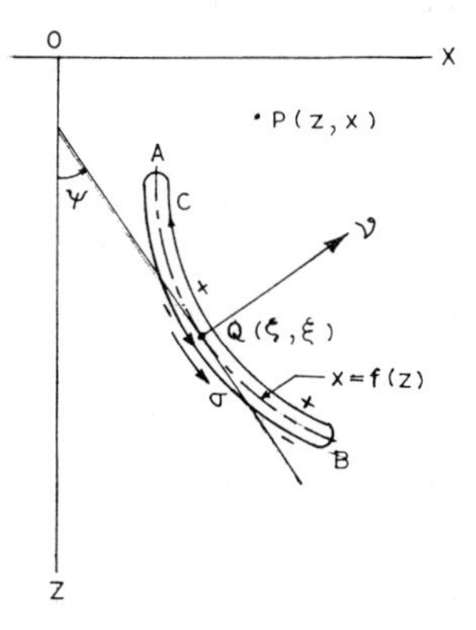

Fig. 1. The crack geometry.

According to the postulate we replace $V(\sigma)$ by $V(s)$. Also, since $\partial/\partial\nu = -\sin\psi\, \partial/\partial\zeta + \cos\psi\, \partial/\partial\xi$, we get by carrying it under the integral of (2),

$$v = v_i - \frac{V(s)}{4\pi}\int_{d_1}^{d_2} d\zeta \int_{-\infty}^{\infty} e^{-s_1|\zeta-z|} e^{-\xi_1|f(\zeta)-x|}$$

$$\times\ [-f'(\zeta)s_1\mathrm{sgn}(\zeta - z) + i\xi_1\mathrm{sgn}(f(\zeta) - x)]\frac{d\xi_1}{s_1} \qquad (3)$$

where $x = f(z)$ is the equation of C and d_1, d_2 are the depths of A and B below OX. Evidently, the inner integral has an infinite discontinuity when (ζ,z) tends to (z,x). Nevertheless, carrying out differentiations, for satisfying the condition $\partial v/\partial n = 0$ on C, we have

$$0 = \frac{\partial v_i}{\partial n} - \frac{V(s)}{4\pi\{1+f'^2(z)\}^{1/2}}\int_{d_1}^{d_2} d\zeta \int_{-\infty}^{\infty} e^{-s_1|\zeta-z|} e^{-i\xi_1|f(\zeta)-x|}[f'(z)f'(\zeta)s_1$$

$$- i\xi_1\{f'(z) + f'(\zeta)\} - s_1 - \frac{k^2}{s_1}]d\xi_1 \qquad (4)$$

Again, excepting the last term in (4), the first three yield non-integrable ζ-integrands. Hence, considering the ζ-integrand as a pseudo-function and taking the Hadamard finite-part of the integral, we get using (2)

$$\frac{iV(s)k^2}{4\{1+f'^2(z)\}^{1/2}}\int_{d_1}^{d_2} H_0^{(2)}[k\{(\zeta - z)^2 + (f(\zeta) - f(z))^2]^{1/2}d\zeta = \frac{\partial v_i}{\partial n} \qquad (5)$$

The integral in (5) is evidently a distribution of SH-wave sources along the length of the crack. With V(s) determined, the diffracted field is easily obtained from equation (1).

When the crack is in a half-space, we have to include the reflected wave v_r in (1) and use G for a half-space which is

$$G = \frac{i\pi}{2}[H_0^{(2)}\{[(\zeta - z)^2 + (\xi - x)^2]^{1/2}\} + H_0^{(2)}\{[(\zeta + z)^2 + (\xi - x)^2]^{1/2}\}] \qquad (6)$$

Carrying out the analysis as before, V(s) and v are given by

$$\frac{iV(s)k^2}{4\{1 + f'^2(z)\}^{1/2}}\int_{d_1}^{d_2}[H_0^{(2)}\{k[(\zeta - z)^2 + (f(\zeta) - f(z))^2]^{1/2}\}$$

$$+ H_0^{(2)}\{k[(\zeta + z)^2 + (f(\zeta) - f(z))^2]^{1/2}\}]d\zeta = \frac{\partial}{\partial n}(v_i + v_r) \qquad (7)$$

$$v = v_i+v_r - \frac{ik}{4}\int_{d_1}^{d_2}[\{f'(\zeta)(\zeta-z)-f(\zeta)+x\}\frac{H_1^{(2)}\ \{k[(\zeta-z)^2+(f(\zeta)-x)^2]\}}{\{(\zeta-z)^2+(f(\zeta)-x)^2\}^{1/2}}$$

$$+ \{f'(\zeta)(\zeta + z) - f(\zeta) + x\}\ \frac{H_1^{(2)}\ \{k[(\zeta+z)^2+(f(\zeta)-x)^2]\}}{\{(\zeta+z)^2+(f(\zeta)-x)^2\}^{1/2}} \qquad (8)$$

On the free surface z = 0, the above simplifies to

$$v = v_i+v_r - \frac{ik}{2}\int_{d_1}^{d_2}[\zeta f'(\zeta) - f(\zeta) + x]\ \frac{H_1^{(2)}\{k[\zeta^2 + (f(\zeta) - x)^2]^{1/2}\}}{[\zeta^2 + (f(\zeta) - x)^2]^{1/2}}V(\sigma(\zeta))d\zeta \qquad (9)$$

COMPUTATIONS AND COMPARISON

In order to deal with the logarithmic singularity in (7), we approximate $f(\zeta)$ in a small interval $[z-\delta,z+\delta]$, by the secent $f(\zeta)-f(z) = m(\zeta-z)$, $m = [f(z+\delta)-f(z-\delta)]/2\delta$. Over this interval the integral becomes

$$\int_{z-\delta}^{z+\delta} H_0^{(2)}\{d_2k[(\zeta - z)^2 + (f(\zeta) - f(z))^2]^{1/2}\}d\zeta$$

$$= \frac{2}{d_2k(1+m^2)^{1/2}}\int_0^{d_2(1+m^2)^{1/2}\delta} H_0^{(2)}(t)dt \qquad (10)$$

and the latter can be expressed in terms of Hankel and Struve functions (Abramowitz and Stegun p.480, [4]):

$$\int_0^x H_0^{(2)}(t)dt = xH_0^{(2)}(x) + \frac{\pi x}{2}[H_0^{(2)}(x)\mathbf{H}_1(x) - \mathbf{H}_1(x)H_0^{(2)}(x)] \tag{11}$$

The Struve functions $\mathbf{H}_0(x)$ and $\mathbf{H}_1(x)$ of orders 0 and 1 can be easily computed by considering their power series expansions for $x \leq 5$ and asymptotic expansions for $x > 5$. The remaining regular integrals in (7) and (8) or (9) can be evaluated by the adaptive quadrature scheme QUANC8 (Forsythe, Malcolm and Moler [5]).

For comparison we consider a normal edge-crack $0 \leq z \leq l$ with $v_i = \exp(-ikx)$. An exact analytical solution is available from Loeber and Sih [6], which however, is known to offer difficulty in accurate numerical computations for higher frequencies. If we consider, the crack opening $z = x = 0$, and compute $|v_d|$ from their equations for a range of values of l/λ (λ = wave length $2\pi/k$) we find an abrupt flatening from $l/\lambda > 3.5$ (Fig. 2 (i)). The same quantity in the present method is $|V(0)/2|$, where from equation (7)

$$\frac{V(0)}{2}\int_0^{kl} H_0^{(2)}(t)dt = -1 \tag{12}$$

and can be computed by using equation (11). The two results are plotted in Fig. 2. The difference between the two is less than 10% for $l/\lambda > 3.0$. Also shown in the same figure, is the curve from Stone, Ghosh and Mal [7] which is based on the Sommerfeld diffraction problem for a half-plane. Equation (12) appears to follow the exact curve from $l/\lambda = 2.0$, than their's and its maximum deviation for $l/\lambda > 3.0$ also appears to be less. By taking a value $l/\lambda = 6.0$, which infact is a very small number in ultrasonics, $|v_d|$ computed from Loeber and Sih [6] and equation (9) for $z = 0$ and increasing $|x|/l$ is presented in Fig. 2(ii). The closeness of the pattern, except for small x/l, is apparant. For larger values of l/λ, we may expect this trend, because for $kl \to \infty$, $V(s) \to -2$ as for $V(0)$ and the integrals in (8) tend to plane waves yielding total shadow for $x > 0$ and a reflected wave for $x < 0$.

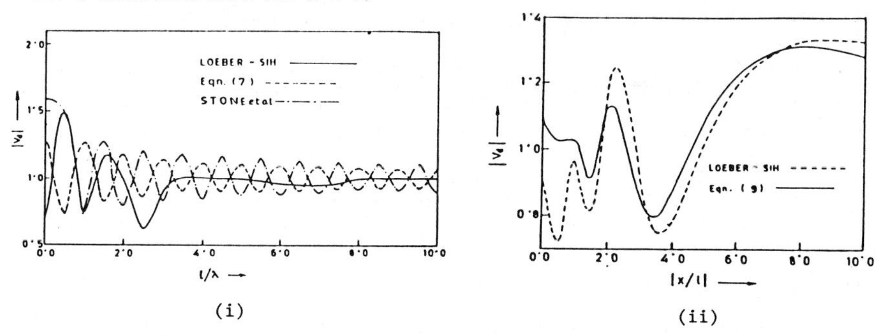

Fig. 2. The normal edge-crack (i) diffraction at 0. (ii) diffraction pattern for $l/\lambda = 6.0$.

Cracks of other orientations and shapes are considered in Fig. 3. In 3(i) an inclind crack at a depth $d_1 = 2l$ is considered with inclinations $\psi = 0, \pm 30, \pm 60$ degrees. in 3(ii) two cracks of equal length l meeting at a depth $d_0 = 3l$ with the first along OZ ($d_1 = 2l$) and the second inclind at $\psi = \pm 60$ degrees to OZ. In 3(iii) a surface parabolic crack with axis vertical, passing through the origin with slope 1/5 and lower tip at a distance $l/4$ to the positive side of OX is considered. Also a similar crack at a depth $d_1 = 2l$ is considered. All the curves have distinctive appearance, particularly when ψ is negative, indicating wave guide action of the crack.

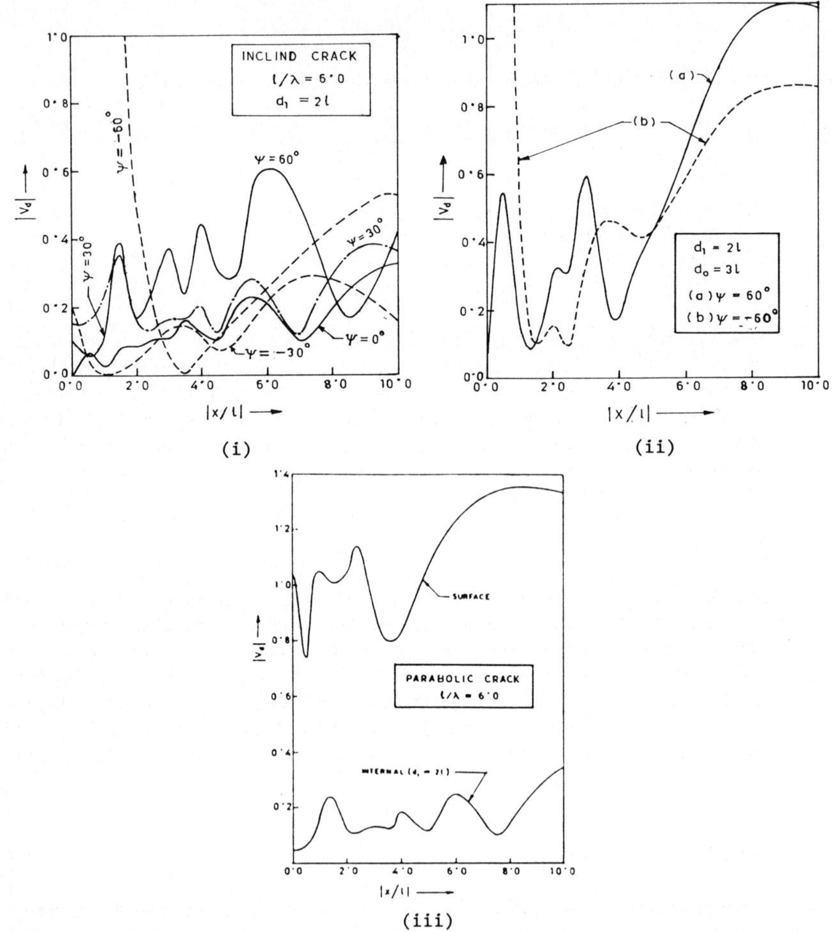

Fig. 3. (i) inclined crack, (ii) pair of converging cracks, and (iii) parabolic crack.

REFERENCES

[1] J.F. Gear, The scattering of a scalar wave by a slender body of revolution, SIAM J. App. Math. 34 (1978) pp. 348-370.

[2] J.F. Gear, Electromagnetic scattering by a slender body of revolution, Axially incident plane wave, SIAM J. App. Math. 38 (1980) pp. 93-102.

[3] S.K. Bose, Dynamic soil stiffness in torsional vibrations of a floating pile, Int. J. Soil Dyn. Earthquake Engg. 7 (1988) pp. 143-148.

[4] M. Abramowitz and I. Stegun, Handbook of mathematical functions, Dover Publications, New York (1965).

[5] G.E. Forsythe, M.A. Malcolm and C.B. Moler, Computer methods for mathematical computations, Prentice Hall, New York (1977).

[6] J.F. Loeber and G.C. Sih, Diffraction of anti-plane shear waves by a finite crack, J. Acoust. Soc. Am. 44 (1968) pp. 90-98.

[7] S.F. Stone, M.L. Ghosh and A.K. Mal, Diffraction of anti-plane shear waves by an edge crack, J. App. Mech. 47 (1980) pp. 359-362.

Elastic Waves and Ultrasonic Nondestructive Evaluation
S.K. Datta, J.D. Achenbach and Y.S. Rajapakse (Editors)

ELASTODYNAMIC SCATTERING BY A RECTANGULAR CRACK

Anders BOSTRÖM

Division of Mechanics
Chalmers University of Technology
Göteborg, Sweden

The scattering of plane elastic waves by a rectangular crack is considered using an integral equation technique. Starting from a double spatial Fourier field representation, the conditions in the plane of the crack lead to an integral equation for the crack opening displacement. The integral equation is discretized by expanding the crack opening displacement in a double series in Chebyshev polynomials which fulfil the right edge conditions.

1. INTRODUCTION

The scattering of ultrasonic waves by various types of cracks is an important problem in nondestructive evaluation of materials and it has therefore been studied in a couple of simple cases. Krenk and Schmidt [1] investigate the penny-shaped crack for general incident waves by a direct integral equation method which does not make use of any Green function. Similar methods can be used also for other types of cracks and in the present contribution this is done for the rectangular crack. This crack has previously been considered by Itou [2] for normal incidence using a method somewhat similar to the present one and by Nishimura and Kobayashi [3] using a boundary integral equation method.

2. THE INTEGRAL EQUATION

Consider a rectangular crack that is situated at $z = 0$, $|x| < a$, $|y| < b$ in an otherwise homogeneous and isotropic elastic solid. The density in the solid is ρ and its Lamé parameters are λ and μ. Only time harmonic conditions are considered with the factor $e^{-i\omega t}$ omitted throughout. The wavenumbers in the solid are k_p and k_s.

Due to space limitations only the symmetric part of the problem is considered; the antisymmetric part can be treated similarly. For the symmetric part the scattered displacement field **u** satisfies the following boundary conditons for $z = 0$:

$$\sigma_{zx} = \sigma_{zy} = 0, \quad \text{all } x \text{ and } y$$

$$u_z = 0, \quad |x| > a \text{ or } |y| > b$$

$$\sigma_{zz} = -\sigma_{zz}^{in} \quad |x| < a \text{ and } |y| < b$$

where σ_{zz}^{in} is the stress component due to the incident field.

For the scattered field for z > 0 a double Fourier transform representation is employed

$$\mathbf{u} = \int_{-\infty}^{\infty}\int_{-\infty}^{\infty} \{f_1 \nabla e^{i(qx+py+h_pz)} + f_2 \nabla \times [\hat{z}\, e^{i(qx+py+h_sz)}]$$

$$+ i f_3 \nabla \times \nabla \times [\hat{z}\, e^{i(qx+py+h_sz)}]\}\, dq\, dp$$

where $h_p = (k_p^2 - q^2 - p^2)^{1/2}$ and $h_s = (k_s^2 - q^2 - p^2)^{1/2}$ with branches chosen so $\mathrm{Im} h_p \geq 0$ and $\mathrm{Im} h_s \geq 0$. $f_i = f_i(q,p)$, $i = 1,2,3$, are clearly the amplitudes for P, SH, and SV waves, respectively. Computing the stress components for z = 0 and using the first boundary conditions yield

$$f_2 = 0$$

$$f_1 = \frac{1}{2h_p}(k_s^2 - 2s^2)\, f_3$$

where $s^2 = q^2 + p^2$. Introducing the normal crack opening displacement ΔU_z the second boundary condition than gives

$$f_3 = \frac{-i}{4\pi^2 k_s^2} \int_{-a}^{a}\int_{-b}^{b} \Delta U_z\, e^{-i(qx+py)}\, dx\, dy$$

The last boundary condition finally gives

$$\mu \int_{-\infty}^{\infty}\int_{-\infty}^{\infty} [-\frac{1}{2h_p}(k_s^2 - 2s^2)^2 - 2s^2 h_s]\, e^{i(qx+py)}\, f_3\, dq\, dp = -\sigma_{zz}^{in}$$

Inserting f_3 this is an integral equation for ΔU_z. This integral equation is nonsingular and it has the virtue of containing the physically interesting quantity ΔU_z as the unknown.

3. DISCUSSION

To discretize the integral equation the ΔU_z is expanded in a double series in Chebyshev polynomials that has the correct square-root behaviour along the edges. No special measures are taken at the corners. After projection of the integral equation on the same set of Chebyshev polynomials a system of linear equations is obtained where the coefficients are double integrals containing Bessel functions. By a two-dimensional stationary phase analysis it is possible to obtain relatively explicit expressions for the scattered far field. See Boström [4] for similar calculations in a two-dimensional problem.

The present method can straightforwardly be adapted to other interesting crack problems such as an interface crack, a crack in a layered structure or a crack in an anisotropic solid.

REFERENCES

[1] Krenk, S. and Schmidt, H., Phil. Trans. Roy. Soc. 308 (1982) 167.

[2] Itou, S., Zeit. ang. Math. Mech. 60 (1980) 317.

[3] Nishimura, N. and Kobayashi, S., An Improved Boundary Integral Equation Method for Crack Problems, in: Cruse, T.A. (ed.), Advanced Boundary Element Methods (Springer, Berlin, 1988) pp. 279-286

[4] Boström, A., J. Appl. Mech. 54 (1986) 503.

Elastic Waves and Ultrasonic Nondestructive Evaluation
S.K. Datta, J.D. Achenbach and Y.S. Rajapakse (Editors)
Elsevier Science Publishers B.V. (North-Holland), 1990

USE OF THE CONNECTION MACHINE TO STUDY ULTRASONIC WAVE PROPAGATION IN MATERIALS

P.P. DELSANTO

Politecnico di Torino
10129 Torino, Italy

Timothy V. WHITCOMBE, Henry H. CHASKELIS and Richard B. MIGNOGNA

Naval Research Laboratory
Washington, DC 20375-5000

1. INTRODUCTION

A detailed understanding of the propagation of ultrasonic waves in materials of industrial interest is essential for the development of refined ultrasonic QNDE techniques. Analytical solutions of the wave equation are well known, but severe difficulties are encountered when dealing with real materials, because of the possible presence of texture, stress, inhomogeneities and other irregularities or defects. A large number of numerical methods and sequential computer models have been devised to overcome these difficulties [1-3]. We present here a different method, based on the use of the Connection Machine* (CM) [4], which is a massively parallel computer with many thousands of independent processors, all working concurrently on the solution of the same problem.

This parallelism allows an opportunity to efficiently reformulate the problem to be studied and modify the approach. Currently, the memory available is limited and requires some care in programming. This limitation should decrease with new CM-type machines. In our approach we divide the material, along the wave propagation path, into "cells" in a one-to-one correspondence with the CM processors. The material property values corresponding to each "cell" are assigned to the respective processors via the CM front-end computer. Variations of the physical parameters of the material are assigned on a one-to-one basis. Time must also be properly discretized. In addition, the source wave, as induced by a transducer, is simulated by inputting the values of the displacements for the time duration of the pulse. Finally, by means of an iteration equation, the displacements of each cell at the time $t + 1$ are evaluated as linear combinations of the displacements of that cell and neighboring ones at the times t and $t - 1$.

2. RESULTS AND DISCUSSION

The treatment, briefly outlined in the introduction, is applicable to the study of ultrasonic wave propagation in the most general case. However, for illustrative purposes we limit ourselves here to the elementary case of one-dimensional wave propagation through a homogeneous material. Results, due to different kinds of source waves, are presented in Figures 1 and 2. Other results are presented in [5], where the formalism for the conversion of the differential equations into difference equations and the adaptation to parallel processing are also described.

From Figures 1 and 2 we see that, in our simulation approach, ultrasonic waves propagate and are reflected by external boundaries, as predicted by the theory. It is also intersting to observe, in Fig. 2b, the interference between arriving and reflected waves.

Although the present results are for a very simple case, similar conclusions about the validity of our approach are expected to hold for more complex situations, due to the reciprocal independence of the processors simulating the material cells.

*Naval Research Laboratory Connection Machine Facility.

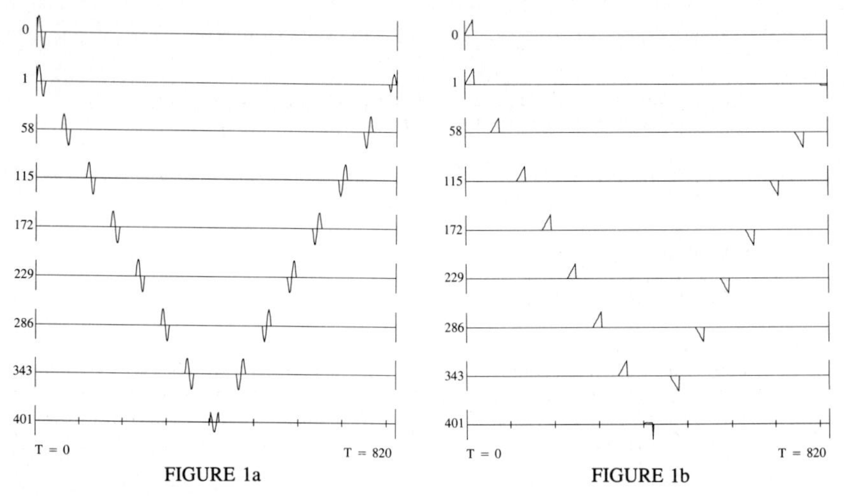

FIGURE 1a FIGURE 1b

(a) Progagation of a sine wave pulse, (b) propagation of a ramp-shaped pulse through a thickness $h = 1.27$ cm. Horizontal (time) scale is in units of $\tau = 5$ ns. Vertical (depth of propagation) scale is in units of $\epsilon = 31.8$ μm. Upper line ($i = 0$) represents the source pulse, next line ($i = 1$) and bottom line represent the top and bottom surface, respectively.

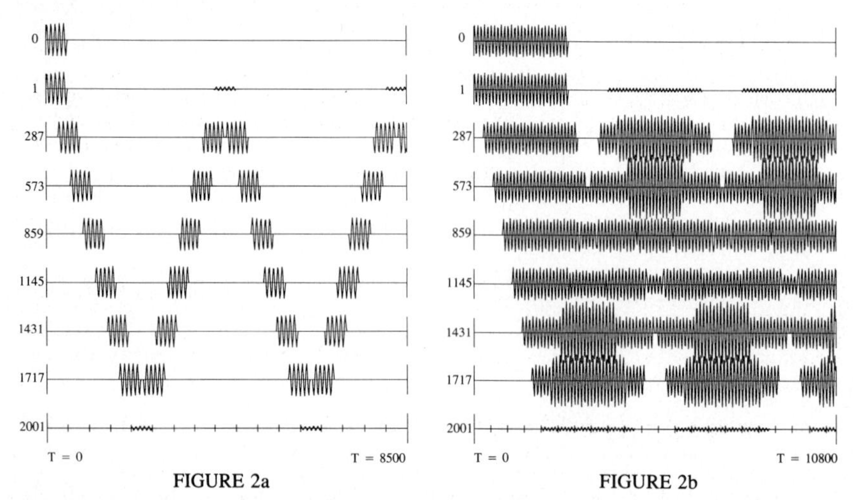

FIGURE 2a FIGURE 2b

Here $\tau = 1$ ns and $\epsilon = 6.4$ μm. Propagation of a sine wave pulse (a) (5 periods), (b) (28 periods).

REFERENCES

[1] Bond, L.J., Numerical Techniques and Their Use to Study Wave Propagation and Scattering, Proc. of this conference.

[2] Temple, A. and Ogilvy J., Numerical Techniques for Wave Propagation and Scattering, ibid.

[3] Fellinger P., et al, Numerical Techniques for Wave Propagation and Scattering, ibid.

[4] Hillis W.D., The Connection Machine (The MIT Press, Cambridge, Mass. 1985).

[5] Delsanto, P.P., Whitcombe, T., Chaskelis, H.H. and Mignogna, R.B., to appear in the Proc. of the Conf. Review of Progress in Quantitative NDE, Brunswick, Maine, July 23-8, 1989.

Elastic Waves and Ultrasonic Nondestructive Evaluation
S.K. Datta, J.D. Achenbach and Y.S. Rajapakse (Editors)

TRANSVERSE CUSP DIFFRACTION CATASTROPHES PRODUCED BY REFLECTING ULTRASONIC TONE BURSTS FROM CURVED SURFACES

Carl K. Frederickson and Philip L. Marston*

Department of Physics
Washington State University
Pullman, Washington 99164

An experiment is summarized in which transverse cusp caustics and associated wavefields are produced by reflecting high-frequency sound and light from smooth curved metal surfaces in water. Structure in the sound field having the form of a Pearcey pattern is clearly observed. It is useful for inverse scattering.

1. INTRODUCTION

Caustics may be classified according to the catastrophe designation of a singularity associated with the merging of rays [1]. The wavefield patterns in regions near caustics are often referred to as "diffraction catastrophes," and while wave amplitude may be large, it does not diverge at caustics as predicted from elementary geometrical optics. The caustic having the generic shape of a cubic cusp curve is associated with the merging of three rays at a cusp point and the disappearance of two rays as the caustic is crossed. The present paper concerns cusp caustics oriented as shown in Fig. 1. The direction of propagation of some outgoing wavefront (produced by reflection in the present study) is nearly parallel to the z axis. The wavefront propagates from the exit plane in water to produce the caustic. The cusp of interest opens up generally transverse to the z axis such that in the uv plane (which is parallel to the exit plane) the caustic coordinates are given by Eq. (1),

$$D_T(u - u_c)^3 = (v - v_c)^2, \qquad D_L(z - z_c)^3 = v^2 \tag{1,2}$$

where for comparison, Eq. (2) shows the form of an axial (or longitudinal) caustic in the zv plane. The coefficients D_T and D_L are cusp opening rates and (u_c, v_c) and $(z_c, v_c = 0)$ are cusp point coordinates in the transverse and axial cases, respectively. The axial caustic is generally associated with aberration of focused cylindrical wavefronts [2] and near the cusp point the Pearcey function

$$P(w_2, w_1) = \int_{-\infty}^{\infty} \exp\left[i\left(s^4 + w_2 s^2 + w_1 s\right)\right] ds, \tag{3}$$

describes the wavefield with w_2 and w_1 proportional to $\pm(z - z_c)$ and $\pm v$, respectively [3]. Near a transverse cusp, Marston [4] has shown how the relevant *two-dimensional* diffraction integral reduces to $P(w_2, w_1)$ with w_2 and w_1 proportional to $\pm(u - u_c)$ and $\pm(v - v_c)$, respectively. Plots of $|P(w_2, w_1)|$ display a diffraction structure known as a Pearcey pattern.

A previous study in which an ultrasonic wavefield near a cusp caustic was scanned by Dong *et al.* [5] failed to resolve a Pearcey pattern. Our experiments are based on the prediction that an outgoing wavefront displaced from the xy plane in the form [4, 6, 7]

$$W(x,y) = -(a_1x^2 + a_2y^2x + a_3y^2 + a_4x + a_5y), \quad a_2 \neq 0, \tag{4}$$

*Research supported by the U.S. Office of Naval Research.

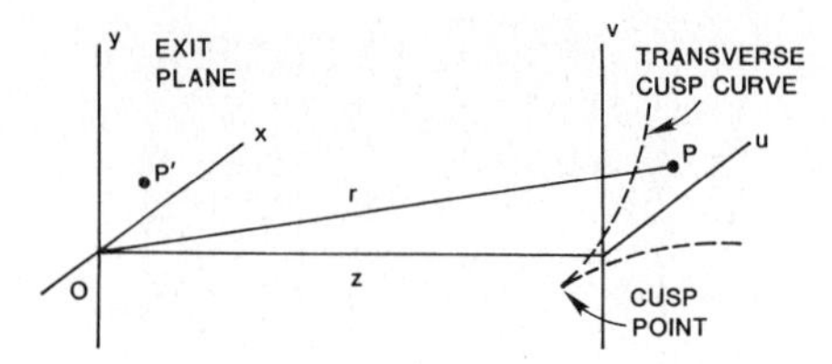

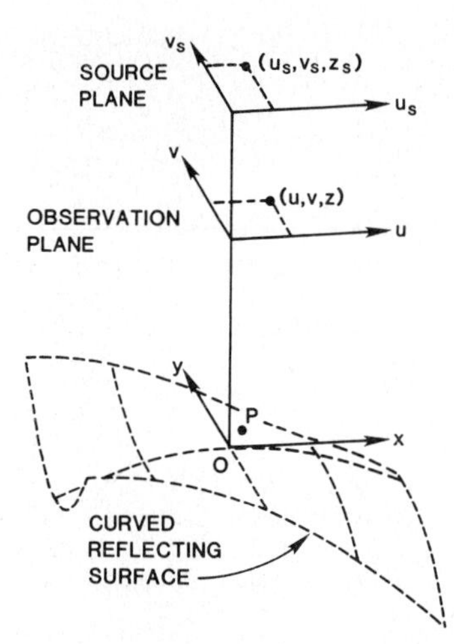

FIGURE 1
Coordinate system for describing a wavefront near the exit plane and the resulting caustics.

FIGURE 2
(Right) A wave from the point source reflects to produce a cusp diffraction catastrophe in the observation plane. The surface has $c_2 < 0$.

propagates to produce a transverse cusp in the plane at z provided $a_1 \neq -(2z)^{-1}$. The relation between the shape parameters a_i and the caustic parameters $D_T(z)$ and (u_c, v_c) is known in the paraxial approximation [4]. Marston [6] calculated that a reflected wavefront having the shape of Eq. (4) is produced by reflecting sound from a point source off of a surface whose height is

$$h(x,y) = c_1x^2 + c_2y^2x + c_3y^2 + c_4x + c_5y, \quad c_2 = -a_2/2 \neq 0, \qquad (5)$$

relative to the xy plane with $a_1 = -2c_1 + (2z_s)^{-1} \neq -(2z)^{-1}$ where (u_s, v_s, z_s) denote source point coordinates as shown in Fig. 2. The relationship between the shape parameters c_i, the source location and the wavenumber $2\pi/\lambda = 2\pi f/c$ with the w_1 and w_2 of the Pearcey pattern and the caustic parameters are calculated in Ref. 6 in the paraxial approximation.

Wavefield data should be useful for inferring the local shape of the reflector which produces the caustic [6]. In related research, optical diffraction catastrophes in scattering from oblate water drops were discovered [8]. Relationships between a scatterer's aspect ratio, wavefront shape, caustic parameters, and diffraction catastrophe patterns have been derived [7, 9-11].

2. SUMMARY OF EXPERIMENTAL METHOD AND RESULTS

A transducer was situated in water which simulated a point source at $z_s = 141$ cm. The duration τ of the tone burst radiated by the transducer was sufficiently long ($\tau \gtrsim 400$ μs) that all of the echoes from the curved surface overlapped in the region of time and space imaged in the reflected field by raster scanning a small hydrophone receiver in the uv observation plane. The duration τ was sufficiently short that the reflected wave could be sampled distinct from the incident wave. The hydrophone voltage was amplified and rectified and then gated and sampled at an appropriate time to simulate a stead-state reflection. The resulting voltage was stored on a Mac II computer along with the current hydrophone position. This was repeated for a large number of positions in the scan. An image was produced by increasing the pixel brightness on a display with the aforementioned voltage for a pixel position corresponding to each hydrophone position. Figure 3 is a representative photograph of the display which clearly shows the Pearcey pattern of a transverse cusp diffraction catastrophe. Dark regions correspond to regions of low reflected pressure amplitude.

One approach to testing the theory would require accurate direct measurements of c_1 and c_2 in Eq. (5). We chose a different method based on comparison with the optical cusp caustic produced by placing an optical point source at the same location as the acoustic source. This was done with the aid of an optical fiber (driven by a laser) with its output end in the water tank. The receiver was replaced by a photo detector. The output voltage was stored and imaged as described above.

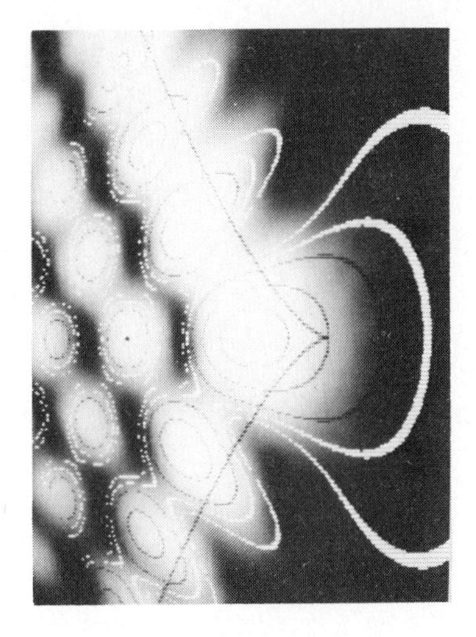

FIGURE 3
Measured ultrasonic wavefield pattern with $z =$ 68 cm and $f = 1.0$ MHz. The displayed spread in u and v are 8.0 and 10.8 cm, respectively.

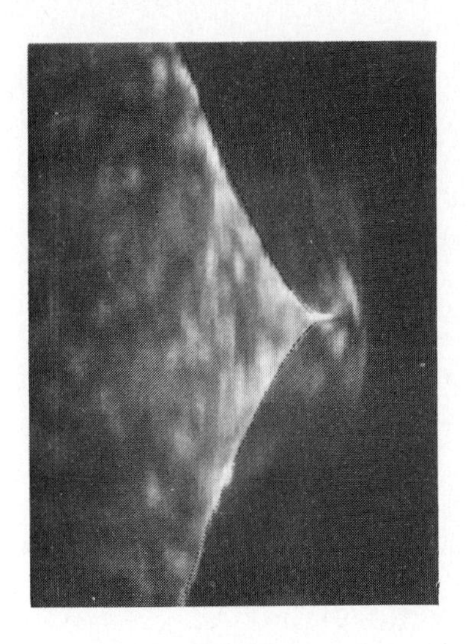

FIGURE 4
Measured pattern of reflected light. Superposed is a faint cusp curve having $D_T = 1.38$/cm which is used in the theoretical contours in Fig. 3.

Figure 4 shows the resulting image of the optical caustic with the same reflecting surface, source location, and observation plane as in Fig. 3 but with λ smaller by a factor > 3000. Superposed on Fig. 4 is a cusp curve given by Eq. (1) in which D_T and the cusp point location (u_c, v_c) were empirically fitted. A cusp curve with the same D_T is superposed on Fig. 3 since from the theory [6] the acoustical D_T is the same. Also shown in Fig. 3 are theoretical contours of constant $|P(w_2, w_1)|$ in which the only remaining free parameter a_1 was varied for agreement between the patterns. From the fitted a_1 and D_T, the surface parameters c_1 and c_2 were inferred and were near the range determined by direct measurement for the surface. The cusp is oriented as predicted. Patterns similar to Fig. 3 were observed for f from 0.5 to 4 MHz where along the u axis the spacing scales as $f^{-1/2}$. The observations not only verify the existence of the transverse Pearcey pattern but support its use in the inverse problem of determining the local surface shape.

REFERENCES

[1] Berry, M. V. and Upstill, C., Progress in Optics 18 (1980) 257.
[2] Solimeno, S., Crosignani, B., and DiPorto, P., Guiding, Diffraction, and Confinement of Optical Radiation (Academic, Orlando, 1986).
[3] Pearcey, T., Philos. Mag. 37 (1946) 311.
[4] Marston, P. L., J. Acoust. Soc. Am. 81 (1987) 226; 83 (1988) 1976.
[5] Dong, R., Adler, L., and Doyle, P. A., J. Appl. Phys. 54 (1983) 2832.
[6] Marston, P. L., Surface Shapes Giving Transverse Cusp Catastrophes in Acoustic or Seismic Echoes, in: Kessler, L. W. (ed.), Acoustical Imaging Vol. 16 (Plenum, New York, 1988) pp. 579-588.
[7] Marston, P. L. in: Schmidt, S. C. and Holmes, N. C. (eds.) Shock Waves in Condensed Matter 1987 (North-Holland, Amsterdam, 1988) pp. 203-206.
[8] Marston, P. L. and Trinh, E. H., Nature 312 (1984) 529.
[9] Marston, P. L., Optics Letters 10 (1985) 588.
[10] Marston, P. L., Dean, C. E., and Simpson, H. J., in: Wang, T. G. (ed.) Proceedings of the Third International Colloquium on Drops and Bubbles (AIP Conf. Proceedings, N.Y., 1989).
[11] Arnott, W. P. and Marston, P. L., J. Acoust. Soc. Am. 85 (1989) 1422.

Elastic Waves and Ultrasonic Nondestructive Evaluation
S.K. Datta, J.D. Achenbach and Y.S. Rajapakse (Editors)

IMPACT WAVES IN TRANSVERSELY ISOTROPIC LAMINATES

E. Rhian GREEN

Department of Engineering
The University
Leicester LE1 7RH
England

1. INTRODUCTION

The problem which is examined here is that of an impulsive line load acting on the top surface of a composite plate, and generating plane wave disturbances travelling in the plate along the direction normal to the line load. The plate is constructed of four layers of a fibre-reinforced material in which the reinforcement of each layer is a family of parallel fibres lying in the plane of the layers. The plies, each of depth h, are bonded together in a symmetric cross-ply configuration. The fibre-reinforced material is modelled as a homogeneous continuum of transversely isotropic elastic material with the axis of transverse isotropy in the fibre direction. A typical material is the ICI product APC-2, formed of carbon fibres embedded in a thermoplastic resin, for which typical dimensions are $h \simeq 125$ μm with the fibre diameter and interface spacing of the order of 6 μm. For wavelengths of the order of 15 μm or less, the continuum model will break down due to diffraction and scattering by the individual fibres. Consequently, the wavelengths of interest are of order $h/3$ or greater. In previous work, e.g. [1] the material was taken to be inextensible in the fibre direction. Here, this constraint of inextensibility has been lifted and results are presented for both extensible and inextensible models.

The method adopted is to solve exactly the system of governing equations appropriate to each layer, matching the solutions across the interfaces and satisfying the appropriate boundary conditions at the outer surfaces of the plate. The method of solution involves taking Laplace transforms in time and Fourier transforms in the in-plane spatial coordinates of the governing equations of the model, and yields the exact solution for the variation of the transforms with depth through the plate. Approximations arise only in the numerical methods for inverting the transforms. All the numerical work reported here refers to $\gamma = 60^0$, γ being the angle between the fibre direction in the inner layers and the direction of propagation.

2. DISCUSSION AND RESULTS

The figures display the variation of the scaled normal displacement at the upper and lower surfaces of the plate as functions of distance from the impact point at a reduced time $T = 200$. T is measured in units of the travel time through one ply thickness, h , for the body wave with speed c_1. This is the fastest body wave speed in the inextensible material, thus the disturbance will not reach a distance of 200h at $T = 200$. Higher speeds exist in the extensible material however, and it can be shown that the disturbance in this material will go further in the same time. For both model materials there exists high frequency motions in the region $110h < x < 140h$. These correspond to shear disturbances whose scaled speeds vary between 0.555 and 0.678. (Velocity is scaled by c_1).

The upper and lower normal displacements are very similar for both the extensible and inextensible models. The distinct step in the inextensible material travels with a scaled speed approximately equal to 0.6. This is associated with the existence of a non-zero low frequency limiting velocity of the fundamental modes of the antisymmetric motion, whereas the corresponding speed in the extensible material is zero which is in agreement with simple plate bending theory.

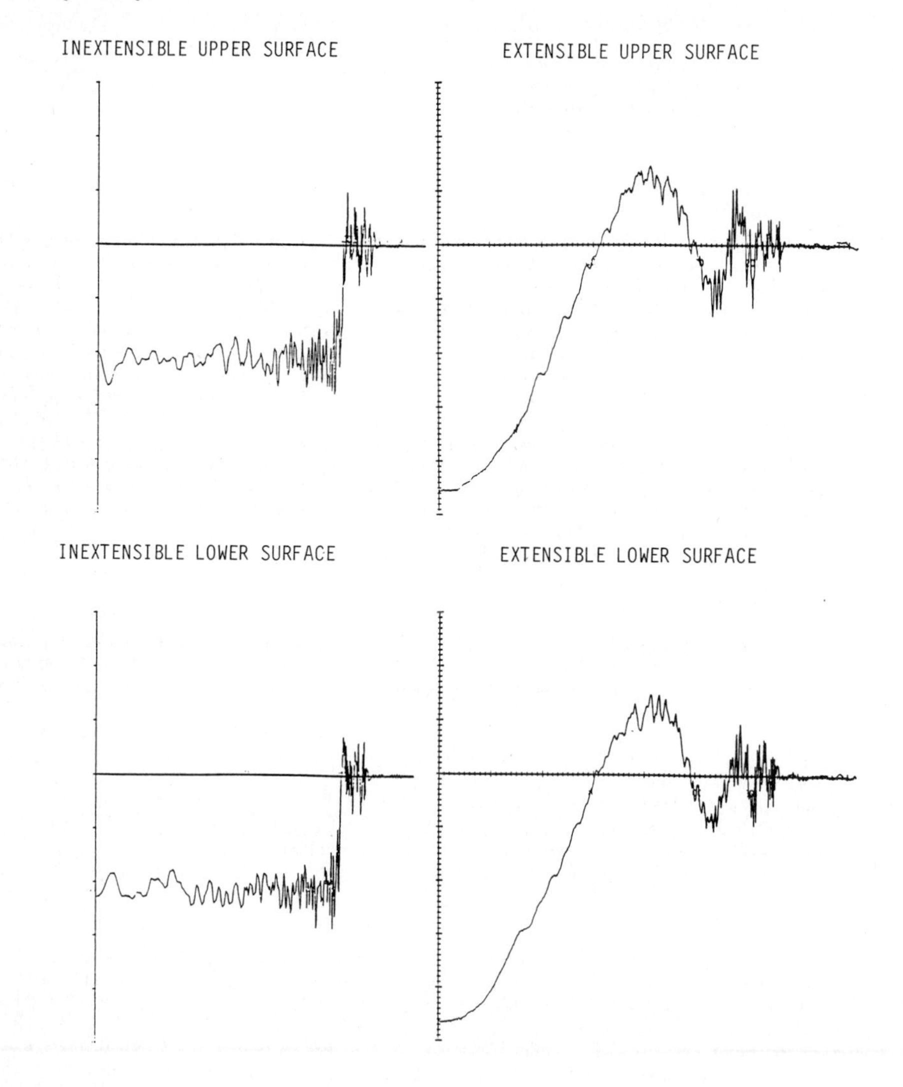

References

[1] Green, W.A. and Baylis, E.R., The Propagation of Impact Stress Waves in anistropic Fibre Reinforced Laminates in: Mal, A.K. and Ting, T.C.T., (eds.), Wave Propagation in Structural Composites, ASME AMD-Vol.90, 1988 pp. 53-68.

Elastic Waves and Ultrasonic Nondestructive Evaluation
S.K. Datta, J.D. Achenbach and Y.S. Rajapakse (Editors)

TRANSIENT STRESS WAVES IN ANISOTROPIC LAMINATES*

W. A. GREEN

Department of Theoretical Mechanics
University of Nottingham
Nottingham NG7 2RD, England

1. INTRODUCTION

The propagator matrix technique, developed for inextensible fibre-reinforced laminates by Green and Baylis [1], is here applied to examine the transmission of transient waves in a six-ply plate with $(-60^{\circ}/60^{\circ}/0)_S$ lay up and in an eight-ply plate with quasi-isotropic, $(90^{\circ}/-45^{\circ}/45^{\circ}/0)_S$ lay up. As in earlier papers, attention is focused on the transients arising due to an impulsive line load acting on the upper surface of the laminate. Such a line load gives rise to plane wave fronts travelling in the plane of the plate along the direction normal to the line. It is the variation of displacement and stress through the composite plate at these wave fronts which is of interest, both from the point of view of providing a standard of comparison for Non-Destructive Evaluation (NDE) studies and as a means of predicting the regions in which high stress levels may arise as a consequence of a surface impact. The results to be presented here relate to the normal displacements at the top and bottom surface of each plate. Calculations are currently in hand to determine the stress levels at these surfaces and at the interfaces within each laminate and these will be reported in due course. Details of the methods of solution are to be found in reference [2].

2. RESULTS

Figures 1 and 2 show the variation of normal displacement at the top and bottom surfaces, as a function of distance from the impact line at a fixed time $t = 200h/c_1$ after the impact. Here h is the thickness of a single ply, which is taken as the unit of distance and c_1 is a typical wave speed in the material.

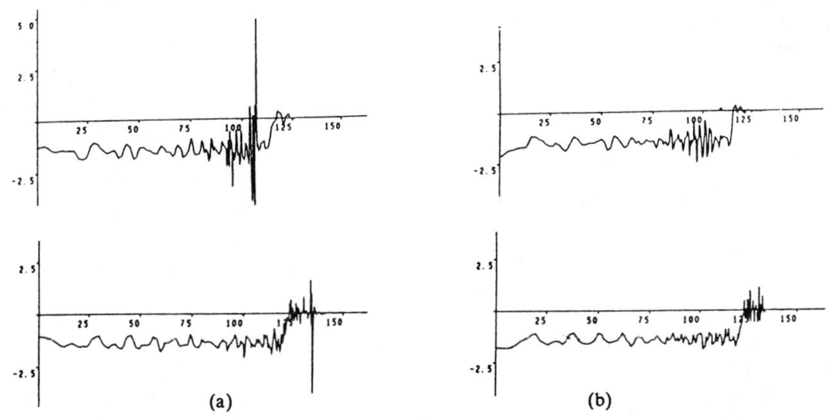

FIGURE 1.(a) Upper and (b) lower surface displacements in 6-ply plate. Top curves $\gamma = 0^{\circ}$, bottom curves $\gamma = 90^{\circ}$.

* Research supported by the United States Air Force under grant AFOSR-88-0353

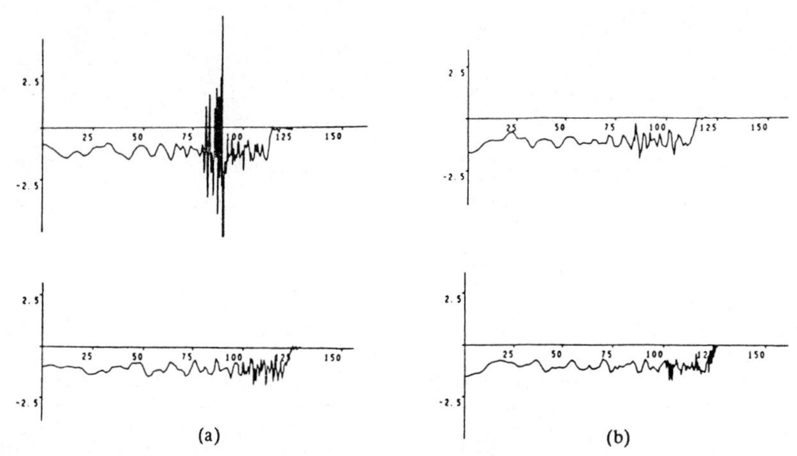

FIGURE 2. (a) Upper and (b) lower surface displacements in 8-ply plate. Top curves $\gamma = 0^0$, bottom curves $\gamma = 90^0$.

Each figure contains two sets of results, corresponding to propagation at two different angles ($\gamma = 0^0$ and $\gamma = 90^0$) relative to the fibre direction of the innermost layer in each plate. Figure 1 relates to the 6-ply plate and clearly shows the existence of a Rayleigh wave on the upper surface for both angles of propagation whereas in Figure 2, which relates to the 8-ply plate, there exists a Rayleigh wave for $\gamma = 0^0$ but not for $\gamma = 90^0$.

The most striking feature of these preliminary results is to be seen in Figure 3. This shows the upper surface normal displacement for a $(90^0/0)_S$ 4-ply plate (ref.[2]) with propagation at $\gamma = 60^0$ and for the 6-ply plate with $\gamma = 90^0$. Both graphs therefore refer to a wave travelling at an angle of 30^0 to the fibre direction in the outer layer. These curves clearly show the presence of a Rayleigh type wave travelling along the upper surface of the 6-ply plate whereas no such wave is apparent in the 4-ply laminate. A more detailed examination of this phenomenon is being carried out.

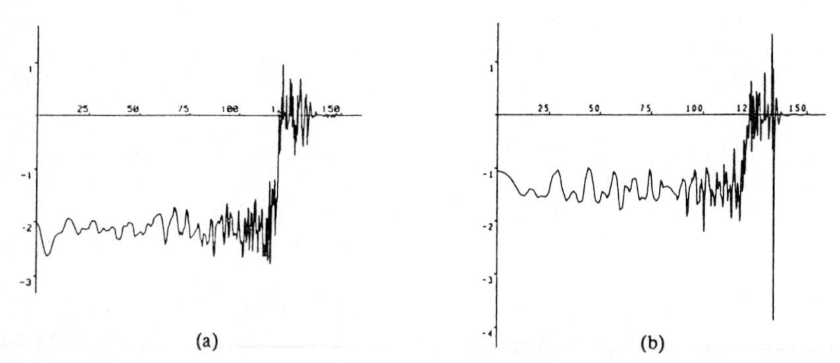

FIGURE 3. Upper surface displacements in (a) 4-ply plate, $\gamma = 60^0$ and (b) 6-ply plate, $\gamma = 90^0$

REFERENCES:

[1] Green, W.A. and Baylis E.R., Archives of Mechanics 31 (1986) 301.
[2] Baylis, E. Rhian and Green W.A., "Impact Stress Waves in Fibre-Reinforced Laminated Plates, "Proc. 3rd Int. Conf. on Recent Advances in Structural Dynamics, AFWAL-TR-88-3034, Vol 1 (1988) pp.171-183.

Elastic Waves and Ultrasonic Nondestructive Evaluation
S.K. Datta, J.D. Achenbach and Y.S. Rajapakse (Editors)

SCATTERING OF WAVES BY A SEMI-CYLINDRICAL GROOVE IN THE SURFACE OF AN ELASTIC HALF-SPACE

R.D. Gregory and D.M. Austin
Department of Mathematics, University of Manchester, Manchester M13 9PL,England

ABSTRACT

An elastic half-space, composed of homogeneous, isotropic material, contains a semi-cylindrical groove in its free surface; a time-harmonic elastic wave (a Rayleigh wave, plane PV wave or plane SV wave) is incident upon the groove, and is scattered by it. What is the scattered elastic field and how is the scattered energy apportioned in the far field? This problem is solved, in the context of the linear theory of elasticity, by a series expansion method. Multipole expansion potentials are used which are singular along the axis of the semi-cylinder, satisfy the free surface boundary condition (except on the groove) and consist of outgoing waves at infinity; these potentials are proved to be complete for the expansion of all time-harmonic elastic wave fields which satisfy the free surface condition on the flat surface of the half-space, and which satisfy "outgoing radiation" conditions at infinity. However, although this natural set of expansion potentials is complete, it is shown that it is not linearly independent; indeed the potentials satisfy infinitely many linear dependencies. We show how to reduce this linearly dependent expansion set to a linearly independent expansion set which is still complete. If the scattered field is now expanded in terms of this reduced set, the expansion coefficients are then unique, a necessary prerequisite for numerical computation. The expansion coefficients are then chosen so as to satisfy (to any desired accuracy) the boundary data on the groove and the far scattered field is then evaluated. Numerical results are presented, showing the effect of the scattering for the various kinds of incident wave.

1 EXPANSION POTENTIALS

These expansion potentials are in two families corresponding to compressional and shear sources of various orders. Since they satisfy the half-space boundary condition exactly, they have a fairly complicated form; the expansion pair $\{\phi_n^P, \psi_n^P\}$ $(-\infty < n < \infty)$ belonging to the (compressional) P-family is given by

$$\phi_n^P = H_n(kr)\left((-1)^n e^{in\theta} - e^{-in\theta}\right) + \int_{C_1} A(t)e^{-nt}\exp(ky\sinh t - ikx\cosh t)\,dt, \qquad (1\cdot 1)$$

$$\psi_n^P = \int_{C_1} B(t)e^{-nt}\exp(-\Delta(t)y - ikx\cosh t)\,dt, \qquad (1\cdot 2)$$

where

$$A(t) = \frac{-8ik^3\Delta(t)\cosh^2 t\sinh t}{\pi f(k\cosh t)}, \qquad (1\cdot 3)$$

$$B(t) = \frac{-4k^2\cosh t\sinh t(2k^2\cosh^2 t - K^2)}{\pi f(k\cosh t)}, \qquad (1\cdot 4)$$

where $\Delta(t) = (k^2 \cosh^2 t - K^2)^{\frac{1}{2}}$, $f(\alpha)$ is the Rayleigh secular determinant, and C_1 is a certain infinite contour in the complex t-plane passing from $-\infty$ to $\infty + \pi i$. The corresponding (shear) S-family of potentials is denoted by $\{\phi_n^S, \psi_n^S\}$.

When a similar expansion is used for the *submerged* cylindrical cavity, the resulting system of linear equations for the coefficients is non-singular. This is not so in the present case owing to infinitely many **linear dependencies between members of the expansion set**. When the redundant potentials are eliminated we obtain:

THEOREM *The potentials $\{\phi, \psi\}$ of any scattered field in the indented half-space can be expanded in the form*

$$\text{even}\,\phi = c\ \text{even}\,\phi_1^P + \sum_{n=0}^{\infty} \left\{c_{2n}\,\text{even}\,\phi_{2n}^P + d_{2n+1}\,\text{even}\,\phi_{2n+1}^S\right\}, \tag{1·5}$$

$$\text{odd}\,\phi = d\ \text{odd}\,\phi_1^S + \sum_{n=0}^{\infty} \left\{c_{2n+1}\,\text{odd}\,\phi_{2n+1}^P + d_{2n}\,\text{odd}\,\phi_{2n}^S\right\}, \tag{1·6}$$

with corresponding expansions for odd ϕ, even ψ.

2 RESULTS

The above method was used to calculate the scattered field when a Rayleigh wave, plane P-wave or plane S-wave was incident upon the groove. The main numerical difficulty is in the calculation of the stress fields corresponding to the expansion potentials $\{\text{even}\,\phi_1^P, \text{odd}\,\psi_1^P\}\ldots$ etc. This involves the evaluation of complex contour integrals similar to those in $(1\cdot1)$,$(1\cdot2)$.

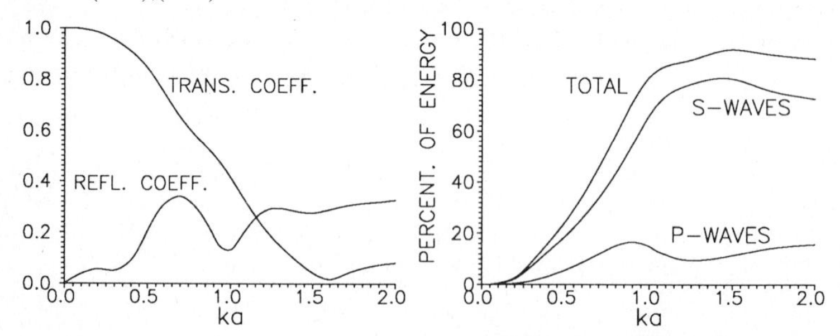

The graphs above show two features of the scattering when a Rayleigh wave, which is propagating in a half-space with Poisson's ratio $\nu = 1/3$, is normally incident on a semi-circular groove of radius a. The left-hand graph shows the (amplitude) reflection and transmission coefficients as functions of ka, where k is the compressional wave number. The right-hand graph shows how the scattered energy is apportioned in the far-field. Further details of the scattering, including scattering of an incident P- or S- wave can be obtained from the first author. All the results obtained were found to be accurately consistent with energy conservation and reciprocity.

Elastic Waves and Ultrasonic Nondestructive Evaluation
S.K. Datta, J.D. Achenbach and Y.S. Rajapakse (Editors)
Elsevier Science Publishers B.V. (North-Holland), 1990

Generation and Propagation of Bulk Heterogeneous Waves

Bernard Hosten and Marc Deschamps
Laboratoire de Mécanique Physique URA C.N.R.S.n°867
351 Cours de la Libération 33405 TALENCE Cédex, France Tél : 56.84.62.18.

In this paper, we present a method to generate bulk heterogeneous wave in a non-absorbing fluid by means of an immerged acoustic prism, made of absorbing material. The heterogeneous structure is verified by measuring the amplitude between equiamplitude planes, with an ultrasonic receptor, scanned along equiphasis planes.

Introduction.

In previous papers [1,2], we show that **bulk** heterogeneous waves are generated through fluid/absorbing solid interfaces and this concept is useful to measure correctly the anisotropic damping of composite materials [2]. In that case, the experimental verification of bulk heterogeneous waves generation is not straightforward. In this paper, we present a direct experimental verification of the structure of **bulk** heterogeneous waves, generated in non absorbing fluid.
The mathematical model for heterogeneous plane waves is given by :
$*\mathbf{U} = *a\ *\mathbf{P}\ \exp i(\omega t - *\mathbf{K}.\mathbf{OM})$, where $*a$ is a complex amplitude, $*\mathbf{K}$ is a complex vector such as : $*\mathbf{K} = \mathbf{K}' - i\,\mathbf{K}''$, ($\mathbf{K}'$: wave vector and $\mathbf{K}''$: damping vector), $*\mathbf{P}$ is a complex polarisation vector and $\mathbf{OM}$ the position vector. The dispersion equation, given by $*\mathbf{K}\ *\mathbf{K} = \mathbf{K}'^2 - \mathbf{K}''^2 - 2i\,\mathbf{K}'.\mathbf{K}'' = (\beta - i\alpha)^2$, links the complex wave vector to the medium properties (β and α) [1].

Bulk heterogeneous waves in non lossy medium.

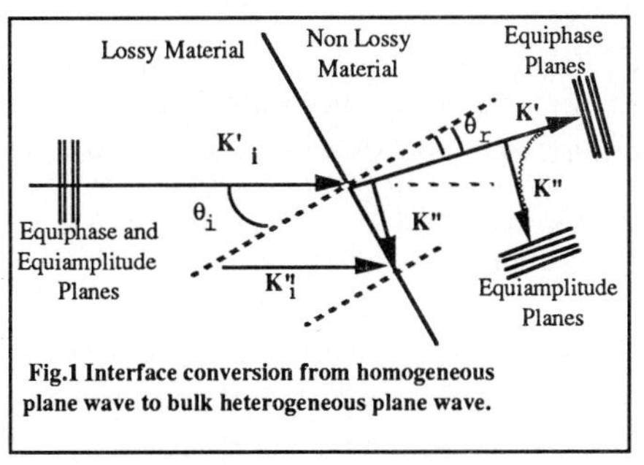

Fig.1 Interface conversion from homogeneous plane wave to bulk heterogeneous plane wave.

Through an interface between lossy and non lossy medium (α=0), an homogeneous damped plane waves is converted in an heterogeneous plane wave of which the structure is given by $\mathbf{K}'.\mathbf{K}'' = 0$ and equiamplitude planes are perpendicular to the equiphase planes.
The Snell's laws generalized to the complex wave vector give the damping vector in the non lossy medium : $k'' = k_i''\ \sin\theta_i\ (\cos\theta_r)^{-1}$.

[1] B.HOSTEN , M.DESCHAMPS . Acustica 59 [1986] , 193-198.
[2] B.HOSTEN, M.DESCHAMPS and B.R.TITTMANN . J.A.S.A. vol.82 (5), Nov.1987, 1763-1770.

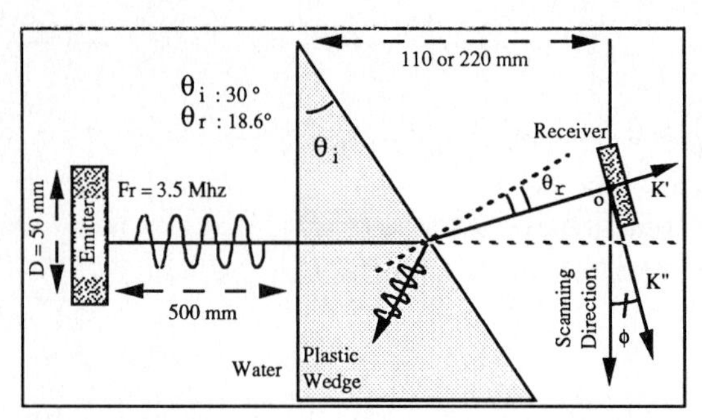

Fig.2 Generator of bulk heterogeneous plane wave. The generator of bulk heterogeneous plane wave [3] is essentialy a prism made with a viscoelastic material (P.V.C.). A damped homogeneous wave is generated in normal incidence through the first fluid/solid interface.

Precise measurements of velocities and dampings (table 1) in prism, let us compute the module of the **K**" vector. With these data, attenuation along the scanning direction is :

$\alpha = K'' \cos(\phi) = 38{,}8\ m^{-1} = 337$ db/m.

Table1	Velocity	Wavelength	K'	K"	K'D
P.V.C.	2340 m/s	0,66 mm	9520 m^{-1}	76 m^{-1}	476
Water.	1478 m/s	0,42 mm	14959 m^{-1}	**40,4 m^{-1}**	748

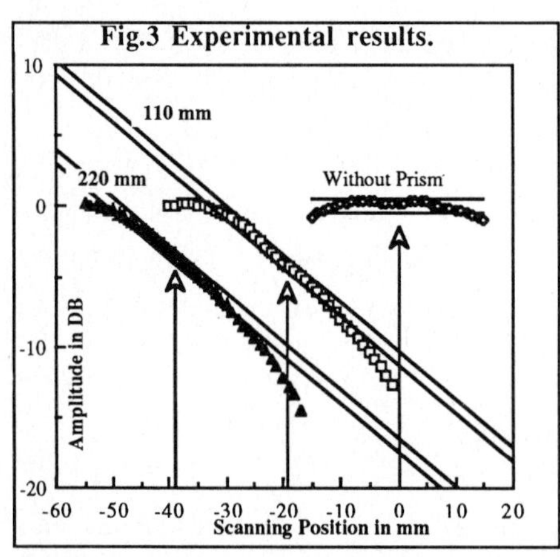

Fig.3 Experimental results.

Amplitude of signals acquired by the receiver is plotted in front of the scanning direction. ◇The prism is removed to verify that incident wave is plane and homogeneous in an area ≈ 20 mm ≈ 50 wavelengths, because the emitter diameter D is very large in front of the wavelength. With the prism, the scanning axis is placed at two positions : □110 mm between the prism input face and scanning axe and ▲ 220 mm (≈ 260 wavelengths further). Computed amplitudes in front of the scanning position are plotted with the maximum variations measured without prism.

Experimental values of **K**" fit very well with predicted values in the region where waves are not influenced by diffraction. In figure 3, arrows indicate the center of this zone calculated by means of Snell's laws .

Conclusions.

In non lossy media, heterogeneous waves are not only propagating at interfaces. The structure of the bulk heterogeneous waves does not change along the propagation and his heterogeneity is only dependent of the generating medium. The generator of bulk heterogeneous waves will now be used to study reflection and transmission of these waves through interfaces.

[3] M.DESCHAMPS , B.HOSTEN. Acustica. Vol.68 No.2 Avril 1989 pp.92/95.

Elastic Waves and Ultrasonic Nondestructive Evaluation
S.K. Datta, J.D. Achenbach and Y.S. Rajapakse (Editors)

AN ULTRASONIC STUDY ON VARIATIONS OF ELASTIC ANISOTROPY WITH PLASTIC DEFORMATIONS

Yukio Iwashimizu

Department of Mechanical Engineering
Ritsumeikan University, Kyoto, Japan

Osami Kobori

Department of Mechanical Engineering
Osaka Sangyo University, Osaka, Japan

Ultrasonic transverse velocities in steel specimens compressed plastically were measured to analyze a transverse isotropy induced by the plastic strain.

1. MEASUREMENTS

Specimens with two sets of parallel faces I and II as illustrated in Fig.1 were made from cylindrical blocks which were cut from a bar of steel S18C and compressed plastically along the bar axis. Several different values were taken for the plastic compressive strain e and face angle θ.
Ultrasonic transverse waves were propagated with the angle θ and (θ =) 90° from the bar axis by the normal incidence on faces I and II, respectively.The thickness normal to these faces was about 20 mm, and to obtain the propagation velocities, the time delay of two echoes was measured by the sing-around method.
In the face I incidence, the vibration direction of the transducer was taken along the x_1- and X_2-axes, and the respective velocities $v_1(\theta)$ and $v_2(\theta)$ were obtained. In the face II incidence, the velocities $v_1(90°)$ and $v_2(90°)$ with the respective polarizations along the X_3- and X_1-axes were obtained.

2. VELOCITIES

Fig.2 shows the results obtained by the face II incidence in all specimens.

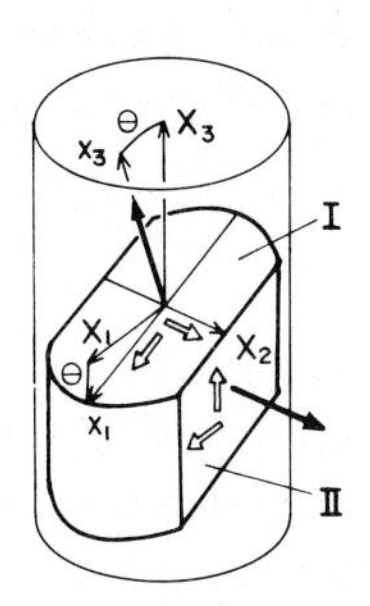

FIGURE 1
Specimen.

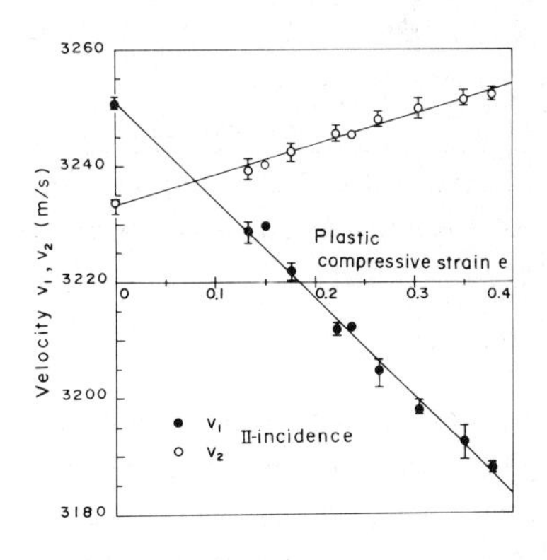

FIGURE 2
Variations in $v_1(90°)$ and $v_2(90°)$ with e.

$v_1(90°)$ and $v_2(90°)$ are varied directly with the plastic compressive strain e, and their difference increased after e = 0.08.
In Fig.3, $v_1(\theta)$ obtained by the face I incidence is plotted with respect to the angle θ from the bar axis to the propagation direction. For the points at $\theta = 90°$, $v_1(90°)$ obtained by the face II incidence in the specimens $\theta = 0°$ are used. At each strain e, $v_1(\theta)$ is well explained by a solid curve

$$v_1(\theta) = a_1 + b_1 \cos 4\theta . \quad (1)$$

Inflection of this curve is reversed in the region e = 0~0.136. This meansthat the initial anisotropy is cancelled by the strain-induced anisotropy.
The relation between $v_2(\theta)$ and θ is also well fitted by the equation

$$v_2(\theta) = a_2 + b_2 \cos 2\theta . \quad (2)$$

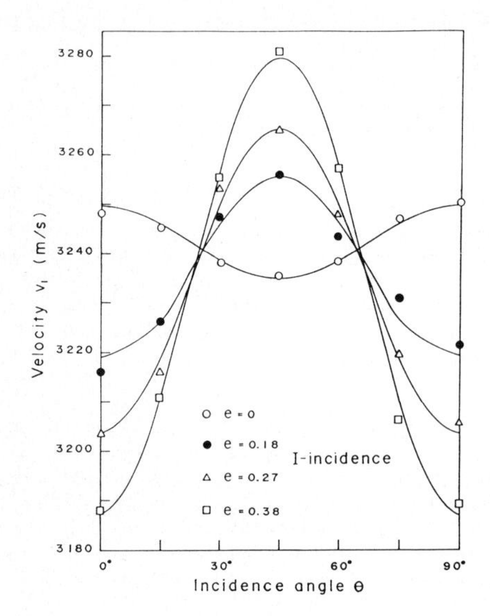

FIGURE 3
Direction dependence of $v_1(\theta)$.

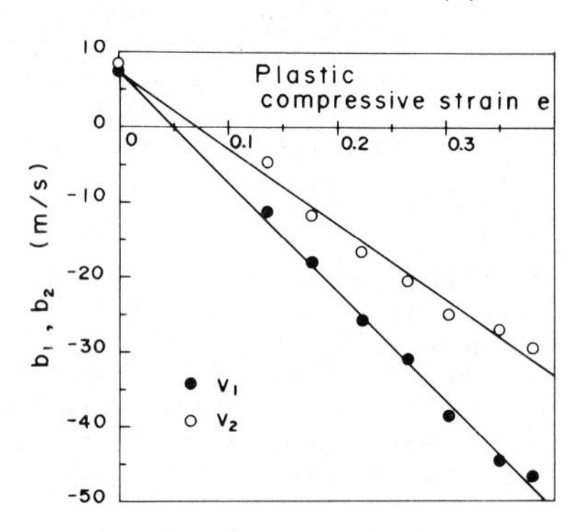

FIGURE 4
Variations in b_1 and b_2 with e.

3. STRAIN-INDUCED TRANSVERSE ISOTROPY

Fig.4 shows that the coefficients b_1 and b_2 in Eqs (1) and (2) are varied directly with the plastic strain e. From this and the similar result for the coefficients a_1 and a_2, we have $a_1 = 3241.7 - 22.8$ e, $a_2 = 3240.8 - 51.3$ e, $b_1 = 7.7 - 147.1$ e, and $b_2 = 7.5 - 101.8$ e (m/s). Since the difference $\pm\{(a_1 + b_1) - (a_2 + b_2)\}/2 = \pm(1-8e)$ is within the measurement error of ± 2 m/s, it is concluded that the initial and the strain-induced anisotropy are transversely isotropic, and that the degree of the latter transverse isotropy is proportional to the strain e [1][2]. If two anisotropic constants [3] are defined by

$$B = B^0 + B'e = C'_{44}, \quad G = G^0 + G'e = (8\,C'_{44} - C'_{33} + 4\,C'_{23})/16,$$

where C'_{KL} is the anisotropic part of the elastic constant C'_{KL}, the initial transverse isotropy (B^0, G^0) and the strain-induced one (B'e, G'e) are given by $B^0 = 4.6$, $G^0 = 3.1$, $B' = -74$, $G' = -60$ ($\times 10^2$ MPa).
It is interesting to note that the ratio $G'/B' = 0.81$ is similar to 7/8 which is obtained by the Voigt estimation for elastic constants of polycrystalline materials.

REFERENCES

[1] Toriyama, T. et al., Jpn Soc. Mech. Eng. Int. J., I , vol.32, no.4 (1989).
[2] Sasaki, K. et al., J. Appl. Mech., in print.
[3] Iwashimizu, Y., J. Soc. Mate. Sci. Jpn, vol.36, p.1409 (1987).

Elastic Waves and Ultrasonic Nondestructive Evaluation
S.K. Datta, J.D. Achenbach and Y.S. Rajapakse (Editors)

SCATTERING CHARACTERISTICS OF NEIGHBORING INCLUSIONS

M.KITAHARA and Y.HIEI

Faculty of Marine Science
and Technology
Shimizu, Shizuoka 424, Japan

K.NAKAGAWA

Total System Institute
Suginami-ku, Tokyo 166, Japan

Characterization of the scattered wavefield from flaws is of great importance to quantify the ultrasonic nondestructive evaluation. The volumetric flaws of inclusions are chosen here for the characterization and attention is paid for two neighboring incusions. The cavities are treated as a special case of inclusions in the present formulation.

1. INTRODUCTION

The three-dimensional elastodynamic multi-scattering problem for two inclusions is formulated by the boundary integral equation (BIE) method and the Born series strategy is adopted to solve the BIE system. The first term of the series corresponds to so-called Born approximation. The second, the third terms, and so forth, represent the mutual interaction effects of the inclusions. The attractive feature of the Born approximation has been described for flaws of cavities and inclusions by Gubernatis et al.[1]. The present treatment of Born series strategy for BIE system is based on Schuster's work for acoustics [2] and somewhat detailed treatment for three dimensional elastodynamics has been shown in Reference [3].

2. FORMULATION

Let $\lambda^{(\alpha)}, \mu^{(\alpha)}$ and $\rho^{(\alpha)}$ $(\alpha = 0, 1, 2)$ be Lamé constants and the mass density for the elastic surrounding matrix $(\alpha = 0)$ and for the first $(\alpha = 1)$ and the second $(\alpha = 2)$ inclusions. The surfaces of inclusions are denoted by S_1 and S_2 for the first and the second inclusions. Then the boundary integral equations can be written as

$$\mathbf{C}^{(\alpha)}(\mathbf{x})\mathbf{u}^{(\alpha)} = \int_{S_\alpha} \mathbf{U}^{(\alpha)}(\mathbf{x},\mathbf{y})\mathbf{t}^{(\alpha)}(\mathbf{y})dS_y - \int_{S_\alpha} \mathbf{T}^{(\alpha)}(\mathbf{x},\mathbf{y})\mathbf{u}^{(\alpha)}(\mathbf{y})dS_y + f_\alpha \mathbf{u}^I(\mathbf{x}) \quad \mathbf{x} \in S_\alpha \ (\alpha = 0,1,2) \tag{1}$$

where the equations for $\alpha = 0, 1, 2$ represent the ones for the surrounding matrix, the first and the second inclusions, respectively. In eq.(1), $f_0 = 1$, $f_1 = f_2 = 0$, and $S_0 = S_1 \cup S_2$. The fundamental solution takes the form:

$$U_{ij}^{(\alpha)}(\mathbf{x},\mathbf{y}) = \frac{1}{4\pi\mu^{(\alpha)}}\left[\frac{e^{ik_T^{(\alpha)}r}}{r}\delta_{ij} + \frac{1}{k_T^{(\alpha)2}}\frac{\partial}{\partial x_i}\frac{\partial}{\partial x_j}\left\{\frac{e^{ik_T^{(\alpha)}r}}{r} - \frac{e^{ik_L^{(\alpha)}r}}{r}\right\}\right] . \tag{2}$$

From the continuity conditions of displacements $\mathbf{u}^{(\alpha)}$ and tractions $\mathbf{t}^{(\alpha)}$ on S_1 and S_2 , and eliminating $\mathbf{t}^{(\alpha)}$, we have the following discretized form of BIE system [3]:

$$\begin{bmatrix} \mathbf{M}_{11} & \mathbf{M}_{12} \\ \mathbf{M}_{21} & \mathbf{M}_{22} \end{bmatrix} \begin{Bmatrix} \mathbf{u}_1 \\ \mathbf{u}_2 \end{Bmatrix} = \begin{Bmatrix} \mathbf{u}_1^I \\ \mathbf{u}_2^I \end{Bmatrix} \tag{3}$$

where $\mathbf{u}_1$ and $\mathbf{u}_2$ are the discretized version of displacements on S_1 and S_2 , respectively. $\mathbf{u}_1^I$ and $\mathbf{u}_2^I$ are incident plane wave fields on S_1 and S_2 . Now we split the equation (3) as

$$(\mathbf{A} - \mathbf{B})\mathbf{u} = \mathbf{u}^I \tag{4}$$

where

$$\mathbf{A}=\begin{bmatrix}\mathbf{M}_{11} & 0\\ 0 & \mathbf{M}_{22}\end{bmatrix},\ \mathbf{B}=\begin{bmatrix}0 & -\mathbf{M}_{12}\\ -\mathbf{M}_{21} & 0\end{bmatrix},\ \mathbf{u}=\{\mathbf{u}_1,\ \mathbf{u}_2\}^{\mathrm{T}},\ \mathbf{u}^I=\{\mathbf{u}_1^I,\ \mathbf{u}_2^I\}^{\mathrm{T}}. \quad (5a,b,c,d)$$

Operating $\mathbf{A}^{-1}$ on both-hand sides of eq.(4), we have

$$(\mathbf{I}-\mathbf{C})\mathbf{u}=\mathbf{A}^{-1}\mathbf{u}^I \quad (6)$$

where

$$\mathbf{A}^{-1}=\begin{bmatrix}\mathbf{M}_{11}^{-1} & 0\\ 0 & \mathbf{M}_{22}^{-1}\end{bmatrix},\ \mathbf{C}=\mathbf{A}^{-1}\mathbf{B}=\begin{bmatrix}\mathbf{M}_{11}^{-1} & 0\\ 0 & \mathbf{M}_{22}^{-1}\end{bmatrix}\begin{bmatrix}0 & -\mathbf{M}_{12}\\ -\mathbf{M}_{21} & 0\end{bmatrix}. \quad (7a,b)$$

The solution of eq.(6) can be written as

$$\mathbf{u}=\sum_{n=0}^{\infty}\mathbf{C}^n\mathbf{u}_0\ , \quad (8)$$

if the condition $\|\mathbf{C}\|=\|\mathbf{B}\|/\|\mathbf{A}\|<1$ is satisfied. The zeroth order term:

$$\mathbf{u}_0=\mathbf{A}^{-1}\mathbf{u}^{\mathrm{I}} \quad (9)$$

of eq.(8) is well known as Born approximation. The convergence check of eq.(8) can be found in Reference [3].

3. RESULTS

The back-scattered displacement fields along the x_3 -axis is shown in Fig.1. Spherical inclusions with same radius d are chosen and the distance of two inclusions is $d/5$.

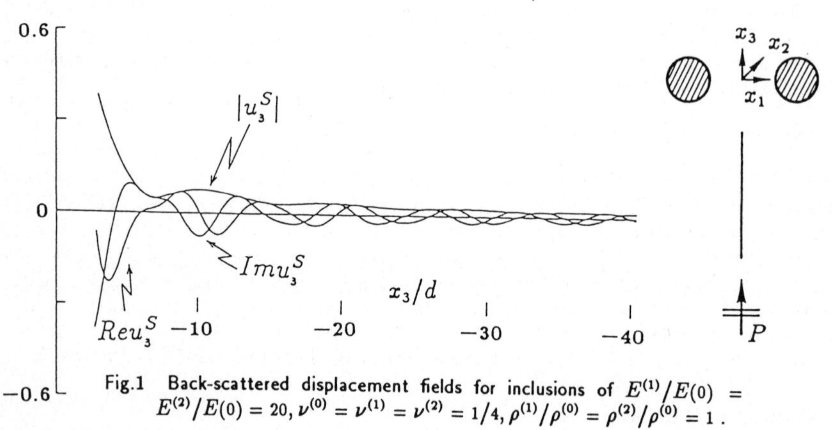

Fig.1 Back-scattered displacement fields for inclusions of $E^{(1)}/E(0) = E^{(2)}/E(0) = 20$, $\nu^{(0)}=\nu^{(1)}=\nu^{(2)}=1/4$, $\rho^{(1)}/\rho^{(0)}=\rho^{(2)}/\rho^{(0)}=1$.

REFERENCES

[1] Gubernatis, J. E., Domany, E., Krumhansl, J. A. and Huberman, M.; The Born approximation in the theory of the scattering of elastic waves by flaws, J. Acoust. Soc. Am., Vol.48(1977), pp.2812-2819.

[2] Schuster, G. T.; A hybrid BIE+Born series modeling scheme : Generalized Born series, J. Acoust. Soc. Am., Vol.77(1985), pp.865-879.

[3] Kitahara, M. and Nakagawa, K.; Born series approach applied to three dimensional elastodynamic inclusion analysis by BIE methods, in: McCarthy, M.F. and Hayes, M.A. (eds.), Elastic Wave Propagation (Elsevier Science Publishers B.V. ,1989) pp.447-452.

Elastic Waves and Ultrasonic Nondestructive Evaluation
S.K. Datta, J.D. Achenbach and Y.S. Rajapakse (Editors)

FREQUENCY SH RESPONSES IN MULTI-LAYERED HALF PLANE

M. K. Kuo

Institute of Applied Mechanics
National Taiwan University
Taipei, Taiwan 10764, R. O. C.

1 Introduction

Direct numerical integration to the transformed solution is efficient to evaluate the responses of multi-layered media, when the total number of layer is large. Traditionally, the transformed solution was constructed by the Thomson-Haskell matrix method. It is well-known that the method suffers serious numerical stability problems when the thickness of layer is large compared to the wavelength. This difficulty can be resolved by using delta matrix technique, Kennett's recursive algorithm [1], etc., though the method become complicated.

2 Problem Statement

Consider the SH motion in a multi-layered half plane consisted of $N-1$ layers overlying on a half-plane. Each layer, including the lower half plane, is homogeneous, isotropic, linearly viscoelastic with complex shear wave speed c_{sn}, complex shear modulus μ_n and thickness H_n. The subscript n refers to the layer number. Point sources are assumed to be monochromatic with angular frequency ω. A common time factor $\exp(i\omega t)$ is assumed and will be suppressed in the sequel. The Fourier transform over the horizontal coordinate x with kernel $\exp(i\xi x)$ will be employed with superposed tilde indicating the transformed quantity. The governing equation of a generic layer are the nonhomogeneous reduced wave equation in terms of z-component of displacement, w_n. The transformed solution is

$$\tilde{w}_n(\xi,\eta_n) = A_n^+ \exp(\gamma_{sn}\eta_n) + A_n^- \exp(-\gamma_{sn}\eta_n) + \tilde{w}_n^*, \tag{1}$$

where A_n^+ and A_n^- are unknown constants, $\gamma_{sn} = \sqrt{\xi^2 - (\omega/c_{sn})^2}$, and η_n is a local depth coordinate from the upper face of the layer, $0 < \eta_n < H_n$, The particular solution $\tilde{w}_n^*$ is due to force in this particular layer and has been so arranged such that it vanishes at both two faces of the source layer for convenience. There are 2 interface conditions along each interface, one free surface condition and one radiation condition at $y = 0$ and $y \to \infty$, respectively. It leads to a system of linear algebraic equations for $A_n^\pm$'s. Schmidt *et al.* [2][3] implemented this system of equation directly in his global matrix approach. It leads to a non-symmetric $(2N-1)\times(2N-1)$ variable band matrix with maximum bandwidth 5.

3 Method of Solutions

The basic idea of the new global matrix method is the use displacement components at both upper face, w_n^u, and lower face, w_n^ℓ, of layer to depict the displacement field

$$\tilde{w}_n(\eta_n) = \frac{\sinh[\gamma_{sn}(H_n-\eta_n)]}{\sinh(\gamma_{sn}H_n)}\, w_n^u + \frac{\sinh(\gamma_{sn}\eta_n)}{\sinh(\gamma_{sn}H_n)}\, w_n^\ell + \tilde{w}_n^*(\eta_n)\ . \tag{2}$$

The continuities of displacements along interfaces can easily be accommodated by letting

$$w_n^u = w_{n-1}^\ell \equiv W_n \quad ; \; w_n^\ell = w_{n+1}^u \equiv W_{n+1} \quad , n = 1,\ldots,(N-1) \quad , \tag{3}$$

where W_n is the displacement at the n-th interface. The relevant transformed stress component in the n-th layer can be deduced from eqns. (2)-(3). The tractions along the upper and lower faces of the n-th layer are denoted as Σ_n^u and Σ_n^ℓ, respectively,

$$\left\{ \begin{array}{c} -\Sigma_n^u \\ \Sigma_n^\ell \end{array} \right\} = \mathbf{K_n} \left\{ \begin{array}{c} W_n \\ W_{n+1} \end{array} \right\} + \frac{i\mu_n f_n}{\sinh(\gamma_{sn} H_n)} \left\{ \begin{array}{c} \sinh\left[\gamma_{sn}(H_n - a_n)\right] \\ \sinh(\gamma_{sn} a_n) \end{array} \right\} , \tag{4}$$

where the last term, local loading vector, in the equation is the consequence of applied force, f_n and a_n are the magnitude and local depth of the point source, respectively, and $\mathbf{K_n}$ is the symmetric 'local stiffness matrix'

$$\mathbf{K_n} = \mu_n \gamma_{sn} \left[\begin{array}{cc} \coth(\gamma_{sn} H_n) & -1/\sinh(\gamma_{sn} H_n) \\ -1/\sinh(\gamma_{sn} H_n) & \coth(\gamma_{sn} H_n) \end{array} \right], \tag{5}$$

which is identical to that derived by Kausel and Roësset [4] starting from propagator matrix. For the lower half-plane, $n = N$, the traction at the $(N-1)$-th interface can be deduced directly from eqns. (4)-(5) by letting $H_N \to \infty$ and the use of radiation condition,

$$-\Sigma_N^u = \mu_{sN} \gamma_{sN} W_N + i\mu_N f_N . \tag{6}$$

Notice that each entry in $\mathbf{K_n}$ is a complex number of finite magnitude even when γ_{sn} and H_n are very large. This eludes the numerical stability problem associated with the Thomson-Haskell formulation entirely. In fact, entries in $\mathbf{K_n}$ will be unbounded only if $\gamma_{sn} H_n = im\pi$, m =integer. These are actually corresponding to the eigenmodes of layer with fixed-fixed face conditions. It simply reflects the fact that the eigenmodes with fixed-fixed face conditions can never be depicted in terms of face displacements. This difficulty can easily removed by including some, even very small, damping in each layer. The boundary condition at $y = 0$,

and conditions of traction continuity along interfaces are ready to conclude a system of N linear equations for W_n, $n = 1, 2, \ldots, N$. The local stiffness matrices and local loading vectors are simply overlapping at corresponding interfaces to form the 'global stiffness matrix' and the 'global loading vector', respectively. It is similar to the 'assembly' process in finite element analysis. It leads to a symmetric $N \times N$ constant band global stiffness matrix with bandwidth 3 (tri-diagonal matrix). Moreover, since only the conditions of traction continuity are used to form the matrix equation, the scaling procedure adopted in the original global matrix formulation which brought the coefficients of displacement equations and traction equations into the same order will not be needed.

4 Conclusion

In this article, an alternative global matrix technique to study frequency response in a multi-layered medium is presented. The essence of the method is the use of displacements at each interface as unknowns. The continuities of displacements along the interfaces of the layered medium are then ensured automatically. The boundary conditions at the outer surface and the continuities of tractions along interfaces lead to a matrix equation for interface displacements at each horizontal wavenumber. The sizes of the matrices are reduced almost by a factor of 4 compared to the original global matrix formulation. This makes it had a big saving on the computational efforts. In addition, numerical stability problems associated with traditional Thomson-Haskell formulation are completely removed automatically.

1 Kennett, B. L. N., *Seismic wave propagation in stratified media* (Cambridge University Press, Cambridge, 1983).
2 Schmidt, H. and Jensen, F. B., J. Acoust. Soc. Am. **77** (1985) 813.
3 Schmidt, H. and Tango, G., Geophys. J. R. astr. Soc. **84** (1986) 331.
4 Kausel, E. and Roësset, J. M., Bull. Seism. Soc. Am. **71** (1981) 1743.

Elastic Waves and Ultrasonic Nondestructive Evaluation
S.K. Datta, J.D. Achenbach and Y.S. Rajapakse (Editors)

PIEZOELECTRIC SURFACE AND INTERFACE WAVES IN NON-CLASSICAL ELASTIC DIELECTRICS

Krystyna MAJORKOWSKA-KNAP and Jürgen LENZ

Institute of Mechanics of Engineering Constructions, Technical University of Warsaw, Płock, Poland

Institut für Theoretische Mechanik, Universität Karlsruhe, Fed. Rep. of Germany

1. INTRODUCTION

The classical theory of piezoelectricity as developed, among others, by W. Voigt and R. A. Toupin, assumes that the stored energy density of a piezoelectric body is a function of mechanical deformation and electric polarization. As an extension of this classical model, Mindlin [1] presented a variational principle in which the stored energy depends additionally on the *polarization gradient*. In contrast to the classical theory, this model accommodates an electromechanical interaction also in *centro-symmetric*, including *isotropic* materials - an experimentally observed physical phenomenon.

Concerning the propagation of electroelastically coupled waves, this non-classical theory leads to various modifications and novel features as compared with the corresponding results obtained within the framework of the classical model of piezo-electricity. For example, in the latter theory it is well-known that the effect of piezoelectricity is to increase the effective elastic constants, and therefore piezo-electricity is said to "stiffen" the crystal, thus bringing about a fractional increase in the phase velocity of waves. We could show [2], on the contrary, that at least for centrosymmetric isotropic media Mindlin's gradient theory delivers the inverse, i. e. piezoelectrically "*softening*" effect. This also holds for the propagation of piezo-electric SH-surface and interface waves in such media, as considered below.

2. RESULTS

In the case of a centrosymmetric cubic crystal and a centrosymmetric isotropic medium the assumption of plane surface waves leads to a *decoupled* set of field equations, i. e. a system of five coupled equations for the displacement and polarization components and the electric potential in the sagittal plane, and a system of two coupled equations for the displacement and polarization in the transverse plane, the latter being identical for cubic and isotropic materials and reading as

$$\left.\begin{aligned} c_{44}\left(\Delta - \frac{\rho}{c_{44}}\partial_t^2\right)u_3 + d_{44}\,\Delta P_3 &= 0 \\ d_{44}\,\Delta u_3 + (\hat{b}_{44}\Delta - a)\,P_3 &= 0 \end{aligned}\right\} \qquad (1)$$

where u_3 and P_3 are the transverse displacement and electric polarization, respectively (ρ: mass density; c_{44}: elastic constant; d_{44}, $\hat{b}_{44}$, a: piezoelectric coefficients).

Restricting ourselves to the propagation of SH-waves, the following wave types are studied:

1) *Love waves* (piezoelectric layer on elastic half-space): As in the classical (elastic and piezoelectric) case, Love waves with a series of modes can exist for *every* angular frequency, provided the shear velocity in the layer (assumed as purely elastic) is less than that in the half-space. But even in the opposite case - and in contrast to the classical model - above certain intrinsic *cut-off frequencies*, additional, hitherto unknown SH-surface wave modes can exist. Whereas the dispersion curves of the "classical" modes differ only insignificantly from those obtained in the purely elastic case (for $d_{44}=0$ the ordinary dispersion relation for Love waves is recovered), the "non-classical" modes turn out to be weakly dispersive [3].

2) *Stoneley waves* (piezoelectric layer between two, in general materially distinct, elastic half-spaces): As in the case of Love waves, in the symmetric (i. e. half-spaces of identical elastic materials) as well as in the non-symmetric problem there again exist "classical" (yet piezoelectrically softened) wave modes for *every* angular frequency if the shear velocity in the layer is smaller than that in the (less stiff) half-space. In the opposite case, like in Love's problem, *novel*, also weakly dispersive *waves* may exist above certain *cut-off frequencies.* In the non-symmetric case - like in the totally elastic model - additional cut-off frequencies arise which can in principle be determined from an analysis of the (analytically expressible, but very complicated) dispersion relation. Numerical calculations show that in the non-symmetric case the waves penetrate deeper into the less stiff medium. Furthermore, the penetration of these interface waves is much more pronounced for the non-classical than for the classical modes. For both, Love and Stoneley waves, the phase velocity increases with growing layer thickness, yet only slightly within the large interval of $(10^{-5} - 1)$ cm of the thickness.

3) *Bleustein-Gulyaev waves* (piezoelectric half-space): Within the classical theory of piezoelectricity, SH-surface waves - without counterpart in purely elastic media - may exist in a homogeneous half-space possessing certain crystal symmetries (e. g. hexagonal 6mm and cubic 23 or $\bar{4}3$m). Since their governing equations formally resemble much the system (1) hope rises for the existence of such waves in non-classical cubic and isotropic piezoelectric media, too. If, however, the inequalities for the material coefficients and certain of their combinations which guarantee the positive definiteness of the stored energy, are introduced into the dispersion relation, no real phase velocity can be found. Thus, such waves *cannot exist* in a half-space of a centrosymmetric cubic or isotropic material of Mindlin's type [3].

4) *Maerfeld-Tournois waves* (piezoelectric half-space attached to elastic half-space): Because of the complex nature of the dispersion relation the existence (known for certain classical piezoelectric crystals) or non-existence could not yet be proved mathematically. However, a painstaking numerical analysis for the (cubic) material combination natrium chloride / silicon did *not* reveal real solutions of the dispersion relation.

The various new phenomena following from this analysis of the propagation of Love and Stoneley waves within the framework of Mindlin's polarization gradient theory, are - qualitatively as well as quantitatively - open for experimental investigations. If the experiments should ascertain the theoretical predictions, a wide class of hitherto unexploited materials could possibly be used in practical applications such as ultrasonic delay lines, radar, sonar, telecommunication, etc.

[1] Mindlin, R.D., Int. J. Solids Struct. 4 (1968) 637.
[2] Majorkowska-Knap, K. and Lenz, J., Int. J. Engng. Sci. 27,8 (1989) 879.
[3] Majorkowska-Knap, K. and Lenz, J., Non-Existence of Bleustein-Gulyaev Waves.

Elastic Waves and Ultrasonic Nondestructive Evaluation
S.K. Datta, J.D. Achenbach and Y.S. Rajapakse (Editors)

LAMB WAVES IN COMPOSITE PLATES

Hidenori MURAKAMI

Department of Applied Mechanics and Engineering Sciences
University of California at San Diego
La Jolla, California 92093

Jonathan S. EPSTEIN

The Center for the Advancement of Computational Mechanics
The Georgia Institute of Technology
Atlanta, Georgia 30332

A high-order mixture continuum model was employed to study the Lamb wave propagation in a cross-ply composite plate. It has been found that the displacement microstructure of the zig-zag shape appears being superposed on the modes of the equivalent homogeneous anisotropic plate. In addition, the Lamb wave in the cross-ply plates exhibits significant dispersion in an acoustic antisymmetric mode.

1. STATEMENT OF THE PROBLEM

Consider a cross-ply laminates with alternating 0° layers (material 1) and 90° layers (material 2), as shown in Figure 1. For notational convenience forms $(\ )^{(\alpha)}$, $\alpha=1,2$ denote quantities associated with material α. Cartesian indicial notation is employed where Latin indices range from 1 to 3, and repeated indices imply the summation convention. In addition, the notation $(\ )_{,i}=\partial(\ )/\partial x_i$ and $(\ )_{,t}=\partial(\ )/\partial t$ will be employed in which t represents time. A dispersive mixture continuum model was employed to investigate the Lamb wave propagation. The governing equations of the mixture model are expressed by average displacements, $U_i^{(\alpha)}$, and average stresses, $\sigma_{ij}^{(\alpha)}$, that are defined for each material on the cell shown in Fig. 1, and the amplitude of zig-zag displacement microstructure, S_i, and moment, M_{ij} [1]:

(a) Equations of motion

$$n^{(\alpha)}\sigma^{(\alpha a)}_{ji,j} + (-1)^{\alpha+1} P_i = n^{(\alpha)}\rho^{(\alpha)} U^{(\alpha)}_{i,tt}, \qquad i=1,3 \quad \alpha=1,2 \tag{1}$$

$$\varepsilon^2 M_{ji,j} + (\sigma^{(2a)}_{3i} - \sigma^{(1a)}_{3i}) = \frac{\varepsilon^2}{12}\sum_{\alpha=1}^{2} n^{(\alpha)}\rho^{(\alpha)} S_{i,tt}, \qquad i=1,3 \tag{2}$$

where $n^{(\alpha)}$ is the volume fraction of material α, and ε is the cell length;

(b) Constitutive relations

$$\sigma^{(\alpha a)}_{ij} = \sigma^{(\alpha a)}_{ij}(U^{(\alpha)}_{k,l}, S_m), \quad M_{ij} = M_{ij}(U^{(\alpha)}_{k}, S_{m,n}) \tag{3}$$

(c) Boundary conditions

$$\sigma_{31}^{(1a)} = \sigma_{31}^{(2a)} = \sigma_{33}^{(1a)} = \sigma_{33}^{(2a)} = M_{31} = M_{33} = 0 \quad \text{on } x_3 = \pm\frac{h}{2} \tag{4}$$

The strategy adopted for the Lamb wave analyses is to perform Fourier transforms with respect to time t and the longitudinal coordinate x_1. This operation applied to the equations of motion expressed in displacements yields a system of ordinary differential equations with respect to x_3. Solving the ODE's and imposing (4), one finds an eigenvalue problem in which each eigenpair consists of Lamb ave phase velocity and the corresponding mode shapes for a given wave number.

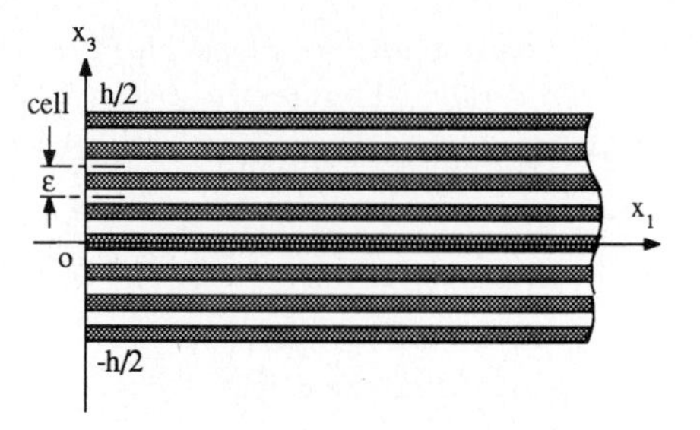

Figure 1 A cross-ply composite plate

2. NUMERICAL RESULTS

Figure 2 illustrates a typical Lamb wave phase velocity spectra of acoustic shear and longitudinal modes of AS4 carbon fiber 3501-6 epoxy cross-ply plates [2]. The phase velocity is nondimensionalized by the bar velocity in 90° layers: $C_0=(E_2/\rho)^{1/2}$, and the nondimensional wave number is defined as $2\pi h/\lambda$ where λ is the harmonic wave length. The corresponding mode shapes exhibits zig-zag shaped displacement microstructure superposed on the mode shapes obtained by the effective modulus theory. This analytical result agrees with the microstructure in the bulk fringes observed experimentally by dynamic Moire interferometry [2].

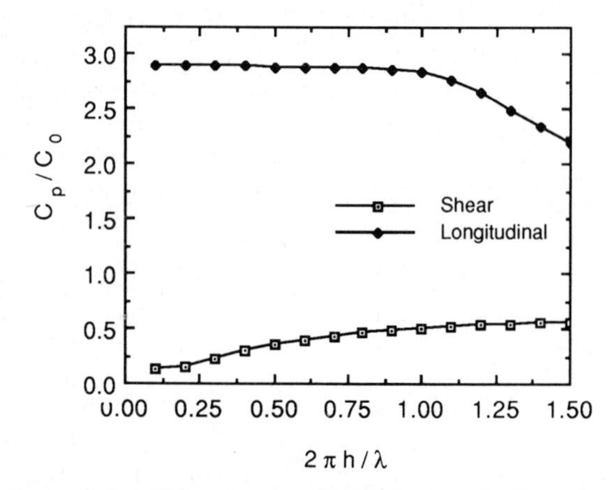

Figure 2 Nondimensional acoustic phase velocity spectra.

REFERENCES

[1] Murakami, H., Maewal, A.,and Hegemier, G. A., Int. J. Solids Struct. 17(1981), pp. 155-173.

[2] Epstein, J. S., Deason, V. A., Abdallah, M. A., and Murakami, H., Proc. Inst. Phys. Conf., (Oxford,1989), pp. 419-432.

Elastic Waves and Ultrasonic Nondestructive Evaluation
S.K. Datta, J.D. Achenbach and Y.S. Rajapakse (Editors)

SURFACE WAVE IN JOINTED HALF-SPACE WITH A SINGLE-SET OF PERIODIC JOINTS

Hidenori MURAKAMI and Kazumi WATANABE*

Department of Applied Mechanics and Engineering Sciences
University of California at San Diego
La Jolla, California 92093

Surface wave propagation in a half space with a single set of periodic rock joints is investigated by using an O(1) homogenization model. It has been found that the anisotropy changes the surface wave velocity, and dramatically modifies the aspect ratios of the ellipses of the wave motion at different depths.

1. STATEMENT OF THE PROBLEM

Consider a half space with a set of regularly spaced parallel joints with uniform spacing ε, as illustrated in Figure 1. A typical interior cell which represents the geometrical microstructure of the medium is also shown in the figure. Two rectangular Cartesian coordinate systems x_i, i=1-3 and x_i, i=1-3 will be adopted. The former is selected with x_3 normal to a joint plane defined by x_1 and x_2, and referred to the joint coordinate system. The latter is chosen to represent the free surface by x_3=0, and referred to the rotated coordinate system. For notational convenience, the usual Cartesian indicial notation, $(\ )_{,i} = \partial(\ )/\partial x_i$, and $(\ )_{,t} = \partial(\ )/\partial t$ will be employed.
With respect to the rotated coordinate system, x_i, i=1-3, the O(1) continuum model is represented by the average displacement U_i and stress N_{ij} defined over the cell [1]. The governing equations for the surface wave propagation under the plane strain in the x_1 and x_3 plane become:
(a) Equations of Motion

$$\hat{N}_{1i,1} + \hat{N}_{3i,3} = \rho \hat{U}_{i,tt} \qquad i = 1, 3 \tag{1}$$

where ρ is mass density of intact layers;
(b) Constitutive Relations

$$\begin{bmatrix} \hat{N}_{11} \\ \hat{N}_{33} \\ \hat{N}_{31} \end{bmatrix} = \hat{C}^{(m)} \begin{bmatrix} \hat{U}_{1,1} \\ \hat{U}_{3,3} \\ \hat{U}_{3,1} + \hat{U}_{1,3} \end{bmatrix} \tag{2}$$

* Faculty of Engineering, Yamagata University, Yonezawa, Yamagata, Japan

where the effective modulus matrix $\mathbf{C}^{(m)}$ are defined by the elastic moduli of the intact layers, the joint moduli, and the angle between the joint plane and the free surface [1];

(c) Traction Free Conditions

$$\hat{N}_{31} = \hat{N}_{33} = 0 \qquad (3)$$

The strategy adopted for the surface wave analyses is to perform Fourier transforms with respect to time t and the space coordinate x_1. This operation applied to the equations of motion expressed in displacements yields a system of ordinary differential equations with respect to x_3. Solving the ODE's and selecting the solutions which decay exponentially as x_3 tends to $-\infty$, a general solution for the half space can be constructed. Finally, the stress free condition is imposed to yield an eigenvalue problem in which each eigen pair consists of surface-wave phase velocity and the corresponding mode shape for a given wave number. In the above eigenvalue problem, the Stroh orthogonality relation [2] was employed to eliminate numerical difficulties.

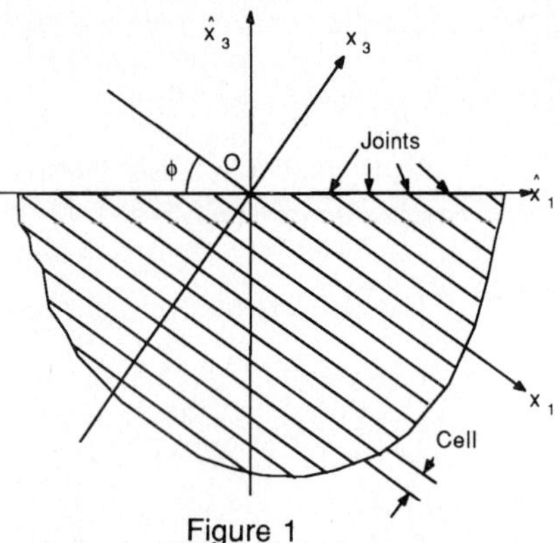

Figure 1
A half-space with a single-set of periodic joints

2. NUMERICAL RESULTS

Figure 2 illustrates the elliptical particle motions of the surface wave in the jointed half-spaces with the angle between the joint plane and the free surface 0° for reasonably stiff joints D_{11}/C_{55}=1.0 and for soft joints D_{11}/C_{55}=0.1. The elliptical particle motion is shown at different nondimensional depths kx_3=0, $\pi/2,\pi,2\pi,3\pi$. When expressed in dimensional quantities kx_3 becomes $2\pi x_3/L$ in which L is the surface wavelength.

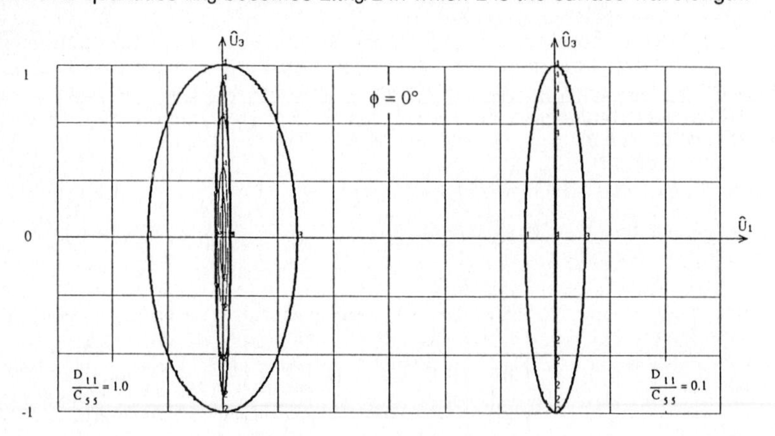

Figure 2 Particle motions at different depths for the half-spaces of φ=0° with D_{11}/C_{55}=1.0 and 0.1.

REFERENCES

[1] Murakami, H., and Hegemier, G. A., Mechanics of Materials 8 (1989), in print.
[2] Stroh, A. N., J. Math. Phys. 41 (1962) pp. 77-103.

Elastic Waves and Ultrasonic Nondestructive Evaluation
S.K. Datta, J.D. Achenbach and Y.S. Rajapakse (Editors)

Distribution of Ultrasonic Energy in Fiber Composite Materials*

Raymond J. Nagem
Department of Aerospace and Mechanical Engineering
Boston University

James H. Williams, Jr.
Department of Mechanical Engineering
Massachusetts Institute of Technology

Recent experiments by Kautz [1] show interesting phenomena associated with wave propagation behavoir in graphite fiber reinforced epoxy composite panels. Kautz introduces a broadband signal into a composite panel, and observes that the response of the composite appears to be divided into a high frequency component traveling at a relatively fast wave speed and a low frequency component traveling at a relatively low wave speed. When the same broadband signal is introduced into a geometrically similar panel consisting only of the epoxy matrix, the response is observed to consist only of the low frequency component. Kautz hypothesizes that ultrasonic energy in the composite is distributed nonuniformly, with higher frequency energy propagating along the fibers at the higher wave speed and lower frequency energy propagating through the matrix at the lower wave speed. This nonuniform distribution of energy, if it in fact exists, would be useful in nondestructive evaluation, since frequency or time partitioning could then be used in ultrasonic tests to extract information about the fibers or the matrix individually.

In order to investigate this possibility of nonuniform energy distribution in unidirectional fiber composites, a single elastic fiber embedded in an infinite elastic medium is considered. The infinite elastic medium is assumed to contain an incident plane SH wave, as shown in Fig. 1. The fiber radius is a, and the wave number of the incident wave is k. Using the known solution for the resulting displacement field within the elastic fiber [2], the total energy (elastic strain energy plus kinetic energy) per unit fiber length is computed analytically, and the resulting expression is time-averaged over one cycle of harmonic oscillation. A nondimensional fiber energy is defined by dividing this time-averaged fiber energy by an appropriate time-averaged matrix energy which would exist, due to the incident SH wave, if the fiber were not present. Thus, the nondimensional fiber energy gives a measure of the fiber energy relative to energy the matrix would contain if there were no fiber present. The explicit expression for the nondimensional fiber energy and the details of its derivation are given in [3]. It is shown in [3] that the nondimensional fiber energy depends on three nondimensional parameters: the product ka of incident wavenumber and fiber radius, the ratio r_μ of fiber shear modulus to matrix shear modulus, and the ratio r_ρ of fiber density to matrix density.

Numerical results for the nondimensional fiber energy as a function of nondimensional wavenumber ka are shown in Figs. 2 and 3. The values of the ratios r_μ and r_ρ which are used in Figs. 2 and 3 are derived from properties of materials commonly used in glass and graphite fiber reinforced polymers. The most interesting features of the fiber energy plots are the relative maxima which the fiber energy attains at certain values of incident wavelength. These maxima indicate a relatively strong fiber response at specific incident wavelength (and, therefore, incident frequency) values. Thus it is reasonable, at least in this single fiber case, to suppose that a broadband energy input would indeed be distributed nonuniformly, with the fiber responding more than the matrix to certain wavelength or frequency components.

*Research supported by NASA Lewis Research Center, Grant NAG3-328.

References

[1] H. E. Kautz, "Ray Propagation Path Analysis of Acousto-Ultrasonic Signals in Composites", NASA Technical Memorandum 100148, July 1987.

[2] Y.-H. Pao and C.-C Mow, *Diffraction of Elastic Waves and Dynamic Stress Concentrations*, Crane Russak and Co., 1973.

[3] J. H. Williams, Jr. and R. J. Nagem, "Energy in Elastic Fiber Embedded in Elastic Matrix Containing Incident SH Wave", NASA Contractor Report 4205, January 1989.

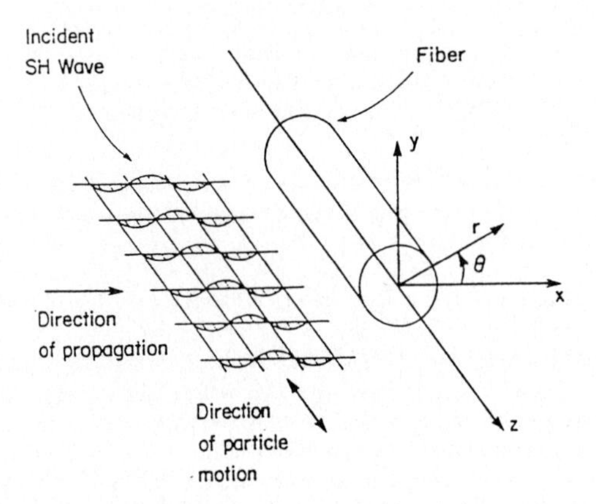

Fig. 1 Elastic fiber embedded in infinite elastic matrix containing incident SH wave.

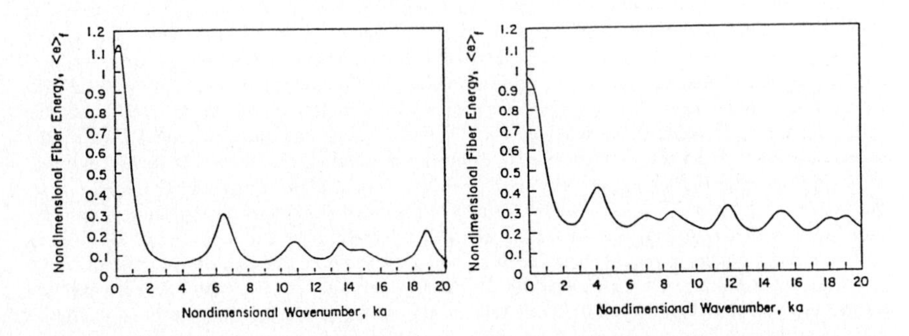

Fig. 2 Nondimensional fiber energy as a function of nondimensional wavenumber ka for $r_\rho = 2.0$, $r_\mu = 25.0$.

Fig. 3 Nondimensional fiber energy as a function of nondimensional wavenumber ka for $r_\rho = 1.5$, $r_\mu = 7.5$.

Elastic Waves and Ultrasonic Nondestructive Evaluation
S.K. Datta, J.D. Achenbach and Y.S. Rajapakse (Editors)

SHEAR WAVE SCATTERING FROM A DEBONDED FIBER

Andrew N. NORRIS and Yang YANG

Department of Mechanics and Materials Science
Rutgers University
Piscataway, New Jersey 08855-0909 USA

A single debond between a circular fiber and matrix is modeled as a circumferential crack along the interface. A system of equations is obtained for the crack opening displacement by representing it as a series of Chebyshev polynomials and the system is solved numerically. It is found that the scattered field exhibits a very strong resonance in the limit as the fiber is almost completely debonded from the matrix. This resonance can occur at arbitrarily low values of ka and is similar in nature to the Helmholtz resonance of an acoustic cavity.

1. INTRODUCTION

Our topic is motivated by the question of ultrasonic detection of debonded fibers in a fiber reinforced composite material. As a first step we here consider the scattering of a plane SH wave from a single fiber in an infinite matrix. Previous work by Coussy [1] obtained the scattered farfield in the long wavelength limit by using known static solutions and it was found that the scattering cross-section is of the order of $k_1^3 a^4$ where k_1 is the incident wave number and a the fiber radius. The present analysis is valid for arbitrary $k_1 a$, and reduces to Coussy's [1] result in the appropriate limit. The vanishingly small amplitude of the long wavelength scattered field suggests that it may be very difficult to quantitatively detect the effect of the debond, and this problem will be aggravated in the real situation of the composite material where multiple scattering from the surrounding bonded fibers could mask the signal from the debonded fibers. However, as the present results indicate, the scattering cross-section may be greatly enhanced if the fiber is almost completely disconnected from the matrix.

2. METHOD OF SOLUTION

The scattering configuration is shown in Fig. 1, where μ_1 and μ_2 are the shear moduli of the matrix and fiber, and ρ_1 and ρ_2 are the corresponding mass densities. The region of debond on the interface $r = a$ is modeled as a traction free crack with nonoverlapping faces. The crack is of angular width 2δ centered at $\theta = 0$. The method of solution adopted is similar to that of Krenk and Schmidt [2] who considered elastic wave scattering by a penny shaped crack. The same technique was used by Boström [3] in solving the problem of SH waves scattered by an interface crack between two half spaces. The method does not require the use of Green's functions and leads to equations for the crack opening displacement (COD) in a direct and simple way.

The total out-of-plane displacement may be expressed as

$$u^{tot}(r,\theta) = \begin{cases} u^{in} + u_1^{(0)} + u_1^{(1)}, & r > a , \\ u_2^{(0)} + u_2^{(1)} , & r < a , \end{cases} \tag{2.1}$$

where u^{in} represents the incident wavefield, $u_1^{(0)}$ and $u_2^{(0)}$ are the fields that would be present if the fiber were perfectly bonded, and $u_1^{(1)}$ and $u_2^{(1)}$ are the additional fields generated by the debond. All these fields are homogeneous solutions to the wave equation in their respective regions

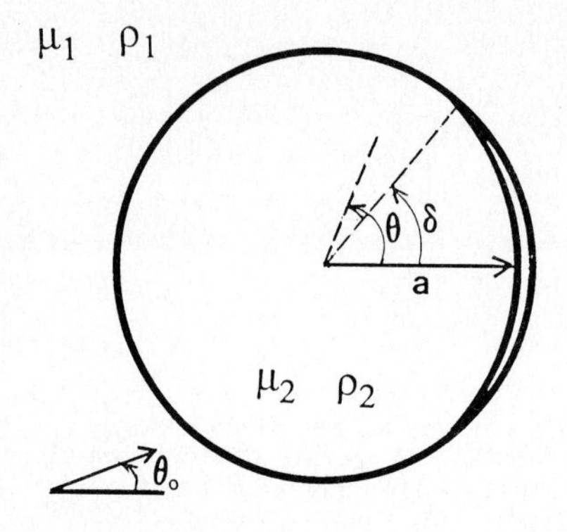

FIGURE 1
Schematic of the fiber scattering configuration.

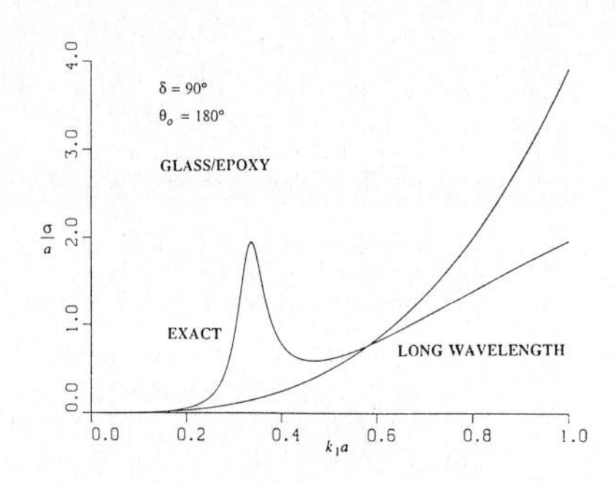

FIGURE 2
Comparison of the exact cross-section and the long wavelength approximation.

$$\nabla^2 u + k_\alpha^2 u = 0, \qquad k_\alpha = \omega(\rho_\alpha/\mu_\alpha)^{1/2}, \qquad \alpha = \begin{cases} 1, & r > a, \\ 2, & r > a. \end{cases} \tag{2.2}$$

The time dependence $\exp(-i\omega t)$ is understood but omitted. The incident plane wave is, see Fig. 1, $u^{in} = A\exp[ik_1 r\cos(\theta-\theta_0)]$, and the fields $u_1^{(0)}$ and $u_2^{(0)}$ can be easily obtained in terms of cylinder functions [4]. The extra fields $u_\alpha^{(1)}$, $\alpha = 1,2$ are found by splitting each into components which are symmetric and antisymmetric about the axis $\theta = 0$. We will describe the symmetric solution only; the antisymmetric fields are found in a similar manner. First expand the symmetric fields $u_{\alpha S}^{(1)}$, $\alpha = 1,2$ in terms of Hankel functions for $\alpha = 1$ and Bessel functions for $\alpha = 2$. The interface condition that the stress be continuous for all θ combined with the orthogonality of $\cos n\theta$, $n = 0,1,2,\ldots$, implies a relation between each of the coefficients in the two expansions.

Let $\Delta U_S(\theta)$ be the symmetric COD, such that

$$u_{1S}^{(1)} - u_{2S}^{(1)} = \begin{cases} 0, & r = a, \quad \delta \le \theta \le \pi, \\ \Delta U_S(\theta), & r = a, \quad 0 \le \theta < \delta. \end{cases} \tag{2.3}$$

Substituting the expansions for $u_{1S}^{(1)}$ and $u_{2S}^{(1)}$ into (2.3), using the relations between the coefficients, and again using the orthogonality of the cosine functions we obtain explicit expressions for the coefficients in terms of integrals of the COD. The final condition to be satisfied is that the total stress must vanish on the crack faces. Since the stress is already guaranteed to be continuous across the interface we need only apply this condition on one side of the crack. Thus for the interior face

$$\frac{\partial u_{2S}^{(0)}}{\partial r} + \frac{\partial u_{2S}^{(1)}}{\partial r} = 0, \qquad r = a, \qquad 0 \le \theta < \delta, \tag{2.4}$$

which becomes, upon substitution of the expansions for $u_{2S}^{(0)}$ and $u_{2S}^{(1)}$,

$$\sum_{p=0}^{\infty} \frac{\epsilon_p}{\pi} \cos p\theta \, J_p'(k_2 a) \frac{H_p^{(1)\prime}(k_1 a)}{D_p} \int_0^\delta \Delta U_S(\psi)\cos p\psi \, d\psi$$

$$= \sum_{p=0}^{\infty} B_p^{(S)} J_p'(k_2 a)\cos p\theta, \qquad 0 \le \theta < \delta, \tag{2.5}$$

where $\epsilon_0 = 1$, $\epsilon_n = 2$, $n > 0$, and

$$D_n = H_n^{(1)\prime}(k_1a)J_n(k_2a) - \frac{\mu_2 k_2}{\mu_1 k_1} H_1^{(1)}(k_1a)J_n'(k_2a) , \qquad B_n^{(S)} = \frac{A\epsilon_n 2\, i^{n+1}}{\pi\, k_1 a\, D_n} \cos n\theta_0$$

Equation (2.5) is an integral equation for the unknown COD which can be converted into an infinite system of linear equations by expanding the COD in a complete set of Chebyshev functions

$$\Delta U_S(\theta) = \sum_{n=1}^{\infty} \alpha_n^{(S)} \phi_n^{(S)}(\theta), \qquad \phi_n^{(S)}(\theta) = \frac{1}{2n-1} \cos[(2n-1)\arcsin(\tfrac{\theta}{\delta})]. \tag{2.6}$$

These functions have the correct square root behavior near the crack edge $\theta = \delta$, and have been used to advantage in other crack scattering problems, e.g. [3]. Substituting (2.6) into (2.5), multiplying both sides by $\phi_m(\theta)$ and integrating over θ using known integrals [3], provides us finally with a system of equations for the COD coefficients

$$\sum_{n=1}^{\infty} Q_{mn}^{(S)} \alpha_n^{(S)} = \frac{\delta}{2} \delta_{n1} J_0'(k_2a)B_0^{(S)} + \sum_{p=1}^{\infty} \frac{B_p^{(S)}}{p} J_{2n-1}(p\delta)J_p'(k_2a) , \quad m = 1,2,.., \tag{2.7}$$

where the symmetric, complex-valued matrix $[Q^{(S)}]$ has elements

$$Q_{mn}^{(S)} = \frac{\delta^2}{8} \delta_{m1}\delta_{n1} \frac{H_0^{(1)\prime}(k_1a)J_0'(k_2a)}{D_0} + \sum_{p=1}^{\infty} \frac{1}{p^2} J_{2m-1}(p\delta)J_{2n-1}(p\delta) \frac{H_p^{(1)\prime}(k_1a)J_p'(k_2a)}{D_p}. \tag{2.8}$$

The system (2.7) must be solved numerically.

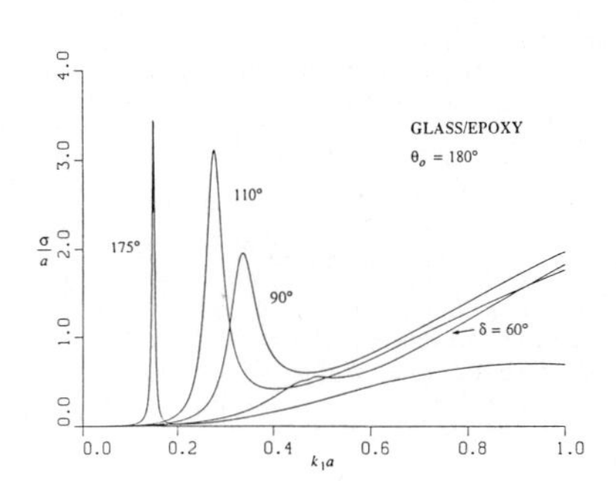

FIGURE 3
The cross-section for different crack sizes.

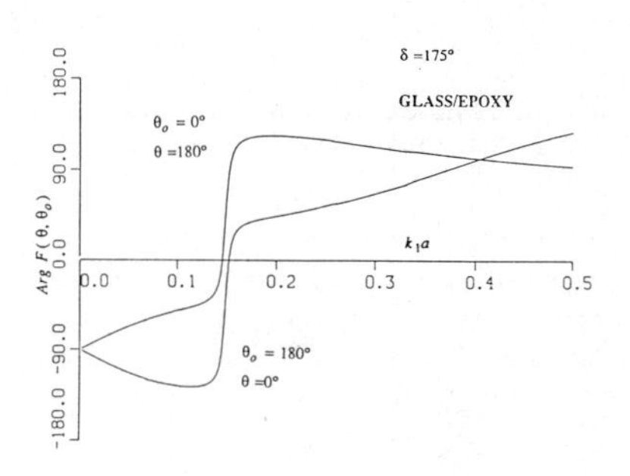

FIGURE 4
The phase of the backscattered farfield amplitude.

3. RESULTS AND DISCUSSION

The farfield radiation pattern $F(\theta,\theta_0)$ is defined by

$$u_1^{(0)} + u_1^{(1)} \sim \left(\frac{8\pi}{k_1 r}\right)^{1/2} e^{i(k_1 r-\pi/4)} F(\theta,\theta_0), \qquad r \to \infty , \tag{3.1}$$

and may be easily expressed in terms of the COD coefficients $\alpha_n^{(S)}$ and $\alpha_n^{(A)}$, $n = 1,2,3,...$, for the symmetric and antisymmetric problems. The associated total scattering cross-section σ is then proportional to the integral of $|F(\theta,\theta_0)|^2$ from 0 to 2π. In the long wavelength limit of $k_1 a \ll 1$, the farfield pattern can be obtained explicitly by suitable asymptotic approximations to the general system (2.7), or more directly by a quasistatic method [1].

Numerical results for finite frequencies have been computed from (2.7) for glass fibers in epoxy, $\mu_1 = 1.28$ GPa, $\mu_2 = 29.9$ GPa, $\rho_1 = 1.25$ and $\rho_2 = 2.55$. Figure 2 shows the scattering cross-section for a fiber that is half debonded. Note the evident failure on the quasistatic approximation near $k_1a = 0.35$. The peak in Fig. 2 becomes more pronounced and occurs at lower frequencies as the crack size is increased, see Fig. 3, until as $\delta \to \pi$, the shape of the peak becomes more and more like that of a sharp resonance. The probable explanation for the resonance is that as $\delta \to \pi$ the narrow neck joining the fiber and matrix permits large relative motion to occur. The resonance frequency is determined by the neck stiffness, which goes to zero, and the inertia of the fiber and matrix, which remain relatively constant independent of δ. There is a simple analogy between this phenomenon and the Helmholtz resonance of an acoustic cavity. The acoustic resonance is caused by the compressibility of the cavity fluid and the inertia of the fluid in the neck joining the cavity to the exterior medium. The roles of the "cavity" material and the material in the neck are reversed in the present situation. Thus, the inertia may be attributed to the gross motion of the fiber, while the spring constant is governed by the elastic strain in the vicinity of the neck.

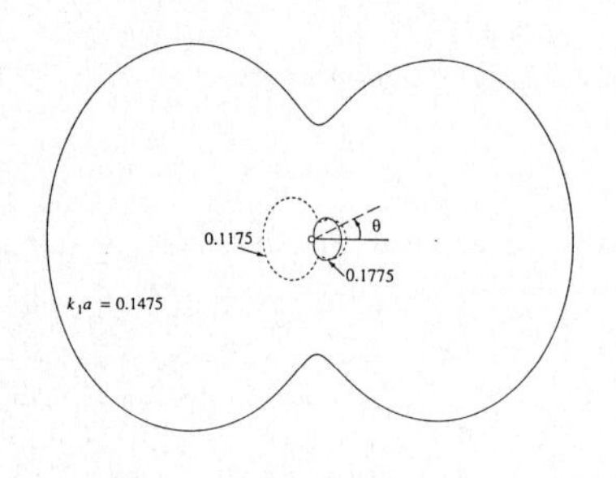

FIGURE 5
The farfield pattern for $\theta_0 = \pi$, $\delta = 175°$. The low frequency resonance occurs at $k_1a = 0.1475$. The curves are normalized by dividing by $(k_1a)^2$, and would be almost identical if the long wavelength approximation held.

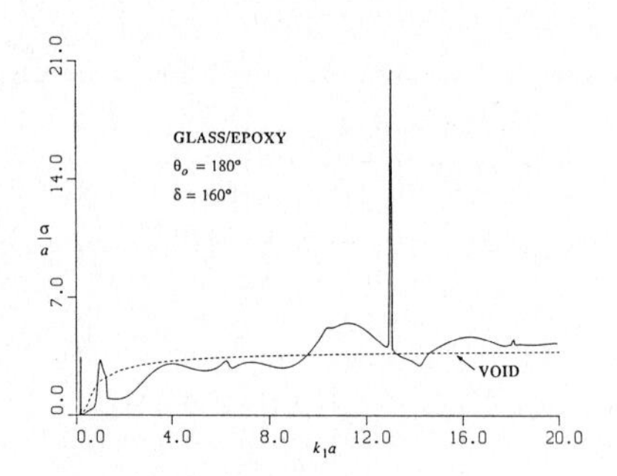

FIGURE 6
The cross-section at high frequency. Note the low frequency resonance similar to those of Fig. 3. The resonance near $k_1a = 13$ is at the lowest frequency for free vibrations of the fiber.

Figures 4 and 5 illustrate the behavior of other quantities near the low frequency resonance. Note in particular the characteristic 180° phase change at resonance in Fig. 4, and the strong dipole pattern in Fig. 5. Future work will concentrate on analyzing the low frequency resonance phenomenon. In particular, we would hope to develop a simple dynamical model that maintains the simplicity of the quasistatic analysis.

REFERENCES

[1] Coussy, O., C.R. Acad. Sci. Paris 295 (1982)1043.

[2] Krenk, S. and Schmidt, H., Phil. Trans. R. Soc. London A308 (1982)167.

[3] Boström, A., J. Appl. Mech. 54 (1987)503.

[4] Pao, Y.H. and Mow, C.C., Diffraction of Elastic Waves and Dynamic Stress Concentrations (Crane Russak, New York, 1973)pp. 113-122.

Elastic Waves and Ultrasonic Nondestructive Evaluation
S.K. Datta, J.D. Achenbach and Y.S. Rajapakse (Editors)

WAVE PROPAGATION DUE TO TIME-HARMONIC VERTICAL MOTION OF A RIGID INCLUSION

Ronald Y.S. Pak and Alain T. Gobert

Dept. of Civil, Envir., & Archit. Engineering
University of Colorado
Boulder, CO 80309-0428, U.S.A.

Abstract: An analysis is presented for the interaction of an infinite solid with an embedded rigid disc inclusion which is undergoing time-harmonic vertical motion. The mixed boundary value problem is reduced to a Fredholm integral equation of the second kind, the solution of which can be evaluated. Results on the response of the inclusion are provided as illustrations.

1. INTRODUCTION

This investigation is concerned with the interaction of an elastic medium with an embedded rigid disc under forced oscillation. On this subject, considerable attention has been given to the forced vibration problem of a disc resting on the surface of an elastic half-space as in Reissner and Sagoci [7], Bycroft [3], Awajobi and Grootenius [1], Robertson [8], and Gladwell [4]. In contrast, treatments of the corresponding problem in which the rigid disc inclusion or inhomogeneity is fully embedded in an elastic matrix as in Selvadurai [9] are somewhat limited. In this paper, an integral equation formulation is presented on the response of a rigid disc inclusion in an infinite medium under forced axial oscillation.

2. ANALYSIS

In this study, one considers a massless thin rigid disc of radius "a" embedded in a homogeneous, isotropic, linearly elastic solid of infinite extent (see Figure 1). With reference to the cylindrical coordinates (r, θ, z) as shown, the disc is assumed to be undergoing a prescribed time-harmonic translation $\Delta e^{i\omega t}$ in the z-direction. Owing to the axial symmetry of the problem, the azimuthal component of the displacement as well as the azimuthal dependence of the solution can be suppressed. In terms of the components of the displacement vector $\vec{u}$ and the stress tensor $\overleftrightarrow{\sigma}$, the interfacial conditions across the plane of the disc can be stated as follows:

$$u_z(r,0) = \Delta e^{i\omega t}, \quad r < a, \tag{1}$$

$$\sigma_{zz}(r,0^+) - \sigma_{zz}(r,0^-) = 0, \quad r > a, \tag{2}$$

$$\sigma_{rz}(r,0^+) - \sigma_{rz}(r,0^-) = 0, \quad r > a, \tag{3}$$

$$\sigma_{zz}(r,0^+) - \sigma_{zz}(r,0^-) = R(r), \quad r < a, \tag{4}$$

where $R(r)$ is the unknown net contact pressure distribution acting on the disc. In addition, the appropriate radiation conditions at infinity (Kupradze [5]) must also be stipulated. With the aid of an integral representation of the displacement field due to an arbitrary distributed

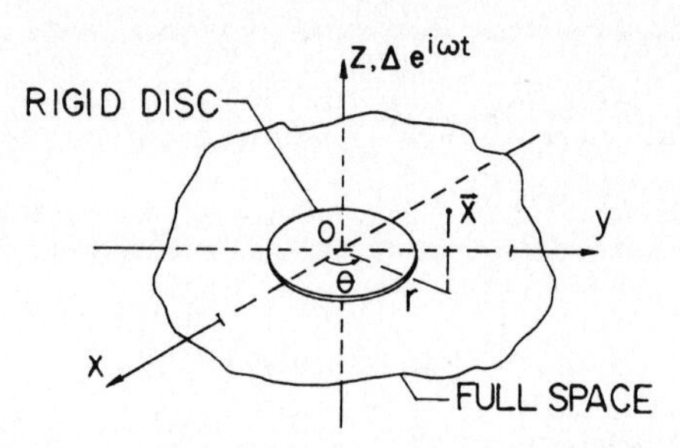

Fig. 1 A rigid disc inclusion in an infinite medium

body-force field in Pak [6], the mixed boundary value problem under consideration can be reduced to a system of dual integral equations. The dual system can be further reduced to a Fredholm integral equation of the second kind which can be expressed as

$$\theta(r) + \int_0^a A(r,\rho;\omega)\theta(\rho)d\rho = \delta \tag{5}$$

where

$$A(r,\rho;\omega) = \frac{2}{\pi}\int_0^\infty H(\xi,\omega)\cos(r\xi)\cos(\rho\xi)d\xi, \tag{6}$$

$$H(\xi,\omega) = \frac{4(1-\nu)}{3-4\nu}\frac{\xi}{k_s^2}\left(-\alpha + \frac{\xi^2}{\beta}\right) - 1, \tag{7}$$

$$\alpha = (\xi^2 - k_d^2)^{1/2}, \quad \beta = (\xi^2 - k_s^2)^{1/2}, \tag{8}$$

$$k_d = \frac{\omega}{C_d}, \quad k_s = \frac{\omega}{C_s}, \tag{9}$$

$$\delta = \frac{8(1-\nu)}{3-4\nu}\Delta, \tag{10}$$

with the time factor $e^{i\omega t}$ henceforth being implicit. In the above, C_d and C_s are the dilatational and the equivoluminal wave speeds of the medium, respectively, and the branches for the radicals α and β are chosen as indicated in Figure 2. The unknown function θ in (5) is related

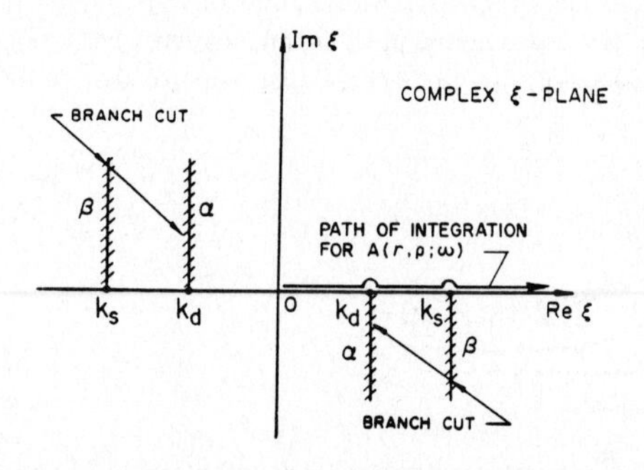

Fig. 2 Branch cuts and path of integration

to the zeroth-order Hankel transform of the contact load distribution $\tilde{R}(\xi)$ through

$$\tilde{R}(\xi) = \mu\sqrt{\frac{2}{\pi}}\xi^{1/2}\int_0^\infty r^{1/2}\theta(r)J_{-1/2}(\xi r)dr. \tag{11}$$

By virtue of (11), the total force required to produce the oscillation of the disc can be computed as

$$P = 4\mu\int_0^a \theta(t)\,dt\ . \tag{12}$$

3. ILLUSTRATIVE RESULTS

By means of the theory of contour integration, the kernel function (6) can be reduced to

$$A(r,\rho;\omega) = -\frac{ik_s 8(1-\nu)}{\pi(3-4\nu)}\left\{\int_0^1 \frac{\tau^3 N(\tau)}{\sqrt{1-\tau^2}}d\tau + \int_0^\kappa \tau\sqrt{\kappa^2-\tau^2}\,N(\tau)d\tau\right\} \tag{13}$$

where

$$N(\tau) = \frac{1}{2}(e^{-ik_s|r+\rho|\tau} + e^{-ik_s|r-\rho|\tau}), \tag{14}$$

$$\kappa = \frac{k_d}{k_s} = \sqrt{\frac{1-2\nu}{2(1-\nu)}}. \tag{15}$$

As the integrals in (13) can be evaluated by quadrature, solutions to the Fredholm integral equation can be computed by standard procedures (Baker [2]). By denoting the dynamic compliance, which is defined as the ratio of Δ to P, as

$$C_{vv} = \frac{3-4\nu}{32(1-\nu)\mu a}[\psi_1 + i\psi_2], \tag{16}$$

its variation with frequency can be described by the functions ψ_1 and ψ_2 which are given in Figure 3 for different Poisson's ratios. As can be seen from the figure, the compliance is in general complex, with its real and imaginary parts representing the in-phase and the 90° out-of-phase components of the response, respectively. With the solution of the Fredholm integral equation (5), the displacement field and the stress field of the medium can also be computed.

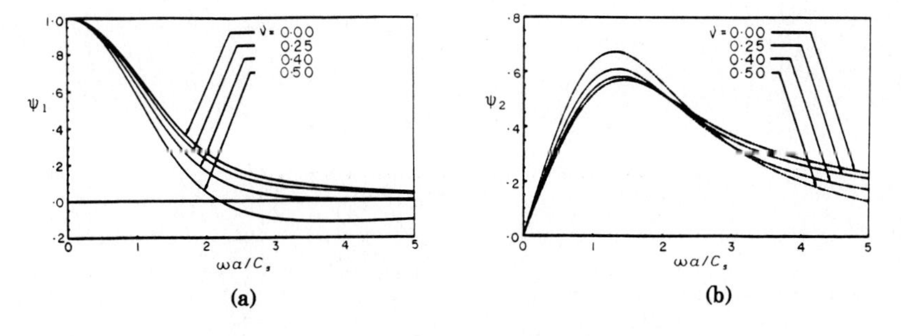

Fig. 3 Compliance functions ψ_1 and ψ_2

ACKNOWLEDGEMENT

The support provided by the U.S. National Science Foundation through Grant No. MSM-8611267 during the course of this study is gratefully acknowledged.

REFERENCES

[1] Awajobi, A.O and Grootenhuis, P. (1965) *Proc. Roy. Soc. Lond.*, A **287**, 27-63.

[2] Baker, C.T.H. (1977) *The Numerical Treatment of Integral Equations*, Camb. Univ. Press, Oxford.

[3] Bycroft, G.N. (1956) *Phil. Trans. Roy. Soc. Lond.*, A **248**, 327-368.

[4] Gladwell, G.M.L. (1968) *Int. J. Engng. Sci.,* **6**, 591-607.

[5] Kupradze, V.D. (1965) *Potential Methods in the Theory of Elasticity*, Daniel Davey and Co., New York.

[6] Pak, R.Y.S. (1987) *J. Appl. Mech.*, **54**, 121-126.

[7] Reissner, E. and Sagoci, H.F. (1944) *J. Appl. Phys.*, **15**, 652-654

[8] Robertson, I.A. (1966) *Proc. Camb. Phil. Soc.*, A **62**, 547-553.

[9] Selvadurai, A.P.S. (1981) *Int. J. Solids Struct.*, **17**, 492-498.

Elastic Waves and Ultrasonic Nondestructive Evaluation
S.K. Datta, J.D. Achenbach and Y.S. Rajapakse (Editors)
Elsevier Science Publishers B.V. (North-Holland), 1990

ACOUSTIC MICROSCOPY FOR NEAR-SURFACE FLAW DETECTION

R. A. ROBERTS

Center for NDE
Iowa State University
Ames, Iowa, U.S.A.

The response of an acoustic microscope to scattering by a near-surface void is modeled rigorously using a boundary integral formulation employing the three-dimensional Green function for contacting fluid and solid half-spaces.

1. INTRODUCTION

Acoustic microscopy has proven to be an effective tool for the detection of near-surface defects in structural materials such as ceramics. This paper presents a model for the prediction of acoustic microscope responses to scattering by near-surface voids. A boundary integral formulation is used in conjunction with appropriate models for the generation and reception of ultrasound by a highly-focused transducer-lens assembly. Numerical results are presented which demonstrate typical acoustic microscope responses to sub-surface flaws.

2. ANALYTICAL FORMULATION

The integral equation for the determination of time harmonic displacements u_i on the surface of a subsurface void is obtained through the application of elastodynamic field reciprocity relations as [1]

$$\frac{1}{2}u_k(x') + \int u_i(x)\, \tau^g_{ij:k}(x|x')\, n_j\, ds = u^p_k(x') \tag{1}$$

where $\tau_{ij:k}{}^g$ is the Green function for a system of contacting fluid and solid half-spaces [2]. The primary field $u_i{}^p$ refers to the field generated in the solid by the focused lens in the absence of the flaw. The primary field in the solid is represented by

$$u^p_i(x) = \int_{-\infty}^{\infty}\int_{-\infty}^{\infty} \hat{p}^{in}(\underline{k})\, T_i(\underline{k})\, \exp(i\, k\cdot x)\, d\underline{k}$$

$$\underline{k} = (k_1, k_2)\,, \quad k_3 = (\, k_w^2 - |\underline{k}|^2\,)^{\frac{1}{2}} \tag{2}$$

where $\hat{p}^{in}$ is the Fourier transform of the pressure field generated by the transducer-lens system in the absence of the solid at the plane $x_3=0$, T_i are plane wave transmission coefficients, and k_w is the wavenumber in water. The pressure field in the water due to scattering by the sub-surface void is given by

$$p^{sc}(x') = \int u_i(x)\, \tau_{ij}^{gw}(x|x')\, n_j\, ds \tag{3}$$

The Green function $\tau_{ij}{}^{gw}$ in eq.(3) refers to a point source acting in the water, whereas the Green function $\tau_{ij:k}{}^{g}$ in eq.(1) refers to a point force acting in the solid. The response of the transducer-lens system is assumed reciprocal in the calculations presented here, hence the reception of the scattered pressure can be modeled as the integral of the product of the scattered pressure p^{sc} and the incident pressure p^{in} over the plane $x_3=0$.

NUMERICAL RESULTS

Results are presented for a 50 mHz center frequency spherically-focused transducer-lens system with a 60 degree aperture. The radial profile of the beam in the focal plane is assumed to be a Gaussian. The response to scattering by a 20 micron diameter spherical void centered 15 microns below the surface of a silicon nitride half-space is considered. Results were obtained by solving eqs.(1-3) using numerical methods. Results are presented in fig. 1 for a) in focus (i.e. focal plane located at the surface $x_3=0$), and b) a defocus of 10 wavelengths in water. The response of a scanning acoustic microscope is simulated by plotting the amplitude of the transducer output (i.e. the peak amplitude of the time harmonic output voltage) versus the distance between the void center and the center of the focused beam.

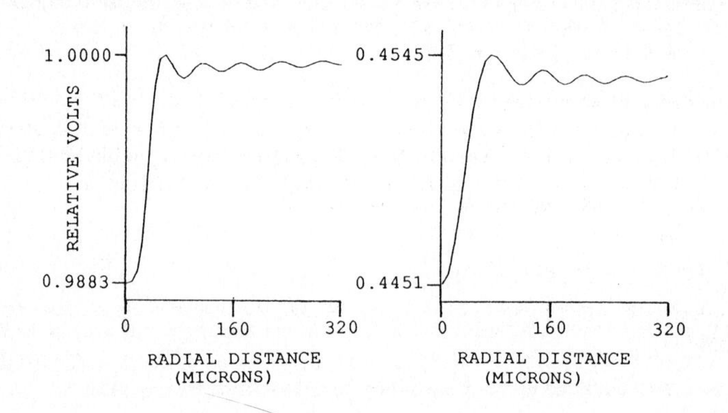

FIGURE 1
Theoretical response of a 50 mHz scanning acoustic microscope to a 20 μm void.

REFERENCES

[1] J. D. Achenbach, Wave Propagation in Elastic Solids, (North-Holland, Amsterdam, 1973)

[2] R. A. Roberts, "Elastodynamic Response of Contacting Fluid and Solid Half-Spaces to a Three-Dimensional Point Load", submitted to Wave Motion.

Elastic Waves and Ultrasonic Nondestructive Evaluation
S.K. Datta, J.D. Achenbach and Y.S. Rajapakse (Editors)

SINUSOIDAL WAVES IN TWINNED CRYSTALS

Nigel H. SCOTT

School of Mathematics
University of East Anglia
Norwich, U.K.

Ericksen [1] has proposed a model for A-15 superconductors near cubic-tetragonal phase transitions treating them as thermoelastic bodies subject to the four internal constraints:

$$e_{23} = 0, \quad e_{31} = 0, \quad e_{12} = 0, \quad e_{11} + e_{22} + e_{33} = 0, \tag{1}$$

where $\mathbf{e}$ is the linear elasticity tensor. (In fact, Ericksen considered nonlinear versions of these constraints but we shall postpone that discussion until another occasion). It has been shown [2,3,4] that a material suffering three or more constraints cannot in general transmit sinusoidal plane waves (or acceleration waves) in any direction $\mathbf{n}$. However, depending on the actual constraints, it is possible that waves may be transmitted in certain exceptional directions. The displacement $\mathbf{u}$ from the rest position $\mathbf{X}$ is assumed to be sinusoidal:

$$\mathbf{u} = \mathbf{U} \exp\{i\omega(t - s\mathbf{n}.\mathbf{X})\}, \tag{2}$$

where $\mathbf{U}$ is the constant amplitude vector, ω is the frequency, t is time, $\mathbf{n}$ is the unit direction of wave propagation and s is the slowness (i.e. $v = s^{-1}$ is the wave speed). Following the analysis of Chadwick *et al.* [3] reveals that wave propagation requires rank $\mathbf{M} < 3$, where

$$\mathbf{M} = \begin{pmatrix} 0 & n_3 & n_2 & n_1 \\ n_3 & 0 & n_1 & n_2 \\ n_2 & n_1 & 0 & n_3 \end{pmatrix}. \tag{3}$$

We define Δ_i to be the determinant of the matrix formed by deleting the i-th column of $\mathbf{M}$ and obtain explicitly

$$\Delta_i = n_i(1 - 2n_i^2), \quad i = 1,2,3, \qquad \Delta_4 = 2n_1n_2n_3. \tag{4}$$

The condition rank $\mathbf{M} < 3$ requires each determinant in (4) to vanish. This happens only for $\mathbf{n}$ parallel to one of the twelve directions

$$(\pm 1,\ \pm 1,\ 0), \quad (\pm 1,\ 0,\ \pm 1), \quad (0,\ \pm 1,\ \pm 1). \tag{5}$$

But these directions correspond also to the so-called twin planes of the tetragonal phase of the crystal and are favoured directions also in experimental and other theoretical discussions [1] of such materials. Thus we have shown that if the constraints (1) apply exactly wave propagation is possible only in the twin planes of the tetragonal phase of the crystal.

The discussion of nearly-constrained materials follows the pattern developed by Rogerson [4]. In terms of $\mathbf{e}$ and its deviatoric part $\mathbf{e}'$ we adopt the strain energy

$$W = \alpha\, e'_{ij}e'_{ij} + (\kappa/2)(e_{kk})^2 + 2\mu\,(e_{23}^2 + e_{31}^2 + e_{12}^2), \tag{6}$$

appropriate to a cubic crystal. If $\mu = 0$ the material is isotropic with shear modulus α and bulk modulus κ, but $\mu > 0$ is a measure of the shear modulus in the coordinate planes (the preferred planes of the cubic crystal). To model a nearly-constrained material we take κ and μ to be large. In order to keep the energy $\dot{W}$ finite this requires each of e_{kk}, e_{23}, e_{31}, e_{12} to be small, thus realising the constraints (1). The acoustic tensor for (6) is

$$\mathbf{Q}(\mathbf{n}) = (\alpha + \mu)\mathbf{I} - 2\mu\,\mathrm{diag}\,\{n_1^2,\, n_2^2,\, n_3^2\} + (\alpha/3 + \kappa + \mu)\,\mathbf{n}\otimes\mathbf{n} \tag{7}$$

in terms of which the usual propagation condition is, letting ρ denote density,

$$\{\mathbf{Q}(\mathbf{n}) - \rho v^2\mathbf{I}\}\,\mathbf{U} = \mathbf{0}, \tag{8}$$

so that the eigenvalues of $\mathbf{Q}(\mathbf{n})$ give the quantities ρv^2. The slowness surface is defined to be the three-sheeted, centro-symmetric surface obtained by marking off on each radial direction $\mathbf{n}$ the three slownesses obtained from the eigenvalues of (8). It can be shown that here the slowness surface is symmetric about each of the three coordinate planes and also about each of the six planes with one of the directions (5) as its normal.

If the constraints are relaxed slightly so that they hold only approximately then it can be shown the slowness surface possesses twelve thin spikes, one in each of the directions (5). This shows that wave propagation is largely, but not completely, confined to these directions for nearly-constrained materials.

The figure bears out these conclusions. Let $\mathbf{i}, \mathbf{j}, \mathbf{k}$ denote unit vectors along the coordinate axes and let $\mathbf{a} = (\mathbf{i} + \mathbf{j})/\sqrt{2}$. In the figure we have taken $\alpha = 1$, $\rho = 1$ and $\kappa = 20$, $\mu = 20$ and plotted the indicated sections of the slowness surface. Even for such relatively small κ and μ the constraints are beginning to operate, with part (a) showing the spikes beginning to form in the directions $(\pm 1, 0, \pm 1)$ and in part (b) the spikes in the directions $(1, 1, 0)$ and $(-1, -1, 0)$.

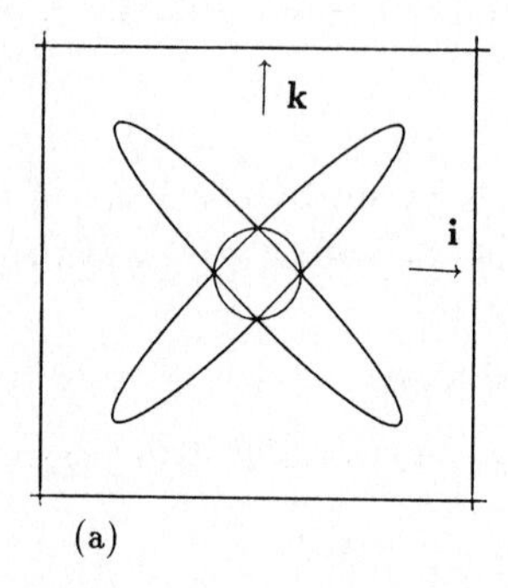

(a)

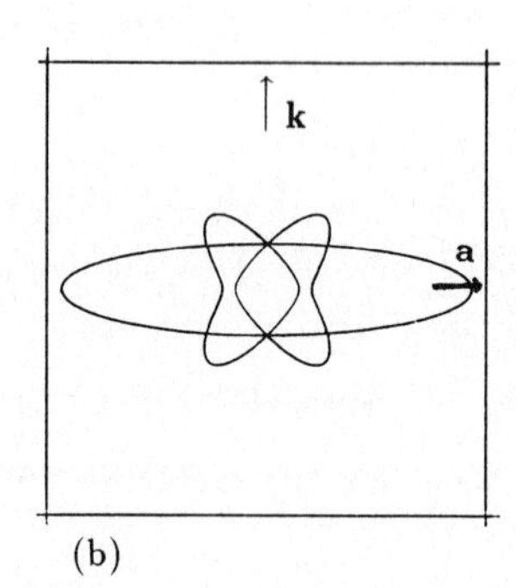

(b)

REFERENCES

[1] Ericksen, J.L., Int. J. Solids Structures **22** (1986), 951–964.
[2] Scott, N.H., Arch. ration. Mech. Analysis **58** (1975), 57–75.
[3] Chadwick, P., Whitworth, A.M. and Borejko, P., *ibid.* **87** (1985), 339–354.
[4] Rogerson, G.A., "Aspects of wave propagation in constrained and nearly constrained elastic bodies". PhD Thesis, University of East Anglia, Norwich, U.K. (1987).

Elastic Waves and Ultrasonic Nondestructive Evaluation
S.K. Datta, J.D. Achenbach and Y.S. Rajapakse (Editors)

STRESS CONCENTRATION DUE TO THE INTERACTION BETWEEN A FINITE CRACK AND A FINITE INCLINED RIGID RIBBON (SH- CASE)

U. SHARMA

Khalsa College, Jallunder, India

and

K. Viswanathan

Defence Science Centre, Metcalfe House, Delhi-110054, India

In this paper we aim at the solution to the problem of interaction between a finite crack and a finite inclined rigid ribbon. We take an incident plane wave and a possible additional load on the crack face. Numerical results for the stress intensity factors at the inside tips of the rigid ribbon caused by the plane wave are given.

1. INTRODUCTION

Understanding stress concentration is vital in many fields such as NDT of materials and seismic fault modeling, (see [1]-[3]). While semi-infinite crack or ribbon can be studied via the Wiener-Hopf technique, it appears that finite defects yield only to numerical approach such as expansion in terms of Chebyshev polynomials. We employ this method here to study the interaction between a finite crack and a finite inclined rigid ribbon. The case of two inclined cracks has recently been treated by So and Huang [4].

2. THE BASIC PROBLEM

We consider a rigid ribbon of length 1 lying along the x_1- axis in the region (d , d+1) and a crack of length c inclined at an angle α_0 to the x_1- axis with its near tip at a distance ρ_0 from the origin. For an incident wave with displacement

$$W^{(in)} = A \exp[-ik(x_1 \cos\theta_0 + x_2 \sin\theta_0) + iwt] \tag{1}$$

the equation for the scattered field may be taken as

$$\begin{aligned} W^{(sc)} = & \int_{-c}^{0} \gamma(\xi_1') \, [\mu \partial G/\partial \xi_2'(\underline{\xi}, \underline{\xi}')]_{\xi_2' = 0} \, d\xi_1' \\ & - \int_{-c}^{0} P_0(\xi_1') \, [G(\underline{\xi}, \underline{\xi}')]_{\xi_2' = 0} \, d\xi_1' \\ & - \int_{d}^{d+1} T(x_1') \, [G(\underline{x}, \underline{x}')]_{x_2' = 0} \, dx_1' \end{aligned} \tag{2}$$

with

$$G(\underline{x},\underline{x}') = (1/2i) \, H_0^{(2)}(k \, |\underline{x} - \underline{x}'|)$$

The boundary conditions are that the total displacement $W = 0$ on the ribbon while the shear stress $\mu \partial W/\partial \xi_2$ takes the value

$$\mu \partial W/\partial \xi_2 = \mp P_0(\xi_1) \exp(iwt) \quad \text{on } \xi_2 = 0, \ -c< \xi_1 <0 \text{ (the crack)}$$

where (ξ_1, ξ_2) denote the local coordinates with axes along and normal to the crack.

The other functions in (2) denote the unknown stress jump $T(x_1)$ across the ribbon, the unknown displacement jump $\gamma(\xi_1)$ across the line of the crack, and the known applied stresses $P_0(\xi_1)$ again on the crack faces (with $\mp$ signs). By defining

$$x_1 = d + (1+\eta_1)\, l/2 \quad ; \quad \xi_1 = (\hat{\xi}_1 - 1)\, c/2$$

$$T(x_1) = (2/l)\, \phi(\eta_1) / (1-\eta_1^2)^{1/2} \quad ; \quad \gamma(\xi_1') = (c/2)\psi(\hat{\xi}_1')(1-\hat{\xi}_1'^2)^{1/2}$$

the equations to solve take the form

$$\int_{-1}^{1} \phi(\eta')[\log|\eta-\eta'| + K_1(\eta, \eta')]\, d\eta' / (1-\eta'^2)^{1/2}$$

$$+ \int_{-1}^{1} \psi(\hat{\xi}_1') K_2(\hat{\xi}_1', \eta)(1-\hat{\xi}_1'^2)^{1/2}\, d\hat{\xi}_1' = g_1(\eta) \quad (-1< \eta< 1)$$

and

$$\int_{-1}^{1} \phi(\eta') K_4(\hat{\xi}_1, \eta')\, d\eta'/(1-\eta'^2)^{1/2} + \int_{-1}^{1} \psi(\hat{\xi}_1')(1-\hat{\xi}_1'^2)^{1/2}.$$

$$[a/|\hat{\xi}_1 - \hat{\xi}_1'|^2 + b \log|\hat{\xi}_1 - \hat{\xi}_1'| + K_3(\hat{\xi}_1, \hat{\xi}_1')]\, d\hat{\xi}_1' = g_2(\hat{\xi}_1), \quad (-1< \hat{\xi}_1 < 1)$$

As noted earlier, these are solved by the Chebyshev polynomial expansion. A typical stress intensity factor curve (for both tips) near the rigid ribbon is computed in Figure 1 shown here. We have taken $d=0$, $c = 0.5$, $k = 0.6$, $\rho_0 = 1$, $\alpha_0 = 30^0$ and only plane wave case.

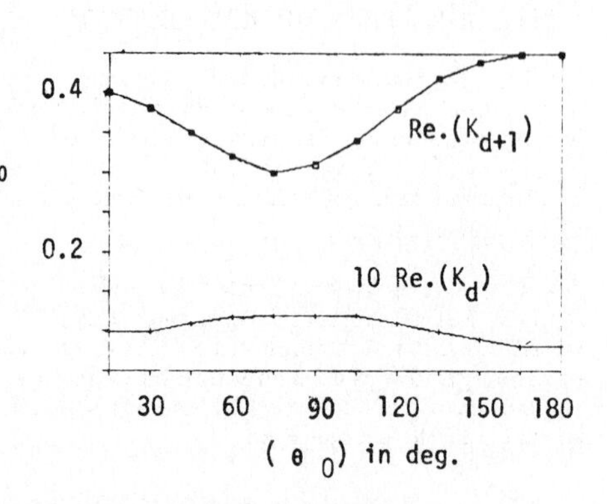

FIGURE 1. Plots of S.I.F. at $x_1 = d$ and $d+l$.

REFERENCES

[1] Achenbach, J.D., Pao, Y.H. and Tiersten, H.F. (editors), Application of Elastic Waves in Electrical Devices, NDT and Seismology(Report of NSF Workshop, Northwestern University, 1976).

[2] Das, S., and Sholtz, C.H., Bull. Seism. Soc. Amer., vol. 71, 1981, pp 1669-1675.

[3] Viswanathan, K. and Sharma, U., Int. J. Engg. Science, vol. 25, 1987, pp 295-306.

[4] So, H. and Huang, J.Y. , Int. J. Engg. Science, vol. 26, 1988, pp 111-119.

Elastic Waves and Ultrasonic Nondestructive Evaluation
S.K. Datta, J.D. Achenbach and Y.S. Rajapakse (Editors)

MAGNETOELASTIC WAVE SCATTERING IN A CRACKED FERROMAGNETIC PLATE*

Yasuhide Shindo and Katsumi Horiguchi†

Department of Mechanical Engineering II
Tohoku University
Sendai 980, Japan

The dynamic analysis of a soft ferromagnetic flat plate with a through crack under a uniform magnetic field is considered. The dynamic moment intensity factor versus frequency is computed and the influence of the magnetic field on the normalized values is displayed graphically.

1. INTRODUCTION

The dynamic behavior of a soft ferromagnetic elastic plate is sufficiently affected by the presence of the magnetic field[1],[2]. In this investigation, the scattering of time harmonic flexural waves by a through crack in a soft ferromagnetic thin plate under a uniform magnetic field is considered. Ferromagnetic martensitic stainless steels such as HT-9 are currently under study as candidate materials for first wall-blanket structures of magnetic fusion reactors[3]. Mindlin's plate theory of magneto-elasticity can be made, of course, but our purpose here is to demonstrate the nature of the magneto-elastic coupling using the simplest theory.

2. FORMULATION OF MAGNETOELASTIC PROBLEM

Let the coordinate axes x and y chosen such that they are in the middle plane of the soft ferromagnetic elastic plate of thickness 2h containing a through crack of length 2a and z-axis is perpendicular to this plane as illustrated in Figure 1. The cracked plate is permeated by a static uniform magnetic field of magnetic induction B_0 normal to the plate surface. The bending and twisting moments per unit length M_x,M_y and M_{xy} may be expressed in terms of the deflection w.

Let a cracked plate be excited by propagating flexural waves. The source that emits the incident waves corresponds to moments applied symmetrically about the crack plane. Fourier transform method is employed to reduce the magneto-elastodynamic problem to the solution of an integro-differential equation for deflection. The integro-differential equation is solved by the standard iteration method and the solution is expressed in terms of a Fredholm integral equation of the second kind.

3. NUMERICAL RESULTS AND DISCUSSION

The dynamic moment intensity factor is defined by

$$K_1 = \lim_{x \to a^+} \{2(x-a)\}^{1/2} M_y(x,0) \tag{1}$$

*This work was supported in part by the Scientific Research Fund of the Ministry of Education for the fiscal year 1988. Partial support was also provided by TEPCO Research Foundation.
† Department of Mechanical Engineering, Tokyo Metropolitan Technical College

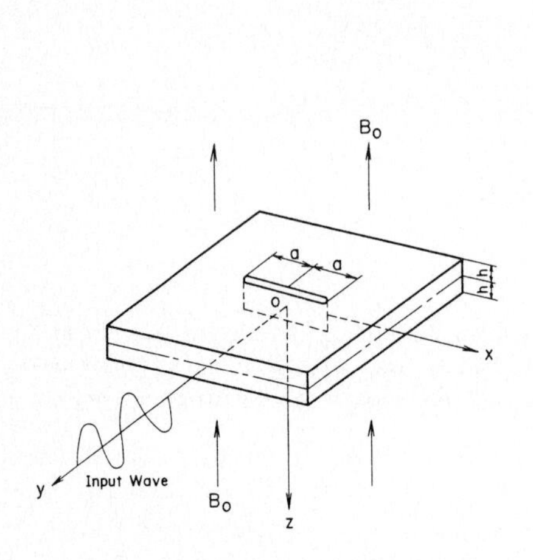

FIGURE 1
Cracked soft ferromagnetic plate.

FIGURE 2
Dynamic moment intensity factor.

The dynamic moment intensity factor K_1 versus circular frequency ω is computed and the influence of the magnetic field on the normalized values is displayed in Figure 2. In Figure 2, M_0, ω_0 and b_c are given by

$$M_0 = D w_0 k_0^4, \qquad k_0^4 = 2\rho h \omega^2 / D, \qquad D = 2Eh^3/3(1-\nu^2),$$

$$\omega_0 = \pi c_2 / 2h, \qquad b_c^2 = B_0^2 / \mu_0 \mu \tag{2}$$

w_0 is the amplitude of the input wave, D is the flexural rigidity of the plate, ω_0 is the cut-off frequency of the plate, $E = 2(1+\nu)\mu$ is the Young's modulus of elasticity, ν is the Poisson's ratio, μ is the shear modulus of elasticity, ρ is the mass density of the material, $c_2 = (\mu/\rho)^{1/2}$ is the shear wave velocity of the material, χ is the magnetic susceptibility, and μ_0 is the magnetic permeability of the vacuum. It is found that the existence of the magnetic field produces higher dynamic singular moments near the crack tip.

Acknowledgements

The authors would like to acknowledge the continuing guidance and encouragement of Emeritus Professor A. Atsumi of Tohoku University.

REFERENCES

[1] Shindo, Y., Trans. JSME 49-44 (1983) 1467.
[2] Shindo, Y., Acta Mechanica 57-1/2 (1985) 99.
[3] Rosenvasser, S. N., Miller, P., Dalessandro, J. A., Rowls, J. M., Toffolo, W. E. and Chen, W., J. Nuclear Materials 85&86 Pt. A (1979) 177.

Elastic Waves and Ultrasonic Nondestructive Evaluation
S.K. Datta, J.D. Achenbach and Y.S. Rajapakse (Editors)

ON THE GENERATION OF DIFFRACTION OF AN SH-WAVE BY A SLIT

Hiroki TODA and Hidekazu FUKUOKA

Faculty of Engineering Science
Osaka University
Osaka, Japan

From the energy flow point of view, we obtained new knowledge understanding the generation mechanism of diffraction of an SH-wave by a slit. The diffraction is caused by the lateral energy flow due to the lateral displacement gradient in both incident and reflected waves.

1. INTRODUCTION

The diffraction of an SH-wave by a semi-infinite slit is analyzed by employing integral transform together with the Wiener-Hopf technique and the Cagniard-de Hoop method [1]. In this paper, in order to study the mechanism of the generation of diffraction, we treat of the energy flow across a wavefront for the diffraction. The time-average energy flow is calculated by means of both the analytical solution by de Hoop and the numerical solution by Huygens-Fresnel's principle.

2. ENERGY FLOW BY THE DIFFRACTION

The time-average energy flow across the wavefront for the diffraction or across the two radial boundaries (one is a plane between an illuminated zone and a shadow zone, and the other is a plane between a reflected zone and the other zone) is calculated by means of the analytical solution given by de Hoop. As an example, the energy flows across each part of the wavefront of the diffraction for the SH-wave of incident angle of 60 degree are shown in Fig.1, and the directions of the energy flow are shown in Fig.2.

Fig.1 and Fig.2 show that the energy source for the diffraction is supplied from both the incident wave and the reflected wave, and the energy conservation law is well maintained for each half-cylindrical wavefront itself, separated by the plane of slit, and that the total energy flow integrated over the whole diffraction wavefront vanishes.

Growth of energy flow with the propagation of the diffraction wavefront reveals that the energy source to generate the diffraction distributes continuously on the two radial boundaries, and that the displacement gradient with respect to the normal to the boundary, which is infinite at the edge of slit and decreases with the propagation, causes the lateral flow of the energy.

The diffraction wave takes an important role to eliminate the displacement jump across the radial boundary. Fig.1 also shows that although the displacement field given by analytical solution is antisymmetric with respect to the plane of slit, the intensities of the energy flow in each half-cylindrical wavefront are different each other. This is because the energy flow is proportional to the square of amplitude of displacement and is affected not only by the displacement of the diffraction but also by the displacements of both incident wave and reflected wave.

3. SIMULATION BY HUYGENS-FRESNEL'S PRINCIPLE

Diffraction by a slit can also be solved approximately by using Huygens-Fresnel's principle. The displacement and the energy flow computed from the simulation for normal angle incidence of a plane harmonic wave are shown in Fig.3 and Fig.4, respectively in comparison with the analytical solution.

The result by Huygens-Fresnel's principle is approximately in good agreement with the analytical solution. The simulation shows that the diffraction by a slit is caused by the spreading of the ultrasonic beam and by the scattering of energy, as it is the case for a radiation of ultrsonic wave from a transducer.

4. CONCLUSIONS

1 The diffraction is caused by the lateral energy flow due to the displacement gradient with respect to the normal to the radial boundary. The energy source to generate diffraction exists not only at the edge of slit, but also distributes continuously on the radial boundary.
2 The energy conservation law is well maintained in each half-cylindrical wavefront of the diffraction itself, and the total energy flow over the full wavefront of the diffraction vanishes. Although the displacement field is antisymmetric with respect to the plane of slit, the intensities of the flow of energy across each half-cylindrical wavefront are different each other.

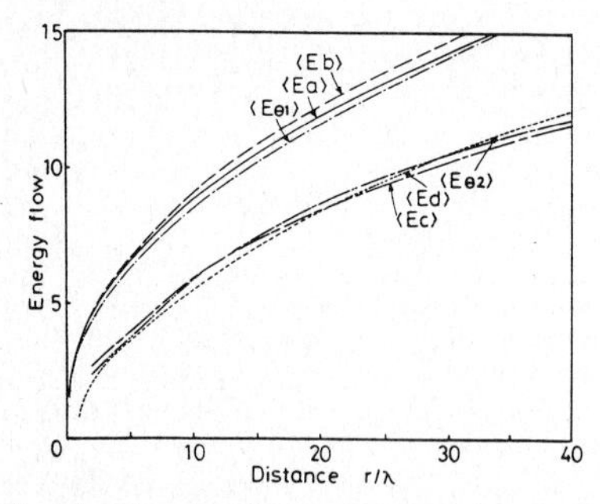

Fig.1 Growth of energy flow across each part of diffraction wavefront or across radial boundary with propagation

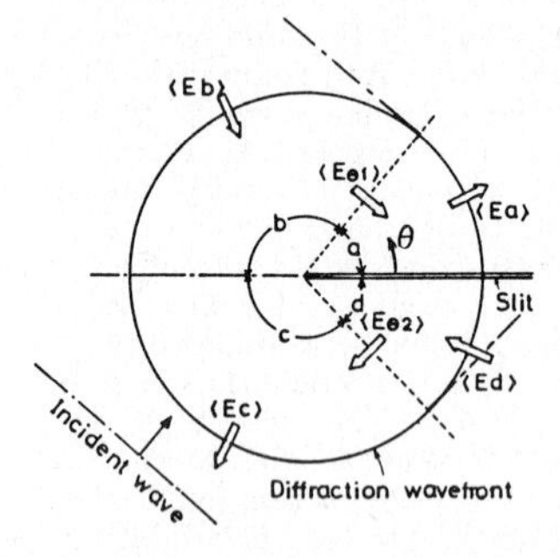

Fig.2 Direction of energy flow derived by difftaction

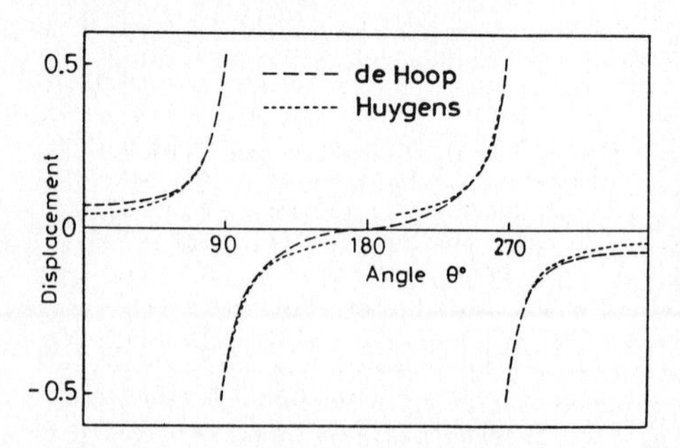

Fig.3 Amplitude of displacement along a cylindrical wavefront

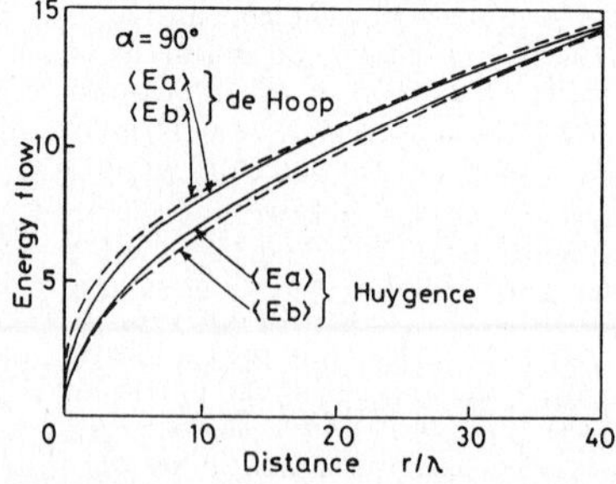

Fig.4 Energy flow across each upper part of diffraction wavefront

Reference [1] Achenbach, J.D., Wave Propagation in Elastic Solids (North-Holland/American Elsevier, 1976) 378.

Elastic Waves and Ultrasonic Nondestructive Evaluation
S.K. Datta, J.D. Achenbach and Y.S. Rajapakse (Editors)

ELASTODYNAMIC WAVES GENERATED BY AN IMPULSIVE POINT SOURCE IN STRATIFIED POROUS MEDIA

Sijtze M. de Vries

Laboratory of Electromagnetic Research, Faculty of Electrical Engineering
Delft University of Technology, P.O. Box 5031, 2600 GA Delft, The Netherlands

The basic ideas that led to the development of a theory to describe the propagation of transient elastodynamic waves in anisotropic, lossless, porous media (fluid saturated poro-elastic solid) are presented. The resulting wave equations have the appearance of the well-known Biot equations, but their interpretation is different, especially as far as the interaction between the two phases is concerned.

In an isotropic elastic solid compressional and shear waves are the two types of waves that can be generated. As soon as the solid becomes porous and the pores are completely filled with an ideal fluid, an extra (slow) compressional wave occurs assuming that the pore sizes are small compared to the wavelength. So the existence of a slow compressional wave is an indication that the solid has become a porous one. Next, we would like to find how the properties of this slow compressional wave depend on the geometrical properties of the pores (e.g., volume fraction occupied by the fluid). To gain insight in the interrelations between the microscopic geometrical and macroscopic elastic properties of the porous medium, a general linearized acoustic wave propagation theory is developed based on a consistent volume averaging technique applied to the fluid phase and the solid phase in the composite [**1,2**]. This technique is simple, and offers great clarity in explaining what kind of averages occur in the macroscopic field quantities and what the different coupling terms in the acoustic wave equations of a two-phase (fluid and solid) model represent. The method also explicitates the kind of average that in a consistent manner occurs in these macroscopic acoustic wave equations. The method circumvents the construction of specific macroscopic energy functions [**3**], the use of variational techniques, Lagrangian formulations, and the two-space method of homogenization, which eventually are complicated intermediate steps in arriving in the end at linearized partial differential equations for describing wave motions of the same general kind as ours. Since we want to analyze the wave shape of acoustic pulses generated by impulsive sources, the corresponding fluid motion is considered at high Reynolds number, i.e., the effects of viscosity are neglected with respect to those of inertia.

The basic idea behind the method is the following. We assume that on the scale of the geometry of the pores the continuum equations for elastic solids and compressible fluids hold. These "microscopic" continuum equations are next spatially averaged over a representative elementary domain of the fluid-solid composite. The diameter of this representative elementary domain should be both sufficiently small and sufficiently large, such that both the macroscopic and the microscopic inhomogeneities do not affect the results of the averaging procedure. At the microscopic level, the continuum equations under consideration have the shape of partial differential equations. Now, for the macroscopic partial differential equations we need partial derivatives of spatially averaged field quantities (acoustic traction and particle velocity in the fluid phase, stress and particle velocity in the solid phase). This poses the question of how the partial derivatives of a spatially averaged field quantity are related to the spatial average of the partial derivatives of this field quantity. The difference between the two is shown to be a surface interaction integral over the microscopic interface. In our theory the only fundamental assumption is that this interaction term is linearly related to the volume averaged acoustic state quantities of the phases involved. Such a general linear interaction is the most general one that is compatible with our ultimate goal: a system of linear partial differential equations that is descriptive of the acoustic waves in the composite at a macroscopic level. To arrive at such a system, the interaction terms of the indicated kind (that contain the fundamental coupling coefficients) can be most straightforwardly be postulated right away. Either explicitly or implicitly, an assumption of this kind is used in all previous treatments, e.g., through the existence of a quadratic energy function in the averaged quantities or in the method of convex linear interpolation (two-space method of homogenization), but at a more complicated intermediate level. The advantage of the procedure is that it directly leads to explicit expressions for the coupling coefficients in terms of the microscopic field quantities along the interface between the two phases, which expressions could be used to draw further conclusions about their magnitude if further techniques of statistical physics were employed. Further, the concept of introducing coupling terms in the pertaining constitutive equations is illustrative for the physics behind all sorts of coupled phenomena, such as in electricity and magnetism the concept of self and mutual inductance of electric circuits and in mechanics the coupled mass-spring systems; the aspect seems, however, to be new in the acoustics of porous media.

In a spatial averaging theory two kinds of averages show up, viz. averages over the single phases (denoted as "intrinsic volume averages") and averages over a time-and shift invariant representative elementary spatial domain (denoted as "volume averages" as such). The two kinds of averages are interconnected by a factor that is the volume fraction of the pertaining phase. Which of the two averages (or combination of them) is to be interpreted as a measurable quantity at a macroscopic level, is still a matter of discussion. It can, however, be argued that the state quantities that describe the acoustic wave phenomena at a macroscopic level should in the corresponding wave equations be treated in an equal manner. In this respect, the Biot theory shows an inconsistency. Biot defined the macroscopic particle displacements (and velocities) to be "intrinsic volume averaged" ones and the macroscopic acoustic pressure and stresses to be "volume averaged" ones. A criterion in selecting one of the two types of averages is the following: the structure of the four coupled partial differential equations should be descriptive for wave propagation (i.e. simultaneous first order hyperbolic equations). If the volume fractions of the porous medium are inhomogeneous and the "intrinsic volume average" is selected we do not end up with hyperbolic partial differential equations since extra terms (that contain the gradient of the volume fractions) appear in the resulting equations. If the "volume average" is selected we do end up with hyperbolic partial differential equations that can be applied to porous media in which the volume fractions are inhomogeneous. Therefore, the "volume averaging" procedure is to be preferred in developing a consistent macroscopic theory of poro-elastic waves, and is taken as the point of departure in the analysis.

Once the final first-order partial differential equations (equations of motion, deformation equations, and constitutive equations) have been obtained, energy theorems are derived. Because the fundamental macroscopic acoustic field quantities are all consistently defined in terms of "volume averages", the kinetic energy density, the deformation (potential) energy density, and the acoustic Poynting vector, too, are expressed in terms of them. In this connection it is remarked that the energy theorems in the Biot theory (from which his first-order equations are deduced) are mutually inconsistent due to the inconsistent use of "intrinsic volume averages" and "volume averages".

The final system of differential equations (with source terms and boundary) conditions serves as the basis for the calculation of the acoustic wave motion generated by impulsive sources in plane layered (porous) media [**1**].

ACKNOWLEDGEMENT

The ideas outlined here have found their shape in many discussions with Professor A.T. de Hoop of our laboratory. His assistance and fruitful cooperation are greatly acknowledged.

REFERENCES

[1] De Vries, S.M., Propagation of transient acoustic waves in porous media (Ph.D.-Thesis), Delft University of Technology, Delft, the Netherlands, (1989) 178pp.

[2] De Vries, S.M., Elastodynamic radiation from an impulsive point source in a poroelastic fluid/solid medium, in: McCarthy, M.F. and Hayes, M.A. (eds.), Elastic Wave Propagation (Elsevier Science Publishers B.V., North-Holland, Amsterdam, 1989) pp. 315–320.

[3] Biot, M.A., Theory of propagation of elastic waves in a fluid-saturated porous solid, J. Acoust. Soc. Am., Vol. 28 (1956), pp. 168–191.

Elastic Waves and Ultrasonic Nondestructive Evaluation
S.K. Datta, J.D. Achenbach and Y.S. Rajapakse (Editors)

MODAL ANALYSIS FOR ELASTIC AND THERMOELASTIC WAVES IN LAYERS OF FINITE THICKNESS

Jörg WAUER

Institut für Technische Mechanik, Universität Karlsruhe
Karlsruhe, West Germany

One of the most elegant but least applied methods of elastodynamics is the expansion of the transient wave function into a series of eigenfunctions. As a generalization, it can also be applied to the interaction of the strain field with the thermal field. In particular, this method can be used advantageously, if vibration problems of finite media described by inhomogeneous boundary value problems are considered and the dynamic behaviour in the time domain is of interest.

The starting point is the set of fundamental equations for thermoelastic waves and corresponding boundary and initial conditions. Typical examples discussed here are the dynamic stress concentration problem of an annular, thick-walled, infinitely long pipe (circular layer: $r \in [R_i, R_a]$, $\varphi \in [0, 2\pi]$, $z \in [-\infty, \infty]$) excited by a circumferentially uniform step pulse at the periphery, and the thermal shock problem for an infinite, plane layer of finite thickness ($x \in [-d, d]$, $y, z \in [-\infty, \infty]$) which is subjected to a sudden change in temperature at the outer surface. Both cases are one-dimensional, i.e., one spatial coordinate is relevant.

In the first case, Lamé-Navier's classical equation without volume forces, specialized to the rotationally symmetric case and formulated in cylindrical coordinates,

$$-\left(\lambda + 2\mu\right) \left(\frac{\partial^2 u}{\partial r^2} + \frac{1}{r}\frac{\partial u}{\partial r} - \frac{u}{r^2}\right) + \rho \frac{\partial^2 u}{\partial t^2} = 0 \tag{1}$$

(λ, μ Lamé's constants, ρ mass density, u radial displacement), supplemented by inhomogeneous boundary conditions (in terms of the radial normal stress σ_r)

$$\sigma_r(R_i, t) = 0, \quad \sigma_r(R_a, t) = \sigma_0 h(t), \text{ where } \sigma_r = (\lambda + 2\mu)\frac{\partial u}{\partial r} + \lambda \frac{u}{r} \tag{2}$$

($h(t)$ Heaviside step function, σ_0 intensity factor) and homogeneous initial conditions

$$u(r, 0_-) = 0, \quad \frac{\partial u}{\partial t}(r, 0_-) = 0 \tag{3}$$

defines the transient problem of radial wave propagation. The initial/boundary value problem (1) to (3) can equivalently be reformulated by introducing a special volume force k on the right side of eq (1),

$$k = \sigma_0\, h(t)\, \delta(r - R_a) \tag{4}$$

($\delta(\cdot)$ Dirac's delta function) and changing the boundary conditions (2) to their homogeneous form.

The second problem requires calculation of the coupled displacement and temperature fields, using

$$(\lambda + 2\mu)\frac{\partial^2 u}{\partial x^2} - \rho\frac{\partial^2 u}{\partial t^2} - \alpha(3\lambda + 2\mu)\frac{\partial T}{\partial x} = 0,$$

$$\frac{\partial^2 T}{\partial x^2} - \frac{1}{\kappa}\frac{\partial T}{\partial t} - \frac{\alpha(3\lambda + 2\mu)T_0}{\kappa\, c_v}\frac{\partial^2 u}{\partial x \partial t} = 0 \tag{5}$$

(α coefficient of expansion, T_0 reference temperature, c_v specific heat, κ thermal diffusivity, u transverse displacement, T temperature change) in a Cartesian coordinate system with the boundary conditions

$$T(\pm d,t) = \Delta T\, h(t), \quad \sigma_x(\pm d,t) = 0 \quad \text{where} \quad \sigma_x = (\lambda + 2\mu)\frac{\partial u}{\partial x} - \alpha T \tag{6}$$

(ΔT temperature step, σ_x transverse normal stress) and the initial conditions

$$T(x,0_-) = 0, \quad u(x,0_-) = 0, \quad \frac{\partial u}{\partial t}(x,0_-) = 0. \tag{7}$$

Due to the symmetry of the problem, two of the boundary conditions (6) can alternatively be replaced by

$$u(0,t) = 0, \quad \frac{\partial T}{\partial x}(0,t) = 0. \tag{8}$$

Also, for the problem of thermoelasticity, a transformation yielding homogeneous boundary conditions exists:

$$T(x,t) = \tau(x,t) + \left(\frac{x}{d}\right)^2 \Delta T\, h(t), \quad u(x,t) = v(x,t) + \frac{\alpha}{\lambda + 2\mu} x\, \Delta T\, h(t). \tag{9}$$

A short calculation leads to the resulting initial/boundary value problem

$$(\lambda + 2\mu)\frac{\partial^2 v}{\partial x^2} - \rho\frac{\partial^2 v}{\partial t^2} - \alpha(3\lambda + 2\mu)\frac{\partial \tau}{\partial x} = \frac{\rho\alpha x}{\lambda + 2\mu}\Delta T \delta^{(1)}(t) + \frac{2\alpha(3\lambda + 2\mu)x}{d^2}\Delta T h(t),$$

$$\frac{\partial^2 \tau}{\partial x^2} - \frac{1}{\kappa}\frac{\partial \tau}{\partial t} - \frac{\alpha(3\lambda + 2\mu)T_0}{\kappa\, c_v}\frac{\partial^2 v}{\partial x \partial t} = \Delta T\left[-\frac{2}{d^2}h(t) + \left(\frac{x}{d}\right)^2 \delta(t) + \frac{\alpha^2(3\lambda + 2\mu)T_0}{\kappa\, c_v(\lambda + 2\mu)}\delta(t)\right], \tag{10}$$

$$\frac{\partial \tau}{\partial x}(0,t) = 0, \quad \tau(d,t) = 0, \quad v(0,t) = 0, \quad \frac{\partial v}{\partial x}(d,t) = 0, \tag{11}$$

$$\tau(x,0_-) = 0, \quad v(x,0_-) = 0, \quad \frac{\partial v}{\partial t}(x,0_-) = 0 \tag{12}$$

($\delta^{(1)}(t)$ denotes the derivative of the δ-function) for τ, v.

Then, the expansion theorem states that the general solution of the problems (1) to (3) with (4) and (10) to (12) can be represented as a series of the governing mode shapes multiplied by unknown time functions

$$u(r,t) = \sum_{k=1}^{N\to\infty} U_k(r)T_k(t) \quad \text{and} \quad \tau(x,t) = \sum_{k=1}^{N\to\infty} \Gamma_k(x)t_k(t), \quad v(x,t) = \sum_{k=1}^{N\to\infty} V_k(x)s_k(t), \tag{13}$$

respectively. The orthonormal eigenfunctions $U_k(r)$ and $\Gamma_k(x)$, $V_k(x)$ have to be determined from the associated eigenvalue problem and the time functions $T_k(t)$ and $t_k(t)$, $s_k(t)$ are solutions of relatively simple initial value problems [1-3].

As an extension to more complicated two-dimensional problems, the thick-walled pipe subjected to a locally concentrated pulse can also be discussed [2]. In this case, radial and tangential displacements u_r, u_φ appear, both depending on the two space coordinates r,φ.

REFERENCES

1. Maaß, M., Dynamische Spannungskonzentrationsprobleme bei allseits berandeten gelochten Scheiben, Dr.-Ing. thesis, University of Karlsruhe, 1986.
2. Maaß, M., Wauer, J., Dynamic Stress Concentration in Annular Disks Under Transient Loading, in: Advances in Fracture Research, Vol. 1, Pergamon Press, 1989, 857-866.
3. Wauer, J., Modalanalysis fur das 1-dimensionale Thermoschockproblem einer elastischen Schicht endlicher Dicke, ZAMM, 70, 1990, to appear.

Elastic Waves and Ultrasonic Nondestructive Evaluation
S.K. Datta, J.D. Achenbach and Y.S. Rajapakse (Editors)

Wave Localization in the Time Domain: Numerical Studies in Two-Dimensions

R.L. Weaver
J. H. Loewenherz
Department of Theoretical and Applied Mechanics
University of Illinois at Urbana-Champaign
104 South Wright Street
Urbana, IL 61801

It is generally held that all modes of a random two-dimensional system are exponentially localized. It has recently been noted, however, Kaveh (1985) that the experimental evidence is not definitive. The possibility that two-dimensional systems manifest a mobility edge separating true exponential localization from mere power law localization remains viable. We explore this question by studying the time-domain behavior of the incoherent diffusive like response of a model random system to a point source tone burst excitation.

Based on the scaling theory of Anderson localization (see for example MacKinnon and Kramer (1983)), we expect all modes of a two-dimensional system to be exponentially localized. MacKinnon and Kramer (1983) have tabulated predicted localization lengths for values of W/V from 2.0 to 15.0. The work of Scher (1983) supports their results to a limited extent, though as we will show, the longest time in his numerical experiments is not long enough to make any definitive conclusions.

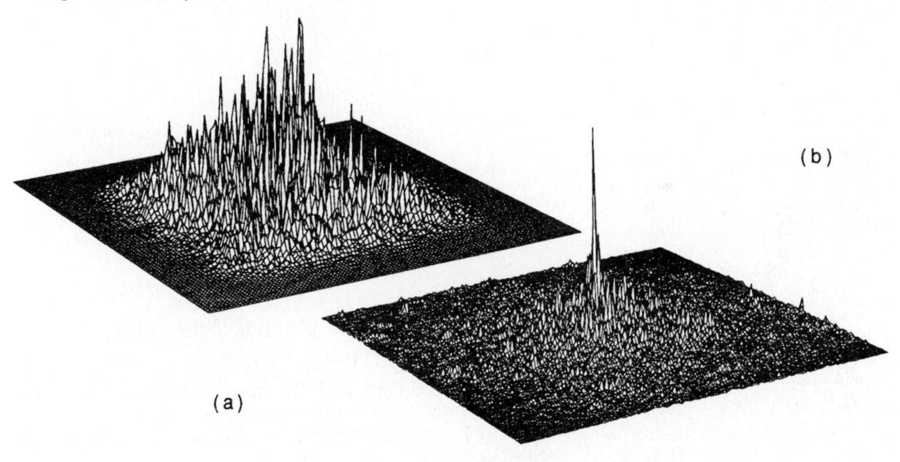

Figure 1. Snapshots of the root mean square displacement. (a) W/V = 2.0, an unlocalized system observed a very short time after application of the excitation. A Gaussian profile is observed, consistent with a process of classical diffusion. (b) W/V = 7.0, in the localized regime (ξ = 18.5, array size = 150x150) observed at the very late time t=50,000.

We have used a model consisting of a 150 x 150 array of point masses, each of which is supported by an ideal spring of random stiffness, and interconnected by massless, constant tension strings. The range over which the spring stiffnesses vary is a direct measure of the amount of disorder in the system and corresponds to the parameter W/V commonly found in the literature. The model is similar to that of Scher (1983). Our system is excited by a tone burst force applied at the center of the array. The tone burst has a central frequency near the middle of the band and a duration of ten cycles. Figure 1 shows snapshots of the array for various values of W/V at times after the application of the excitation.

In Figure 2 the second moment of the energy density as measured from the point of application is plotted. The classical theory of diffusion would predict an $L^2 = 4Dt$, increasing linearly with time with a diffusion constant D. We clearly observe localization for W/V values of 7.0 and 8.0 (Fig. 2), with an L^2_{max} of 740 and 310 respectively. These values are in agreement with the predictions for localization lengths ξ of 18.5 and 11.1 from MacKinnon and Kramer (1983), if one notes that $L^2_{max} = 2\xi^2$ in accordance with the hydrodynamical model of Vollhardt and Wolfle (1980). It should be noted that the time used in our experiments corresponds to Scher's time multiplied by a factor of six, so that his maximum time of 2200 corresponds to a time of 13,200 on our plots. As can be seen from Fig.2, at this time $L^2(t)$ has not yet achieved its asymptotic value, indicating that the transport is still underway. Kaveh's (1985) conclusion based on Scher's curves that the localization is merely power-law, and that the transport continues indefinitely, is therefore seen to have been unwarranted.

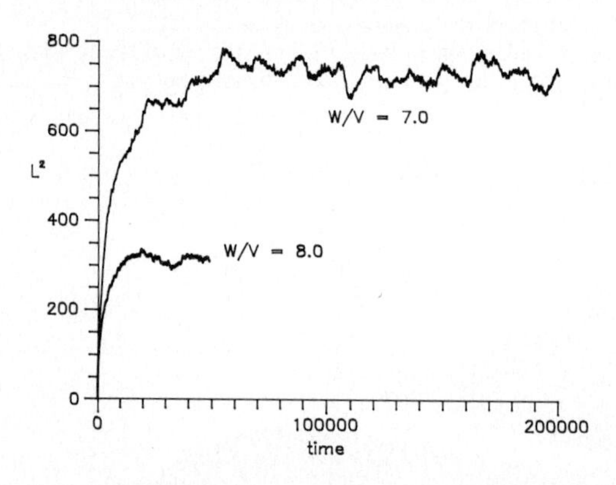

Figure 2. The second moment of the energy density is plotted as a function of time for two different values of the disorder W/V.

A plot of dL^2/dt vs. $\log_e(L^2)$ for a W/V value of 7.0 (Fig 3) is further evidence in support of the standard theory, which predicts (see e.g. Weaver,1989) that the diffusivity scales logarithmically with the length scale on length scales shorter than the localization length ξ by

$$D = D_o - (1/2\pi^2\rho)\log_e(L/L_o) ,$$

where ρ is the modal density and D_o is the diffusivity on the bare, micro, length scale L_o . If we identify the diffusivity at scale L with the slope $(1/4)dL^2/dt$ this formula predicts the slope of a plot of dL^2/dt vs. $\log_e(L^2)$. We see that for early times the slope of these curves is a constant, and compares very well with the theoretical slope.

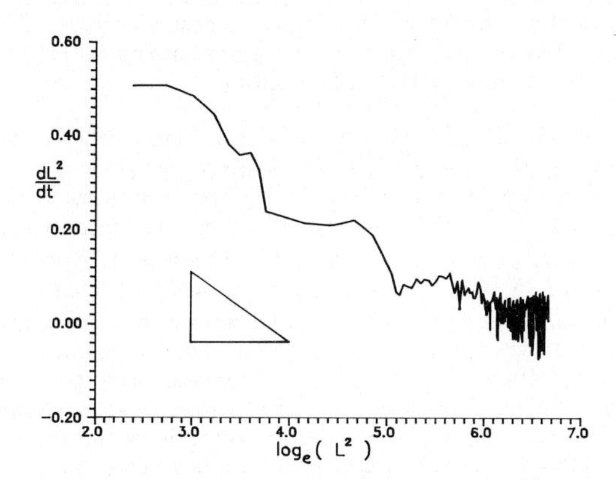

Figure 3. dL^2/dt versus $\log_e(L^2)$ for a W/V value of 7.0. The triangle corresponds to the theoretical slope of -0.15 obtained from an independent assessment of the modal density.

This work was supported by the National Science Foundation through grant number MSM 87-22413.

References

A. MacKinnon and B. Kramer, "The Scaling Theory of Electrons in Disordered Solids: Additional Numerical Results", Z. Phys. B - Condensed Matter 53, 1-13 (1983).

G. M. Scher, "Anderson Transition and Minimum Metallic Conductivity", J. Non-Crys. Solids 59, 60, 33-36 (1983).

M. Kaveh, "Power-Law Localization in Two-dimensional Systems, Theory and Experiment", Philos. Mag.B 52, 521-540 (1985).

R. Weaver, "Anderson Localization of Ultrasound" in press, Wave Motion (1989).

D. Vollhardt and P. Wolfle, "Anderson Localization in $d \leq 2$ Dimensions: A Self Consistent Diagrammatic Theory", Phys. Rev. Lett. 45, 842-846 (1980).

Elastic Waves and Ultrasonic Nondestructive Evaluation
S.K. Datta, J.D. Achenbach and Y.S. Rajapakse (Editors)

FINITE ELEMENT STUDY OF DIFFRACTION TOMOGRAPHY

Zhongqing YOU and William LORD

Department of Electrical Engineering and Computer Engineering
Iowa State University
Ames, Iowa 50011

1. INTRODUCTION

A finite element model for elastic wave propagation in solids has been developed [1,2] which can predict displacement for all the nodes in the solid at every instant of time in a given time interval. The model displacement field predictions on the surface of the solid can serve as the raw data for inverse algorithms which discover any inhomogeneity in the material. Even though these raw data are usually from experiments in practical situations, the numerical techniques are superior to experiments in the sense that material parameters and geometries are easily varied. The objective of this research, therefore, is to utilize such a numerical model as a test bed for diffraction tomography [3,4]. This paper gives results which show that the normal component of the displacements on the measurement surface can be used as the prime data for a diffraction tomographic algorithm derived form the scalar wave equation. Other results and detailed derivation of the algorithm are given in [5].

2. FINITE ELEMENT GENERATION OF RAW DATA

Finite element modeling is summarized in the following steps. Detailed formulations are discussed in [1,2].

1. Discretize space into small elements.
2. Interpolate field value by shape function.
3. Minimize energy functional of the system to derive elemental mass and stiffness matrices.
4. Assemble all elemental matrices to form the global equation.
5. Approximate the second time derivative term by central difference formula.
6. Obtain the field values for all nodes at a given time by solving equations simultaneously. A lumped mass matrix is used to avoid matrix inversion.
7. Iterate over a given time interval.

The code is used for both acoustical and elastic problems where a small rectangular flaw is located in the otherwise homogeneous medium. The reflected wave from the flaw is sensed on the front surface of the block. The CPU-time for a 65,025 element mesh is 1.6 seconds per iteration for the acoustic case, while 3.2 seconds are used for the elastic problem because of the doubled number of variables. All computations are performed on the NAS AS/9160 computer. A typical distribution of the acoustic potential at $t = 14\mu s$ is illustrated in Figure 1 where a plane incident wave propagating along the positive z-axis produces a strong reflection when it passes by a small rectangular flaw.

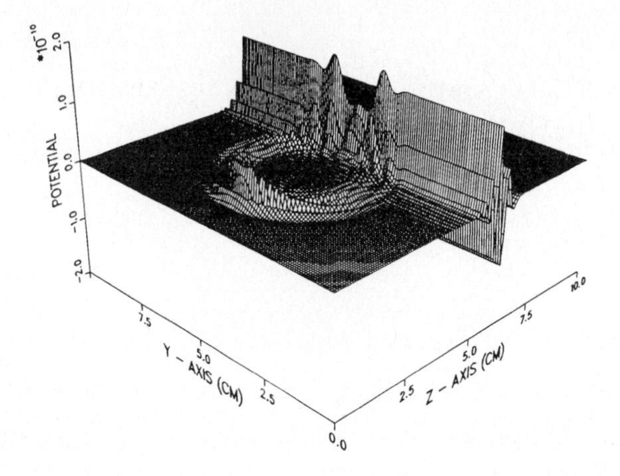

Figure 1. Interaction of plane acoustic wave with a rectangular crack.

3. RECONSTRUCTION ALGORITHM AND THE RECONSTRUCTED IMAGE

Based on the Born approximation, a 2-D frequency diversity diffraction tomographic algorithm (FDDTA) in reflection mode can be derived as follows [5]:

$$O_{\Lambda P}(\underline{r}) = \frac{j}{(2\pi)^3}\int_0^{\infty}\frac{dk}{k^2F(\omega)}\int_{-\infty}^{\infty}\left|k+\sqrt{k^2-k_y^2}\right|\hat{\phi}_M(k_y,z=d,\omega)$$
$$e^{-jkz}e^{j(z-d)\sqrt{k^2-k_y^2}}u(k^2-k_y^2)e^{jk_y y}\,dk_y \qquad (1)$$

where Λ is the shape of the space filter involved in the derivation, k is the wave number of the host medium at frequency ω, k_y and k_z are the Fourier transformed variables corresponding to y and z respectively, u is a step function, and $\hat{\phi}_M$ is the 1-D Fourier transformation of the measured data ϕ_M at the $z = d$ line with respect to y. Fourier transformation of the A-scan potential data on the front wall with respect to time t provides the measured data $\phi_M(x, z = 0, \omega)$ in the frequency domain. Further Fourier transformation of the ϕ_M with respect to x gives $\hat{\phi}_M(k_x, z = 0, \omega)$ as required in Equation (1). The reconstructed image by FDDTA in reflection mode is shown in Figure 2. This image shows good agreement with the original crack. The same algorithm now is applied for the elastic case where the normal component of the displacement data measured in a straight line is used as the measured potential. The same process reconstructs the image shown in Figure 3 for a similar geometry to that of Figure 2. Obviously, these two images have about the same resolution. This is also true for other examples such as a two-crack geometry, a weak volumetric flaw, etc. Results are not presented here because of the limited space.

4. CONCLUSIONS

The normal component of the displacement data in the measure line can be used as the raw data for the FDDTA to reconstruct the filtered object function. Finite element models can provide an effective tool for evaluating inverse algorithms.

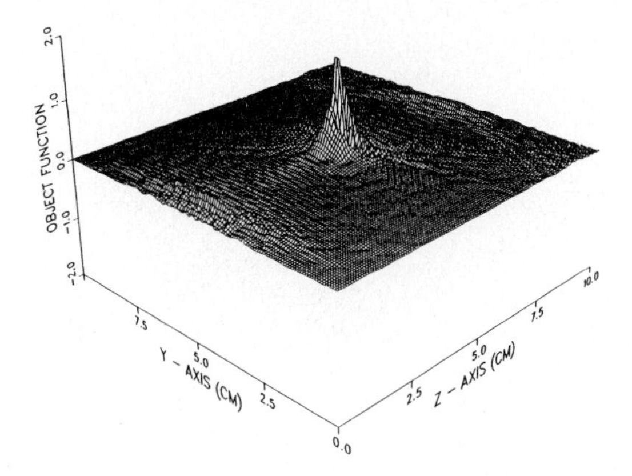

Figure 2. Reconstructed image from the reflected acoustic potentials.

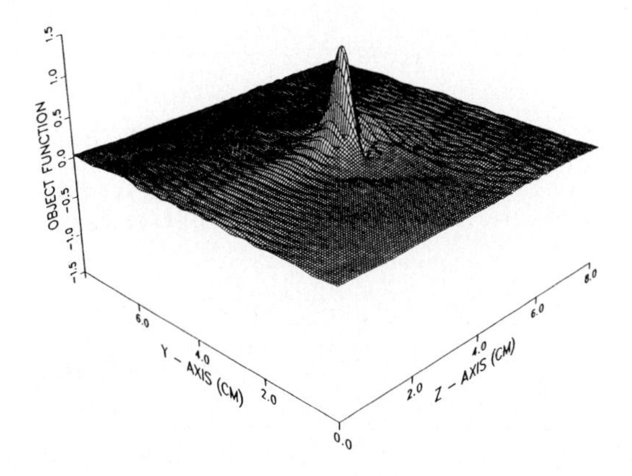

Figure 3. Reconstructed image from the normal component of the reflected L-wave displacements.

REFERENCES

[1] R. Ludwig and W. Lord, IEEE Transactions on Ultrasonics, Ferroelectrics and Frequency Control, Vol. 35., No. 6 (1988) 809.

[2] Z. You and W. Lord, Finite Element Study of Elastic Wave Interaction with Cracks, in: D. O. Thompson and D. E. Chimenti, (eds.), Review of Progress in Quantitative NDE, Vol. 8A (Plenum Press, New York, 1989) pp. 109-116.

[3] K. J. Langenberg, Applied Inverse Problems for Acoustic, Electromagnetic and Elastic Wave Scattering, in: D. C. Sabatico, (ed.), Tomography and Inverse Problems (Adam Hilger, 1987) pp. 125-467.

[4] A. J. Devaney, Ultrasonic Imaging, Vol. 4 (1982) 336.

[5] Z. You and W. Lord, A Finite Element Test Bed for Diffraction Tomography, in: D. O. Thompson and D. E. Chimenti, (eds.), Review of Progress in Quantitative NDE, Vol. 9, in print.

 **Mile-High IUTAM** 

by

L.B. Felsen

Summarizing the IUTAM Symposium on Elastic Wave Propagation

The motion of elastic waves
Is IUTAM's concern.
Last year, we met in Ireland
The latest status to discern.

This year, again elastic waves,
But more to NDE applied.
For flaw detection, there are means
That ultrasonic waves provide.

Waves carry signals that are launched
From there to be detected here.
Elastic waves transmit themselves
Both with compression and with shear.

Participants came well prepared:
They were compressed as well as sheared.
After long flights in crowded planes,
They inelastically appeared.

The hoped, in vain, as they arrived,
Blue Colorado sky to see.
It poured -- thereby preparing them
For fluid-loaded NDE.

But soon, initial gloom subsides.
Quick registration, juice and snacks,
A neat blue briefcase, fully packed --
You're done, and now you can relax.

The talks began right on the theme.
Scatterers -- fractures, holes and cracks --
Were chased by ultrasonic waves
Along an awesome maze of tracks.

Bulk waves showed up, and surface waves
On layers that are dipped and straight.
A favored combination dish
Was laminated sandwich plate.

When plates or shells in fluids sit,
The waves and fluid interact.
What was confined within a shell
Can now leak out for new impact.

Tortures that man concocts have been
Transferred to waves. It's not farfetched:
Waves creep, they leak, and they are trapped;
They are diffracted, shrunk and stretched.

And as the meeting days progress,
The wave pathology takes hold.
The audience, elastic first,
Moves strongly damped, is much less bold.

When from the lectures they emerge,
One sees much creeping, even shock.
The tortured waves take their revenge.
Those who've abused are on the block.

But remedy is near at hand.
For meeting blues, it does prescribe
Long coffee breaks, where you can stretch
While fruit or pastry you imbibe.

Even the skies did soon clear up,
And therefore it is safe to say
That, at the closing, there will reign
Good cheer as order of the day.

I could not stay there to the end.
I missed Banquet and Bar-B-Que.
I guess, that those who skipped some talks
Found much in Boulder they could do.

The days were packed. But single track
of sessions gave each one a chance
To hear reports, across the board,
On latest status and advance.

We owe much thanks to our hosts
Who labored hard for this event.
That it succeeded is for them
The proof their efforts were well spent.

Author Index